Social Sciences

Life Sciences

General Interest

Applied

Finite

Mathematics

Applied

Finite

Mathematics

Second Edition

S. T. Tan
Stonehill College

Prindle,
Weber &
Schmidt
Boston

PWS PUBLISHERS

Prindle, Weber & Schmidt • 🖋• Duxbury Press • ♠• PWS Engineering •🔺• Breton Publishers •⚙
20 Park Plaza • Boston, Massachusetts 02116

PWS Publishers is a division of Wadsworth, Inc.

Library of Congress Cataloging-in-Publication Data

Tan, Soo Tang.
  Applied finite mathematics.

  Rev. ed. of: Applied finite mathematics for the managerial and social sciences. c1983.
    Includes index.
    1. Mathematics—1961–   .   I. Tan, Soo Tang.
Applied finite mathematics for the managerial and social sciences.   II. Title.
QA39.2.T34   1987   510   86-22500
ISBN 0-87150-074-4

The author is grateful for permission to reprint the following Chapter Opening photos courtesy of: H. Armstrong Roberts (p. 2, p. 229); U.S. Environmental Protection Agency (p. 71); American Airlines (p. 149); Public Service Information Department (p. 187); Camerique Stock Photos (p. 275); and W. Atlee Burpee Company (p. 487).

Printed in the United States of America.
87 88 89 90 91 — 10 9 8 7 6 5 4 3 2 1

Sponsoring Editor: *Chuck Glaser*
Production Coordinator/Designer: *Elise Kaiser*
Production: *Del Mar Associates*
Composition: *Weimer Typesetting Company, Inc.*
Cover Printer: *Phoenix Color Corp.*
Text Printer/Binder: *Halliday Lithograph*

Cover art *"La Mer du Soir"* by Kasai Masahiro; a silkscreen print used with the permission of the artist.

To Pat, Bill, and Michael

Preface

This book, which treats the standard topics in finite mathematics, is directed toward the student in the managerial, social, and life sciences. The only prerequisite for understanding this book is a year of high school algebra. The objective of *Applied Finite Mathematics* is two-fold: first, to provide the student with the background in the quantitative techniques that are necessary to better understand and appreciate the courses normally taken in one's undergraduate training; and, second, to lay the foundation for more advanced courses, such as statistics and operations research. We have hoped to accomplish this by striking a careful balance between theory and applications.

Our approach is intuitive and we state the results informally. However, we took special care to ensure that this does not compromise the mathematical content and accuracy. The applications are drawn from many fields, and we made every effort to make them interesting, relevant, and up to date. Numerous examples and solved problems are used to amplify each new concept or result in order to facilitate the student's comprehension of the new material. Each section is accompanied by an extensive set of exercises, which contains an ample set of problems of a routine computational nature to help the student master new techniques, followed by an extensive set of applications-oriented problems to test his or her mastery of the topics. Each chapter of the text also ends with a set of review exercises. Answers to odd-numbered exercises appear in the back of the book.

Since the book contains more than ample material for a one-semester or two-quarter course, the instructor may be flexible in choosing the topics most suitable for his or her course. The following chart on chapter dependency is provided to help the instructor design a course that is most suitable for the intended audience.

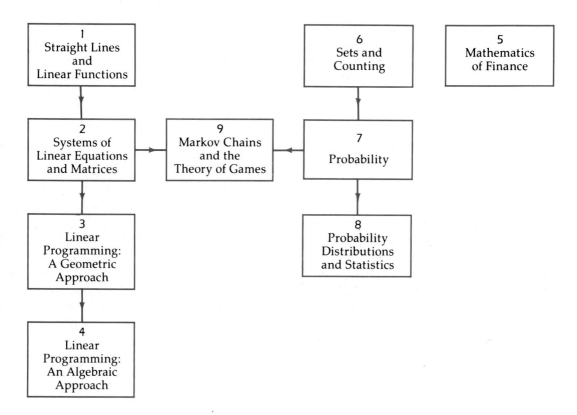

Changes in the Second Edition

This edition contains an improved treatment of many of the topics in the previous edition, as well as a few additions and changes:

* The addition of Self-Check Exercises at the end of each section. These exercises, with completely worked out solutions appearing at the end of each exercise set, give students a chance to test themselves on their understanding of the material.

* The addition of many more examples and exercises. The examples, drawn from the fields of business, economics, social and behavioral sciences, life and physical sciences, and other fields of general interest, provide further illustrations of the concepts. The exercises are of varied degrees of difficulty ranging from rote to more challenging problems.

* Chapter 2 has been rewritten. A more traditional approach has been adopted with systems of linear equations appearing before matrices.

* Chapter 5, "Mathematics of Finance," has been moved forward. Also, the section on arithmetic and geometric progressions has been relegated to the end of the chapter and is now optional.

* An appendix on logic has been added for those who wish to cover this material in their course.

An answer book containing the answers to the even-numbered exercises and a computerized test generator are available for instructors. A student supplement has been prepared to accompany the text and includes detailed worked out solutions for every third problem that appears in the text.

Acknowledgments

I wish to express my personal appreciation to each of the following reviewers whose many suggestions have helped make a much improved second edition: Frank E. Bennett, Mount Saint Vincent University; Patricia Hickey, Baylor University; Lavon B. Page, North Carolina State University; James Perkins, Piedmont Virginia Community College; and Robert H. Rodine, Northern Illinois University. I also wish to thank Reverend Robert J. Kruse, Academic Dean of Stonehill College, for his enthusiastic support of this project. My thanks also go to the editorial and production staffs of Prindle, Weber & Schmidt: Chuck Glaser and Elise Kaiser, for their assistance and cooperation in the development and production of this book. Finally, I wish to thank Nancy Sjoberg of Del Mar Associates and Andrea Olshevsky for doing an excellent job ensuring the accuracy and readability of this second edition.

S. T. Tan

Contents

Applied

Finite

Mathematics

Straight Lines
and Linear
Functions

Which road to take? Two towns are connected by two different routes: one runs along the coast and the other includes a stretch of mountain highway. The coastal route is faster but longer than the inland route. In Example 2, page 8, we will solve the problem of getting from one town to the other in the shortest time.

► CHAPTER

ONE

1.1

The Cartesian Coordinate System

▶ The Cartesian Coordinate System

▶ The Distance Formula

▶ Application

▶ The Cartesian Coordinate System

The system of real numbers plays a fundamental role throughout this book. This system is made up of the set of real numbers together with the usual operations of addition, subtraction, multiplication, and division. We will assume that you are familiar with the rules governing these algebraic operations (see Appendix A).

It is convenient and fruitful to have a geometrical representation of the set of real numbers. Such a representation is called a **number line** and is constructed as follows: Arbitrarily select a point on a straight line to represent the number zero. This point is called the **origin.** If the line is horizontal, then a point at a convenient distance to the right of the origin is chosen to represent the number one. This determines the scale for the number line. Each positive real number x lies x units to the right of zero, and each negative real number $-x$ lies x units to the left of zero.

In this manner, a one-to-one correspondence is set up between the set of real numbers and the set of points on the number line, with all the positive numbers lying to the right of the origin and all the negative numbers lying to the left of the origin (see Figure 1.1).

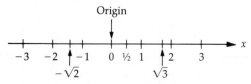

Figure 1.1   The real number line

A similar representation for points in a plane (a two-dimensional space) is realized through the **Cartesian coordinate system,** which is constructed as follows: Take two perpendicular lines, one of which is normally chosen to be horizontal. These lines intersect at a point O, called the **origin** (see Figure 1.2). The horizontal line is called the **axis of abscissas,** or more simply, the **x-axis.** The vertical line is called the **axis of ordinates,** or the **y-axis.** A number scale is set up along the x-axis with the positive numbers lying to the right of the origin and the negative numbers lying to the left of the origin. Similarly, a number scale is set up along the y-axis with the positive numbers lying above the origin and the negative numbers lying below the origin.

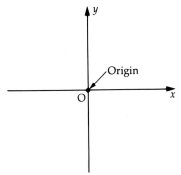

Figure 1.2   The Cartesian coordinate system

The number scales on the two axes need not be the same. Indeed, in many applications different quantities are represented by x and y. For example, x may represent the number of typewriters sold and y the total revenue resulting from the sales. In such cases it is often desirable to choose different number scales to represent the different quantities. Note, however, that the zeros of both number scales coincide at the origin of the two-dimensional coordinate system.

A point in the plane can now be represented uniquely in this coordinate system by an **ordered pair** of numbers, that is, a pair (x, y) where x is the first number and y the second. To see this, let P be any point in the plane (see Figure 1.3). Draw perpendiculars from P to the x-axis and y-axis, respectively. Then the number x is precisely the number corresponding to the point on the x-axis at which the perpendicular through P hits the x-axis. Similarly, y is the number

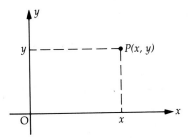

Figure 1.3

corresponding to the point on the y-axis at which the perpendicular through P crosses the y-axis.

Conversely, given an ordered pair (x, y) with x as the first number and y the second, a point P in the plane is uniquely determined as follows: Locate the point on the x-axis represented by the number x and draw a line through that point parallel to the y-axis. Next, locate the point on the y-axis represented by the number y and draw a line through that point parallel to the x-axis. The point of intersection of these two lines is the point P (see Figure 1.3).

In the ordered pair (x, y), x is called the **abscissa,** or **x-coordinate,** y is called the **ordinate,** or **y-coordinate,** and x and y together are referred to as the **coordinates** of the point P.

The points $A = (2, 3)$, $B = (-2, 3)$, $C = (-2, -3)$, $D = (2, -3)$, $E = (3, 2)$, $F = (4, 0)$, and $G = (0, -5)$ are plotted in Figure 1.4. The fact that, in general, $(x, y) \neq (y, x)$ is clearly illustrated by points A and E.

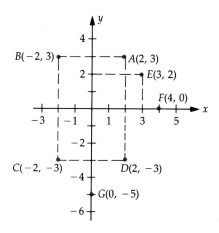

Figure 1.4

The axes divide the plane into four quadrants. Quadrant I consists of the points P with coordinates x and y, denoted by $P(x, y)$, satisfying $x > 0$ and $y > 0$; Quadrant II, the points $P(x, y)$ where $x < 0$ and $y > 0$; Quadrant III, the points $P(x, y)$ where $x < 0$ and $y < 0$; and Quadrant IV, the points $P(x, y)$ where $x > 0$ and $y < 0$ (see Figure 1.5).

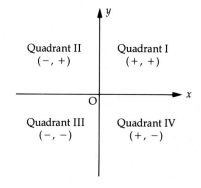

Figure 1.5

▶ The Distance Formula

One immediate benefit that arises from using the Cartesian coordinate system is that the distance between any two points in the plane may be expressed solely in terms of their coordinates. Suppose, for example, that (x_1, y_1) and (x_2, y_2) are any two points in the plane (see Figure 1.6).

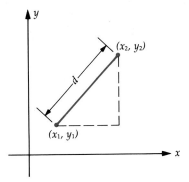

Figure 1.6

Then the distance d between these two points is, by the Pythagorean Theorem,

$$d = \sqrt{(x_2 - x_1)^2 + (y_2 - y_1)^2}. \qquad (1)$$

For a proof of this result see Exercises 1.1, Problem 33.

Distance Formula

The distance d between two points (x_1, y_1) and (x_2, y_2) in the plane is given by

$$d = \sqrt{(x_2 - x_1)^2 + (y_2 - y_1)^2}.$$

EXAMPLE

1 Find the distance between the points $(-4, 3)$ and $(2, 6)$.

SOLUTION We have, by the distance formula (1)

$$
\begin{aligned}
d &= \sqrt{[2 - (-4)]^2 + (6 - 3)^2} \\
  &= \sqrt{6^2 + 3^2} \\
  &= \sqrt{45} \\
  &= 3\sqrt{5}.
\end{aligned}
$$

◀

▶ Application

EXAMPLE

2 Towns A, B, C, and D are located as shown in Figure 1.7. Two highways connect towns B and D. Route 1, from town B to town D via town A, includes a stretch of coastal highway joining A to D. Route 2, from town B to town D via town C, includes a stretch of mountain highway joining town C to town D. If a man wishes to drive from town B to town D and can average 52 mph on route 1 and 39 mph on route 2, which route should he take in order to arrive in the shortest time?

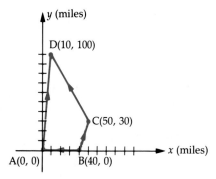

Figure 1.7

SOLUTION

If the man took route 1, he would have to drive a distance of

$$d_1 = 40 + \sqrt{10^2 + 100^2}$$
$$= 40 + \sqrt{10,100} \approx 140.50,$$

or 140.50 miles. The time required to complete the trip would be

$$\frac{140.50}{52} \approx 2.7 \quad \text{(distance covered divided by the average speed)}$$

or 2.7 hours.
If he took route 2, he would have to drive a distance of

$$d_2 = \sqrt{(50 - 40)^2 + (30 - 0)^2} + \sqrt{(10 - 50)^2 + (100 - 30)^2}$$
$$= \sqrt{1000} + \sqrt{6500}$$
$$\approx 112.25,$$

or 112.25 miles. The time required to complete the trip would then be

$$\frac{112.25}{39} \approx 2.9,$$

or 2.9 hours.
Therefore, the man should take route 1.

◀

Self-Check Exercises

1.1

1. Plot the points (1, 3) and (−1, −3).
2. Find the distance between the points (−1, 2) and (3, −2).
3. Cities A, B, and C are located as shown in the following figure:

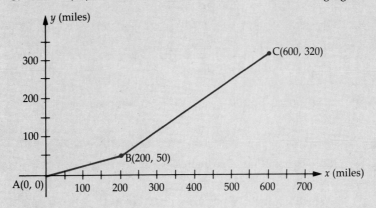

Suppose that a pilot wishes to fly from city A to city C but must make a mandatory stopover in city B. If the single-engine light plane has a range of 650 miles, can she make the trip without refueling in city B?

Solutions to Self-Check Exercises 1.1 can be found on page 12.

Exercises 1.1

In Exercises 1–6, refer to the following figure and determine the coordinates of the given point and the quadrant in which it is located.

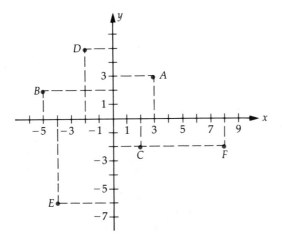

Figure for
Exercises 1–6

1. A   *2. B*   *3. C*   *4. D*   *5. E*   *6. F*

In Exercises 7–11, refer to the following figure.

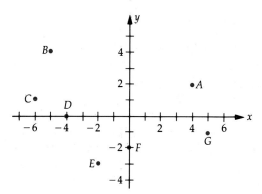

Figure for
Exercises 7–11

7. Which point has coordinates (4, 2)?

8. What are the coordinates of point *B*?

9. Which points have negative *y*-coordinates?

10. Which point has a negative *x*-coordinate and a negative *y*-coordinate?

11. Which point has an *x*-coordinate that is equal to zero?

In Exercises 12–19, sketch a set of coordinate axes and plot the given point.

12. (1, 3)   13. (−2, 5)   14. (3, −4)   15. (3, −1)

16. (−5/2, 3/2)   17. (8, −7/2)   18. (1.2, −3.4)   19. (4.5, −4.5)

In Exercises 20–23, find the distance between the given points.

20. (1, 0) and (4, 4)   21. (1, 3) and (4, 7)

22. (−2, 1) and (10, 6)   23. (−1, 3) and (4, 9)

24. Find the coordinates of the points that are 5 units away from the origin and have an *x*-coordinate equal to 3.

25. Find the coordinates of the points that are 10 units away from the origin and have a *y*-coordinate equal to −6.

26. Show that the triangle with vertices (−5, 2), (−2, 5), and (5, −2) is a right triangle.

27. Show that the points (3, 4), (−3, 7), (−6, 1), and (0, −2) form the vertices of a square.

28. A furniture store offers free set-up and delivery services to all points within a 25-mile radius of its warehouse distribution center. If you live 20 miles east and 14 miles south of the warehouse, will you incur a delivery charge? Justify your answer.

29. A grand tour of four cities begins at city A and makes successive stops at cities B, C, and D before returning to city A. If the cities are located as shown in the following figure, find the total distance covered on the tour.

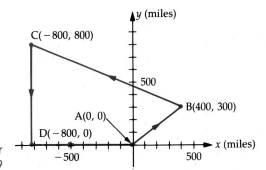

Figure for
Exercise 29

30. Mr. Barclay wishes to determine which antenna he should purchase for his home. The TV store has supplied him with the following information:

	Range in Miles		
VHF	UHF	Model	Price
30	20	A	$40.00
45	35	B	$50.00
60	40	C	$60.00
75	55	D	$70.00

Barclay wishes to get channel 17 (VHF), which is located 25 miles east and 35 miles north of his home, and channel 38 (UHF), which is located 20 miles south and 32 miles west of his home. Which model will allow him to receive both channels at the least cost? (Assume that the terrain between Barclay's home and both broadcasting stations is flat.)

31. Towns A, B, C, and D are located as shown in the following figure. Two highways link town A to town D. Route 1 runs from town A to town D via town B, and route 2 runs from town A to town D via town C. If a salesman wishes to drive from town A to town D and traffic conditions are such that he could expect to average the same speed on either route, which highway should he take in order to arrive in the shortest time?

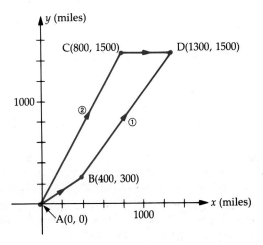

Figure for
Exercises 31 and 32

32. Refer to the figure for Exercise 31. Suppose a fleet of 100 automobiles are to be shipped from an assembly plant in town A to town D. They may be shipped either by freight train along route 1 at a cost of 11 cents per mile per automobile or by truck along route 2 at a cost of 10½ cents per mile per automobile. Which means of transportation minimizes the shipping cost? What is the net savings?

33. Let (x_1, y_1) and (x_2, y_2) be two points lying in the xy-plane. Show that the distance between the two points is given by

$$d = \sqrt{(x_2 - x_1)^2 + (y_2 - y_1)^2}.$$

[*Hint:* Refer to the following figure and use the Pythagorean Theorem.]

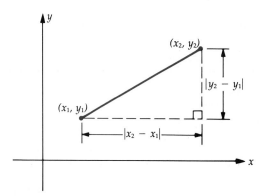

Figure for
Exercise 33

Solutions to Self-Check Exercises

1.1

1. The points are plotted in the following figure:

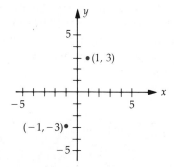

Figure for
Exercise 1

2. Using the distance formula (1) with $x_1 = -1$, $y_1 = 2$ and $x_2 = 3$, $y_2 = -2$, we find that the required distance is

$$d = \sqrt{[3 - (-1)]^2 + [-2 - 2]^2}$$
$$= \sqrt{4^2 + (-4)^2}$$
$$= \sqrt{32}$$
$$= 4\sqrt{2}.$$

3. The distance between city A and city B is

$$d_1 = \sqrt{200^2 + 50^2} \approx 206,$$

or 206 miles. The distance between city B and city C is

$$d_2 = \sqrt{[600 - 200]^2 + [320 - 50]^2}$$
$$= \sqrt{400^2 + 270^2} \approx 483,$$

or 483 miles. Therefore, the total distance the pilot would have to cover is 689 miles, so she must refuel in city B.

1.2
Straight Lines

▶ Slope of a Line

▶ Equations of Lines

▶ Applications

▶ General Equation of a Line

▶ Slope of a Line

In this section we will recall the properties of straight lines and the fact that a straight line in the xy-plane may be represented by a linear equation in x and y.

First, we define the slope of a straight line. Let L denote the unique straight line that passes through the two distinct points (x_1, y_1) and (x_2, y_2). If $x_1 = x_2$, then L is a vertical line and the slope is undefined (see Figure 1.8). If $x_1 \neq x_2$, we define the slope of L as follows:

Slope of a Nonvertical Line

If (x_1, y_1) and (x_2, y_2) are distinct points on a nonvertical line L, then the slope m of L is given by

$$m = \frac{\Delta y}{\Delta x} = \frac{y_2 - y_1}{x_2 - x_1} \tag{2}$$

(see Figure 1.9).

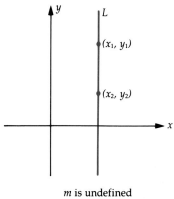

m is undefined

Figure 1.8

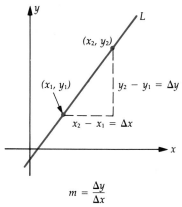

$$m = \frac{\Delta y}{\Delta x}$$

Figure 1.9

Thus, the slope of a straight line is a constant whenever it is defined. The number $\Delta y = y_2 - y_1$ (Δy is read "delta y") is a measure of the vertical change in y, and $\Delta x = x_2 - x_1$ is a measure of the horizontal change in x as shown in Figure 1.9. From this figure we can see that the slope m of a straight line L is a measure of the *rate of change of y with respect to x*.

Figure 1.10(a) shows a straight line L_1 with slope 2. Observe that L_1 has the property that a unit increase in x results in a two-unit increase in y. To see this, let $\Delta x = 1$ in Equation (2) so that $m = \Delta y$. Since $m = 2$, we conclude that $\Delta y = 2$. Similarly, Figure 1.10(b) shows a line L_2 with slope -1. Observe that a straight line with positive slope slants upward from left to right (y increases as x increases), whereas a line with negative slope slants downward from left to right (y decreases as x increases). Finally, Figure 1.11 shows a family of straight lines passing through the origin with indicated slopes.

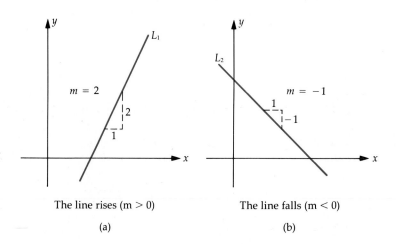

The line rises (m > 0)

(a)

The line falls (m < 0)

(b)

Figure 1.10

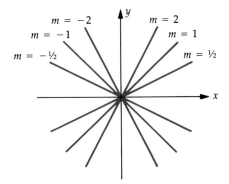

Figure 1.11

EXAMPLE

3 Sketch the graph of the straight line passing through the point $(-2, 5)$ and having slope $-4/3$.

SOLUTION First plot the point $(-2, 5)$ [see Figure 1.12]. Next, recall that the slope of $-4/3$ indicates that an increase of 1 unit in the x-direction produces a *decrease* of 4/3 units in the y-direction, or equivalently, a 3-unit increase in the x-direction produces a 3(4/3), or 4-unit decrease, in the y-direction. Using this information, we plot the point $(1, 1)$. Finally, draw the line through the two points.

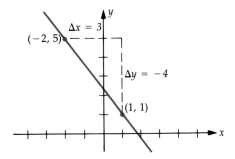

Figure 1.12

EXAMPLE

4 Find the slope m of the line passing through the points $(-1, 1)$ and $(5, 3)$.

SOLUTION Choose (x_1, y_1) to be the point $(-1, 1)$ and (x_2, y_2) to be the point $(5, 3)$. Then, with $x_1 = -1$, $y_1 = 1$, $x_2 = 5$, and $y_2 = 3$, we find, using Equation (2),

$$m = \frac{y_2 - y_1}{x_2 - x_1} = \frac{3 - 1}{5 - (-1)} = \frac{1}{3}$$

(see Figure 1.13).

Try to verify that the result obtained would have been the same had we chosen the point $(-1, 1)$ to be (x_2, y_2) and the point $(5, 3)$ to be (x_1, y_1).

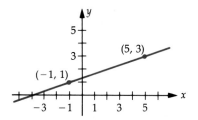

Figure 1.13

EXAMPLE

5 Find the slope of the line passing through the points $(-2, 5)$ and $(3, 5)$.

SOLUTION   The slope of the required line is given by

$$m = \frac{5 - 5}{3 - (-2)} = \frac{0}{5} = 0$$

(see Figure 1.14).

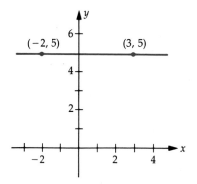

Figure 1.14

We can use the slope of a straight line to determine whether a line is parallel or perpendicular to another line.

Parallel Lines

Two distinct lines are parallel if and only if their slopes are equal or their slopes are undefined.

Perpendicular Lines

If L_1 and L_2 are two distinct nonvertical lines with slopes m_1 and m_2, respectively, then L_1 is perpendicular to L_2 (written $L_1 \perp L_2$) if and only if

$$m_1 = -\frac{1}{m_2}.$$

If the line L_1 is vertical (so that its slope is undefined), then L_1 is perpendicular to another line, L_2, if and only if L_2 is horizontal (so that its slope is zero). For a proof of these results, see Exercises 1.2, Problem 43.

EXAMPLE

6 Let L_1 be a line that passes through the points $(-2, 9)$ and $(1, 3)$, and let L_2 be the line that passes through the points $(-4, 10)$ and $(3, -4)$. Determine whether L_1 and L_2 are parallel.

SOLUTION   The slope m_1 of L_1 is given by

$$m_1 = \frac{3 - 9}{1 - (-2)} = -2.$$

The slope m_2 of L_2 is given by

$$m_2 = \frac{-4 - 10}{3 - (-4)} = -2.$$

Since $m_1 = m_2$, the lines L_1 and L_2 are in fact parallel (see Figure 1.15).

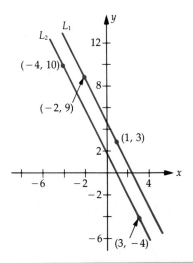

Figure 1.15

▶ Equations of Lines

We will now show that every straight line lying in the xy-plane may be represented by an equation involving the variables x and y. One immediate benefit of this is that problems involving straight lines may be solved algebraically.

Let L be a straight line parallel to the y-axis (perpendicular to the x-axis) [see Figure 1.16]. Then L crosses the x-axis at some point $(a, 0)$ with the x-coordinate given by $x = a$, where a is some real number. The vertical line L is described by the sole condition

$$x = a,$$

and this is accordingly the equation of L. For example, the equation $x = -2$ represents a vertical line 2 units to the left of the y-axis, and the equation $x = 3$ represents a vertical line 3 units to the right of the y-axis (see Figure 1.17).

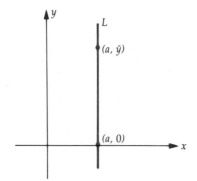

Figure 1.16

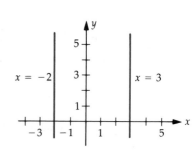

Figure 1.17

Next, suppose that L is a nonvertical line so that it has a well-defined slope m. Suppose (x_1, y_1) is a fixed point lying on L and (x, y) is a variable point on L distinct from (x_1, y_1) [see Figure 1.18]. Using Equation (2) with (x_1, y_1) and $(x_2, y_2) = (x, y)$, we find that the slope of L is given by

$$m = \frac{y - y_1}{x - x_1}.$$

Upon multiplying both sides of the equation by $x - x_1$, we obtain Equation (3).

Point-Slope Form

The equation of the line with slope m and passing through the point (x_1, y_1) is given by

$$y - y_1 = m(x - x_1). \tag{3}$$

Equation (3) is called the point-slope form of a line, since it utilizes a given point (x_1, y_1) on a line and the slope m of the line.

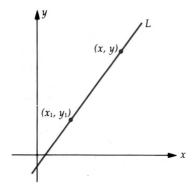

Figure 1.18

EXAMPLE

7 Find an equation of the line passing through the point (1, 3) and having slope 2.

SOLUTION Using the point-slope form of the equation of a line with $P_1(1, 3)$ and $m = 2$, we obtain

$$y - 3 = 2(x - 1),$$

which, when simplified, becomes

$$2x - y + 1 = 0$$

(see Figure 1.19).

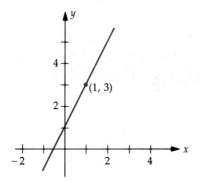

Figure 1.19

◀

EXAMPLE

8 Find an equation of the line passing through the point (3, 1) and perpendicular to the line of Example 7.

SOLUTION   Since the slope of the line in Example 7 is 2, the slope of the required line is given by $m = -1/2$, the negative reciprocal of 2. Using the point-slope form of the equation of a line, we obtain

$$y - 1 = -\frac{1}{2}(x - 3)$$

$$2y - 2 = -x + 3$$

or $\qquad\qquad x + 2y - 5 = 0$

(see Figure 1.20).

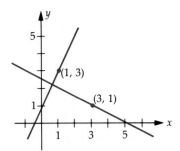

Figure 1.20

EXAMPLE

9   Find an equation of the line passing through the points $(-3, 2)$ and $(4, -1)$.

SOLUTION   The slope of the line is given by

$$m = \frac{-1 - 2}{4 - (-3)} = -\frac{3}{7}.$$

Using the point-slope form of the equation of a line with the point $(4, -1)$ and the slope $m = -3/7$, we have

$$y + 1 = -\frac{3}{7}(x - 4)$$

$$7y + 7 = -3x + 12$$

or $\qquad\qquad 3x + 7y - 5 = 0$

(see Figure 1.21).

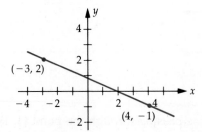

Figure 1.21

A straight line L that is neither horizontal nor vertical cuts the x-axis and the y-axis at, say, points $(a, 0)$ and $(0, b)$, respectively (see Figure 1.22). The numbers a and b are called the **x-intercept** and **y-intercept,** respectively, of L.

Now, let L be a line with slope m and y-intercept b. Using the point-slope form of the equation of a line, Equation (3), with the point given by $(0, b)$ and the slope given by m, we have

$$y - b = m(x - 0)$$

or
$$y = mx + b.$$

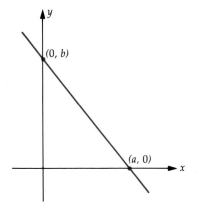

Figure 1.22

Slope-Intercept Form

The equation of the line with slope m and intersecting the y-axis at the point $(0, b)$ is given by

$$y = mx + b. \tag{4}$$

EXAMPLE

10 Find the equation of the line with slope 3 and y-intercept -4.

SOLUTION Using Equation (4) with $m = 3$ and $b = -4$, we obtain the required equation

$$y = 3x - 4. \qquad \blacktriangleleft$$

EXAMPLE

11 Determine the slope and y-intercept of the line whose equation is $3x - 4y = 8$.

SOLUTION   Rewrite the given equation in the slope-intercept form and obtain

$$y = \frac{3}{4}x - 2.$$

Comparing this result with Equation (4) we find $m = 3/4$ and $b = -2$, and we conclude that the slope and y-intercept of the given line are 3/4 and -2, respectively.   ◀

▶ Applications

EXAMPLE

12   The sales manager of a local sporting goods store plotted sales versus time for the last five years and found the points to lie approximately along a straight line (see Figure 1.23). By using the points corresponding to the first and fifth years, find the equation of the trend line. What sales figure can be predicted for the sixth year?

SOLUTION   Using Equation (2) with the points $(1, 20)$ and $(5, 60)$, we find that the slope of the required line is given by

$$m = \frac{60 - 20}{5 - 1} = 10.$$

Next, using the point-slope form of the equation of a line with the point $(1, 20)$ and $m = 10$, we obtain

$$y - 20 = 10(x - 1)$$

or

$$y = 10x + 10,$$

as the required equation.

The sales figure for the sixth year is obtained by letting $x = 6$ in the above equation, giving

$$y = 70,$$

or \$70,000.

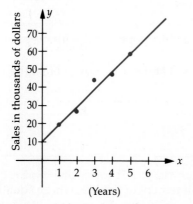

Figure 1.23

EXAMPLE

13 Suppose an art object purchased for $50,000 is expected to appreciate in value at a constant rate of $5,000 per year for the next five years. Use Equation (4) to write an equation predicting the value of the art object in the next several years. What will its value be three years from the date of purchase?

SOLUTION Let x denote the time (in years) that has elapsed since the date the object was purchased and let y denote its value (in dollars). Then $y = 50,000$ when $x = 0$. Furthermore, the slope of the required equation is given by $m = 5,000$, since each unit increase in x (one year) implies an increase of 5,000 units (dollars) in y. Using Equation (4) with $m = 5,000$ and $b = 50,000$ we obtain

$$y = 5,000x + 50,000.$$

Three years from the date of purchase, the value of the object will be given by

$$y = 5,000(3) + 50,000,$$

or $65,000. ◀

▶ General Equation of a Line

We have considered several forms of the equation of a straight line in the plane. These different forms of the equation are equivalent to each other. In fact, each is a special case of the following equation.

General Form of a Linear Equation

The equation

$$Ax + By + C = 0 \qquad (5)$$

where A, B, and C are constants and A and B are not both zero, is called the general form of a linear equation in the variables x and y.

We will now state (without proof) an important result concerning the algebraic representation of straight lines in the plane:

Theorem

An equation of a straight line is a linear equation; conversely, every linear equation represents a straight line.

This result justifies the use of the adjective *linear* in describing Equation (5).

EXAMPLE

14 Sketch the straight line represented by the equation

$$3x - 4y - 12 = 0.$$

SOLUTION Since every straight line is uniquely determined by two distinct points, we need find only two such points through which the line passes in order to sketch it. For convenience let us compute the points at which the line crosses the x and y axes. Setting $y = 0$, we find $x = 4$, so the line crosses the x-axis at the point $(4, 0)$. Setting $x = 0$ gives $y = -3$, so the line crosses the y-axis at the point $(0, -3)$. A sketch of the line appears in Figure 1.24.

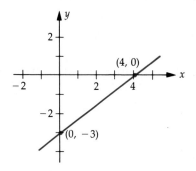

Figure 1.24

Following is a summary of the common forms of the equations of straight lines discussed in this section.

Equations of Straight Lines

Vertical line:	$x = a$
Horizontal line:	$y = b$
Point-slope form:	$y - y_1 = m(x - x_1)$
Slope-intercept form:	$y = mx + b$
General form:	$Ax + By + C = 0$

Self-Check Exercises
1.2

1. Determine the number a so that the line passing through the points $(a, 2)$ and $(3, 6)$ is parallel to a line with slope 4.

2. Find an equation of the line passing through the point $(3, -1)$ and perpendicular to a line with slope $-\frac{1}{2}$.

3. Does the point $(3, -3)$ lie on the line with equation $2x - 3y - 12 = 0$? Sketch the graph of the line.

4. The percentage of people over 65 who have high school diplomas is summarized in the following table:

Year (x)	1960	1965	1970	1975	1980	1985
Percent with Diplomas (y)	20	25	30	36	42	47

a. Plot the percentages of people over 65 who have high school diplomas (y) versus the year (x).

b. Draw the straight line L through the points $(1960, 20)$ and $(1985, 47)$.

c. Find an equation of the line L.

d. If the trend continues, estimate the percentage of people over 65 who will have high school diplomas by the year 2000.

Solutions to Self-Check Exercises 1.2 can be found on page 30.

Exercises 1.2

In Exercises 1–4, find the slope of the line shown in each figure.

1.

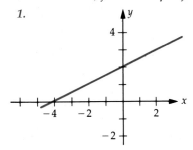

2.

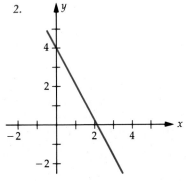

3.

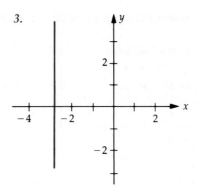

4.

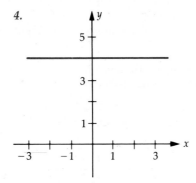

In Exercises 5–10, find the slope of the line passing through the given pair of points.

5. $(4, 3)$ and $(5, 8)$

6. $(4, 5)$ and $(3, 8)$

7. $(-2, 3)$ and $(4, 8)$

8. $(-2, -2)$ and $(4, -4)$

9. (a, b) and (c, d)

10. $(-a + 1, b - 1)$ and $(a + 1, -b)$

11. Given the equation $y = 4x - 3$, answer the following questions.

    a. If x increases by 1 unit, what is the corresponding change in y?

    b. If x decreases by 2 units, what is the corresponding change in y?

12. Given the equation $2x + 3y = 4$, answer the following questions.

    a. Is the slope of the line described by this equation positive or negative?

    b. As x increases in value, does y increase or decrease?

    c. If x decreases by 2 units, what is the corresponding change in y?

In Exercises 13–15, determine whether the lines through the given pairs of points are parallel.

13. $A(1, -2)$, $B(-3, -10)$ and $C(1, 5)$, $D(-1, 1)$

14. $A(2, 3)$, $B(2, -2)$ and $C(-2, 4)$, $D(-2, 5)$

15. $A(-2, 4)$, $B(2, 4)$ and $C(1, -2)$, $D(3, -2)$

In Exercises 16–17, determine whether the lines through the given pairs of points are perpendicular.

16. $A(-2, 5)$, $B(4, 2)$ and $C(-1, -2)$, $D(3, 6)$

17. $A(2, 0)$, $B(1, -2)$ and $C(4, 2)$, $D(-8, 4)$

18. If the line passing through the points $(1, a)$ and $(4, -2)$ is parallel to the line passing through the points $(2, 8)$ and $(-7, a + 4)$, what is the value of a?

In Exercises 19 and 20, determine whether the given points lie on a straight line.

19. $A(-2, 1)$, $B(1, 7)$, and $C(4, 13)$

20. $A(-1, 7)$, $B(2, -2)$, and $C(5, -9)$

21. Find an equation of the horizontal line passing through $(-4, -3)$.

22. Find an equation of the vertical line passing through $(0, 5)$.

23. Find an equation of the line passing through the given point and having the given slope m.

    a. $(3, -4)$; $m = 2$ b. $(2, 4)$; $m = -1$

    c. $(-3, 2)$; $m = 0$

24. Find an equation of the line passing through the given points.

    a. $(2, 4)$ and $(3, 7)$ b. $(1, 2)$ and $(-3, -2)$

    c. $(-1, -2)$ and $(3, -4)$ d. $(2, 1)$ and $(2, 5)$

    e. $(3, 4)$ and $(5, 4)$

25. Find an equation of the line having slope m and y-intercept b.

    a. $m = 3$; $b = 4$ b. $m = -2$; $b = -1$

    c. $m = 0$; $b = 5$

26. Write each of the following equations in the slope-intercept form and then find the slope and y-intercept of the corresponding lines.

    a. $2x - 3y - 9 = 0$ b. $3x - 4y + 8 = 0$

    c. $2x + 4y = 14$ d. $5x + 8y - 24 = 0$

    e. $x - 2y = 0$ f. $y - 2 = 0$

27. Find an equation of the line that passes through the point $(-2, 2)$ and that is parallel to the line $2x - 4y - 8 = 0$.

28. Find an equation of the line that passes through the point $(2, 4)$ and that is perpendicular to the line $3x + 4y - 22 = 0$.

29. Find an equation of the line satisfying the given conditions.

    a. The line parallel to the x-axis and 6 units below it.

    b. The line passing through the origin and parallel to the line joining the points $(2, 4)$ and $(4, 7)$.

    c. The line passing through the point (a, b) with slope equal to zero.

30. Find an equation of the line satisfying the given conditions.

    a. The line that contains the x-axis.

    b. The line passing through $(-3, 4)$ and parallel to the x-axis.

    c. The line passing through $(-5, -4)$ and parallel to the line joining $(-3, 2)$ and $(6, 8)$.

    d. The line passing through (a, b) with undefined slope.

31. Given that the point $P(-3, 5)$ lies on the line $kx + 3y + 9 = 0$, find k.

32. Show that an equation of a line through the points $(a, 0)$ and $(0, b)$ with $a \neq 0$ and $b \neq 0$ can be written on the form

$$\frac{x}{a} + \frac{y}{b} = 1.$$

(Recall that the numbers a and b are the x- and y-intercepts, respectively, of the line. Accordingly, this form of an equation of a line is called the **intercept form.**)

*33. Find an equation of the line whose x- and y-intercepts are:

    a. 3; 4   *b.* -2; -4

    c. $-\dfrac{1}{2}; \dfrac{3}{4}$ *d.* $4; -\dfrac{1}{2}$

    [*Hint:* See Exercise 32.]

34. Which of the following are linear equations?

    a. $3x + 2y^2 = 4$ *b.* $3(x + 2) + 4(y - 3) = 4$

    c. $x(2 - y) + 3(x - 1) = 0$ *d.* $xy = 4$

    e. $y = 0$ *f.* $\dfrac{3 - x}{2} = \dfrac{y - x}{4}$

35. For each of the following, sketch the straight line defined by the given linear equation by finding the x- and y-intercepts. [*Hint:* See Example 14.]

    a. $3x - 2y + 6 = 0$ *b.* $2x - 5y + 10 = 0$

    c. $x + 2y - 4 = 0$ *d.* $2x + 3y - 15 = 0$

    e. $y + 5 = 0$ *f.* $-2x - 8y + 24 = 0$

36. Show that two lines with equations $a_1x + b_1y + c_1 = 0$ and $a_2x + b_2y + c_2 = 0$, respectively, are parallel if and only if $a_1b_2 - b_1a_2 = 0$. [*Hint:* Write each equation in the point-slope form and compare.]

37. During the period 1986–1989, social security contributions by employees are scheduled to be 7.15 percent of the employee's wages (up to the maximum taxable wage base).

    a. Find an equation expressing the relationship between the wages earned (x) and the social security taxes paid (y) by an employee.

    b. For each additional dollar that an employee earns, by how much is his or her social security contribution increased? (Assume that the employee's wages are less than the maximum taxable wage base.)

    c. What social security contributions will an employee who earns $35,000 be required to make?

38. Using data compiled by the Admissions Office at Faber University, college admission officers estimate that 55 percent of the students who are offered admission to the freshman class at the university will actually enroll.

    a. Find an equation expressing the relationship between the number of students who actually enroll (y) and the number of students who are offered admission to the university (x).

---

*An asterisk indicates a more difficult exercise.

b. If the desired freshman class size for the upcoming academic year is 1100 students, how many students should be admitted?

39. The equation $W = 3.51L - 192$, expressing the relationship between the length L (in feet) and the expected weight W (in British tons) of adult blue whales, was adopted in the late 1960s by the International Whaling Commission.

a. What is the expected weight of an 80-foot blue whale?

b. Sketch the straight line that represents the equation.

40. A manufacturer obtained the following data relating the cost y (in dollars) to the number of units (x) of a commodity produced:

Number of Units Produced (x)	0	20	40	60	80	100
Cost in Dollars (y)	200	208	222	230	242	250

a. Plot the cost (y) versus the quantity produced (x).

b. Draw the straight line through the points $(0, 200)$ and $(100, 250)$.

c. Derive an equation of the straight line of part (b).

d. Taking this equation to be an approximation of the relationship between the cost and the level of production, estimate the cost of producing 54 units of the commodity.

41. The Venus Health Club for Women provides its members with the following table, which gives the average desirable weight for women of a certain height:

Height x *(in Inches)*	60	63	66	69	72
Weight y *(in Pounds)*	108	118	129	140	152

a. Plot the weight (y) versus the height (x).

b. Draw a straight line L through the points corresponding to heights of 5 feet and 6 feet.

c. Derive an equation of the line L.

d. Using the equation of part (c), estimate the average desirable weight for a woman who is 5 feet 5 inches tall.

42. The annual sales (in millions of dollars) of the Metro Department Store during the past five years were:

Annual Sales	5.8	6.2	7.2	8.4	9.0
Year	1	2	3	4	5

a. Plot the annual sales (y) versus the year (x).

b. Draw a straight line L through the points corresponding to the first and fifth years.

c. Derive an equation of the line L.

d. Using the equation found in (*c*), estimate the annual sales for the Metro Department Store four years from now.

43. Prove that if a line L_1 with slope m_1 is perpendicular to a line L_2 with slope m_2, then $m_1 m_2 = -1$. [*Hint:* Refer to the following figure. Show that $m_1 = b/a$ and $m_2 = c/a$. Next, apply the Pythagorean Theorem and the distance formula to the triangle OBA to show that $a^2 = -bc$.]

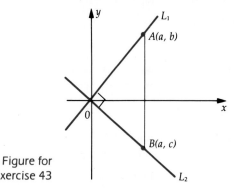

Figure for
Exercise 43

Solutions to Self-Check Exercises

1.2

1. The slope of the line passing through the points $(a, 2)$ and $(3, 6)$ is

$$m = \frac{6 - 2}{3 - a} = \frac{4}{3 - a}.$$

Since this line is parallel to a line with slope 4, m must be equal to 4; that is,

$$\frac{4}{3 - a} = 4$$

or, upon multiplying both sides of the equation by $3 - a$,

$$4 = 4(3 - a)$$
$$4 = 12 - 4a$$
$$4a = 8$$

and $\qquad\qquad a = 2.$

2. Since the required line L is perpendicular to a line with slope $-\frac{1}{2}$, the slope of L is

$$-\frac{1}{-\frac{1}{2}} = 2.$$

Next, using the point-slope of the equation of a line, we have

$$y - (-1) = 2(x - 3)$$
$$y + 1 = 2x - 6$$

or

$$y = 2x - 7.$$

3. Substituting $x = 3$ and $y = -3$ into the left-hand side of the given equation, we find

$$2(3) - 3(-3) - 12 = 3,$$

which is not equal to zero (the right-hand side). Therefore, $(3, -3)$ does not lie on the line with equation $2x - 3y - 12 = 0$.

Setting $x = 0$, we find $y = -4$, the y-intercept. Next, setting $y = 0$ gives $x = 6$, the x-intercept. We now draw the line passing through the points $(0, -4)$ and $(6, 0)$ as shown.

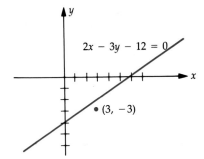

Figure for
Exercise 3

4. *a* and *b*. See the accompanying figure.

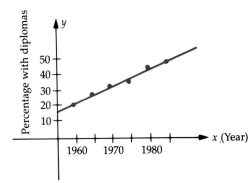

Figure for
Exercises 4a and 4b

c. The slope of L is

$$m = \frac{47 - 20}{1985 - 1960} = \frac{27}{25}.$$

Using the point-slope form of the equation of a line with the point $(1960, 20)$, we find

$$y - 20 = \frac{27}{25}(x - 1960)$$

$$= \frac{27}{25}x - \frac{10584}{5}$$

or

$$y = \frac{27}{25}x - \frac{10484}{5}.$$

d. To estimate the percentage of people over 65 who will have high school diplomas by the year 2000, let $x = 2000$ in the equation obtained in part (c). Thus, the required estimate is

$$y = \frac{27}{25}(2000) - \frac{10484}{5}$$

$$\approx 63.2,$$

or approximately 63 percent.

▶ 1.3

Linear Functions and Their Applications

▶ Functions

▶ Simple Interest and Simple Discount

▶ Simple Depreciation

▶ Linear Cost, Revenue, and Profit Functions

▶ Linear Demand and Supply Curves

▶ Functions

In many mathematical applications we are concerned with the relationship between two quantities. This relationship is conveniently described by the concept of a **function.** The following definition is suitable for our purpose:

Function

A function $y = f(x)$ is a rule that assigns to each value of x one and only one value of y.

The number y is normally denoted by $f(x)$, read "f of x," emphasizing the dependency of y upon x.

An example of a function may be drawn from the familiar relationship between the area of a circle and its radius. Letting x and y denote the radius and area of a circle, respectively, we have from elementary geometry,

$$y = \pi x^2.$$

This equation defines y as a function of x, since for each admissible value of x, that is, a nonnegative number representing the radius of a certain circle, there corresponds precisely one number $y = \pi x^2$ giving the area of the circle. This *area function* may be written as

$$f(x) = \pi x^2. \tag{6}$$

For example, to compute the area of a circle with a radius of 5 inches, we simply replace x in Equation (6) by the number 5. Thus, the area of the circle is

$$f(5) = \pi 5^2 = 25\pi,$$

or 25π square inches.

Suppose we are given the function $y = f(x)$.[†] The variable x is referred to as the **independent variable,** and the variable y is called the **dependent variable.** The set of all values that may be assumed by x is called the **domain** *of the function* f, and the set comprising all the values assumed by $y = f(x)$ as x takes on all possible values in its domain is called the **range** *of the function* f. For the area function (6), the domain of f is the set of all nonnegative numbers x, and the range of f is the set of all nonnegative numbers y.

We now focus our attention on an important class of functions known as linear functions. Recall that a linear equation in x and y is of the form $Ax + By + C = 0$, where A, B, and C are constants and A and B are not both zero. If $B \neq 0$, the equation can always be solved for y in terms of x; in fact, as we saw in Section 1.2, the equation may be cast in the slope-intercept form:

$$y = mx + b \quad (m, b \text{ constants}) \tag{7}$$

Equation (7) defines y as a function of x. The domain (and range) of this function is the set of all real numbers. Furthermore, the graph of the function, as we saw in Section 1.2, is a straight line in the plane. For this reason, the function $f(x) = mx + b$ is called a *linear function.*

Linear Function

The function f defined by

$$f(x) = mx + b,$$

where m and b are constants, is called a **linear function.**

[†]It is customary to refer to a function f as $f(x)$.

Linear functions play an important role in the quantitative analysis of business and economic problems. First, many problems arising in these and other fields are *linear* in nature or are *linear* in the intervals of interest and thus can be formulated in terms of linear functions. Second, because linear functions are relatively easy to work with, assumptions involving linearity are often made in the formulation of problems. In many cases these assumptions are justified, and acceptable mathematical models are obtained that approximate real-life situations.

▶ Simple Interest and Simple Discount

A natural application of linear functions to the business world is found in computations of **simple interest**—interest that is computed on the original principal only. Thus, if I denotes the interest on a principal P (in dollars) at an interest rate of r per year for n years, then we have

$$I = Prn.$$

The **accumulated value** A, the sum of the principal and interest after n years, is given by

$$A = P + I = P + Prn$$

or

$$A = P(1 + rn)$$

and is a linear function of n (see Problem 10, Exercises 1.3). In a business application we are normally interested only in the case where n is positive, so only that part of the line that lies in Quadrant I is of interest to us (see Figure 1.25).

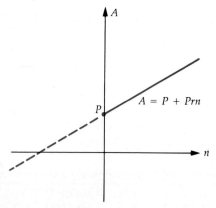

Figure 1.25

Simple Interest Formulas

Interest: $\qquad I = Prn$

Accumulated Amount: $\quad A = P(1 + rn)$ $\qquad$ (8)

EXAMPLE

15 A bank pays simple interest at the rate of 8 percent per year for certain deposits. If a customer deposits $1000 and makes no withdrawals for three years, what is the total amount on deposit at the end of three years? What is the interest earned in that period of time?

SOLUTION The total amount on deposit at the end of three years is given by

$$A = P(1 + rn)$$
$$= 1000[1 + (0.08)(3)]$$
$$= 1240,$$

or $1240.
　　　The interest earned over the three-year period is given by

$$I = Prn$$
$$= 1000(0.08)(3)$$
$$= 240,$$

or $240.　　　　　　　　　　　　　　　　　　　　　　　　　　◄

EXAMPLE

16 An amount of $2000 is invested in a ten-year trust fund that pays 6 percent annual simple interest. What is the total amount of the trust fund at the end of ten years?

SOLUTION The total amount of the trust fund at the end of ten years is given by

$$A = P(1 + rn)$$
$$= 2000[1 + (0.06)(10)]$$
$$= 3200,$$

or $3200.　　　　　　　　　　　　　　　　　　　　　　　　　◄

For small loans made over relatively short periods of time a **simple discount** procedure is normally employed. Simple discount is also referred to as "banker's discount" or "interest paid in advance." In this procedure the borrower agrees to pay a certain sum A at a certain date, say in n years' time. The lender deducts a certain percentage of this amount and gives the borrower the remainder. The amount deducted, D, is the *amount discounted* and is the (simple) interest on the amount A at the *discount rate* of d(percent) per year for n years. Thus,

$$D = Adn.$$

If P denotes the amount the borrower receives, called the **present value** of the amount A, then

$$P = A - D = A - Adn$$
$$= A(1 - dn)$$

and is a linear function of n (see Problem 14, Exercises 1.3).

Simple Discount Formulas

Amount discounted: $D = Adn$

Present value: $P = A(1 - dn)$ (9)

EXAMPLE

17   Mr. Brown signs a bank note in which he agrees to pay back $1000 in four months. The bank's discount rate is 9 percent. How much does Mr. Brown receive from the bank?

SOLUTION   Mr. Brown receives

$$P = 1000\left[1 - (0.09)\left(\frac{1}{3}\right)\right]$$
$$= 970,$$

or $970.   ◄

EXAMPLE

18   Ms. Smith receives $95.50 from a bank and promises to repay $100 six months later. What is the corresponding discount rate?

SOLUTION   We have

$$95.5 = 100\left(1 - \frac{d}{2}\right)$$

$$0.955 = 1 - \frac{d}{2}$$

$$\frac{d}{2} = 1 - 0.955$$

or $d = 0.09,$

giving a discount rate of 9 percent per year.   ◄

▶ Simple Depreciation

In computing income tax, business firms are allowed by law to depreciate certain assets such as buildings, machines, furniture, automobiles, and so on, over a period of time. The **linear depreciation,** or *straight line method,* is often used for this purpose. The following example illustrates how to derive an equation describing the book value of an asset being depreciated linearly.

EXAMPLE

19 A printing machine has an initial value of $100,000 and is to be depreciated linearly over five years with a $30,000 scrap value. Find an expression giving the book value at the end of the second year. What is the rate of depreciation of the printing machine?

SOLUTION Let V denote the book value of the asset at the end of the nth year. Since the depreciation is linear, V is a linear function of n; that is,

$$V = an + b \quad (a, b \text{ constants})$$

To determine the values of a and b, observe that the initial value of the machine, $100,000, translates into the condition $V = 100,000$ when $n = 0$. Using this condition, we find

$$100,000 = a(0) + b,$$

or $b = 100,000$. Next, the scrap value of $30,000 at the end of five years tells us that $V = 30,000$ when $n = 5$. Using this condition and the value of b found earlier gives

$$30,000 = a(5) + 100,000$$
$$5a = -70,000,$$

or $a = -14,000$. Therefore, the required expression is

$$V = -14,000n + 100,000.$$

The book value at the end of the second year is given by

$$V = -14,000(2) + 100,000 = 72,000,$$

or $72,000. The rate of depreciation of the machine is given by the negative of the slope of the (straight) depreciation line. Since the slope of the line is $a = -14,000$, the rate of depreciation is $14,000 per year. ◀

▶ Linear Cost, Revenue, and Profit Functions

Whether a business is a sole proprietorship or a large corporation, the owner or chief executive must constantly keep track of operating costs, revenue resulting from the sale of products or services, and perhaps most important, the profits realized. Three functions provide management with a measure of these quantities: the total cost function, the total revenue function, and the total profit function.

Cost, Revenue, and Profit Functions

Let x denote the number of units of a product manufactured or sold. Then, the **total cost function** is

$$C(x) = \text{total cost of manufacturing } x \text{ units of the product.}$$

The **total revenue function** is

$$R(x) = \text{total revenue realized from the sale of } x \text{ units of the product.}$$

The **total profit function** is

$$P(x) = \text{total profit realized from manufacturing and selling } x \text{ units of the product.}$$

[*Note:* Profit = Revenue − Cost, or $P(x) = R(x) - C(x)$.]

Generally speaking, the total cost, total revenue, and total profit functions associated with a company are more likely than not to be nonlinear (these functions are best studied using the tools of calculus). But *linear* cost, revenue, and profit functions do arise in practice, and we will consider such functions in this section. Before deriving explicit forms of these functions, we need to recall some common terminology.

The costs incurred in operating a business are usually classified into two categories. Costs that remain more or less constant regardless of the level of activity of the firm are called **fixed costs.** Examples of fixed costs are rental fees and executive salaries. On the other hand, costs that vary with production or sales are called **variable costs.** Examples of variable costs are wages and costs for raw materials.

Suppose a firm has a fixed cost of F dollars, a production cost of c dollars per unit, and a selling price of s dollars per unit. Then the *cost function* $C(x)$, the *revenue function* $R(x)$, and the *profit function* $P(x)$ for the firm are given by

$$C(x) = cx + F$$
$$R(x) = sx$$

and
$$P(x) = R(x) - C(x) \quad (\text{revenue} - \text{cost})$$
$$= (s - c)x - F,$$

where x denotes the number of units of the commodity produced and sold. The functions C, R, and P are linear functions of x.

EXAMPLE

20   Puritron, a manufacturer of water filters, has a monthly fixed cost of $20,000, a production cost of $20 per unit, and a selling price of $30 per unit. Find the cost function, the revenue function, and the profit function for the company.

SOLUTION   Let x denote the number of units produced and sold. Then

$$C(x) = 20x + 20{,}000$$
$$R(x) = 30x$$

and
$$P(x) = R(x) - C(x) = 30x - (20x + 20{,}000)$$
$$= 10x - 20{,}000.$$ ◀

▶ Linear Demand and Supply Curves

In a free market economy, consumer demand for a particular commodity depends on the commodity's unit price. A **demand equation** expresses this relationship between the unit price and the quantity demanded. The corresponding graph of the demand equation is called a **demand curve.** In general, the quantity demanded of a commodity decreases as its unit price increases, and vice versa. Accordingly, a **demand function,** defined by $p = f(x)$, where p measures the unit price and x measures the number of units of the commodity, is generally characterized as a decreasing function of x; that is, $p = f(x)$ decreases as x increases.

The simplest demand function is defined by a linear equation in x and p. Its graph is a straight line having a negative slope, and both x and p assume only nonnegative values. Thus, the demand curve in this case is that part of the graph of a straight line that lies in the first quadrant (Figure 1.26). The p-intercept, p_0, is the highest price that anyone would pay for the commodity—there is no demand for the commodity at all if the unit price is greater than or equal to p_0. The x-intercept, x_0, is the quantity demanded when the commodity is free—the total demand never exceeds x_0.

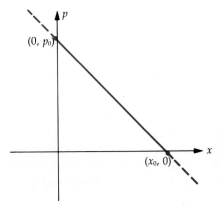

Figure 1.26

EXAMPLE

21 The demand for a certain model of clock is 48,000 units when the unit price is $8. At $12 per unit, the demand drops to 32,000 units. Given that it is linear, find the demand equation. What is the highest price anyone would pay for a clock? What is the maximum quantity demanded?

SOLUTION   Let p denote the unit price of a clock (in dollars) and let x (measured in units of 1000) denote the quantity demanded when the unit price of the clocks is p. When $p = 8$, $x = 48$ and the point (48, 8) lies on the demand curve. Similarly, when $p = 12$, $x = 32$ and the point (32, 12) also lies on the demand curve. Given that the demand equation is linear, its graph is a straight line. The slope of the required line is given by

$$m = \frac{12 - 8}{32 - 48} = \frac{4}{-16} = -\frac{1}{4},$$

so using the point-slope form of the equation of a line, we find that

$$p - 8 = -\frac{1}{4}(x - 48)$$

or

$$p = -\frac{1}{4}x + 20$$

is the required equation. The demand curve is shown in Figure 1.27. The highest price anyone would pay for a clock, $20, is given by the p-intercept. The maximum quantity demanded, 80,000 units, is given by the x-intercept.

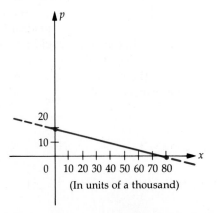

Figure 1.27

In a competitive market, a relationship also exists between the unit price of a commodity and its availability in the market. In general, an increase in a commodity's unit price will induce the manufacturer to increase the supply of that commodity. Conversely, a decrease in the unit price generally leads to a drop in the supply. The equation that expresses the relationship between the unit price and the quantity supplied is called a **supply equation,** and the corresponding graph is called a **supply curve.** A **supply function,** defined by $p = f(x)$, is generally characterized by an increasing function of x, that is, $p = f(x)$ increases as x increases.

As in the case of a demand equation, the simplest supply equation is a linear equation in p and x, where p and x have the same meaning as before but the line has a positive slope. The supply curve corresponding to a linear supply function is that part of the straight line that lies in the first quadrant (Figure 1.28). The p-intercept, p_0, is the lowest price at which the supplier will offer the commodity.

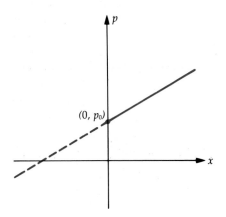

Figure 1.28

EXAMPLE

22 The supply equation for a commodity is given by $4p - 5x = 120$, where p is measured in dollars and x is measured in units of 100.

a. Sketch the corresponding curve.

b. What is the lowest price at which the supplier will make the commodity available in the market?

c. How many units will be marketed when the unit price is $55?

SOLUTION

a. Setting $x = 0$ and $p = 0$, we find that the p- and x-intercepts are 30 and -24, respectively. The supply curve is sketched in Figure 1.29.

b. From the solution to (a) we see that the p-intercept is 30; therefore the lowest price at which the commodity will be made available is $30.

c. Substituting $p = 55$ in the supply equation, we have $4(55) - 5x = 120$, or $x = 20$, so the amount marketed will be 2000 units.

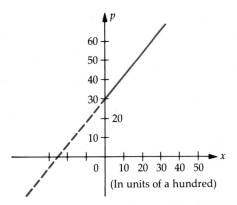

Figure 1.29

Self-Check Exercises

1.3

1. Find the accumulated amount at the end of 15 months on a $1000 deposit in a bank paying simple interest at the rate of 8 percent per year.

2. A manufacturer has a monthly fixed cost of $60,000 and a production cost of $10 for each unit produced. The product sells for $15 per unit.

   a. What is the cost function?

   b. What is the revenue function?

   c. What is the profit function?

   d. Compute the profit (loss) corresponding to production levels of 10,000 and 14,000 units.

3. There is no demand for a certain make of 30" x 52" area rugs when the unit price is $120 or more. But for each $20 decrease in the unit price, the quantity demanded increases by 500 units. Find the demand equation and sketch its graph.

Solutions to Self-Check Exercises 1.3 can be found on page 45.

▶ Exercises 1.3

1. Find the simple interest on a $500 investment made for two years at an interest rate of 8 percent per year. What is the accumulated amount?

2. Find the accumulated amount after 180 days on a $10,000 investment at a simple interest rate of 10 percent per year. (Assume 360 days in a year; interest computed under this assumption is called *ordinary simple interest.*)

3. Find the accumulated amount at the end of nine months on an $800 deposit in a bank paying simple interest at a rate of 6 percent per year.

4. If the accumulated amount is $1160 at the end of two years and the simple rate of interest is 8 percent per year, what is the principal?

5. How many days will it take for a sum of $1000 to earn $15 interest if it is deposited in a bank paying ordinary simple interest at the rate of 5 percent per year?

6. How much time will it take for a sum of money to double when it earns (simple) interest at the rate of 6 percent per year?

7. Mr. Mitchell has been given the option of either paying his $300 bill now or settling it for $306 after 30 days. If he chooses to pay after 30 days, find the simple interest rate at which he would be charged.

8. A bank deposit paying simple interest at the rate of 5 percent per year grew to a sum of $3100 in ten months. Find the principal.

9. An automobile including all optional equipment costs $8400. The trade-in is valued at $3400. The (simple) interest rate is 8 percent per year. Compute the monthly payment on a three-year contract.

10. Write Equation (8) in the slope-intercept form and interpret the meaning of the slope and the A-intercept in terms of r and P.

11. A man receives $10,000 from a bank and promises to repay $12,500 two years later. What is the simple discount rate used by the bank?

12. Find the present value of $3000 due at the end of two years if the (simple) discount rate is 8 percent per year.

13. What is the present value of a note for $10,000 due in 60 days if the simple discount rate is 9 percent per year? (Use a 360-day year.)

14. Write Equation (9) in the slope-intercept form and interpret the meaning of the slope and the P-intercept in terms of d and A. Compare your results with Exercise 10.

15. An office building worth $1,000,000 when completed in 1975 is being depreciated linearly over 50 years. What will the book value of the building be in 1985? in 1990?

16. An automobile purchased for the use of the manager of a firm at a price of $14,000 is to be depreciated using the straight line method over five years. What will the book value of the automobile be at the end of three years?

17. The consumption function in a certain economy is given by the equation

$$C(y) = 0.75y + 6$$

where $C(y)$ is the personal consumption expenditure in billions of dollars and y is the disposable personal income in billions of dollars. Find $C(0)$, $C(50)$, $C(100)$.

18. In a certain state, the sales tax T on the amount of taxable goods is 6 percent of the value of the goods purchased (x) where both T and x are measured in dollars.

a. Express T as a function of x.
b. Find $f(200)$, $f(5.65)$.

19. Social security recipients receive an automatic cost-of-living adjustment (COLA) once each year. Their monthly benefit is increased by the amount that consumer prices have increased during the preceding year. Suppose that consumer prices have increased by 5.3 percent during the preceding year.

a. Express the adjusted monthly benefit of a social security recipient as a function his current monthly benefit.

b. If Mr. Harrington's monthly social security benefit is now $620, what will his adjusted monthly benefit be?

20. In 1980, the National Textile Company installed a new machine in one of its factories at a cost of $250,000. The machine is depreciated linearly over ten years with a scrap value of $10,000.

    a. Find an expression for the book value of the machine in the nth year of use $(0 \leq n \leq 10)$.

    b. Sketch the graph of the function of part (*a*).

    c. Find the book value of the machine in 1984.

    d. Find the rate at which the machine is being depreciated.

21. A minicomputer purchased at a cost of $60,000 in 1978 had a residual (scrap) value of $12,000 at the end of four years. If the straight line method of depreciation is used:

    a. Find the rate of depreciation.

    b. Find the linear equation expressing the book value of the minicomputer at the end of n years.

    c. Sketch the graph of the function of part (*b*).

    d. Find the book value of the minicomputer at the end of the third year.

22. Auto-time, a manufacturer of 24-hour variable timers, has a monthly fixed cost of $48,000 and a production cost of $8 for each timer manufactured. The timers sell for $14 each.

    a. What is the cost function?

    b. What is the revenue function?

    c. What is the profit function?

    d. Compute the profit (loss) corresponding to production levels of 4,000, 6,000, and 10,000 timers, respectively.

23. The management of T.M.I. Inc. finds that the monthly fixed costs attributable to the blank tape division of the company amount to $12,100. If the cost for producing each reel of tape is $.60 and each reel of tape sells for $1.15, find the cost function, the revenue function, and the profit function for the company.

For each of the demand equations in Exercises 24–27, where x represents the quantity demanded in units of 1000 and p is the unit price in dollars:

    a. Sketch the demand curve.

    b. Determine the highest price that the consumer is willing to pay for a unit of the commodity.

    c. Determine the largest amount of the commodity that will be demanded.

24. $2x + 3p - 18 = 0$ **25.** $5p + 4x - 80 = 0$

26. $p = -3x + 60$ **27.** $p = -0.4x + 120$

28. At a unit price of $55, the quantity demanded of a certain commodity is 1000 units. At a unit price of $85, the demand drops to 600 units. Given that it is linear, find the demand equation. Above what price will there be no demand? What is the maximum quantity demanded?

29. There is no demand for a certain commodity when the unit price is $100 or more. But for each $10 decrease in price, the quantity demanded increases by 200 units. Find the demand equation and sketch its graph.

30. Assume that the demand equation for a certain commodity has the form $p = ax + b$, where x is the quantity demanded and p is the unit price in dollars. Determine the demand equation if the highest price a customer is willing to pay for a unit of the commodity is $10 and the largest amount of the commodity that will be demanded is 10,000 units. What is the quantity demanded when the unit price is $7.50?

31. The demand equation for the Sicard wristwatch is

$$p = -0.025x + 50$$

where x is the quantity demanded per week and p is the unit price in dollars. Sketch the graph of the demand equation. What is the highest price anyone would pay for the watch?

For each supply equation in Exercises 32–35, where x is the quantity supplied in units of 1000 and p is the unit price in dollars:

    a. Sketch the supply curve.

    b. Determine the lowest price at which the supplier will make the commodity available in the market.

32. $3x - 4p + 24 = 0$ 33. $\frac{1}{2}x - \frac{2}{3}p + 12 = 0$

34. $p = 2x + 10$ 35. $p = \frac{1}{2}x + 20$

36. Suppliers of 15-inch black and white television sets will not make any available on the market if the unit price is $40. However, at a unit price of $50, 20,000 units will be made available. Assuming that the relationship between the unit price and the quantity supplied is linear, derive the supply equation. Sketch the supply curve and determine the quantity suppliers will make available when the unit price is $70.

37. Producers will make 2000 refrigerators available when the unit price is $330. At a unit price of $390, 6000 refrigerators will be marketed. Find the equation relating the unit price of a refrigerator to the quantity supplied if the equation is known to be linear. How many refrigerators will be marketed when the unit price is $450? What is the lowest price at which a refrigerator will be marketed?

Solutions to Self-Check Exercises

1.3

1. Here $P = 1000$, $r = 0.08$, and $n = \dfrac{15}{12}$. Therefore, the accumulated amount at the end of 15 months is given by

$$A = 1000\left[1 + (0.08)\left(\frac{15}{12}\right)\right]$$
$$= 1000(1 + 0.1)$$
$$= 1100,$$

or $1100.

2. Let x denote the number of units produced and sold. Then

    a. $C(x) = 10x + 60,000$.

    b. $R(x) = 15x$.

    c. $P(x) = R(x) - C(x) = 15x - (10x + 60,000)$
               $= 5x - 60,000$.

d. $P(10,000) = 5(10,000) - 60,000$
$$= -10,000,$$

or a loss of $10,000.

$$P(14,000) = 5(14,000) - 60,000$$
$$= 10,000,$$

or a profit of $10,000.

3. Let p denote the price of a rug (in dollars) and let x denote the quantity of rugs demanded when the unit price is p. The condition that there is no demand when the unit price is $120 or more ($x = 0$ when $p = 120$) tells us that the demand curve passes through the point (0, 120). Next, the condition that for each $20 decrease in the unit price the quantity demanded increases by 500 tells us that the demand curve is linear and has (constant) slope given by $-20/500$ or $-1/25$. Therefore, letting $m = -1/25$ in the demand equation

$$p = mx + b$$

we find

$$p = -\frac{1}{25}x + b.$$

To determine b, use the fact that the straight line passes through (0, 120) to obtain

$$120 = -\frac{1}{25}(0) + b,$$

or $b = 120$. Therefore, the required equation is

$$p = -\frac{1}{25}x + 120.$$

The demand curve is sketched in the accompanying figure.

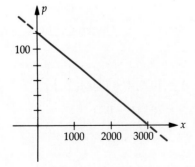

Figure for
Exercise 3

1.4

Intersection of Straight Lines

▶ Finding the Point of Intersection

▶ Break-Even Analysis

▶ Market Equilibrium

▶ Finding the Point of Intersection

The solution of certain practical problems sometimes involves finding the point of intersection of two straight lines. To see how such a problem may be solved algebraically, suppose we are given two straight lines L_1 and L_2 with equations

$$y = m_1x + b_1 \quad \text{and} \quad y = m_2x + b_2$$

(where m_1, b_1, m_2, and b_2 are constants) that intersect at the point $P(x, y)$ [see Figure 1.30].

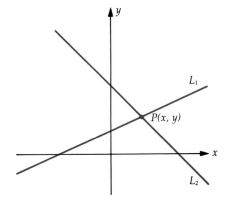

Figure 1.30

The point $P(x, y)$ lies on the line L_1 and therefore satisfies the equation $y = m_1x + b_1$. At the same time, it also lies on the line L_2 and satisfies the equation $y = m_2x + b_2$. Therefore, to find the point of intersection $P(x, y)$ of the lines L_1 and L_2, we solve the equations

$$y = m_1x + b_1 \quad \text{and} \quad y = m_2x + b_2$$

simultaneously for x and y.

EXAMPLE

23 Find the point of intersection of the straight lines with equations $y = x + 1$ and $y = -2x + 4$.

SOLUTION We solve the given equations simultaneously. Substituting the value y as given in the first equation into the second, we obtain

$$x + 1 = -2x + 4$$
$$3x = 3$$

or

$$x = 1.$$

Substituting this value of x into either one of the given equations gives $y = 2$. Therefore, the required point of intersection is $(1, 2)$ [see Figure 1.31].

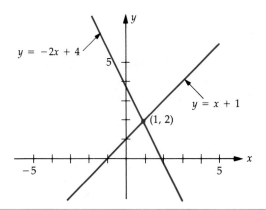

Figure 1.31

We now turn to some applications involving the intersections of pairs of straight lines.

▶ Break-Even Analysis

Consider a firm with (linear) cost function $C(x)$, revenue function $R(x)$, and profit function $P(x)$ given by

$$C(x) = cx + F$$
$$R(x) = sx$$

and

$$P(x) = R(x) - C(x) = (s - c)x - F$$

where c denotes the unit cost of production, s denotes the selling price per unit, F denotes the fixed cost incurred by the firm, and x denotes the level of production and sales. The level of production at which the firm neither makes a profit nor sustains a loss is called the **break-even level of operation** and may be determined by solving the equations $p = C(x)$ and $p = R(x)$ simultaneously. For, at this level of production, x_0, the profit is zero, so $P(x_0) = R(x_0) - C(x_0) = 0$ and $R(x_0) = C(x_0)$. The point $P_0(x_0, p_0)$, the solution of the simultaneous equations $p = R(x)$ and $p = C(x)$, is referred to as the **break-even point;** the number x_0 and the number p_0 are called the **break-even quantity** and the **break-even revenue,** respectively.

Geometrically, the break-even point $P_0(x_0, p_0)$ is just the point of intersection of the straight lines representing the cost and revenue functions, respectively. This follows because $P_0(x_0, p_0)$, being the solution of the simultaneous equations $p = R(x)$ and $p = C(x)$, must lie on both these lines simultaneously (Figure 1.32). Note that if $x < x_0$, then $R(x) < C(x)$ so that $P(x) = R(x) - C(x) < 0$ and thus the firm sustains a loss at this level of production. On the other hand, if $x > x_0$, then $P(x) > 0$ and the firm operates at a profitable level.

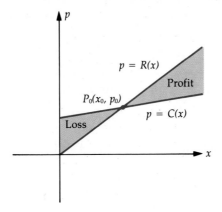

Figure 1.32

EXAMPLE

24 Prescott, Inc. manufactures its products at a cost of $4 per unit and sells them for $10 per unit. If the fixed cost for the firm is $12,000 per month, determine the break-even point for the firm.

SOLUTION The cost function C and the revenue function R are given by $C(x) = 4x + 12,000$ and $R(x) = 10x$, respectively (Figure 1.33).

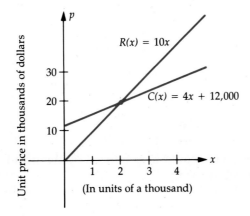

Figure 1.33

Setting $R(x) = C(x)$, we obtain

$$10x = 4x + 12{,}000$$
$$6x = 12{,}000$$
$$x = 2{,}000.$$

Substituting this value of x into $R(x) = 10x$ gives $R(2000) = (10)(2000) = 20{,}000$. So, for a break-even operation, the firm should manufacture 2,000 units of its product, resulting in a break-even revenue of $20,000. ◀

EXAMPLE

25 Using the data supplied in Example 24, answer the following questions:

a. What is the loss sustained by the firm if only 1500 units are produced and sold per month?

b. What is the profit if 3000 units are produced and sold per month?

c. What is the number of units the firm should produce in order to realize a minimum monthly profit of $9000?

SOLUTION The profit function P is given by the rule

$$P(x) = R(x) - C(x)$$
$$= 10x - (4x + 12{,}000)$$
$$= 6x - 12{,}000.$$

a. If 1500 units are produced and sold per month, we have

$$P(1{,}500) = 6(1{,}500) - 12{,}000$$
$$= -3{,}000,$$

so the firm will sustain a loss of $3,000 per month.

b. If 3,000 units are produced and sold per month, we have

$$P(3{,}000) = 6(3{,}000) - 12{,}000$$
$$= 6{,}000,$$

or a monthly profit of $6,000.

c. Substituting 9,000 for $P(x)$ in the equation $P(x) = 6x - 12{,}000$, we obtain

$$9{,}000 = 6x - 12{,}000$$
$$6x = 21{,}000$$
$$x = 3{,}500.$$

Thus, at least 3,500 units should be produced by the firm in order to realize a $9,000 minimum monthly profit. ◀

EXAMPLE

26 The management of a firm must decide between two manufacturing processes for a certain product. The monthly cost of the first process is given by $C_1(x) = 20x + 10{,}000$, where x is the number of units produced, whereas the monthly cost of the second process is given by $C_2(x) = 10x + 30{,}000$. Both costs are measured in dollars. If the projected sales are 800 units at a unit price of $40, which process should the management choose?

SOLUTION The break-even level of operation using the first process is obtained by solving the equation

$$40x = 20x + 10{,}000$$

or
$$20x = 10{,}000$$

whence
$$x = 500,$$

giving an output of 500 units. Next, we solve the equation

$$40x = 10x + 30{,}000$$

or
$$30x = 30{,}000$$

whence
$$x = 1{,}000,$$

giving an output of 1,000 units for a break-even operation using the second process. Since the projected sales are 800 units, we conclude that the management should choose the first process, which would give the firm a profit. ◀

EXAMPLE

27 Referring to Example 26, which process should the management choose if the projected sales are (a) 1500 units and (b) 3000 units?

SOLUTION In both cases, the production is past the break-even level. Since the revenue is the same regardless of which process is employed, the decision will be based on how much each process costs.

a. If $x = 1500$, then

$$C_1(x) = (20)(1500) + 10{,}000 = 40{,}000$$
$$C_2(x) = (10)(1500) + 30{,}000 = 45{,}000.$$

Hence, management should choose the first process.

b. If $x = 3000$, then

$$C_1(x) = (20)(3000) + 10{,}000 = 70{,}000$$
$$C_2(x) = (10)(3000) + 30{,}000 = 60{,}000.$$

In this case, management should choose the second process. ◀

▶ Market Equilibrium

Under pure competition, the price of a commodity will eventually settle at a level dictated by the condition that the supply of the commodity be equal to the demand for it. If the price is too high, the consumer will not buy, and if the price is too low, the supplier will not produce. **Market equilibrium** is said to prevail when the quantity produced is equal to the quantity demanded. The quantity produced at market equilibrium is called the **equilibrium quantity,** and the corresponding price is called the **equilibrium price.**

Market equilibrium corresponds to the point at which the demand curve and the supply curve intersect. In Figure 1.34, x_0 represents the equilibrium quantity and p_0 the equilibrium price. The point (x_0, p_0) lies on the supply curve and therefore satisfies the supply equation. At the same time, it also lies on the demand curve and therefore satisfies the demand equation. Thus, to find the point (x_0, p_0), and hence the equilibrium quantity and price, one solves the demand and supply equations simultaneously for x and p. For meaningful solutions, x and p must both be positive.

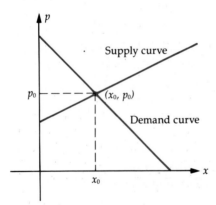

Figure 1.34

EXAMPLE

28   There is no demand for a certain model of loudspeaker when the unit price is $500 or higher. However, at a $200 unit price the quantity demanded is 10,000 units. The manufacturer will not market any loudspeakers if the price is $100 or lower. On the other hand, for each $50 increase in the unit price above $100, the manufacturer will market an additional 1,000 units. Both the demand and the supply equations are known to be linear.

   a. Find the demand equation.

   b. Find the supply equation.

   c. Find the equilibrium quantity and price.

SOLUTION   Let p denote the unit price in hundreds of dollars, and let x denote the number of units of loudspeakers in thousands.

a. Since the demand equation is linear, it has the form $p = mx + b$. When $x = 0$, $p = 5$, and when $x = 10$, $p = 2$, leading us to the equations

$$5 = b$$

and

$$2 = 10m + b.$$

Solving the system of equations, we find $b = 5$ and $m = -0.3$. Thus, the demand equation is $p = -0.3x + 5$ (Figure 1.35).

b. Since the supply equation is also linear, it has the form $p = mx + b$. When $x = 0$, $p = 1$, and when $x = 1$, $p = 1.5$, leading to the equations

$$1 = b$$

and

$$1.5 = m + b.$$

Solving the above system of equations, we find $b = 1$ and $m = 0.5$. Thus, the supply equation is $p = 0.5x + 1$ (Figure 1.35).

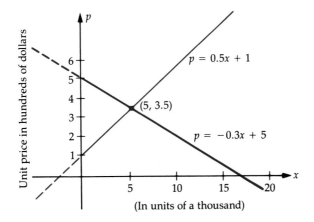

Figure 1.35

c. To find the market equilibrium, we solve simultaneously the system comprising the demand and supply equations obtained in parts (a) and (b), respectively; that is, the system

$$p = -0.3x + 5$$
$$p = 0.5x + 1.$$

Subtracting the first equation from the second gives

$$0.8x - 4 = 0,$$

whence $x = 5$. Substituting this value of x in the second equation gives $p = 3.5$. Thus, the equilibrium quantity is 5000 units and the equilibrium price is \$350 (Figure 1.35). ◀

Self-Check Exercises

1.4

1. Find the point of intersection of the straight lines with equations $2x + 3y = 6$ and $x - 3y = 4$.

2. There is no demand for a certain make of videocassette tape when the unit price is $12. However, when the unit price is $8, the quantity demanded is 8000 a week. The suppliers will not market any tapes if the unit price is $2 or lower. At $4 per tape, however, the manufacturer will make 5000 tapes available per week. Both the demand and supply functions are known to be linear.

    a. Find the demand equation.

    b. Find the supply equation.

    c. Find the equilibrium quantity and price.

Solutions to Self-Check Exercises 1.4 can be found on page 56.

▶

Exercises 1.4

In Exercises 1–6, find the point of intersection of each of the given pairs of straight lines.

1. $y = 3x + 4$
   $y = -2x + 14$

2. $y = -4x - 7$
   $-y = 5x + 10$

3. $2x - 3y = 6$
   $3x + 6y = 16$

4. $2x + 4y = 11$
   $-5x + 3y = 5$

5. $y = \dfrac{1}{4}x - 5$
   $2x - \dfrac{3}{2}y = 1$

6. $y = \dfrac{2}{3}x - 4$
   $x + 3y + 3 = 0$

In Exercises 7–10, find the break-even point for the firm whose cost function C, and revenue function R, are given.

7. $C(x) = 5x + 10{,}000,$ $R(x) = 15x$

8. $C(x) = 15x + 12{,}000,$ $R(x) = 21x$

9. $C(x) = 0.2x + 120,$ $R(x) = 0.4x$

10. $C(x) = 150x + 20{,}000,$ $R(x) = 270x$

11. Auto-time, a manufacturer of 24-hour variable timers, has a monthly fixed cost of $48,000 and a production cost of $8 for each timer manufactured. The units sell for $14 each.

    a. Sketch the graphs of the cost function and the revenue function, and hence find the break-even point graphically.

    b. Find the break-even point algebraically.

    c. Sketch the graph of the profit function.

    d. At what point does the graph of the profit function cross the *x*-axis? Interpret your result.

12. A division of Carter Enterprises produces "Personal Income Tax" diaries. Each diary sells for $8. The monthly fixed costs incurred by the division are $25,000, and the variable cost of producing each diary is $3.

   a. Find the break-even point for the division.

   b. What should the level of sales be in order for the division to realize a 15 percent sales profit? (Discount income tax.)

13. A division of the Gibson Corporation manufactures bicycle pumps. Each pump sells for $9, and the variable cost of producing each unit is 40 percent of the selling price. The monthly fixed costs incurred by the division are $50,000. What is the break-even point for the division?

14. The Ace Truck Leasing Company leases a certain size truck for $30 a day and $.15 a mile, whereas the Acme Truck Leasing Company leases the same size truck for $25 a day and $.20 a mile.

   a. Find the functions describing the daily cost of leasing from each company.

   b. Sketch the graphs of the two functions on the same set of axes.

   c. If a customer plans to drive at most 70 miles, which company should he rent a truck from for one day?

15. A product may be made using machine I or machine II. The manufacturer estimates that the monthly fixed costs of using machine I are $18,000, whereas the monthly fixed costs of using machine II are $15,000. The variable costs of manufacturing one unit of the product using machine I and machine II are $15 and $20, respectively. The product sells for $50 each.

   a. Obtain expressions for the cost functions associated with the use of each machine.

   b. Sketch the graphs of the cost functions of part (a) and the revenue functions on the same set of axes.

   c. Which machine should management choose if the projected sales are 450 units? 550 units? 650 units?

   d. What is the profit for each case in part (c)?

For each pair of supply and demand equations in Exercises 16–19, where x represents the quantity demanded in units of 1000 and p is the unit price in dollars, find the equilibrium quantity and the equilibrium price.

16. $4x + 3p - 59 = 0$ and $5x - 6p + 14 = 0$

17. $2x + 7p - 56 = 0$ and $3x - 11p + 45 = 0$

18. $p = -2x + 22$ and $p = 3x + 12$

19. $p = -0.3x + 6$ and $p = 0.15x + 1.5$

20. There is no demand for a certain brand of videocassette recorder when the unit price is $1000. However, for each $150 decrease below $1000, the quantity demanded increases by 300 units. The suppliers will not market any recorders if the unit price is $300 or lower. But at a unit price of $525, they are willing to make available 2500 units in the market. Both the demand and supply equations are known to be linear.

   a. Find the demand equation.

   b. Find the supply equation.

   c. Find the equilibrium quantity and price.

Solutions to Self-Check Exercises

1.4

1. The point of intersection of the two straight lines is found by solving the system of linear equations

$$2x + 3y = 6$$
$$x - 3y = 4.$$

Solving the first equation for y in terms of x we obtain

$$y = -\frac{2}{3}x + 2.$$

Substituting this expression for y into the second equation we obtain

$$x - 3\left(-\frac{2}{3}x + 2\right) = 4$$
$$x + 2x - 6 = 4$$
$$3x = 10,$$

or $x = \frac{10}{3}$. Substituting this value of x into the expression for y obtained earlier, we find

$$y = -\frac{2}{3}\left(\frac{10}{3}\right) + 2$$
$$= -\frac{2}{9}.$$

Therefore, the point of intersection is $\left(\frac{10}{3}, -\frac{2}{9}\right)$.

2. a. Let p denote the price per cassette and x the quantity demanded. The given conditions imply that $x = 0$ when $p = 12$, and $x = 8000$ when $p = 8$. Since the demand equation is linear, it has the form

$$p = mx + b.$$

Now, the first condition implies that

$$12 = m(0) + b, \text{ or } b = 12.$$

Therefore,

$$p = mx + 12.$$

Using the second condition, we find

$$8 = 8000m + 12$$

or $$m = -\frac{4}{8000} = -0.0005.$$

Therefore, the required demand equation is

$$p = -0.0005x + 12.$$

b. Let p denote the price per cassette and x the quantity made available at that price. Then, since the supply equation is linear, it also has the form

$$p = mx + b.$$

The first condition implies that $x = 0$ when $p = 2$, so we have

$$2 = m(0) + b \text{ or } b = 2.$$

Therefore

$$p = mx + 2.$$

Next, using the second condition, $x = 5000$ when $p = 4$, we find

$$4 = 5000m + 2,$$

giving $m = 0.0004$. So the required supply equation is

$$p = 0.0004x + 2.$$

c. The equilibrium quantity and price are found by solving the system of linear equations

$$p = -0.0005x + 12$$
$$p = 0.0004x + 2.$$

Equating the two equations yields

$$-0.0005x + 12 = 0.0004x + 2$$
$$0.0009x = 10,$$

or $x = 11,111$. Substituting this value of x into either equation in the system yields

$$p = 6.44.$$

Therefore, the equilibrium quantity is 11,111 and the equilibrium price is $6.44.

1.5

The Method of Least Squares (Optional)

▶ The Method of Least Squares

▶ Application

▶ The Method of Least Squares

In Example 12 we saw how a linear equation may be used to approximate the sales trend for a local sporting goods store. The **trend line,** as we have seen, may be used to predict the future sales of the store. Recall that the trend line in this particular case was obtained by requiring that it pass through two data points; the rationale for this lies in the fact that such a line seems to *fit* the data reasonably well.

In this section we describe a general method known as the **method of least squares** for determining a straight line that, in some sense, best fits a set of data points when the points are scattered about a straight line. In order to illustrate the principle behind the method of least squares, suppose, for simplicity, that we are given five data points

$$P_1(x_1, y_1), \qquad P_2(x_2, y_2), \qquad P_3(x_3, y_3), \qquad P_4(x_4, y_4), \quad \text{and} \quad P_5(x_5, y_5),$$

describing the relationship between the two variables x and y. By plotting these data points we obtain a graph called a **scatter diagram** (see Figure 1.36).

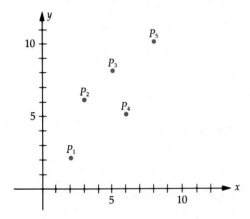

Figure 1.36

Since the five data points need not be collinear, there may not be a straight line passing through all or, for that matter, any of them. Thus, if we try to *fit* a straight line to these data points, the line will miss the first, second, third, fourth, and fifth data points by the amounts $d_1, d_2, d_3, d_4,$ and d_5, respectively (Figure 1.37).

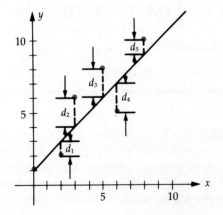

Figure 1.37

The **principle of least squares** states that the straight line L that fits the data points *best* is the one that should be chosen by requiring that the sum of the squares of $d_1, d_2, \ldots, d_5$, that is,

$$d_1^2 + d_2^2 + d_3^2 + d_4^2 + d_5^2$$

be made as small as possible. If we think of the amount d_1 as the error made when the value y_1 is approximated by the corresponding value of y lying on the straight line L, then it can be seen that the least squares criterion calls for minimizing the sum of the squares of the errors. The line L obtained in this manner is called the **least-squares line** or *regression line*.

The method for computing the least-squares line that best fits a set of data points is contained in the following result, which we state without proof.

The Method of Least Squares

Suppose we are given n data points

$$P_1(x_1, y_1), P_2(x_2, y_2), P_3(x_3, y_3), \ldots, P_n(x_n, y_n).$$

Then, the least-squares (regression) line for the data is given by the linear equation (function)

$$y = f(x) = mx + b$$

where the constants m and b satisfy the equations

$$nb + (x_1 + x_2 + \cdots + x_n)m = y_1 + y_2 + \cdots + y_n \tag{10}$$

and

$$(x_1 + x_2 + \cdots + x_n)b + (x_1^2 + x_2^2 + \cdots + x_n^2)m$$
$$= x_1 y_1 + x_2 y_2 + \cdots + x_n y_n \tag{11}$$

simultaneously. Equations (10) and (11) are called **normal equations.**

EXAMPLE

29 Find the least-squares line for the data

$$P_1(1, 1), \qquad P_2(2, 3), \qquad P_3(3, 4), \qquad P_4(4, 3), \quad \text{and} \quad P_5(5, 6).$$

SOLUTION

Here we have $n = 5$ and

$$x_1 = 1, \qquad x_2 = 2, \qquad x_3 = 3, \qquad x_4 = 4, \qquad x_5 = 5,$$
$$y_1 = 1, \qquad y_2 = 3, \qquad y_3 = 4, \qquad y_4 = 3, \qquad y_5 = 6,$$

so Equation (10) becomes, in this case,

$$5b + (1 + 2 + 3 + 4 + 5)m = 1 + 3 + 4 + 3 + 6$$

or
$$5b + 15m = 17, \tag{12}$$

while Equation (11) becomes

$$(1 + 2 + 3 + 4 + 5)b + (1 + 4 + 9 + 16 + 25)m = 1 + 6 + 12 + 12 + 30$$

or
$$15b + 55m = 61. \tag{13}$$

Solving Equation (12) for b gives

$$b = -3m + \frac{17}{5}, \tag{14}$$

which upon substitution into (13) gives

$$15\left(-3m + \frac{17}{5}\right) + 55m = 61$$

$$-45m + 51 + 55m = 61$$

$$10m = 10$$

or
$$m = 1.$$

Substituting this value of m into Equation (14) gives

$$b = -3 + \frac{17}{5}$$

or
$$b = \frac{2}{5} = 0.4.$$

Therefore, the required least-squares line is

$$y = x + 0.4.$$

The scatter diagram and the least-squares line are shown in Figure 1.38.

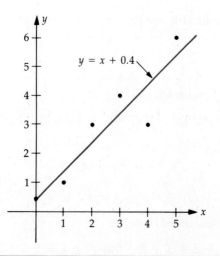

Figure 1.38

▶ Application

EXAMPLE

30 The proprietor of the Leisure Travel Service compiled the following data relating the annual profit of the firm to its annual advertising expenditure (both measured in thousands of dollars):

Annual Advertising Expenditure	(x)	12	14	17	21	26	30
Annual Profit	(y)	60	70	90	100	100	120

a. Determine the equation of the least-squares line for these data.

b. Draw a scatter diagram and the least-squares line for these data.

c. Use the result obtained in part (*a*) to predict the annual profit of the firm if the annual advertising budget is $20,000.

SOLUTION *a.* The calculations required for obtaining the normal equations may be summarized as follows:

	x	y	x^2	xy
	12	60	144	720
	14	70	196	980
	17	90	289	1530
	21	100	441	2100
	26	100	676	2600
	30	120	900	3600
Sum	120	540	2646	11,530

The normal equations are

$$6b + 120m = 540 \tag{15}$$

and $$120b + 2646m = 11{,}530. \tag{16}$$

Solving Equation (15) for b gives

$$b = -20m + 90, \tag{17}$$

which, upon substitution into Equation (16) gives

$$120(-20m + 90) + 2646m = 11{,}530$$
$$-2400m + 10{,}800 + 2646m = 11{,}530$$
$$246m = 730$$

or $$m = 2.97.$$

Substituting this value of m into Equation (17) gives

$$b = -20(2.97) + 90$$

or

$$b = 30.6.$$

Therefore, the required least-squares line is given by

$$y = f(x) = 2.97x + 30.6.$$

b. The scatter diagram and the least-squares line are shown in Figure 1.39.

c. The predicted annual profit of the firm corresponding to an annual budget of $20,000 is given by

$$f(20) = 2.97(20) + 30.6$$
$$= 90,$$

or $90,000.

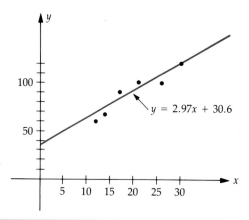

Figure 1.39

Self-Check Exercises

1.5

1. Find an equation of the least-squares line for the data

x	1	3	4	5	7
y	4	10	11	12	16

2. In a market research study for the Century Communications Company, the following data were provided based on the projected monthly sales x (in thousands) of a videocassette version of a box-office-hit adventure movie with a proposed wholesale unit price of p dollars.

p	38	36	34.5	30	28.5
x	2.2	5.4	7.0	11.5	14.6

Find the demand equation if the demand curve is the least-squares line for these data.

Solutions to Self-Check Exercises 1.5 can be found on page 66.

▶

Exercises 1.5

In Exercises 1–6, (a) find the equation of the least-squares line for the given data and (b) draw a scatter diagram for the given data and graph the least-squares line for this data.

1.

x	1	2	3	4
y	4	6	8	11

2.

x	1	3	5	7	9
y	9	8	6	3	2

3.

x	1	2	3	4	4	6
y	4.5	5	3	2	3.5	1

4.

x	1	1	2	3	4	4	5
y	2	3	3	3.5	3.5	4	5

5. $P_1(1, 3)$, $P_2(2, 5)$, $P_3(3, 5)$, $P_4(4, 7)$, $P_5(5, 8)$

6. $P_1(1, 8)$, $P_2(2, 6)$, $P_3(5, 6)$, $P_4(7, 4)$, $P_5(10, 1)$

7. The following data were compiled by the admissions office at Faber College during the past five years. The data relate the number of college brochures and follow-up letters (x) sent to a preselected list of high-school juniors who had taken the PSAT and the number of completed applications (y) received from these students (both measured in units of 1000).

x	4	4.5	5	5.5	6
y	0.5	0.6	0.8	0.9	1.2

a. Determine the equation of the least-squares line for this data.

b. Draw a scatter diagram and the least-squares line for this data.

c. Use the result obtained in part (a) to predict the number of completed applications that might be expected if 6400 brochures and follow-up letters are sent out during the next year.

8. The management of Kaldor, Inc., a manufacturer of electric motors, submitted the following data in the annual report to its stockholders. The table shows the net sales (in millions of dollars) during the five years that have elapsed since the new management team took over. (The first year the firm operated under the new management corresponds to the time period $x = 1$, and the four subsequent years correspond to $x = 2, 3, 4, 5$.)

Year (x)	1	2	3	4	5
Net Sales (y)	426	437	460	473	477

a. Determine the equation of the least-squares line for this data.

b. Draw a scatter diagram and the least-squares line for this data.

c. Use the result obtained in part (a) to predict the net sales for the upcoming year.

9. The following data were compiled by the superintendent of schools in a large metropolitan area. The table shows the average SAT verbal scores of high-school seniors during the five years since the district implemented the "back-to-basics" program.

Year (x)	1	2	3	4	5
Average Score (y)	436	438	428	430	426

a. Determine the equation of the least-squares line for these data.

b. Draw a scatter diagram and the least-squares line for these data.

c. Use the result obtained in part (a) to predict the average SAT verbal score of high-school seniors two years from now.

10. Shown below are figures compiled by Clarke, Kingsley, and Company, a consulting firm that specializes in auto operating costs, relating the annual mileage (in thousands of miles) that an average new compact car is driven to the cost per mile (in cents) of operating the car.

Annual Mileage (x)	5	10	15	20	25	30
Cost Per Mile (y)	50.3	34.8	30.1	27.4	25.6	23.5

a. Determine the equation of the least-squares line for these data.

b. Draw a scatter diagram and the least-squares line for these data.

c. Use the result obtained in part (a) to estimate the cost per mile of operating a new compact car if it is driven 8000 miles during the first year of ownership.

11. The highest price paid (in thousands of dollars) for a seat on the New York Stock Exchange from 1979 to 1983 is given in the following table:

Year (x)	1979	1980	1981	1982	1983
Highest Price Paid (y)	210	275	285	340	425

a. Find the equation of the least-squares line for these data. (Let $x = 1$ represent 1979.)

b. Use your result to estimate the highest price paid for a seat on the New York Stock Exchange in 1985.

12. The average new car price (in thousands of dollars) in the United States from 1980 to 1984 is given in the following table:

Year (x)	1980	1981	1982	1983	1984
Price (y)	7.1	8.1	9.1	9.9	10.7

a. Find the equation of the least-squares line for these data. (Let $x = 1$ represent 1980.)

b. Use your result to estimate the average new car price in 1986.

13. The Department of Agriculture's index of the amount spent by farmers in operating their farms over a six-year period is given in the following table:

Year (x)	1978	1979	1980	1981	1982	1983
Index (y)	115	132	138	150	156	160

a. Find the equation of the least-squares line for these data. (Let $x = 1$ represent 1978.)

b. Estimate the value of the index in 1990.

14. The average price per 1000 cubic feet of natural gas (in dollars) paid by consumers in the United States from 1978 to 1983 is given in the following table:

Year (x)	1978	1979	1980	1981	1982	1983
Price (y)	2.45	2.86	3.23	3.83	4.16	4.60

a. Find the equation of the least-squares line for these data. (Let $x = 1$ represent 1978.)

b. Estimate what the average price of gas will be in the United States in 1988.

15. The social security (FICA) wage base (in thousands of dollars) from 1979 to 1984 is given in the following table:

Year (x)	1979	1980	1981	1982	1983	1984
Wage Base (y)	22.9	25.9	29.7	32.4	35.7	37.8

a. Find the equation of the least-squares line for these data. (Let $x = 1$ represent 1979.)

b. Use your result to estimate what the FICA wage base will be in the year 1990.

Solutions to Self-Check Exercises

1.5

1. The calculations required for obtaining the normal equations may be summarized as follows:

x	y	x^2	xy
1	4	1	4
3	10	9	30
4	11	16	44
5	12	25	60
7	16	49	112
Sum 20	53	100	250

The normal equations are

$$5b + 20m = 53$$
$$20b + 100m = 250.$$

Solving the first equation for b gives

$$b = -4m + \frac{53}{5},$$

which upon substitution into the second equation yields

$$20\left(-4m + \frac{53}{5}\right) + 100m = 250$$
$$-80m + 212 + 100m = 250$$
$$20m = 38$$

or

$$m = 1.9.$$

Substituting this value of m into the expression for b found earlier, we find

$$b = -4(1.9) + \frac{53}{5}$$

or
$$b = 3.$$

Therefore, the required least-squares line is

$$y = 1.9x + 3.$$

2. The calculations required for obtaining the normal equations may be summarized as follows:

	x	p	x^2	xp
	2.2	38	4.84	83.6
	5.4	36	29.16	194.4
	7.0	34.5	49	241.5
	11.5	30	132.25	345
	14.6	28.5	213.16	416.1
Sum	40.7	167	428.41	1280.6

The normal equations are

$$5b + 40.7m = 167$$
$$40.7b + 428.41m = 1280.6.$$

Solving this system of linear equations simultaneously, we find that

$$m = -0.81 \quad \text{and} \quad b = 39.99.$$

Therefore, the required least-squares line is given by

$$p = f(x) = -0.81x + 39.99,$$

which is the required demand equation, provided $0 \le x \le 49.37$.

▶

Chapter 1 Review Exercises

1. Find an equation of the straight line passing through the point (2, 3) and parallel to the line with equation $3x + 4y - 8 = 0$.

2. Find an equation of the straight line passing through the point $(-1, 3)$ and parallel to the line joining the points $(-3, 4)$ and $(2, 1)$.

3. Find an equation of the line passing through the point $(-2, -4)$ that is perpendicular to the line with equation $2x - 3y - 24 = 0$.

4. Sketch the graph of the equation $3x - 4y = 24$.

5. Sales of a certain clock/radio are approximated by the relationship $S(x) = 6{,}000x + 30{,}000$ $(0 \le x \le 5)$, where $S(x)$ denotes the number of clock/radios sold in year x

($x = 0$ corresponds to the year 1982). Find the number of clock/radios expected to be sold in the year 1987.

6. The total sales of a company (in millions of dollars) are approximately linear as a function of time (in years). Sales in 1978 were $2.4 million, whereas sales in 1983 amounted to $7.4 million.

    a. Find an equation giving the sales of the company as a function of time.

    b. What were the sales in 1981?

7. Show that the triangle with vertices $A(1, 1)$, $B(5, 3)$, and $C(4, 5)$ is a right triangle.

8. Find the accumulated amount at the end of eight months on a $1500 bank deposit paying simple interest at a rate of 9 percent per year.

9. Find the present value of a $20,000 note due in ninety days if the simple discount rate is 9 percent per year. (Use a 360-day year.)

10. A manufacturer installed a new machine in her factory in 1980 at a cost of $300,000. The machine is depreciated linearly over 12 years with a scrap value of $30,000.

    a. Find an expression for the book value of the machine in year n ($0 \le n \le 12$).

    b. What is the rate of depreciation of the machine per year?

11. A company has a fixed cost of $30,000 and a production cost of $6 for each unit it manufactures. A unit sells for $10.

    a. What is the cost function?

    b. What is the revenue function?

    c. What is the profit function?

    d. Compute the profit (loss) corresponding to production levels of 6,000, 8,000, and 12,000 units, respectively.

12. There is no demand for a certain commodity when the unit price is $200 or more, but for each $10 decrease in price below $200, the quantity demanded increases by 200 units. Find the demand equation and sketch its graph.

13. Suppliers of bicycles will make 200 bicycles available in the market per month when the unit price is $50 and 2000 bicycles available per month when the unit price is $100. Find the supply equation if it is known to be linear. What is the lowest price at which a bicycle will be marketed?

14. Find the point of intersection of the two straight lines with equations $y = \frac{3}{4}x + 6$ and $3x - 2y + 3 = 0$.

15. The cost function and the revenue function for a certain firm are given by $C(x) = 12x + 20,000$ and $R(x) = 20x$, respectively. Find the break-even point for the company.

16. Given the demand equation $3x + p - 40 = 0$ and the supply equation $2x - p + 10 = 0$, where p is the unit price in dollars and x represents the quantity in units of a thousand, determine the equilibrium quantity and the equilibrium price.

Systems of Linear

Equations and

Matrices

How fast is the traffic moving? The flow of downtown traffic is controlled by traffic lights installed at each of the six intersections. One of the roads is to be resurfaced. In Example 10, page 91, we will see how the flow patterns must be altered in order to ensure a smooth flow of traffic even during rush hour.

▶CHAPTER

TWO

2.1

Systems of Linear Equations: Introduction

▶ Systems of Equations

▶ Application

▶ Solution(s) of Systems of Equations

▶ Systems of Equations

When we first encountered a system of linear equations (Section 1.4), we were required to solve a system of two linear equations in two variables. Recall that such a system may be written in the general form

$$ax + by = h$$
$$cx + dy = k \tag{1}$$

where a, b, c, d, h, and k are real constants and a and b (c and d) are not both zero.

Now let us study the nature of the solution of the system (1) in more detail. Recall that each equation in (1) has for its graph a straight line in the plane, so that geometrically the solution to the system is the point(s) of intersection of the two straight lines L_1 and L_2, represented by the first and second equations of the system.

It is clear that the two lines L_1 and L_2 may

a. intersect at exactly one point,

b. be parallel and coincident, or

c. be parallel and distinct

(see Figure 2.1).

Note that one and only one of the above cases must occur. In the first case, the system (1) has a unique solution corresponding to the single point of intersection of the two lines. In the second case, the system (1) has infinitely many solutions corresponding to the points lying simultaneously on both lines. Fi-

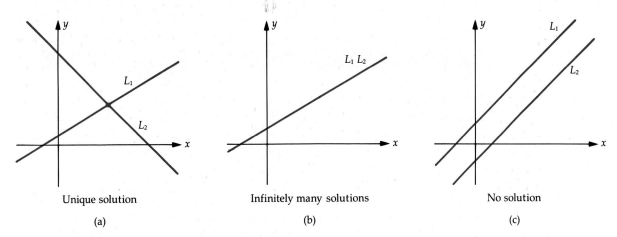

Unique solution

(a)

Infinitely many solutions

(b)

No solution

(c)

Figure 2.1

nally, in the third case, the system (1) has no solution, since the two lines do not intersect.

Let us illustrate each of these possibilities by considering some specific examples. We begin by looking at an example of a system with exactly one solution. Consider the system

$$2x - y = 1$$
$$3x + 2y = 12.$$

Solving the first equation for y in terms of x, we obtain the equation

$$y = 2x - 1.$$

Substituting this expression for y into the second equation yields

$$3x + 2(2x - 1) = 12$$
$$3x + 4x - 2 = 12$$
$$7x = 14$$
$$x = 2.$$

Finally, substituting this value of x into the expression for y obtained earlier gives

$$y = 2(2) - 1 = 3.$$

Therefore, the (unique) solution of the system is given by $x = 2$ and $y = 3$. Geometrically, the two lines represented by the two linear equations that make up the system intersect at the point $(2, 3)$ [see Figure 2.2]. Incidentally, we can check our result by substituting the values $x = 2$ and $y = 3$ into the equations. Thus

$$2(2) - (3) = 1 \quad \text{(Check)}$$
$$3(2) + 2(3) = 12 \quad \text{(Check)}$$

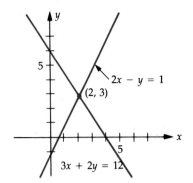

Figure 2.2

From the geometrical point of view we have just verified that the point (2, 3) lies on both lines.

Next, let us investigate a situation in which a system has infinitely many solutions. Consider the system

$$2x - y = 1$$
$$6x - 3y = 3.$$

Solving the first equation for y in terms of x, we obtain the equation

$$y = 2x - 1.$$

Substituting this expression for y into the second equation gives

$$6x - 3(2x - 1) = 3$$
$$6x - 6x + 3 = 3$$

or $$0 = 0.$$

This result simply tells us that the second equation is equivalent to the first. (To see this, just multiply both sides of the first equation by 3.) Our computations have revealed that the system of two equations is in fact equivalent to the single equation, say, $2x - y = 1$. Thus, any ordered pair of numbers (x, y) satisfying the equation $2x - y = 1$ (or $y = 2x - 1$) constitutes a solution to the system.

In particular, by assigning the value t to x, where t is any real number, we find that $y = 2t - 1$, so the ordered pair $(t, 2t-1)$ is a solution of the system. The variable t is called a **parameter.** For example, setting $t = 0$ gives the point $(0, -1)$ as a solution, and setting $t = 1$ gives the point $(1, 1)$ as another solution of the system. Since t represents any real number, there are infinitely many solutions to the system. Geometrically, the two equations in the system represent the same line, and all solutions of the system are points lying on the line (see Figure 2.3). Such a system is said to be **dependent.**

Finally, let us study a system that does not have a solution. Consider the system

$$2x - y = 1$$
$$6x - 3y = 12.$$

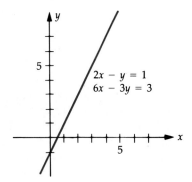

Figure 2.3

The first equation is equivalent to $y = 2x - 1$. Substituting this expression for y into the second equation gives

$$6x - 3(2x - 1) = 12$$
$$6x - 6x + 3 = 12$$

or $\qquad\qquad\qquad\qquad\qquad 0 = 9,$

which is clearly impossible. Thus, there is no solution to the system of equations. To interpret this situation geometrically, cast both equations in the slope-intercept form, obtaining

$$y = 2x - 1$$

and $\qquad\qquad\qquad\qquad y = 2x - 4.$

We see at once that the lines represented by these equations are parallel (each has slope 2) and distinct, since the first has intercept -1 and the second has intercept -4 (see Figure 2.4). Systems with no solutions, such as this one, are said to be **inconsistent.**

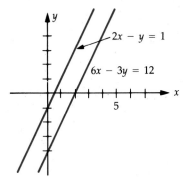

Figure 2.4

In Section 1.4 we presented some real-world applications of systems involving two linear equations in two variables. Here is an example involving a system of three linear equations in three variables.

▶ **Application**

EXAMPLE

1 The Ace Novelty Company wishes to produce three types of souvenirs: types A, B, and C. To manufacture a type-A souvenir requires 2 minutes on machine I, 1 minute on machine II, and 2 minutes on machine III. A type-B souvenir requires 1 minute on machine I, 3 minutes on machine II, and 1 minute on machine III. A type-C souvenir requires 1 minute on machine I and 2 minutes each on machines II and III. There are 3 hours available on machine I, 5 hours available on machine II, and 4 hours available on machine III for processing the order. How many souvenirs of each type should Ace Novelty make in order to use all of the available time? Formulate, but do not solve, the problem.

SOLUTION The given information may be tabulated as follows:

	Type A	Type B	Type C	Time Available
Machine I	2	1	1	180
Machine II	1	3	2	300
Machine III	2	1	2	240

Let x, y, and z denote the respective numbers of type-A, type-B, and type-C souvenirs to be made. The total amount of time that machine I is used is given by $2x + y + z$ minutes and must equal 180 minutes. This leads to the equation

$$2x + y + z = 180.$$

Similar considerations on the use of machines II and III lead to the following equations:

$$x + 3y + 2z = 300$$
$$2x + y + 2z = 240.$$

Since the variables x, y, and z must satisfy simultaneously the three conditions represented by the three equations, the solution to the problem is found by solving the following system of linear equations.

$$2x + y + z = 180$$
$$x + 3y + 2z = 300$$
$$2x + y + 2z = 240$$

◀

▶ **Solution(s) of Systems of Equations**

We will complete the solution of the problem posed in Example 1 later on (see page 91). For the moment let us look at the geometrical interpretation of a system of linear equations such as the system in Example 1 in order to gain some insight into the nature of the solution.

Now, a linear system comprising three linear equations in three variables x, y, and z has the general form

$$a_1x + b_1y + c_1z = d_1$$
$$a_2x + b_2y + c_2z = d_2$$
$$a_3x + b_3y + c_3z = d_3. \tag{2}$$

Just as a linear equation in two variables represents a straight line in the plane, it can be shown that a linear equation $ax + by + cz = d$ (a, b, and c not simultaneously equal to zero) in three variables represents a plane in three-dimensional space. Thus, each equation in system (2) represents a *plane* in three-dimensional space, and the *solution(s) of the system* is precisely the point(s) of intersection of the three planes defined by the three linear equations that make up the system. As before, the system has one and only one solution, infinitely many solutions, or no solution, depending on whether and how the planes intersect each other. Figure 2.5 illustrates each of these possibilities.

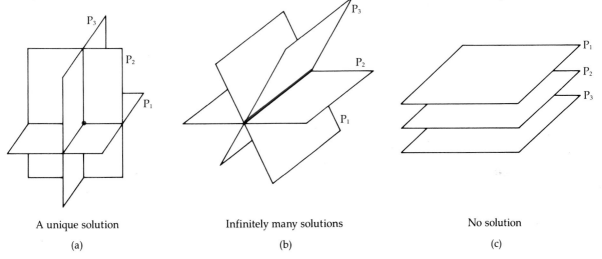

| A unique solution | Infinitely many solutions | No solution |
| (a) | (b) | (c) |

Figure 2.5

In Figure 2.5(a) the three planes intersect at a point corresponding to the situation in which system (2) has a unique solution. Figure 2.5(b) depicts the situation in which there are infinitely many solutions to the system. Here, the three planes intersect along a line, and the solutions are represented by the infinitely many points lying on this line. In Figure 2.5(c), the three planes are parallel and distinct, so there is no point in common to all three planes, and system (2) has no solution in this case.

The situations depicted in Figure 2.5 are by no means exhaustive. You may consider various other orientations of the three planes that would illustrate the three possible outcomes in solving a system of linear equations involving three variables.

Linear Equation in n Variables

A linear equation in n variables $x_1, x_2, \ldots, x_n$ is one of the form

$$a_1x_1 + a_2x_2 + \cdots + a_nx_n = c,$$

where $a_1, a_2, \ldots, a_n$ (not all zero) and c are constants.

For example, the equation

$$3x_1 + 2x_2 - 4x_3 + 6x_4 = 8$$

is a linear equation in the four variables $x_1, x_2, x_3,$ and x_4.

When the number of variables involved in a linear equation exceeds three, we no longer have the geometrical interpretation we had for the lower-dimensional spaces. Nevertheless, the algebraic concepts of the lower-dimensional spaces generalize to higher dimensions. For this reason, a linear equation in n variables $a_1x_1 + a_2x_2 + \cdots + a_nx_n = c$, where $a_1, a_2, \ldots, a_n$ are not all zero, is referred to as an ***n*-dimensional hyperplane.** We may interpret the solution(s) to a system comprising a finite number of such linear equations to be the *point(s) of intersection* of the hyperplanes defined by the equations that make up the system. As in the case of systems involving two or three variables, it can be shown that only three possibilities exist regarding the nature of the solution of such a system: (a) **a unique solution,** (b) **infinitely many solutions,** and (c) **no solution.**

Self-Check Exercises

2.1

1. Determine whether the system of linear equations

$$2x - 3y = 12$$
$$x + 2y = 6$$

has (a) a unique solution, (b) infinitely many solutions, or (c) no solution. Find all solutions whenever they exist. Make a sketch of the set of lines described by the system.

2. A farmer has 200 acres of land suitable for cultivating crops A, B, and C. The cost of cultivating crops A, B, and C is $40 per acre, $60 per acre, and $80 per acre, respectively. The farmer has $12,600 available for cultivation. Each acre of crop A requires 20 hours of labor, each acre of crop B requires 25 hours of labor, and each acre of crop C requires 40 hours of labor. The farmer has a maximum of 5,950 hours of labor available. If he wishes to use all of his cultivatable land, the entire budget, and all the labor available, how many acres of each crop should he plant? Formulate, but do not solve, the problem.

Solutions to Self-Check Exercises 2.1 can be found on page 80.

Exercises 2.1

In Exercises 1–8, determine whether each system of linear equations has (a) one and only one solution, (b) infinitely many solutions, or (c) no solution. Find all solutions whenever they exist.

1. $x - 3y = -1$
   $4x + 3y = 11$

2. $x + 4y = 7$
   $(1/2)x + 2y = 5$

3. $2x - 5y = 10$
   $6x - 15y = 30$

4. $4x - 5y = 14$
   $2x + 3y = -4$

5. $2x - 3y = 6$
   $6x - 9y = 12$

6. $x + 2y = 7$
   $2x - y = 4$

7. $x + 2y = 6$
   $3x + y = 8$
   $2x - 3y = 2$

8. $3x - y = 8$
   $2x + 4y = 3$
   $x + y = 2$

In Exercises 9–14, formulate, but do not solve, the problem. You will be asked to solve some of them in the next section.

9. Michael has a total of $2000 on deposit with two savings institutions. One pays simple interest at the rate of 6 percent per year, whereas the other pays simple interest at the rate of 8 percent per year. If Michael earned a total of $144 in interest during a single year, how much does he have on deposit in each institution?

10. Cantwell Associates, a real estate developer, is planning to build a new apartment complex consisting of one-bedroom units and two- and three-bedroom townhouses. A total of 192 units is planned, and the number of family units (two- and three-bedroom townhouses) will equal the number of one-bedroom units. If the number of one-bedroom units will be three times the number of three-bedroom units, find how many units of each type will be in the complex.

11. The annual interest on Mr. Carrington's three investments amounted to $21,600: 6 percent on a savings account, 8 percent on mutual funds, and 12 percent on money market certificates. If the amount of Carrington's investment in money market certificates was twice the amount of his investment in the savings account, and the interest earned from his investment in money market certificates was equal to the dividends he received from his investment in mutual funds, find how much money he placed in each type of investment.

12. The management of a private investment club has a fund of $200,000 earmarked for investment in stocks and bonds. To arrive at an acceptable overall level of risk, the stocks the company is considering have been classified into three categories: high risk, medium risk, and low risk. Management estimates that high-risk stocks will return 15 percent; medium-risk stocks, 10 percent; and low-risk stocks, 6 percent. The investment in low-risk stocks is to be twice the sum of the investments in stocks of the other two categories. If the investment goal is to have an average return of 9 percent on the total investment, determine how much the club should invest in each type of stock.

13. The management of Hartman Rent-A-Car has allocated $1 million to buy a fleet of new automobiles consisting of compact, intermediate, and full-size cars. Compacts cost $8,000 each, intermediate-size cars cost $12,000 each, and full-size cars cost $16,000 each. If Hartman purchases twice as many compacts as intermediate-size cars, and the total number of cars to be purchased is 100, determine how many cars of each type will be purchased. (Assume that the entire budget will be used.)

14. A theater has a seating capacity of 900 and charges $2 for children, $3 for students, and $4 for adults. At a certain screening with full attendance there were half as many adults as children and students combined. The receipts totaled $2800. How many children attended the show?

Solutions to Self-Check Exercises

2.1

1. Solving the first equation for y in terms of x, we obtain

$$y = \frac{2}{3}x - 4.$$

Next, substituting this result into the second equation of the system, we find

$$x + 2\left(\frac{2}{3}x - 4\right) = 6$$

$$x + \frac{4}{3}x - 8 = 6$$

$$\frac{7}{3}x = 14$$

or

$$x = 6.$$

Substituting this value for x into the expression for y obtained earlier, we arrive at

$$y = \frac{2}{3}(6) - 4 = 0.$$

Therefore, the system has the unique solution $x = 6$ and $y = 0$. The lines are shown in the following figure.

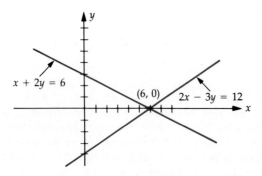

2. Let x, y, and z denote the number of acres of crop A, crop B, and crop C, respectively, to be cultivated. Then, the condition that all the cultivatable land be used translates into the equation

$$x + y + z = 200.$$

Next, the total cost incurred in cultivating all three crops is $40x + 60y + 80z$ dollars, and since the entire budget is to be expended, we have

$$40x + 60y + 80z = 12,600.$$

Finally, the amount of labor required to cultivate all three crops is $20x + 25y + 40z$ hours, and since all the available labor is to be used, we have

$$20x + 25y + 40z = 5,950.$$

Thus, the solution is found by solving the following system of linear equations

$$
\begin{aligned}
x + y + z &= 200 \\
40x + 60y + 80z &= 12,600 \\
20x + 25y + 40z &= 5,950.
\end{aligned}
$$

2.2

Solving Systems of Linear Equations—
The Gauss-Jordan Method

▶ The Gauss-Jordan Method

▶ Augmented Matrices

▶ Solution(s) of Systems of Equations

▶ Applications

▶ The Gauss-Jordan Method

The method of substitution used in Section 2.1 is well suited to solving a system of linear equations when the number of linear equations and variables is small. But for large systems, the steps involved in the procedure become difficult to manage.

A suitable technique for solving systems of linear equations of any size is the **Gauss-Jordan elimination method.** One advantage of this technique is its adaptability to the electronic computer. This method involves a sequence of operations on the system of linear equations to obtain at each stage an **equivalent system**—in other words, a system having the same solution as the original system. The reduction is complete when the original system has been transformed so that it has a certain standard form from which the solution can be easily read.

The operations of the Gauss-Jordan elimination method are:

1. Interchange any two equations.
2. Multiply any equation by a nonzero constant.

3. Replace an equation by adding a constant multiple of any other equation to it.

To illustrate the Gauss-Jordan elimination method for solving systems of linear equations, let us apply it to the solution of the system defined by the following equations:

$$\begin{aligned} x + 2y + 3z &= 1 \\ 2x + 3y + 2z &= 2 \\ -x + y + 2z &= -8 \end{aligned} \tag{3}$$

First, we transform the above system into an equivalent one in which the variable x is eliminated from all equations except the first. To accomplish this, we use Operation 3 twice: we replace the second equation of (3) by an equation obtained by adding -2 times the first equation to it, obtaining the equivalent system

$$\begin{aligned} x + 2y + 3z &= 1 \\ -y - 4z &= 0 \\ -x + y + 2z &= -8 \end{aligned} \qquad \begin{matrix} -2 \text{ times first} \\ \text{second} \end{matrix} \qquad \begin{aligned} -2x - 4y - 6z &= -2 \\ \underline{2x + 3y + 2z = 2} \\ -y - 4z = 0 \end{aligned} \tag{4}$$

and we replace the third equation of (4) by an equation obtained by adding the first equation of (4) to the third equation of (4), giving

$$\begin{aligned} x + 2y + 3z &= -1 \\ -y - 4z &= 0 \\ 3y + 5z &= -7 \end{aligned} \qquad \begin{matrix} \text{first} \\ \text{third} \end{matrix} \qquad \begin{aligned} x + 2y + 3z &= 1 \\ \underline{-x + y + 2z = -8} \\ 3y + 5z = -7 \end{aligned} \tag{5}$$

By multiplying the second equation of (5) by -1 (Operation 2), we see that (5) is equivalent to

$$\begin{aligned} x + 2y + 3z &= 1 \\ y + 4z &= 0 \\ 3y + 5z &= -7 \end{aligned} \qquad -1 \text{ times second} \qquad y + 4z = 0 \tag{6}$$

We now eliminate the variable y from all equations except the second. Again, this may be accomplished by using Operation 3. We replace the first equation in (6) by one obtained by adding the first equation to -2 times the second equation of (6). This gives

$$\begin{aligned} x \quad - 5z &= 1 \\ y + 4z &= 0 \\ 3y + 5z &= -7 \end{aligned} \qquad \begin{matrix} \text{first} \\ -2 \text{ times second} \end{matrix} \qquad \begin{aligned} x + 2y + 3z &= 1 \\ \underline{-2y - 8z = 0} \\ x \quad - 5z = 1 \end{aligned} \tag{7}$$

Next, we replace the third equation in (7) by one obtained by adding -3 times the second equation of (7) to it. This leads to the (equivalent) system

$$\begin{aligned} x \quad - 5z &= 1 \\ y + 4z &= 0 \\ -7z &= -7 \end{aligned} \qquad \begin{matrix} -3 \text{ times second} \\ \text{third} \end{matrix} \qquad \begin{aligned} -3y - 12z &= 0 \\ \underline{3y + 5z = -7} \\ -7z = -7 \end{aligned} \tag{8}$$

Multiplying the third equation of (8) by $-1/7$ gives

$$
\begin{aligned}
x \quad -5z &= 1 \\
y + 4z &= 0 \\
z &= 1
\end{aligned}
\qquad -\frac{1}{7}\text{ times third} \qquad z = 1
\tag{9}
$$

Finally, we eliminate z from the first two equations of (9) by (a) replacing the first equation of (9) by a sum involving that equation and 5 times the third equation of (9), and (b) replacing the second equation of (9) by a sum involving that equation and -4 times the third equation of (9). We obtain the equivalent system

$$
\begin{aligned}
x \quad &= \quad 6 \\
y \quad &= \; -4 \\
z &= \quad 1
\end{aligned}
\tag{10}
$$

in standard form, from which we can read off the solution to system (3) as $x = 6$, $y = -4$, and $z = 1$, or more concisely, the point $(6, -4, 1)$ in three-dimensional space.

▶ Augmented Matrices

Observe from the preceding example that in each step of the reduction process the variables x, y, and z play no significant role except as a reminder of the position of each coefficient in the system. With the aid of **matrices,** rectangular arrays of numbers, we can eliminate writing the variables at each step of the reduction and thus save ourselves a great deal of work. To this end, note that the system

$$
\begin{aligned}
x + 2y + 3z &= \quad 1 \\
2x + 3y + 2z &= \quad 2 \\
-x + \;\; y + 2z &= -8
\end{aligned}
\tag{11}
$$

may be represented by the matrix

$$
\left[
\begin{array}{ccc|c}
1 & 2 & 3 & 1 \\
2 & 3 & 2 & 2 \\
-1 & 1 & 2 & -8
\end{array}
\right]
\tag{12}
$$

The **submatrix,** the first three columns of the matrix (12), is called the **matrix of coefficients** of system (11). The matrix itself (12) is referred to as the **augmented matrix** of system (11), since it is obtained by joining the matrix of coefficients to the column (matrix) of constants. The vertical line separates the column of constants from the matrix of coefficients.

The next example shows how much work can be saved by using matrices instead of the standard representation of the systems of linear equations.

EXAMPLE

2   Write the augmented matrix corresponding to each equivalent system given in (3) through (10).

SOLUTION   The required sequence of augmented matrices is given by:

Augmented Matrix	Equivalent System

a.
$$\begin{bmatrix} 1 & 2 & 3 & 1 \\ 2 & 3 & 2 & 2 \\ -1 & 1 & 2 & -8 \end{bmatrix}$$
$$\begin{aligned} x + 2y + 3z &= 1 \\ 2x + 3y + 2z &= 2 \\ -x + y + 2z &= -8 \end{aligned}$$

b.
$$\begin{bmatrix} 1 & 2 & 3 & 1 \\ 0 & -1 & -4 & 0 \\ -1 & 1 & 2 & -8 \end{bmatrix}$$
$$\begin{aligned} x + 2y + 3z &= 1 \\ -y - 4z &= 0 \\ -x + y + 2z &= -8 \end{aligned}$$

c.
$$\begin{bmatrix} 1 & 2 & 3 & 1 \\ 0 & -1 & -4 & 0 \\ 0 & 3 & 5 & -7 \end{bmatrix}$$
$$\begin{aligned} x + 2y + 3z &= 1 \\ -y - 4z &= 0 \\ 3y + 5z &= -7 \end{aligned}$$

d.
$$\begin{bmatrix} 1 & 2 & 3 & 1 \\ 0 & 1 & 4 & 0 \\ 0 & 3 & 5 & -7 \end{bmatrix}$$
$$\begin{aligned} x + 2y + 3z &= 1 \\ y + 4z &= 0 \\ 3y + 5z &= -7 \end{aligned}$$

e.
$$\begin{bmatrix} 1 & 0 & -5 & 1 \\ 0 & 1 & 4 & 0 \\ 0 & 3 & 5 & -7 \end{bmatrix}$$
$$\begin{aligned} x - 5z &= 1 \\ y + 4z &= 0 \\ 3y + 5z &= -7 \end{aligned}$$

f.
$$\begin{bmatrix} 1 & 0 & -5 & 1 \\ 0 & 1 & 4 & 0 \\ 0 & 0 & -7 & -7 \end{bmatrix}$$
$$\begin{aligned} x - 5z &= 1 \\ y + 4z &= 0 \\ -7z &= -7 \end{aligned}$$

g.
$$\begin{bmatrix} 1 & 0 & -5 & 1 \\ 0 & 1 & 4 & 0 \\ 0 & 0 & 1 & 1 \end{bmatrix}$$
$$\begin{aligned} x - 5z &= 1 \\ y + 4z &= 0 \\ z &= 1 \end{aligned}$$

h.
$$\begin{bmatrix} 1 & 0 & 0 & 6 \\ 0 & 1 & 0 & -4 \\ 0 & 0 & 1 & 1 \end{bmatrix}$$
$$\begin{aligned} x &= 6 \\ y &= -4 \\ z &= 1 \end{aligned}$$

(13)

◀

The augmented matrix in (13) is an example of a matrix in row-reduced form. In general, a matrix with m rows and n columns (called an $m \times n$ matrix) is in row-reduced form if it has the following properties.

Row-Reduced Form of a Matrix

1. Any rows that have only zero entries lie below the rows that have nonzero entries.

2. The first nonzero entry in each row is 1, and the column containing such an entry has zeros elsewhere.

3. The first nonzero entry 1 in each nonzero row lies to the right of the first nonzero entry 1 of any row above it.

EXAMPLE

3 The following matrices are in row-reduced form:

$$\begin{bmatrix} 1 & 0 & 0 & | & 0 \\ 0 & 1 & 0 & | & 0 \\ 0 & 0 & 1 & | & 3 \end{bmatrix} \quad \begin{bmatrix} 1 & 0 & 0 & | & 4 \\ 0 & 1 & 0 & | & 3 \\ 0 & 0 & 0 & | & 0 \end{bmatrix} \quad \begin{bmatrix} 1 & 2 & 0 & | & 3 \\ 0 & 0 & 1 & | & 2 \\ 0 & 0 & 0 & | & 3 \end{bmatrix} \quad \begin{bmatrix} 1 & 0 & | & 2 \\ 0 & 1 & | & 4 \\ 0 & 0 & | & 0 \end{bmatrix} \quad \blacktriangleleft$$

The foregoing discussion suggests the following adaptation of the Gauss-Jordan elimination method in solving systems of linear equations using matrices. First, the three operations on the equations of the system (see page 81) translate into the following row operations on the corresponding augmented matrices.

Row Operations

1. Interchange any two rows.
2. Multiply any row by a nonzero constant.
3. Replace any row by adding a constant multiple of any other row to it.

The sequence of *equivalent augmented matrices* in Example 2 was obtained by employing the appropriate row operations on the augmented matrices corresponding to the operations on the equivalent system of equations.

The Gauss-Jordan elimination method may be summarized as follows:

The Gauss-Jordan Elimination Method

1. Write the augmented matrix corresponding to the given linear system.

2. Consider the first nonzero column. Using row operations, if necessary, transform the augmented matrix so that the uppermost entry in this column is a 1. Then use multiples of this first row to obtain an augmented matrix with zeros in the rest of this column.

3. Proceed to the next nonzero column that contains at least one nonzero entry other than the first. Using row operations involving only rows other than the first, if necessary, transform the augmented matrix so that the second entry (from the top) in this column is a 1. Then, use multiples of this second row to obtain an augmented matrix with zeros in the rest of this column.

4. Continue until the resulting matrix is in row-reduced form.

A word of caution: Before writing the augmented matrix, be sure to write all equations with the variables on the left and constant terms on the right of the equals sign. Also, make sure that the variables are in the same order in all equations.

Before going on, we need to introduce some notation for the three types of row operations:

Notation for Row Operations

Letting R_i denote the ith row of a matrix, we write:

1. $R_i \leftrightarrow R_j$ to mean: Interchange row i with row j (Operation 1).
2. cR_i to mean: Replace row i with c times row i (Operation 2).
3. $R_i + aR_j$ to mean: Replace row i with the sum of row i and a times row j (Operation 3).

▶ Solution(s) of Systems of Equations

EXAMPLE

4 Solve the system of linear equations given by

$$3x - 2y + 8z = 9$$
$$-2x + 2y + z = 3$$
$$x + 2y - 3z = 8 \qquad (14)$$

SOLUTION   Using the Gauss-Jordan elimination method, we obtain the following sequence of equivalent augmented matrices:

$$\begin{bmatrix} 3 & -2 & 8 & | & 9 \\ -2 & 2 & 1 & | & 3 \\ 1 & 2 & -3 & | & 8 \end{bmatrix} \xrightarrow{R_1 + R_2} \begin{bmatrix} 1 & 0 & 9 & | & 12 \\ -2 & 2 & 1 & | & 3 \\ 1 & 2 & -3 & | & 8 \end{bmatrix} \xrightarrow[R_3 - R_1]{R_2 + 2R_1} \begin{bmatrix} 1 & 0 & 9 & | & 12 \\ 0 & 2 & 19 & | & 27 \\ 0 & 2 & -12 & | & -4 \end{bmatrix}$$

$$\xrightarrow{R_2 \leftrightarrow R_3} \begin{bmatrix} 1 & 0 & 9 & | & 12 \\ 0 & 2 & -12 & | & -4 \\ 0 & 2 & 19 & | & 27 \end{bmatrix} \xrightarrow{\frac{1}{2}R_2} \begin{bmatrix} 1 & 0 & 9 & | & 12 \\ 0 & 1 & -6 & | & -2 \\ 0 & 2 & 19 & | & 27 \end{bmatrix} \xrightarrow{R_3 - 2R_2}$$

$$\begin{bmatrix} 1 & 0 & 9 & | & 12 \\ 0 & 1 & -6 & | & -2 \\ 0 & 0 & 31 & | & 31 \end{bmatrix} \xrightarrow{\frac{1}{31}R_3} \begin{bmatrix} 1 & 0 & 9 & | & 12 \\ 0 & 1 & -6 & | & -2 \\ 0 & 0 & 1 & | & 1 \end{bmatrix} \xrightarrow[R_2 + 6R_3]{R_1 - 9R_3} \begin{bmatrix} 1 & 0 & 0 & | & 3 \\ 0 & 1 & 0 & | & 4 \\ 0 & 0 & 1 & | & 1 \end{bmatrix}$$

The solution to system (14) is given by $x = 3$, $y = 4$, and $z = 1$ and may be verified by substitution into system (14) as follows:

$$3(3) - 2(4) + 8(1) = 9 \quad \text{(Check)}$$
$$-2(3) + 2(4) + 1 = 3 \quad \text{(Check)}$$
$$3 + 2(4) - 3(1) = 8 \quad \text{(Check)} \quad \blacktriangleleft$$

The following example illustrates the situation in which a system of linear equations has infinitely many solutions.

EXAMPLE

5   Solve the system of linear equations given by

$$x + 2y - 3z = -2$$
$$3x - y - 2z = 1$$
$$2x + 3y - 5z = -3 \tag{15}$$

SOLUTION

Using the Gauss-Jordan elimination method, we obtain the following sequence of equivalent augmented matrices:

$$\begin{bmatrix} 1 & 2 & -3 & | & -2 \\ 3 & -1 & -2 & | & 1 \\ 2 & 3 & - & | & -3 \end{bmatrix} \xrightarrow[R_3 - 2R_1]{R_2 - 3R_1} \begin{bmatrix} 1 & 2 & -3 & | & -2 \\ 0 & -7 & 7 & | & 7 \\ 0 & -1 & 1 & | & 1 \end{bmatrix} \xrightarrow{-\frac{1}{7}R_2} \begin{bmatrix} 1 & 2 & -3 & | & -2 \\ 0 & 1 & -1 & | & -1 \\ 0 & -1 & 1 & | & 1 \end{bmatrix}$$

$$\xrightarrow[R_3 + R_2]{R_1 - 2R_2} \begin{bmatrix} 1 & 0 & -1 & | & 0 \\ 0 & 1 & -1 & | & -1 \\ 0 & 0 & 0 & | & 0 \end{bmatrix}$$

The last augmented matrix is in row-reduced form. Interpreting it as a system of linear equations gives

$$x - z = 0$$
$$y - z = -1,$$

a system of two equations in the three variables x, y, and z.

   Let us now single out one variable, say z, and solve for x and y in terms of it. We obtain

$$x = z$$
$$y = z - 1.$$

If we assign a particular value to z, say $z = 0$, we obtain $x = 0$ and $y = -1$, giving the solution $(0, -1, 0)$ to system (15). By setting $z = 1$ we obtain the solution $(1, 0, 1)$. In general, if we set $z = t$ where t represents some real number, we obtain a solution given by $(t, t-1, t)$. Since the parameter t may be any real number, we see that the system (15) has infinitely many solutions. Geometrically, the solutions of the system (15) lie on the straight line in three-dimensional space given by the intersection of the three planes determined by the three equations in the system. $\blacktriangleleft$

The next example shows what happens in the elimination procedure when the system does not have a solution.

EXAMPLE

6 Solve the system of linear equations given by

$$\begin{aligned} x + y + z &= 1 \\ 3x - y - z &= 4 \\ x + 5y + 5z &= -1 \end{aligned} \tag{16}$$

SOLUTION Using the Gauss-Jordan elimination method, we obtain the following sequence of equivalent augmented matrices:

$$\begin{bmatrix} 1 & 1 & 1 & | & 1 \\ 3 & -1 & -1 & | & 4 \\ 1 & 5 & 5 & | & -1 \end{bmatrix} \xrightarrow[\substack{R_2 - 3R_1 \\ -R_3 + R_1}]{} \begin{bmatrix} 1 & 1 & 1 & | & 1 \\ 0 & -4 & -4 & | & 1 \\ 0 & -4 & -4 & | & 2 \end{bmatrix} \xrightarrow{R_3 - R_2} \begin{bmatrix} 1 & 1 & 1 & | & 1 \\ 0 & -4 & -4 & | & 1 \\ 0 & 0 & 0 & | & 1 \end{bmatrix}$$

Observe that the third row in the last matrix reads $0x + 0y + 0z = 1$, that is, $0 = 1$! We conclude therefore that system (16) is inconsistent and has no solution. Geometrically, we have a situation in which two of the planes intersect in a straight line but the third plane is parallel to this line of intersection of the two planes and does not contain it. Consequently, there is no point of intersection of the three planes. ◀

Example 6 illustrates the following more general result of using the Gauss-Jordan elimination procedure.

Systems with No Solution

If there is a row in the augmented matrix containing all zeros to the left of the vertical line and a nonzero entry to the right of the line, then the system of equations has no solution.

It may have dawned on you that in all the previous examples we have dealt only with systems involving exactly the same number of linear equations as there are variables. However, systems in which the number of equations is different from the number of variables also occur in practice. Indeed, we will see an example of this in Chapter 3.

The following theorem provides us with some preliminary information on a system of linear equations.

Theorem

    a. If there are at least as many equations as there are variables in a linear system, then one of the following is true:

        i. The system has no solution

        ii. The system has exactly one solution

        iii. The system has infinitely many solutions.

    b. If there are fewer equations than variables in a linear system, then the system either has no solution or it has infinitely many solutions.

    Although we will not prove this theorem, you should recall that we have illustrated geometrically part (a) for the case in which there are exactly as many equations (three) as there are variables. To show the validity of part (b), let us once again consider the case in which a system has three variables. Now, if there is only one equation in the system, then it is clear that there are infinitely many solutions corresponding geometrically to all the points lying on the plane represented by the equation.

    Next, if there are two equations in the system, then *only* the following possibilities exist:

1. The two planes are parallel and distinct.

2. The two planes intersect in a straight line.

3. The two planes are coincident (the two equations define the same plane) [see Figure 2.6].

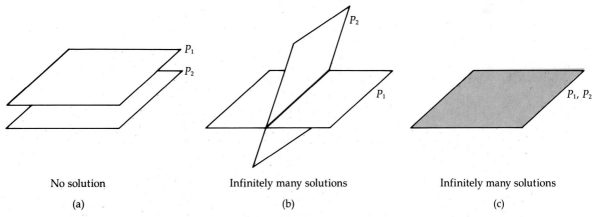

    No solution   Infinitely many solutions   Infinitely many solutions

      (a)   (b)   (c)

Figure 2.6

    Thus, either there is no solution or there are infinitely many solutions corresponding to the points lying on a line (of intersection) of the two planes or on a single plane determined by the two equations.

EXAMPLE

7 Solve the following system of linear equations:

$$x + 2y = 4$$
$$x - 2y = 0$$
$$4x + 3y = 12$$

SOLUTION We obtain the following sequence of equivalent augmented matrices:

$$\begin{bmatrix} 1 & 2 & | & 4 \\ 1 & -2 & | & 0 \\ 4 & 3 & | & 12 \end{bmatrix} \xrightarrow[R_3-4R_1]{-R_2+R_1} \begin{bmatrix} 1 & 2 & | & 4 \\ 0 & 4 & | & 4 \\ 0 & -5 & | & -4 \end{bmatrix} \xrightarrow{\frac{1}{4}R_2} \begin{bmatrix} 1 & 2 & | & 4 \\ 0 & 1 & | & 1 \\ 0 & -5 & | & -4 \end{bmatrix} \xrightarrow[R_3+5R_2]{R_1-2R_2}$$

$$\begin{bmatrix} 1 & 0 & | & 2 \\ 0 & 1 & | & 1 \\ 0 & 0 & | & 1 \end{bmatrix}$$

The last row of the row-reduced augmented matrix implies that $0 = 1$, which is impossible, so we conclude that the given system has no solution. Geometrically, the three lines defined by the three equations in the system do not intersect at a point. You are urged to draw the graphs of these equations. ◀

EXAMPLE

8 Solve the following system of linear equations:

$$x_1 + 2x_2 - 3x_3 = -2$$
$$3x_1 - x_2 - 2x_3 = 1$$
$$2x_1 + 3x_2 - 5x_3 = -3$$
$$x_1 + 7x_2 - 8x_3 = -7$$

SOLUTION We obtain the following sequence of equivalent augmented matrices:

$$\begin{bmatrix} 1 & 2 & -3 & | & -2 \\ 3 & -1 & -2 & | & 1 \\ 2 & 3 & -5 & | & -3 \\ 1 & 7 & -8 & | & -7 \end{bmatrix} \xrightarrow[\substack{R_2-3R_1 \\ R_3-2R_1 \\ R_4-R_1}]{} \begin{bmatrix} 1 & 2 & -3 & | & -2 \\ 0 & -7 & 7 & | & 7 \\ 0 & -1 & 1 & | & 1 \\ 0 & 5 & -5 & | & -5 \end{bmatrix} \xrightarrow[\substack{7R_3-R_2 \\ R_4+5R_3}]{-\frac{1}{7}R_2} \begin{bmatrix} 1 & 2 & -3 & | & -2 \\ 0 & 1 & -1 & | & -1 \\ 0 & 0 & 0 & | & 0 \\ 0 & 0 & 0 & | & 0 \end{bmatrix}$$

$$\xrightarrow{R_1-2R_2} \begin{bmatrix} 1 & 0 & -1 & | & 0 \\ 0 & 1 & -1 & | & -1 \\ 0 & 0 & 0 & | & 0 \\ 0 & 0 & 0 & | & 0 \end{bmatrix}$$

The last augmented matrix is in row-reduced form. Observe that the given system is equivalent to the system

$$x_1 - x_3 = 0$$
$$x_2 - x_3 = -1$$

of two equations in three variables. Letting $x_3 = t$, we find that the required solutions may be represented by the 3-tuple $(t, t-1, t)$, where t is any real number. ◀

▶ Applications

EXAMPLE

9 Complete the solution to Example 1, page 76.

SOLUTION To complete the solution of the problem posed in Example 1, recall that the mathematical formulation of the problem led to the following system of linear equations:

$$2x + y + z = 180$$
$$x + 3y + 2z = 300$$
$$2x + y + 2z = 240$$

where x, y, and z denote the respective numbers of type-A, type-B, and type-C souvenirs to be made.

Solving the above system of linear equations by the Gauss-Jordan elimination method we obtain the following sequence of equivalent augmented matrices:

$$\begin{bmatrix} 2 & 1 & 1 & | & 180 \\ 1 & 3 & 2 & | & 300 \\ 2 & 1 & 2 & | & 240 \end{bmatrix} \xrightarrow{R_1 \leftrightarrow R_2} \begin{bmatrix} 1 & 3 & 2 & | & 300 \\ 2 & 1 & 1 & | & 180 \\ 2 & 1 & 2 & | & 240 \end{bmatrix} \xrightarrow[-R_3+2R_1]{-R_2+2R_1} \begin{bmatrix} 1 & 3 & 2 & | & 300 \\ 0 & 5 & 3 & | & 420 \\ 0 & 5 & 2 & | & 360 \end{bmatrix}$$

$$\xrightarrow{\frac{1}{5}R_2} \begin{bmatrix} 1 & 3 & 2 & | & 300 \\ 0 & 1 & \frac{3}{5} & | & 84 \\ 0 & 5 & 2 & | & 360 \end{bmatrix} \xrightarrow[-R_3+5R_2]{R_1-3R_2} \begin{bmatrix} 1 & 0 & \frac{1}{5} & | & 48 \\ 0 & 1 & \frac{3}{5} & | & 84 \\ 0 & 0 & 1 & | & 60 \end{bmatrix}$$

$$\xrightarrow[R_2-\frac{3}{5}R_3]{R_1-\frac{1}{5}R_3} \begin{bmatrix} 1 & 0 & 0 & | & 36 \\ 0 & 1 & 0 & | & 48 \\ 0 & 0 & 1 & | & 60 \end{bmatrix}$$

Thus, $x = 36$, $y = 48$, and $z = 60$; that is, Ace Novelty should make 36 type-A souvenirs, 48 type-B souvenirs, and 60 type-C souvenirs in order to use all the machine time available. ◀

The following example illustrates a situation in which a system of linear equations has infinitely many solutions.

EXAMPLE

10 Figure 2.7 shows the flow of downtown traffic during the rush hours on a typical weekday. The arrows indicate the direction of the flow of traffic on each one-way road, and the average number of vehicles entering and leaving each intersection per hour appears beside each road. Fifth and Sixth Avenues can handle up to 2000 vehicles per hour without causing congestion, whereas the maximum capacity of each of the two streets is 1000 vehicles per hour. The flow of traffic is controlled by traffic lights installed at each of the four intersections.

a. Write a general expression involving the rates of flow, x_1, x_2, x_3, x_4, and suggest two possible flow patterns that would ensure that there is no traffic congestion.

b. Suppose that the part of 4th Street between 5th Avenue and 6th Avenue is to be resurfaced and that traffic flow between the two junctions has to be slowed down to at most 300 vehicles per hour. Find two possible flow patterns that would result in a smooth flow of traffic.

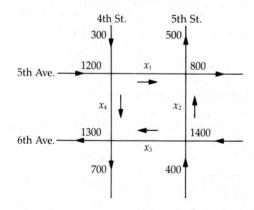

Figure 2.7

SOLUTION

a. To avoid congestion, all traffic entering an intersection must also leave that intersection. Applying this condition to each of the four intersections in a clockwise direction beginning with the 5th Avenue and 4th Street intersection, we obtain the following equations:

$$1500 = x_1 + x_4$$
$$1300 = x_1 + x_2$$
$$1800 = x_2 + x_3$$
$$2000 = x_3 + x_4$$

This system of four linear equations in the four variables x_1, x_2, x_3, x_4 may be rewritten in the more standard form

$$\begin{aligned}
x_1 \quad\quad\quad + x_4 &= 1500 \\
x_1 + x_2 \quad\quad\quad &= 1300 \\
x_2 + x_3 \quad\quad &= 1800 \\
x_3 + x_4 &= 2000.
\end{aligned}$$

Using the Gauss-Jordan elimination method to solve the system, we obtain

$$\begin{bmatrix} 1 & 0 & 0 & 1 & | & 1500 \\ 1 & 1 & 0 & 0 & | & 1300 \\ 0 & 1 & 1 & 0 & | & 1800 \\ 0 & 0 & 1 & 1 & | & 2000 \end{bmatrix} \xrightarrow{R_2 - R_1} \begin{bmatrix} 1 & 0 & 0 & 1 & | & 1500 \\ 0 & 1 & 0 & -1 & | & -200 \\ 0 & 1 & 1 & 0 & | & 1800 \\ 0 & 0 & 1 & 1 & | & 2000 \end{bmatrix} \xrightarrow{R_3 - R_2}$$

$$\begin{bmatrix} 1 & 0 & 0 & 1 & | & 1500 \\ 0 & 1 & 0 & -1 & | & -200 \\ 0 & 0 & 1 & 1 & | & 2000 \\ 0 & 0 & 1 & 1 & | & 2000 \end{bmatrix} \xrightarrow{R_4 - R_3} \begin{bmatrix} 1 & 0 & 0 & 1 & | & 1500 \\ 0 & 1 & 0 & -1 & | & -200 \\ 0 & 0 & 1 & 1 & | & 2000 \\ 0 & 0 & 0 & 0 & | & 0 \end{bmatrix}.$$

The last augmented matrix is in row-reduced form and is equivalent to a system of three linear equations in the four variables x_1, x_2, x_3, x_4. Thus, we may express three of the variables, say, x_1, x_2, x_3, in terms of x_4. Setting $x_4 = t$ (t, a parameter), we may write the infinitely many solutions of the system as

$$x_1 = 1500 - t$$
$$x_2 = -200 + t$$
$$x_3 = 2000 - t$$
$$x_4 = t.$$

Observe that for a meaningful solution, $200 \le t \le 1000$. (Why?) For example, picking $t = 300$ gives the flow pattern

$$x_1 = 1200, \ x_2 = 100, \ x_3 = 1700, \ x_4 = 300.$$

Selecting $t = 500$ gives the flow pattern as

$$x_1 = 1000, \ x_2 = 300, \ x_3 = 1500, \ x_4 = 500.$$

b. In this case, x_4 must not exceed 300. Again, using the results of part (a), we find upon setting $x_4 = t = 300$ the flow pattern

$$x_1 = 1200, \ x_2 = 100, \ x_3 = 1700, \ x_4 = 300$$

obtained earlier. Picking $t = 250$ gives the flow pattern

$$x_1 = 1250, \ x_2 = 50, \ x_3 = 1750, \ x_4 = 250. \qquad \blacktriangleleft$$

Self-Check Exercises

2.2

1. The following augmented matrix in row-reduced form is equivalent to the augmented matrix of a certain system of linear equations. Use this result to solve the system of equations.

$$\begin{bmatrix} 1 & 0 & -1 & | & 3 \\ 0 & 1 & 5 & | & -2 \\ 0 & 0 & 0 & | & 0 \end{bmatrix}$$

2. Solve the system of linear equations

$$2x - 3y + z = 6$$
$$x + 2y + 4z = -4$$
$$x - 5y - 3z = 10$$

using the Gauss-Jordan elimination method.

3. A farmer has 200 acres of land suitable for cultivating crops A, B, and C. The cost of cultivating crop A, crop B, and crop C is \$40 per acre, \$60 per acre, and \$80 per acre, respectively. The farmer has \$12,600 available for land cultivation. Each acre of crop A requires 20 hours of labor, each acre of crop B requires 25 hours of labor, and each acre of crop C requires 40 hours of labor. The farmer has a maximum of 5,950 hours of labor available. If he wishes to use all of his cultivatable land, the entire budget, and all the labor available, how many acres of each crop should he plant?

Solutions to Self-Check Exercises 2.2 can be found on page 98.

▶

Exercises 2.2

In Exercises 1–10, indicate whether the matrix is in row–reduced form.

1.
$$\begin{bmatrix} 1 & 0 & 3 \\ 0 & 1 & -2 \end{bmatrix}$$

2.
$$\begin{bmatrix} 1 & 1 & 3 \\ 0 & 0 & 0 \end{bmatrix}$$

3.
$$\begin{bmatrix} 0 & 1 & 3 \\ 1 & 0 & 5 \end{bmatrix}$$

4.
$$\begin{bmatrix} 0 & 1 & 3 \\ 0 & 0 & 5 \end{bmatrix}$$

5.
$$\begin{bmatrix} 1 & 0 & 0 & 3 \\ 0 & 1 & 0 & 4 \\ 0 & 0 & 1 & 5 \end{bmatrix}$$

6.
$$\begin{bmatrix} 1 & 0 & 0 & -1 \\ 0 & 1 & 0 & -2 \\ 0 & 0 & 2 & -3 \end{bmatrix}$$

7.
$$\begin{bmatrix} 1 & 0 & 1 & 3 \\ 0 & 1 & 0 & 4 \\ 0 & 0 & -1 & 6 \end{bmatrix}$$

8.
$$\begin{bmatrix} 1 & 0 & -10 \\ 0 & 1 & 2 \\ 0 & 0 & 0 \end{bmatrix}$$

9.
$$\begin{bmatrix} 1 & 0 & 0 & 2 \\ 0 & 0 & 1 & 3 \end{bmatrix}$$

10.
$$\begin{bmatrix} 1 & 0 & 0 & 3 \\ 0 & 1 & 0 & 6 \\ 0 & 0 & 0 & 4 \\ 0 & 0 & 1 & 5 \end{bmatrix}$$

In Exercises 11 and 12, fill in the missing entries by performing the indicated row operations to obtain the row-reduced matrices.

11.
$$\begin{bmatrix} 1 & 3 & 1 & 3 \\ 3 & 8 & 3 & 7 \\ 2 & -3 & 1 & -10 \end{bmatrix} \xrightarrow[R_3 - 2R_1]{R_2 - 3R_1} \begin{bmatrix} 1 & 3 & 1 & 3 \\ \cdot & \cdot & \cdot & \cdot \\ \cdot & \cdot & \cdot & \cdot \end{bmatrix} \xrightarrow{-R_2} \begin{bmatrix} 1 & 3 & 1 & 3 \\ \cdot & \cdot & \cdot & \cdot \\ 0 & -9 & -1 & -16 \end{bmatrix}$$

$$\xrightarrow[R_3 + 9R_2]{R_1 - 3R_2} \begin{bmatrix} \cdot & \cdot & \cdot & \cdot \\ 0 & 1 & 0 & 2 \\ \cdot & \cdot & \cdot & \cdot \end{bmatrix} \xrightarrow[-R_3]{R_1 + R_3} \begin{bmatrix} 1 & 0 & 0 & -1 \\ 0 & 1 & 0 & 2 \\ 0 & 0 & 1 & -2 \end{bmatrix}$$

12. $\begin{bmatrix} 0 & 1 & 3 & | & -4 \\ 1 & 2 & 1 & | & 7 \\ 1 & -2 & 0 & | & 1 \end{bmatrix} \xrightarrow{R_1 \leftrightarrow R_2} \begin{bmatrix} \cdot & \cdot & \cdot & | & \cdot \\ \cdot & \cdot & \cdot & | & \cdot \\ 1 & -2 & 0 & | & 1 \end{bmatrix} \xrightarrow{R_3 - R_1} \begin{bmatrix} 1 & 2 & 1 & | & 7 \\ 0 & 1 & 3 & | & -4 \\ \cdot & \cdot & \cdot & | & \cdot \end{bmatrix}$

$\xrightarrow[R_3 + 4R_2]{R_1 + \frac{1}{2}R_3} \begin{bmatrix} \cdot & \cdot & \cdot & | & \cdot \\ 0 & 1 & 3 & | & -4 \\ \cdot & \cdot & \cdot & | & \cdot \end{bmatrix} \xrightarrow{\frac{1}{11}R_3} \begin{bmatrix} 1 & 0 & 1/2 & | & 4 \\ 0 & 1 & 3 & | & -4 \\ \cdot & \cdot & \cdot & | & \cdot \end{bmatrix} \xrightarrow[R_2 - 3R_3]{R_1 - \frac{1}{2}R_3}$

$\begin{bmatrix} 1 & 0 & 0 & | & 5 \\ 0 & 1 & 0 & | & 2 \\ 0 & 0 & 1 & | & -2 \end{bmatrix}$

13. Write a system of linear equations for the augmented matrix of Exercise 11. Using the results of Exercise 11, determine the solution of the system.

14. Repeat Exercise 13 for the augmented matrix of Exercise 12.

In Exercises 15–28, given that the augmented matrix in row-reduced form is equivalent to the augmented matrix of some system of linear equations, (a) determine whether the system has a solution and (b) find the solution or solutions to the system if they exist.

15. $\begin{bmatrix} 1 & 0 & 0 & | & 3 \\ 0 & 1 & 0 & | & -1 \\ 0 & 0 & 1 & | & 2 \end{bmatrix}$

16. $\begin{bmatrix} 1 & 0 & 0 & | & 3 \\ 0 & 1 & 0 & | & 1 \\ 0 & 0 & 0 & | & 4 \end{bmatrix}$

17. $\begin{bmatrix} 1 & 0 & 0 & | & 1 \\ 0 & 1 & 0 & | & 2 \\ 0 & 0 & 0 & | & 0 \end{bmatrix}$

18. $\begin{bmatrix} 1 & 0 & 0 & | & 2 \\ 0 & 1 & 1 & | & 2 \\ 0 & 0 & 0 & | & 0 \end{bmatrix}$

19. $\begin{bmatrix} 1 & 0 & | & 2 \\ 0 & 1 & | & 4 \\ 0 & 0 & | & 0 \end{bmatrix}$

20. $\begin{bmatrix} 1 & 0 & 1 & | & 4 \\ 0 & 1 & 0 & | & -2 \end{bmatrix}$

21. $\begin{bmatrix} 1 & 0 & 0 & 0 & | & 3 \\ 0 & 1 & 0 & 0 & | & -2 \\ 0 & 0 & 1 & 0 & | & 1 \\ 0 & 0 & 0 & 1 & | & 2 \end{bmatrix}$

22. $\begin{bmatrix} 1 & 0 & 0 & 0 & | & 2 \\ 0 & 1 & 0 & 0 & | & 1 \\ 0 & 0 & 1 & 0 & | & 3 \\ 0 & 0 & 0 & 0 & | & -5 \end{bmatrix}$

23. $\begin{bmatrix} 1 & 0 & 0 & 0 & | & 3 \\ 0 & 1 & 1 & 0 & | & -1 \\ 0 & 0 & 0 & 1 & | & 2 \end{bmatrix}$

24. $\begin{bmatrix} 1 & 0 & 0 & | & 4 \\ 0 & 1 & 0 & | & -1 \\ 0 & 0 & 1 & | & 3 \\ 0 & 0 & 0 & | & 2 \end{bmatrix}$

25. $\begin{bmatrix} 1 & 0 & 0 & 0 & | & 2 \\ 0 & 1 & 0 & 0 & | & -1 \\ 0 & 0 & 1 & 1 & | & 2 \\ 0 & 0 & 0 & 0 & | & 0 \end{bmatrix}$

26. $\begin{bmatrix} 1 & 0 & 0 & 0 & | & 0 \\ 0 & 1 & 0 & 0 & | & 1 \\ 0 & 0 & 0 & 0 & | & 0 \\ 0 & 0 & 0 & 0 & | & 0 \end{bmatrix}$

27. $\begin{bmatrix} 1 & 0 & 0 & 0 & | & 2 \\ 0 & 0 & 0 & 0 & | & 0 \\ 0 & 0 & 0 & 0 & | & 0 \\ 0 & 0 & 0 & 0 & | & 0 \end{bmatrix}$

28. $\begin{bmatrix} 1 & 0 & 0 & 0 & | & 0 \\ 0 & 1 & 0 & 0 & | & 0 \\ 0 & 0 & 1 & 0 & | & 0 \\ 0 & 0 & 0 & 1 & | & 0 \end{bmatrix}$

In Exercises 29–50, solve the system of linear equations using the Gauss-Jordan elimination method.

29. $x - 2y = 8$
$3x + 4y = 6$

30. $3x + y = 1$
$-7x - 2y = -1$

31. $2x - y = 3$
$x + 2y = 4$
$2x + 3y = 7$

32. $2x + 3y = 2$
$x + 3y = -2$
$x - y = 3$

33. $x + y + z = 0$
$2x - y + z = 1$
$x + y - 2z = 2$

34. $2x + y - 2z = 4$
$x + 3y - z = -3$
$3x + 4y - z = 7$

35. $2x_1 + 2x_2 + x_3 = 9$
$x_1 + x_3 = 4$
$4x_2 - 3x_3 = 17$

36. $x + 2y + z = -2$
$-2x - 3y - z = 1$
$2x + 4y + 2z = -4$

37. $2x + 3y - 2z = 10$
$3x - 2y + 2z = 0$
$4x - y + 3z = -1$

38. $2x_1 - x_2 + x_3 = -4$
$3x_1 - (3/2)x_2 + (3/2)x_3 = -6$
$-6x_1 + 3x_2 - 3x_3 = 12$

39. $x + y - 2z = -3$
$2x - y + 3z = 7$
$x - 2y + 5z = 0$

40. $x_1 - 2x_2 + x_3 = -3$
$2x_1 + x_2 - 2x_3 = 2$
$x_1 + 3x_2 - 3x_3 = 5$

41. $4x + y - z = 4$
$8x + 2y - 2z = 8$

42. $3x_1 - 2x_2 + x_3 = -2$
$x_1 + x_2 + 2x_3 = 1$

43. $2x_1 - x_2 + 3x_3 - 2x_4 = -2$
$x_1 - 2x_2 + x_3 - 3x_4 = 2$
$x_1 - 5x_2 + 2x_3 + 3x_4 = -6$
$-3x_1 + 3x_2 - 4x_3 - 4x_4 = 9$

44. $2x_1 + 3x_2 + 2x_3 + x_4 = -1$
$x_1 - x_2 + x_3 - 2x_4 = -8$
$5x_1 + 6x_2 - 2x_3 + 2x_4 = 11$
$x_1 + 3x_2 + 8x_3 + x_4 = -14$

45. $x_1 - 2x_2 - 2x_3 + x_4 = 1$
$2x_1 - x_2 + 2x_3 + 3x_4 = -2$
$-x_1 - 5x_2 + 7x_3 - 2x_4 = 3$
$3x_1 - 4x_2 + 3x_3 + 4x_4 = -4$

46. $x_1 - x_2 + 3x_3 - 6x_4 = 2$
$x_1 + x_2 + x_3 - 2x_4 = 2$
$-2x_1 - x_2 + x_3 + 2x_4 = 0$

47. $x_1 + 2x_2 - x_3 + x_4 = 1$
$2x_1 + x_2 + x_3 - 4x_4 = 2$
$x_1 + 3x_2 + 2x_3 - 3x_4 = 5$

48. $2x_1 - x_2 - x_3 = 0$
$3x_1 - 2x_2 - x_3 = -1$
$-x_1 + 2x_2 - x_3 = 3$
$2x_2 - 2x_3 = 4$

49. $x_1 + x_2 - x_3 - x_4 = -1$
$x_1 - x_2 + x_3 + 4x_4 = -6$
$3x_1 + x_2 - x_3 + 2x_4 = -4$
$5x_1 + x_2 - 3x_3 + x_4 = -9$

50. $2x_1 + x_2 + 3x_3 - x_4 = 9$
$-x_1 - 2x_2 - 3x_4 = -1$
$x_1 - 3x_3 + x_4 = 10$
$x_1 - x_2 - x_3 - x_4 = 8$

51. Cantwell Associates, a real estate developer, is planning to build a new apartment complex consisting of one-bedroom units and two- and three-bedroom townhouses. A total of 192 units is planned, and the number of family units (two- and three-bedroom townhouses) will equal the number of one-bedroom units. If the number of

one-bedroom units will be three times the number of three-bedroom units, find how many units of each type will be in the complex.

52. The annual interest on Mr. Carrington's three investments amounted to $21,600: 6 percent on a savings account, 8 percent on mutual funds, and 12 percent on money market certificates. If the amount of Carrington's investment in money market certificates was twice the amount of his investment in the savings account, and the interest earned from his investment in money market certificates was equal to the dividends he received from his investment in mutual funds, find how much money he placed in each type of investment.

53. The management of a private investment club has a fund of $200,000 earmarked for investment in stocks and bonds. To arrive at an acceptable overall level of risk, the stocks the company is considering have been classified into three categories: high risk, medium risk, and low risk. Management estimates that high-risk stocks will return 15 percent; medium-risk stocks, 10 percent; and low-risk stocks, 6 percent. The investment in low-risk stocks is to be twice the sum of the investments in stocks of the other two categories. If the investment goal is to have an average return of 9 percent on the total investment, determine how much the club should invest in each type of stock.

54. The management of Hartman Rent-A-Car has allocated $1 million to buy a fleet of new automobiles consisting of compact, intermediate, and full-size cars. Compacts cost $8,000 each, intermediate-size cars cost $12,000 each, and full-size cars cost $16,000 each. If Hartman purchases twice as many compacts as intermediate-size cars, and the total number of cars to be purchased is 100, determine how many cars of each type will be purchased. (Assume that the entire budget will be used.)

55. A manufacturer of women's blouses makes three types of blouses: sleeveless, short-sleeve, and long-sleeve. The time required by each department to produce a dozen blouses of each type is shown in the following table.

Department	Sleeveless	Short-sleeve	Long-sleeve
Cutting	9 min.	12 min.	15 min.
Sewing	22 min.	24 min.	28 min.
Packaging	6 min.	8 min.	8 min.

The cutting, sewing, and packaging departments have a maximum of 80, 160, and 48 labor hours, respectively, available per day. How many dozens of each type of blouse can be produced each day if the plant is operated at full capacity?

*56. The following figure shows the flow of traffic near a city's Civic Center during the rush hours on a typical weekday. Each road can handle a maximum of 1000 cars per hour without causing congestion. The flow of traffic is controlled by traffic lights at each of the five intersections.

a. Set up a system of linear equations describing the traffic flow.

b. Solve the system devised in (a) and suggest two possible traffic flow patterns that would ensure that there is no traffic congestion.

c. Suppose that 7th Avenue between 3rd and 4th Streets is soon to be closed for road repairs, and find one possible flow pattern that would result in a smooth flow of traffic.

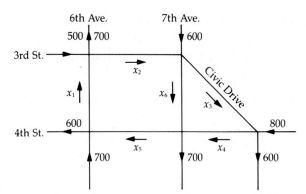

Figure for
Exercise 56

*57. The following figure shows the flow of downtown traffic during the rush hours on a typical weekday. Each avenue can handle up to 1500 vehicles per hour without causing congestion, whereas the maximum capacity of each street is 1000 vehicles per hour. The flow of traffic is controlled by traffic lights at each of the six intersections.

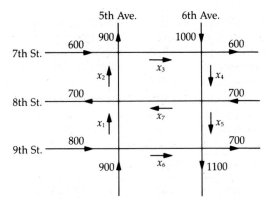

Figure for
Exercise 57

a. Set up a system of linear equations describing the traffic flow.

b. Solve the system devised in (a) and suggest two possible traffic flow patterns that would ensure there is no traffic congestion.

c. Suppose that the traffic flow along 9th Street between 5th Avenue and 6th Avenue, x_6, is restricted due to sewer construction. What is the minimum permissible traffic flow along this road that will not result in traffic congestion?

Solutions to Self-Check Exercises

2.2

1. Let x, y, and z denote the variables. Then the given row-reduced augmented matrix tells us that the system of linear equations is equivalent to the two equations

$$x \quad\quad - z = 3$$
$$y + 5z = -2.$$

Letting $z = t$, where t is a parameter, we find the infinitely many solutions given by

$$x = \quad t + 3$$
$$y = -5t - 2$$
$$z = \quad t.$$

2. We obtain the following sequence of equivalent augmented matrices:

$$\begin{bmatrix} 2 & -3 & 1 & | & 6 \\ 1 & 2 & 4 & | & -4 \\ 1 & -5 & -3 & | & 10 \end{bmatrix} \xrightarrow{R_1 \leftrightarrow R_2} \begin{bmatrix} 1 & 2 & 4 & | & -4 \\ 2 & -3 & 1 & | & 6 \\ 1 & -5 & -3 & | & 10 \end{bmatrix}$$

$$\xrightarrow[R_3 - R_1]{R_2 - 2R_1} \begin{bmatrix} 1 & 2 & 5 & | & -4 \\ 0 & -7 & -7 & | & 14 \\ 0 & -7 & -7 & | & 14 \end{bmatrix} \xrightarrow{-\frac{1}{7}R_2}$$

$$\begin{bmatrix} 1 & 2 & 5 & | & -4 \\ 0 & 1 & 1 & | & -2 \\ 0 & -7 & -7 & | & 14 \end{bmatrix} \xrightarrow[R_3 + 7R_2]{R_1 - 2R_2} \begin{bmatrix} 1 & 0 & 3 & | & 0 \\ 0 & 1 & 1 & | & -2 \\ 0 & 0 & 0 & | & 0 \end{bmatrix}.$$

The last augmented matrix, which is in row-reduced form, tells us that the given system of linear equations is equivalent to the following system of two equations

$$x \quad\quad + 3z = 0$$
$$y + z = -2.$$

Letting $z = t$, where t is a parameter, we see that the infinitely many solutions are given by

$$x = -3t$$
$$y = -t - 2$$
$$z = \quad t.$$

3. Referring to the solution of Exercise 2 of Self-Check Exercises 2.1, we see that the problem reduces to solving the following system of linear equations

$$x + y + z = 200$$
$$40x + 60y + 80z = 12600$$
$$20x + 25y + 40z = 5950.$$

Using the Gauss-Jordan elimination method, we have

$$\begin{bmatrix} 1 & 1 & 1 & | & 200 \\ 40 & 60 & 80 & | & 12600 \\ 20 & 25 & 40 & | & 5950 \end{bmatrix} \xrightarrow[R_3 - 20R_1]{R_2 - 40R_1} \begin{bmatrix} 1 & 1 & 1 & | & 200 \\ 0 & 20 & 40 & | & 4600 \\ 0 & 5 & 20 & | & 1950 \end{bmatrix} \xrightarrow{\frac{1}{20}R_2} \begin{bmatrix} 1 & 1 & 1 & | & 200 \\ 0 & 1 & 2 & | & 230 \\ 0 & 5 & 20 & | & 1950 \end{bmatrix} \xrightarrow[R_3 - 5R_2]{R_1 - R_2}$$

$$\begin{bmatrix} 1 & 0 & -1 & | & -30 \\ 0 & 1 & 2 & | & 230 \\ 0 & 0 & 10 & | & 800 \end{bmatrix} \xrightarrow{\frac{1}{10}R_3} \begin{bmatrix} 1 & 0 & -1 & | & -30 \\ 0 & 1 & 2 & | & 230 \\ 0 & 0 & 1 & | & 80 \end{bmatrix} \xrightarrow[R_2 - 2R_3]{R_1 + R_3} \begin{bmatrix} 1 & 0 & 0 & | & 50 \\ 0 & 1 & 0 & | & 70 \\ 0 & 0 & 1 & | & 80 \end{bmatrix}.$$

From the last augmented matrix in reduced form we see that $x = 50$, $y = 70$, and $z = 80$. Therefore, the farmer should plant 50 acres of crop A, 70 acres of crop B, and 80 acres of crop C.

2.3

Matrices

▶ Definition

▶ Application

▶ Sum and Difference

▶ Scalar Multiplication

▶ Definition

Many practical problems are solved by using arithmetic operations on the data associated with the problems. By properly organizing the data into *blocks* of numbers, we can use **matrix theory** to help us carry out these arithmetic operations in an orderly and efficient manner. In particular, this systematic approach enables us to use the electronic computer to full advantage.

Let us begin by considering how the monthly output data of a manufacturer may be organized. The Acrosonic Company manufactures four different loudspeaker systems in three separate locations. The output of the company for the month of May is summarized in Table 2.1.

	Model A	*Model B*	*Model C*	*Model D*
Location I	320	280	460	280
Location II	480	360	580	0
Location III	540	420	200	880

Table 2.1

Now, if we agree to preserve the location of each entry in Table 2.1, the set of data may be summarized further as follows:

$$\begin{bmatrix} 320 & 280 & 460 & 280 \\ 480 & 360 & 580 & 0 \\ 540 & 420 & 200 & 880 \end{bmatrix}$$

The array of numbers displayed above is an example of a *matrix*. Observe that the numbers in the first row give the output of models A, B, C, and D of Acrosonic loudspeaker systems manufactured in location I; similarly, the numbers in the second and third rows give the respective outputs of these loudspeaker systems in locations II and III. The numbers in each column of the matrix give the outputs of a particular model of loudspeaker system manufactured in each of the company's three manufacturing locations.

More generally, a matrix is an ordered rectangular array of real numbers. For example, each of the following arrays is a matrix:

$$A = \begin{bmatrix} 3 & 0 & -1 \\ 2 & 1 & 4 \end{bmatrix} \qquad B = \begin{bmatrix} 3 & 2 \\ 0 & 1 \\ -1 & 4 \end{bmatrix} \qquad C = \begin{bmatrix} 1 \\ 2 \\ 4 \\ 0 \end{bmatrix} \qquad D = [1 \quad 3 \quad 0 \quad 1]$$

Another example of a matrix is the *augmented matrix,* encountered in Section 2.2.

The real numbers that make up the array are called the **entries,** or *elements, of the matrix.* The entries in a row in the array are referred to as a **row** of the matrix, whereas the entries in a column in the array are referred to as a **column** of the matrix. Matrix A, for example, has two rows and three columns, which may be identified as follows:

$$\begin{array}{cccc} & Column\ 1 & Column\ 2 & Column\ 3 \\ Row\ 1 & \begin{bmatrix} 3 & 0 & -1 \\ 2 & 1 & 4 \end{bmatrix} \\ Row\ 2 \end{array}$$

The **order,** or *size, of a matrix* is described in terms of the number of rows and columns of the matrix. For example, matrix A has two rows and three columns and is said to have order 2 by 3, denoted by 2×3. In general, a matrix having m rows and n columns is said to have order $m \times n$.

Matrix

A **matrix** is an ordered rectangular array of numbers. A matrix with m rows and n columns has order **m × n**. The entry in the ith row and jth column is denoted by a_{ij}.

A matrix of order $1 \times n$—a matrix having one row and n columns—is referred to as a **row matrix,** or *row vector* (of dimension n). For example, matrix D is a row vector of dimension 4. Similarly, a matrix having m rows and one column is referred to as a **column matrix,** or *column vector* (of dimension m). The matrix C is a column vector of dimension 4. Finally, an $n \times n$ matrix, that is, a matrix having the same number of rows as columns, is called a **square matrix.** For example, the matrix

$$\begin{bmatrix} -3 & 8 & 6 \\ 2 & \frac{1}{4} & 4 \\ 1 & 3 & 2 \end{bmatrix}$$

is a square matrix of order 3×3, or simply of order 3.

▶ **Application**

EXAMPLE

11 Consider the matrix

$$P = \begin{bmatrix} 320 & 280 & 460 & 280 \\ 480 & 360 & 580 & 0 \\ 540 & 420 & 200 & 880 \end{bmatrix}$$

representing the output of loudspeaker systems of the Acrosonic Company discussed earlier (see Table 2.1).

a. What is the order of the matrix P?

b. Find a_{24} (the entry in the second row and fourth column of the matrix P) and give an interpretation of this number.

c. Find the sum of the entries that make up the first row of P and interpret the result.

d. Find the sum of the entries that make up the fourth column of P and interpret the result.

SOLUTION

a. The matrix P has three rows and four columns and hence has order 3×4.

b. The required entry lies in the second row and fourth column and is the number zero. This means that no model D loudspeaker system was manufactured in location II in the month of May.

c. The required sum is given by

$$320 + 280 + 460 + 280 = 1340,$$

which gives the total number of loudspeaker systems manufactured in location I in the month of May, or 1340 units.

d. The required sum is given by

$$280 + 0 + 880 = 1160,$$

giving the output of model D loudspeaker systems in the month of May, or 1160 units. ◀

▶ Sum and Difference

Two matrices are said to be *equal* if they have the same order and their corresponding entries are equal. For example,

$$\begin{bmatrix} 2 & 3 & 1 \\ 4 & 6 & 2 \end{bmatrix} = \begin{bmatrix} (3-1) & 3 & 1 \\ 4 & (4+2) & 2 \end{bmatrix},$$

but

$$\begin{bmatrix} 1 & 3 & 5 \\ 2 & 4 & 3 \end{bmatrix} \neq \begin{bmatrix} 1 & 2 \\ 3 & 4 \\ 5 & 3 \end{bmatrix},$$

since the matrix on the left has order 2×3, whereas the matrix on the right has order 3×2,

and

$$\begin{bmatrix} 2 & 3 \\ 4 & 6 \end{bmatrix} \neq \begin{bmatrix} 2 & 3 \\ 4 & 7 \end{bmatrix},$$

since the corresponding elements in the second row and second column of the two square matrices are not equal.

Equality of Matrices

Two matrices are equal if they have the same order and their corresponding entries are equal.

EXAMPLE

12 Solve the following matrix equation for x, y, and z:

$$\begin{bmatrix} 1 & x & 3 \\ 2 & y-1 & 2 \end{bmatrix} = \begin{bmatrix} 1 & 4 & z \\ 2 & 1 & 2 \end{bmatrix}$$

SOLUTION   Since the corresponding elements of the two matrices must be equal, we find that $x = 4$, $z = 3$, and $y - 1 = 1$, or $y = 2$.   ◀

Two matrices A and B of the *same order* can be added or subtracted to produce a matrix of the same order. This is done by adding or subtracting the corresponding entries in the two matrices. For example,

$$\begin{bmatrix} 1 & 3 & 4 \\ -1 & 2 & 0 \end{bmatrix} + \begin{bmatrix} 1 & 4 & 3 \\ 6 & 1 & -2 \end{bmatrix} = \begin{bmatrix} 1+1 & 3+4 & 4+3 \\ -1+6 & 2+1 & 0+(-2) \end{bmatrix} = \begin{bmatrix} 2 & 7 & 7 \\ 5 & 3 & -2 \end{bmatrix}$$

and

$$\begin{bmatrix} 1 & 2 \\ -1 & 3 \\ 4 & 0 \end{bmatrix} - \begin{bmatrix} 2 & -1 \\ 3 & 2 \\ -1 & 0 \end{bmatrix} = \begin{bmatrix} 1-2 & 2-(-1) \\ -1-3 & 3-2 \\ 4-(-1) & 0-0 \end{bmatrix} = \begin{bmatrix} -1 & 3 \\ -4 & 1 \\ 5 & 0 \end{bmatrix}.$$

Addition and Subtraction of Matrices

If A and B are two matrices of the same order:

1. The *sum* $A + B$ is the matrix obtained by adding the corresponding entries in the two matrices.

2. The *difference* $A - B$ is the matrix obtained by subtracting the corresponding entries in B from A.

EXAMPLE

13 The total output of the Acrosonic Company for the month of June is summarized in Table 2.2.

	Model A	Model B	Model C	Model D
Location I	210	180	330	180
Location II	400	300	450	40
Location III	420	280	180	740

Table 2.2

Find the total output of the company for May and June.

SOLUTION The production matrix for the Acrosonic Company for the month of May (see page 100) is given by

$$A = \begin{bmatrix} 320 & 280 & 460 & 280 \\ 480 & 360 & 580 & 0 \\ 540 & 420 & 200 & 880 \end{bmatrix}.$$

Next, from Table 2.2, we see that the production matrix for June is given by

$$B = \begin{bmatrix} 210 & 180 & 330 & 180 \\ 400 & 300 & 450 & 40 \\ 420 & 280 & 180 & 740 \end{bmatrix}.$$

Finally, the total output of the Acrosonic Company for May and June is given by the matrix

$$A + B = \begin{bmatrix} 320 & 280 & 460 & 280 \\ 480 & 360 & 580 & 0 \\ 540 & 420 & 200 & 880 \end{bmatrix} + \begin{bmatrix} 210 & 180 & 330 & 180 \\ 400 & 300 & 450 & 40 \\ 420 & 280 & 180 & 740 \end{bmatrix}$$

$$= \begin{bmatrix} 530 & 460 & 790 & 460 \\ 880 & 660 & 1030 & 40 \\ 960 & 700 & 380 & 1620 \end{bmatrix}. \qquad \blacktriangleleft$$

The following laws hold for matrix addition. If A, B, and C are matrices of the same order, then

$$A + B = B + A \qquad \text{(Commutative law)}$$

and $\quad (A + B) + C = A + (B + C) = A + B + C \quad$ (Associative law)

A *zero matrix* is one in which all entries are zero. The zero matrix O has the property that $A + O = O + A = A$ for any matrix A having the same order as that of O. For example, the zero matrix of order 3×2 is the matrix

$$O = \begin{bmatrix} 0 & 0 \\ 0 & 0 \\ 0 & 0 \end{bmatrix},$$

and if A is any 3×2 matrix, then

$$A + O = \begin{bmatrix} a_{11} & a_{12} \\ a_{21} & a_{22} \\ a_{31} & a_{32} \end{bmatrix} + \begin{bmatrix} 0 & 0 \\ 0 & 0 \\ 0 & 0 \end{bmatrix} = \begin{bmatrix} a_{11} & a_{12} \\ a_{21} & a_{22} \\ a_{31} & a_{32} \end{bmatrix} = A,$$

where a_{ij} denotes the entry in the ith row and jth column of the matrix A.

The matrix obtained by interchanging the rows and columns of a given matrix A is called the *transpose* of A and is denoted A^T. For example, if

$$A = \begin{bmatrix} 1 & 2 & 3 \\ 4 & 5 & 6 \\ 7 & 8 & 9 \end{bmatrix},$$

then

$$A^T = \begin{bmatrix} 1 & 4 & 7 \\ 2 & 5 & 8 \\ 3 & 6 & 9 \end{bmatrix}.$$

Transpose of a Matrix

If A is an $m \times n$ matrix with elements a_{ij}, then the **transpose** of A is the $n \times m$ matrix A^T with elements a_{ji}.

▶ Scalar Multiplication

A matrix A may be multiplied by a real number, called a **scalar** in the context of matrix algebra. The **scalar product**, denoted by cA, is a matrix obtained by multiplying each entry of A by c. For example, the scalar product of the matrix

$$A = \begin{bmatrix} 3 & -1 & 2 \\ 0 & 1 & 4 \end{bmatrix}$$

by the scalar 3 is the matrix

$$3A = 3\begin{bmatrix} 3 & -1 & 2 \\ 0 & 1 & 4 \end{bmatrix} = \begin{bmatrix} 9 & -3 & 6 \\ 0 & 3 & 12 \end{bmatrix}.$$

Scalar Product

If A is a matrix and c is a real number, then the scalar product cA is the matrix obtained by multiplying each entry of A by c.

EXAMPLE

14 Given

$$A = \begin{bmatrix} 3 & 4 \\ -1 & 2 \end{bmatrix} \quad \text{and} \quad B = \begin{bmatrix} 3 & 2 \\ -1 & 2 \end{bmatrix},$$

find the matrix X satisfying the *matrix equation* $2X + B = 3A$.

SOLUTION From the given equation $2X + B = 3A$ we find that

$$2X = 3A - B$$

$$= 3 \begin{bmatrix} 3 & 4 \\ -1 & 2 \end{bmatrix} - \begin{bmatrix} 3 & 2 \\ -1 & 2 \end{bmatrix}$$

$$= \begin{bmatrix} 9 & 12 \\ -3 & 6 \end{bmatrix} - \begin{bmatrix} 3 & 2 \\ -1 & 2 \end{bmatrix} = \begin{bmatrix} 6 & 10 \\ -2 & 4 \end{bmatrix}$$

so

$$X = \frac{1}{2} \begin{bmatrix} 6 & 10 \\ -2 & 4 \end{bmatrix} = \begin{bmatrix} 3 & 5 \\ -1 & 2 \end{bmatrix}.$$

◄

EXAMPLE

15 The management of the Acrosonic Company has decided to increase its July production of loudspeaker systems by 10 percent (over its June output). Find a matrix giving the targeted production for the month of July.

SOLUTION From the results of Example 13 we see that Acrosonic's total output for the month of June may be represented by the matrix

$$B = \begin{bmatrix} 210 & 180 & 330 & 180 \\ 400 & 300 & 450 & 40 \\ 420 & 280 & 180 & 740 \end{bmatrix}.$$

The required matrix is given by

$$(1.1)B = 1.1 \begin{bmatrix} 210 & 180 & 330 & 180 \\ 400 & 300 & 450 & 40 \\ 420 & 280 & 180 & 740 \end{bmatrix} = \begin{bmatrix} 231 & 198 & 363 & 198 \\ 440 & 330 & 495 & 44 \\ 462 & 308 & 198 & 814 \end{bmatrix}$$

and is interpreted in the usual manner.

◄

Self-Check Exercises

2.3

1. Perform the indicated operations:

$$\begin{bmatrix} 1 & 3 & 2 \\ -1 & 4 & 7 \end{bmatrix} - 3\begin{bmatrix} 2 & 1 & 0 \\ 1 & 3 & 4 \end{bmatrix}$$

2. Solve the following matrix equation for x, y, and z:

$$\begin{bmatrix} x & 3 \\ z & 2 \end{bmatrix} + \begin{bmatrix} 2-y & z \\ 2-z & -x \end{bmatrix} = \begin{bmatrix} 3 & 7 \\ 2 & 0 \end{bmatrix}$$

3. Jack owns two gas stations—one downtown and the other in the Wilshire district. Over two consecutive days his gas stations recorded gasoline sales represented by the following matrices:

		Premium	Regular	Super-Unleaded	Unleaded
$A =$	Downtown	400	700	1100	1300
	Wilshire	300	800	900	1100

and

		Premium	Regular	Super-Unleaded	Unleaded
$B =$	Downtown	350	600	1000	1200
	Wilshire	250	700	400	1200

Find a matrix representing the total sales of the two gas stations over the two-day period.

Solutions to Self-Check Exercises 2.3 can be found on page 110.

▶

Exercises 2.3

In Exercises 1–4, refer to the following matrices:

$$A = \begin{bmatrix} 2 & -3 & 9 & -4 \\ -11 & 2 & 6 & 7 \\ 6 & 0 & 2 & 9 \\ 5 & 1 & 5 & -8 \end{bmatrix}, B = \begin{bmatrix} 3 & -1 & 2 \\ 0 & 1 & 4 \\ 3 & 2 & 1 \\ -1 & 0 & 8 \end{bmatrix}, C = [1 \quad 0 \quad 3 \quad 4 \quad 5]$$

1. What is the order of A, B, and C?

2. Find a_{14}, a_{21}, a_{31}, and a_{43}.

3. Find b_{13}, b_{31}, and b_{43}.

4. Identify the row matrix. What is its transpose?

In Exercises 5–8, refer to the following matrices:

$$A = \begin{bmatrix} -1 & 2 \\ 3 & -2 \\ 4 & 0 \end{bmatrix}, B = \begin{bmatrix} 2 & 4 \\ 3 & 1 \\ -2 & 2 \end{bmatrix}, C = \begin{bmatrix} 3 & -1 & 0 \\ 2 & -2 & 3 \\ 4 & 6 & 2 \end{bmatrix}, D = \begin{bmatrix} 2 & -2 & 4 \\ 3 & 6 & 2 \\ -2 & 3 & 1 \end{bmatrix}$$

5. Compute $A + B$.

6. Compute $2A - 3B$.

7. Compute $C - D$.

8. Compute $4D - 2C$.

In Exercises 9–11, perform the indicated operations.

9. $\begin{bmatrix} 1 & 4 & -5 \\ 3 & -8 & 6 \end{bmatrix} + \begin{bmatrix} 4 & 0 & -2 \\ 3 & 6 & 5 \end{bmatrix} - \begin{bmatrix} 2 & 8 & 9 \\ -11 & 2 & -5 \end{bmatrix}$

10. $3\begin{bmatrix} 1 & 1 & -3 \\ 3 & 2 & 3 \\ 7 & -1 & 6 \end{bmatrix} + 4\begin{bmatrix} -2 & -1 & 8 \\ 4 & 2 & 2 \\ 3 & 6 & 3 \end{bmatrix}$

11. $\dfrac{1}{2}\begin{bmatrix} 1 & 0 & 0 & -4 \\ 3 & 0 & -1 & 6 \\ -2 & 1 & -4 & 2 \end{bmatrix} + \dfrac{4}{3}\begin{bmatrix} 3 & 0 & -1 & 4 \\ -2 & 1 & -6 & 2 \\ 8 & 2 & 0 & -2 \end{bmatrix} - \dfrac{1}{3}\begin{bmatrix} 3 & -9 & -1 & 0 \\ 6 & 2 & 0 & -6 \\ 0 & 1 & -3 & 1 \end{bmatrix}$

In Exercises 12–15, solve for u, x, y, and z in the matrix equation.

12. $\begin{bmatrix} 2x - 2 & 3 & 2 \\ 2 & 4 & y - 2 \\ 2z & -3 & 2 \end{bmatrix} = \begin{bmatrix} 3 & u & 2 \\ 2 & 4 & 5 \\ 4 & -3 & 2 \end{bmatrix}$

13. $\begin{bmatrix} x & -2 \\ 3 & y \end{bmatrix} + \begin{bmatrix} -2 & z \\ -1 & 2 \end{bmatrix} = \begin{bmatrix} 4 & -2 \\ 2u & 4 \end{bmatrix}$

14. $\begin{bmatrix} 1 & x \\ 2y & -3 \end{bmatrix} - 4\begin{bmatrix} 2 & -2 \\ 0 & 3 \end{bmatrix} = \begin{bmatrix} 3z & 10 \\ 4 & -u \end{bmatrix}$

15. $\begin{bmatrix} 1 & 2 \\ 3 & 4 \\ x & -1 \end{bmatrix} - 3\begin{bmatrix} y - 1 & 2 \\ 1 & 2 \\ 4 & 2z + 1 \end{bmatrix} = 2\begin{bmatrix} -4 & -u \\ 0 & -1 \\ 4 & 4 \end{bmatrix}$

16. Let

$$A = \begin{bmatrix} 2 & -4 & 3 \\ 4 & 2 & 1 \end{bmatrix}, \quad B = \begin{bmatrix} 4 & -3 & 2 \\ 1 & 0 & 4 \end{bmatrix}, \quad \text{and} \quad C = \begin{bmatrix} 1 & 0 & 2 \\ 3 & -2 & 1 \end{bmatrix}.$$

Verify by direct computation the validity of the commutative law for matrix addition and the associative law for matrix addition for this special case.

In Exercises 17–18, let

$$A = \begin{bmatrix} 3 & 1 \\ 2 & 4 \\ -4 & 0 \end{bmatrix} \quad \text{and} \quad B = \begin{bmatrix} 1 & 2 \\ -1 & 0 \\ 3 & 2 \end{bmatrix}.$$

Verify by direct computation each of the given equations:

17. *a.* $(3 + 5)A = 3A + 5A$ *b.* $2(4A) = (2 \cdot 4)A = 8A$

18. *a.* $4(A + B) = 4A + 4B$ *b.* $2(A - 3B) = 2A - 6B$

In Exercises 19–22 find the transpose of the given matrix.

19. $\begin{bmatrix} 3 & 2 & -1 & 5 \end{bmatrix}$

20. $\begin{bmatrix} 4 & 2 & 0 & -1 \\ 3 & 4 & -1 & 5 \end{bmatrix}$

21. $\begin{bmatrix} 1 & -1 & 2 \\ 3 & 4 & 2 \\ 0 & 1 & 0 \end{bmatrix}$

22. $\begin{bmatrix} 1 & 2 & 6 & 4 \\ 2 & 3 & 2 & 5 \\ 6 & 2 & 3 & 0 \\ 4 & 5 & 0 & 2 \end{bmatrix}$

23. Mr. Cross, Mr. Jones, and Mr. Smith each suffer from coronary heart disease. As part of their treatment, they were put on special low-cholesterol diets: Mr. Cross on diet I, Mr. Jones on diet II, and Mr. Smith on diet III. Progressive records of each patient's cholesterol level were kept. At the beginning of the first, second, third, and fourth months the cholesterol levels of the three patients were:

Cross: 220, 215, 210, and 205

Jones: 220, 210, 200, and 195

Smith: 215, 205, 195, and 190

Represent this information in a 3 × 4 matrix.

24. The Campus Bookstore's inventory of books is:

Hardcover: Textbooks, 5280; Fiction, 1680; Nonfiction, 2320; Reference, 1890.

Paperback: Fiction, 2810; Nonfiction, 1490; Reference, 2070; Textbooks, 1940.

The College Bookstore's inventory of books is:

Hardcover: Textbooks, 6340; Fiction, 2220; Nonfiction, 1790; Reference, 1980.

Paperback: Fiction, 3100; Nonfiction, 1720; Reference, 2710; Textbooks, 2050.

a. Represent Campus's inventory as matrix *A*.

b. Represent College's inventory as matrix *B*.

c. If the two companies decide to merge, write a matrix *C* that would represent the total inventory of the newly amalgamated company.

25. The numbers of three types of bank accounts on January 1 in the Central Bank and its branches are represented by matrix *A*.

	Checking accounts	Savings accounts	Fixed-deposit accounts
Main office	2820	1470	1120
A = West side branch	1030	520	480
East side branch	1170	540	460

The number and types of accounts opened during the first quarter are represented by matrix *B*, and the number and types of accounts closed during the same period are represented by matrix *C*. Thus,

$$B = \begin{bmatrix} 260 & 120 & 110 \\ 140 & 60 & 50 \\ 120 & 70 & 50 \end{bmatrix}$$

$$C = \begin{bmatrix} 120 & 80 & 80 \\ 70 & 30 & 40 \\ 60 & 20 & 40 \end{bmatrix}.$$

a. Find matrix *D*, which represents the number of each type of account at the end of the first quarter at each location.

b. Because a new manufacturing plant is opening in the immediate area, it is anticipated that there will be a 10 percent increase in the number of accounts at each location during the second quarter. Write a matrix *E* to reflect this anticipated increase.

Solutions to Self-Check Exercises

2.3

1.
$$\begin{bmatrix} 1 & 3 & 2 \\ -1 & 4 & 7 \end{bmatrix} - 3\begin{bmatrix} 2 & 1 & 0 \\ 1 & 3 & 4 \end{bmatrix} = \begin{bmatrix} 1 & 3 & 2 \\ -1 & 4 & 7 \end{bmatrix} - \begin{bmatrix} 6 & 3 & 0 \\ 3 & 9 & 12 \end{bmatrix}$$

$$= \begin{bmatrix} -5 & 0 & 2 \\ -4 & -5 & -5 \end{bmatrix}$$

2. We are given

$$\begin{bmatrix} x & 3 \\ z & 2 \end{bmatrix} + \begin{bmatrix} 2-y & z \\ 2-z & -x \end{bmatrix} = \begin{bmatrix} 3 & 7 \\ 2 & 0 \end{bmatrix}.$$

Performing the indicated operation on the left-hand side, we obtain

$$\begin{bmatrix} 2+x-y & 3+z \\ 2 & 2-x \end{bmatrix} = \begin{bmatrix} 3 & 7 \\ 2 & 0 \end{bmatrix}.$$

By the equality of matrices, we have

$$2 + x - y = 3$$
$$3 + z = 7$$
$$2 - x = 0,$$

from which we deduce that $x = 2$, $y = 1$, and $z = 4$.

3. The required matrix is

$$A + B = \begin{bmatrix} 400 & 700 & 1100 & 1300 \\ 300 & 800 & 900 & 1100 \end{bmatrix} + \begin{bmatrix} 350 & 600 & 1000 & 1200 \\ 250 & 700 & 400 & 1200 \end{bmatrix}$$

$$= \begin{bmatrix} 750 & 1300 & 2100 & 2500 \\ 550 & 1500 & 1300 & 2300 \end{bmatrix}.$$

2.4

Multiplication of Matrices

▶ Matrix Product

▶ Application

▶ Matrix Representation

▶ Matrix Product

In Section 2.3 we saw how matrices of the same order may be added or subtracted and how a matrix may be multiplied by a scalar (real number), an operation referred to as scalar multiplication. In this section we will see how, under certain restrictions, one matrix may be multiplied by another matrix.

In order to define matrix multiplication, let us consider the following problem: On a certain day, Al's Service Station sold 800 gallons of premium, 1200 gallons of regular, 1000 gallons of super-unleaded, and 1600 gallons of regular-unleaded gasoline. If the price of gasoline on this day was $1.20 per gallon for premium, $0.90 for regular, $1.10 for super-unleaded, and $1.00 for regular-unleaded gasoline, find the total revenue realized by Al's for that day.

The day's sale of gasoline may be represented by the matrix

$$A = [800 \quad 1200 \quad 1000 \quad 1600] \quad \text{Row matrix } (1 \times 4)$$

Next, we let the unit selling price of premium, regular, super-unleaded, and regular-unleaded gasoline be the entries in the matrix

$$B = \begin{bmatrix} 1.20 \\ 0.90 \\ 1.10 \\ 1.00 \end{bmatrix} \quad \text{Column matrix } (4 \times 1)$$

The first entry in matrix A gives the number of gallons of premium gasoline sold, and the first entry in matrix B gives the selling price for each gallon of premium gasoline, so their product $(800)(1.20)$ gives the revenue realized from the sale of premium gasoline for the day. A similar interpretation of the second, third, and fourth entries in the two matrices suggests that we multiply the corresponding entries to obtain the respective revenues realized from the sale of regular, super-unleaded, and regular-unleaded gasoline. Finally, the total revenue realized by Al's from the sale of gasoline is given by adding these products to obtain

$$(800)(1.20) + (1200)(0.90) + (1000)(1.10) + (1600)(1.00) = 4740,$$

or $4740.

This example suggests that if we have a row matrix of order $1 \times n$,

$$A = [a_1 \quad a_2 \quad a_3 \quad \cdots \quad a_n],$$

and a column matrix of order $n \times 1$,

$$B = \begin{bmatrix} b_1 \\ b_2 \\ b_3 \\ \vdots \\ b_n \end{bmatrix},$$

we may define the *product* of A and B, written AB, by

$$AB = [a_1 \quad a_2 \quad a_3 \quad \cdots \quad a_n] \begin{bmatrix} b_1 \\ b_2 \\ b_3 \\ \vdots \\ b_n \end{bmatrix} = a_1b_1 + a_2b_2 + a_3b_3 + \cdots + a_nb_n.$$

Observe that the number of columns of the row matrix A is *equal* to the number of rows of the column matrix B. Observe further that the product matrix AB has order 1×1 (a real number may be thought of as a 1×1 matrix). Schematically,

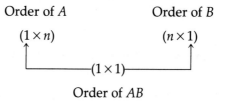

Order of AB

More generally, if A is a matrix of order $m \times n$ and B is a matrix of order $n \times p$ (the number of columns of A equals the number of rows of B), then the *matrix product* of A and B, AB, is defined and is a matrix of order $m \times p$. Schematically,

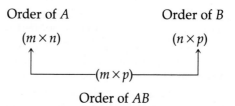

Order of AB

We now illustrate the mechanics of matrix multiplication by computing the product of a 2×3 matrix A and a 3×4 matrix B. Suppose

$$A = \begin{bmatrix} a_{11} & a_{12} & a_{13} \\ a_{21} & a_{22} & a_{23} \end{bmatrix} \quad \text{and} \quad B = \begin{bmatrix} b_{11} & b_{12} & b_{13} & b_{14} \\ b_{21} & b_{22} & b_{23} & b_{24} \\ b_{31} & b_{32} & b_{33} & b_{34} \end{bmatrix}.$$

From the schematic

```
                    ┌─────same─────┐
                    ↓              ↓
Order of A (2×3)              (3×4) Order of B
        ↑                        ↑
        └────────(2×4)───────────┘
```

Order of AB

we see that the matrix product $C = AB$ is defined (the number of columns of A equals the number of rows of B) and has order 2×4. Thus

$$C = \begin{bmatrix} c_{11} & c_{12} & c_{13} & c_{14} \\ c_{21} & c_{22} & c_{23} & c_{24} \end{bmatrix}.$$

The entries of C are computed as follows: The entry c_{11} (the entry in the *first* row, *first* column of C) is the product of the row matrix comprising the entries from the *first* row of A and the column matrix comprising the *first* column of B. Thus,

$$c_{11} = [a_{11} \quad a_{12} \quad a_{13}] \begin{bmatrix} b_{11} \\ b_{21} \\ b_{31} \end{bmatrix} = a_{11}b_{11} + a_{12}b_{21} + a_{13}b_{31}.$$

The entry c_{12} (the entry in the *first* row, *second* column of C) is the product of the row matrix comprising the *first* row of A and the column matrix comprising the *second* column of B. Thus,

$$c_{12} = [a_{11} \quad a_{12} \quad a_{13}] \begin{bmatrix} b_{12} \\ b_{22} \\ b_{32} \end{bmatrix} = a_{11}b_{12} + a_{12}b_{22} + a_{13}b_{32}.$$

The other entries in C are computed in a similar manner.

EXAMPLE

16 Let

$$A = \begin{bmatrix} 3 & 1 & 4 \\ -1 & 2 & 3 \end{bmatrix} \quad \text{and} \quad B = \begin{bmatrix} 1 & 3 & -3 \\ 4 & -1 & 2 \\ 2 & 4 & 1 \end{bmatrix}.$$

Compute AB.

SOLUTION

The order of matrix A is 2×3, and the order of matrix B is 3×3. Since the number of columns of matrix A is equal to the number of rows of matrix B, the matrix product $C = AB$ is defined. Furthermore, the order of matrix C is 2×3. Thus,

$$\begin{bmatrix} 3 & 1 & 4 \\ -1 & 2 & 3 \end{bmatrix} \begin{bmatrix} 1 & 3 & -3 \\ 4 & -1 & 2 \\ 2 & 4 & 1 \end{bmatrix} = \begin{bmatrix} c_{11} & c_{12} & c_{13} \\ c_{21} & c_{22} & c_{23} \end{bmatrix}.$$

It remains now to determine the entries $c_{11}, c_{12}, c_{13}, c_{21}, c_{22},$ and c_{23}. We have

$$c_{11} = [3 \quad 1 \quad 4] \begin{bmatrix} 1 \\ 4 \\ 2 \end{bmatrix} = (3)(1) + (1)(4) + (4)(2) = 15$$

$$c_{12} = [3 \quad 1 \quad 4] \begin{bmatrix} 3 \\ -1 \\ 4 \end{bmatrix} = (3)(3) + (1)(-1) + (4)(4) = 24$$

$$c_{13} = [3 \quad 1 \quad 4] \begin{bmatrix} -3 \\ 2 \\ 1 \end{bmatrix} = (3)(-3) + (1)(2) + (4)(1) = -3$$

$$c_{21} = [-1 \quad 2 \quad 3] \begin{bmatrix} 1 \\ 4 \\ 2 \end{bmatrix} = (-1)(1) + (2)(4) + (3)(2) = 13$$

$$c_{22} = \begin{bmatrix} -1 & 2 & 3 \end{bmatrix} \begin{bmatrix} 3 \\ -1 \\ 4 \end{bmatrix} = (-1)(3) + (2)(-1) + (3)(4) = 7$$

and
$$c_{23} = \begin{bmatrix} -1 & 2 & 3 \end{bmatrix} \begin{bmatrix} -3 \\ 2 \\ 1 \end{bmatrix} = (-1)(-3) + (2)(2) + (3)(1) = 10,$$

so the required product AB is given by

$$AB = \begin{bmatrix} 15 & 24 & -3 \\ 13 & 7 & 10 \end{bmatrix}. \qquad \blacktriangleleft$$

EXAMPLE

17 If

$$A = \begin{bmatrix} 3 & 2 & 1 \\ -1 & 2 & 3 \\ 3 & 1 & 4 \end{bmatrix}, \quad B = \begin{bmatrix} 1 & 3 & 4 \\ 2 & 4 & 1 \\ -1 & 2 & 3 \end{bmatrix}$$

then

$$AB = \begin{bmatrix} 3 \cdot 1 & + 2 \cdot 2 + 1 \cdot (-1) & 3 \cdot 3 & + 2 \cdot 4 + 1 \cdot 2 & 3 \cdot 4 & + 2 \cdot 1 + 1 \cdot 3 \\ (-1) \cdot 1 + 2 \cdot 2 + 3 \cdot (-1) & (-1) \cdot 3 + 2 \cdot 4 + 3 \cdot 2 & (-1) \cdot 4 + 2 \cdot 1 + 3 \cdot 3 \\ 3 \cdot 1 & + 1 \cdot 2 + 4 \cdot (-1) & 3 \cdot 3 & + 1 \cdot 4 + 4 \cdot 2 & 3 \cdot 4 & + 1 \cdot 1 + 4 \cdot 3 \end{bmatrix}$$

$$= \begin{bmatrix} 6 & 19 & 17 \\ 0 & 11 & 7 \\ 1 & 21 & 25 \end{bmatrix}$$

$$BA = \begin{bmatrix} 1 \cdot 3 & + 3 \cdot (-1) + 4 \cdot 3 & 1 \cdot 2 & + 3 \cdot 2 + 4 \cdot 1 & 1 \cdot 1 & + 3 \cdot 3 + 4 \cdot 4 \\ 2 \cdot 3 & + 4 \cdot (-1) + 1 \cdot 3 & 2 \cdot 2 & + 4 \cdot 2 + 1 \cdot 1 & 2 \cdot 1 & + 4 \cdot 3 + 1 \cdot 4 \\ (-1) \cdot 3 + 2 \cdot (-1) + 3 \cdot 3 & (-1) \cdot 2 + 2 \cdot 2 + 3 \cdot 1 & (-1) \cdot 1 + 2 \cdot 3 + 3 \cdot 4 \end{bmatrix}$$

$$= \begin{bmatrix} 12 & 12 & 26 \\ 5 & 13 & 18 \\ 4 & 5 & 17 \end{bmatrix}. \qquad \blacktriangleleft$$

As the last example shows, in general $AB \neq BA$ for any two square matrices A and B. However, the following properties are valid for matrix multiplication:

$$A(BC) = (AB)C \qquad \text{(Associative law)}$$

and
$$A(B + C) = AB + AC \qquad \text{(Distributive law)}$$

whenever the products are defined.

A square matrix of order n having ones along the main diagonal and zeros elsewhere is called an **identity matrix** of order n.

Identity Matrix

The identity matrix of order n is given by

$$I_n = \begin{bmatrix} 1 & 0 & \cdots & & 0 \\ 0 & 1 & \cdots & & 0 \\ \vdots & & \ddots & & \vdots \\ & & & & \\ 0 & 0 & \cdots & & 1 \end{bmatrix} \quad (n \text{ rows})$$

(n columns)

The identity matrix has the property that $I_n A = A$ for any $n \times r$ matrix A, and $BI_n = B$ for any $s \times n$ matrix B. In particular, if A is a square matrix of order n, then

$$I_n A = AI_n = A.$$

EXAMPLE

18 Let

$$A = \begin{bmatrix} 1 & 3 & 1 \\ -4 & 3 & 2 \\ 1 & 0 & 1 \end{bmatrix}.$$

Then $\quad I_3 A = \begin{bmatrix} 1 & 0 & 0 \\ 0 & 1 & 0 \\ 0 & 0 & 1 \end{bmatrix} \begin{bmatrix} 1 & 3 & 1 \\ -4 & 3 & 2 \\ 1 & 0 & 1 \end{bmatrix} = \begin{bmatrix} 1 & 3 & 1 \\ -4 & 3 & 2 \\ 1 & 0 & 1 \end{bmatrix} = A$

$AI_3 = \begin{bmatrix} 1 & 3 & 1 \\ -4 & 3 & 2 \\ 1 & 0 & 1 \end{bmatrix} \begin{bmatrix} 1 & 0 & 0 \\ 0 & 1 & 0 \\ 0 & 0 & 1 \end{bmatrix} = \begin{bmatrix} 1 & 3 & 1 \\ -4 & 3 & 2 \\ 1 & 0 & 1 \end{bmatrix} = A,$

so $I_3 A = AI_3$, confirming the above result for this special case. ◀

▶ Application

EXAMPLE

19 The Ace Novelty Company received an order from Magic World Amusement Park for 900 "Giant Pandas," 1200 "Saint Bernards," and 2000 "Big Birds." Ace's management decided that 500 Giant Pandas, 800 Saint Bernards and 1300 Big

Birds could be manufactured in their Los Angeles plant, and that the balance of the order could be filled by their Seattle plant. Each Panda requires 1.5 square yards of plush, 30 cubic feet of stuffing, and 5 pieces of trim; each Saint Bernard requires 2 square yards of plush, 35 cubic feet of stuffing, and 8 pieces of trim; and each Big Bird requires 2.5 square yards of plush, 25 cubic feet of stuffing, and 15 pieces of trim. The plush costs $4.50 per square yard, the stuffing costs $.10 per cubic foot, and the trim costs $.25 per unit.

a. Find how much of each type of material must be purchased for each plant.

b. What is the total cost of materials incurred by each plant and the total cost of materials incurred by Ace Novelty in filling the order?

SOLUTION

The quantities of each type of stuffed animal to be produced at each plant location may be expressed as a 2×3 *production matrix P.* Thus

$$P = \begin{matrix} \text{L.A.} \\ \text{Seattle} \end{matrix} \begin{matrix} \text{Pandas} & \text{St. Bernards} & \text{Birds} \\ \begin{bmatrix} 500 & 800 & 1300 \\ 400 & 400 & 700 \end{bmatrix} \end{matrix}.$$

Similarly, we may represent the amount and type of material required to manufacture each type of animal by a 3×3 *activity matrix A.* Thus

$$A = \begin{matrix} \text{Pandas} \\ \text{St. Bernards} \\ \text{Birds} \end{matrix} \begin{matrix} \text{Plush} & \text{Stuffing} & \text{Trim} \\ \begin{bmatrix} 1.5 & 30 & 5 \\ 2 & 35 & 8 \\ 2.5 & 25 & 15 \end{bmatrix} \end{matrix}.$$

Finally, the unit cost for each type of material may be represented by the 3×1 *cost matrix C*

$$C = \begin{matrix} \text{Plush} \\ \text{Stuffing} \\ \text{Trim} \end{matrix} \begin{bmatrix} 4.50 \\ 0.10 \\ 0.25 \end{bmatrix}.$$

a. The amount of each type of material required for each plant is given by the matrix *PA*. Thus,

$$PA = \begin{bmatrix} 500 & 800 & 1300 \\ 400 & 400 & 700 \end{bmatrix} \begin{bmatrix} 1.5 & 30 & 5 \\ 2 & 35 & 8 \\ 2.5 & 25 & 15 \end{bmatrix}$$

$$= \begin{matrix} \text{L.A.} \\ \text{Seattle} \end{matrix} \begin{matrix} \text{Plush} & \text{Stuffing} & \text{Trim} \\ \begin{bmatrix} 5,600 & 75,500 & 28,400 \\ 3,150 & 43,500 & 15,700 \end{bmatrix} \end{matrix}.$$

b. The total cost of materials for each plant is given by the matrix *PAC*:

$$PAC = \begin{bmatrix} 5,600 & 75,500 & 28,400 \\ 3,150 & 43,500 & 15,700 \end{bmatrix} \begin{bmatrix} 4.50 \\ 0.10 \\ 0.25 \end{bmatrix}$$

$$= \begin{matrix} \text{L.A.} \\ \text{Seattle} \end{matrix} \begin{bmatrix} 39,850 \\ 22,450 \end{bmatrix}$$

or $39,850 for the L.A. plant and $22,450 for the Seattle plant. Thus the total cost of materials incurred by Ace Novelty is $62,300. ◀

▶ Matrix Representation

Example 20 shows how a system of linear equations may be written in a compact form with the help of matrices. (This matrix equation representation will prove useful in Section 2.5.)

EXAMPLE

20   Write the system of linear equations

$$\begin{aligned} 2x - 4y + z &= 6 \\ -3x + 6y - 5z &= -1 \\ x - 3y + 7z &= 0 \end{aligned}$$

in matrix form.

SOLUTION

Let us write

$$A = \begin{bmatrix} 2 & -4 & 1 \\ -3 & 6 & -5 \\ 1 & -3 & 7 \end{bmatrix}, \quad X = \begin{bmatrix} x \\ y \\ z \end{bmatrix}, \quad \text{and} \quad B = \begin{bmatrix} 6 \\ -1 \\ 0 \end{bmatrix}.$$

Note that A is just the 3×3 matrix of coefficients of the system, X is the 3×1 column matrix of unknowns (variables), and B is the 3×1 column matrix of constants. We will now show that the required matrix representation of the system of linear equations is

$$AX = B.$$

To see this, observe that

$$AX = \begin{bmatrix} 2 & -4 & 1 \\ -3 & 6 & -5 \\ 1 & -3 & 7 \end{bmatrix} \begin{bmatrix} x \\ y \\ z \end{bmatrix} = \begin{bmatrix} 2x - 4y + z \\ -3x + 6y - 5z \\ x - 3y + 7z \end{bmatrix}.$$

Equating this 3×1 matrix with matrix B now gives

$$\begin{bmatrix} 2x - 4y + z \\ -3x + 6y - 5z \\ x - 3y + 7z \end{bmatrix} = \begin{bmatrix} 6 \\ -1 \\ 0 \end{bmatrix},$$

which, by matrix equality, is easily seen to be equivalent to the given system of linear equations. ◀

Self-Check Exercises

2.4

1. Compute

$$\begin{bmatrix} 1 & 3 & 0 \\ 2 & 4 & -1 \end{bmatrix} \begin{bmatrix} 3 & 1 & 4 \\ 2 & 0 & 3 \\ 1 & 2 & -1 \end{bmatrix}.$$

2. Write the following system of linear equations in matrix form:

$$\begin{aligned} y - 2z &= 1 \\ 2x - y + 3z &= 0 \\ x \qquad + 4z &= 7 \end{aligned}$$

3. On July 1, the stock holdings of Ash and Joan Robinson were given by the matrix

$$A = \begin{array}{c} \\ \text{Ash} \\ \text{Joan} \end{array} \begin{array}{cccc} \text{AT\&T} & \text{GTE} & \text{IBM} & \text{TWA} \\ \begin{bmatrix} 2000 & 1000 & 500 & 5000 \\ 1000 & 500 & 2000 & 0 \end{bmatrix}, \end{array}$$

and the closing prices of AT&T, GTE, IBM, and TWA were \$24, \$47, \$150, and \$14 per share, respectively. Use matrix multiplication to determine the separate values of Ash's and Joan's stock holdings as of that date.

Solutions to Self-Check Exercises 2.4 can be found on page 123.

▶

Exercises 2.4

In Exercises 1–4, determine the order of AB and BA whenever they are defined.

1. *A* is of order 2×3 and *B* is of order 3×5.

2. *A* is of order 3×4 and *B* is of order 4×3.

3. *A* is of order 1×7 and *B* is of order 7×1.

4. *A* is of order 4×4 and *B* is of order 4×4.

*5. Let A be a matrix of order $m \times n$ and B be a matrix of order $s \times t$. Find conditions on m, n, s, and t so that both the matrix products AB and BA are defined.

6. Find condition(s) on the order of a matrix A so that A^2 (that is, AA) is defined.

In Exercises 7–17, compute the indicated products.

7. $\begin{bmatrix} 1 & 2 \\ 3 & 0 \end{bmatrix} \begin{bmatrix} 1 \\ -1 \end{bmatrix}$

8. $\begin{bmatrix} -1 & 2 \\ 3 & 1 \end{bmatrix} \begin{bmatrix} 2 & 4 \\ 3 & 1 \end{bmatrix}$

9. $\begin{bmatrix} 0.1 & 0.9 \\ 0.2 & 0.8 \end{bmatrix} \begin{bmatrix} 1.2 & 0.4 \\ 0.5 & 2.1 \end{bmatrix}$

10. $\begin{bmatrix} 1 & 3 \\ -1 & 2 \end{bmatrix} \begin{bmatrix} 1 & 3 & 0 \\ 3 & 0 & 2 \end{bmatrix}$

11. $\begin{bmatrix} 2 & 1 & 2 \\ 3 & 2 & 4 \end{bmatrix} \begin{bmatrix} -1 & 2 \\ 4 & 3 \\ 0 & 1 \end{bmatrix}$

12. $\begin{bmatrix} -1 & 2 \\ 4 & 3 \\ 0 & 1 \end{bmatrix} \begin{bmatrix} 2 & 1 & 2 \\ 3 & 2 & 4 \end{bmatrix}$

13. $\begin{bmatrix} 6 & -3 & 0 \\ -2 & 1 & -8 \\ 4 & -4 & 9 \end{bmatrix} \begin{bmatrix} 1 & 0 & 0 \\ 0 & 1 & 0 \\ 0 & 0 & 1 \end{bmatrix}$

14. $\begin{bmatrix} 3 & 0 & -2 & 1 \\ 1 & 2 & 0 & -1 \end{bmatrix} \begin{bmatrix} 2 & 1 & -1 \\ -1 & 2 & 0 \\ 0 & 0 & 1 \\ -1 & -2 & 2 \end{bmatrix}$

15. $4\begin{bmatrix} 1 & -2 & 0 \\ 2 & -1 & 1 \\ 3 & 0 & -1 \end{bmatrix} \begin{bmatrix} 1 & 3 & 1 \\ 1 & 4 & 0 \\ 0 & 1 & -2 \end{bmatrix}$

16. $\begin{bmatrix} 1 & 0 \\ 0 & 1 \end{bmatrix} \begin{bmatrix} 4 & -3 & 2 \\ 7 & 1 & -5 \end{bmatrix} \begin{bmatrix} 1 & 0 & 0 \\ 0 & 1 & 0 \\ 0 & 0 & 1 \end{bmatrix}$

17. $2\begin{bmatrix} 3 & 2 & -1 \\ 0 & 1 & 3 \\ 2 & 0 & 3 \end{bmatrix} \begin{bmatrix} 1 & 0 & 0 \\ 0 & 1 & 0 \\ 0 & 0 & 1 \end{bmatrix} \begin{bmatrix} 1 & 2 & 0 \\ 0 & -1 & -2 \\ 1 & 3 & 1 \end{bmatrix}$

18. Let

$$A = \begin{bmatrix} 1 & 0 & -2 \\ 1 & -3 & 2 \\ -2 & 1 & 1 \end{bmatrix}, \quad B = \begin{bmatrix} 3 & 1 & 0 \\ 2 & 2 & 0 \\ 1 & -3 & -1 \end{bmatrix}, \quad \text{and} \quad C = \begin{bmatrix} 2 & -1 & 0 \\ 1 & -1 & 2 \\ 3 & -2 & 1 \end{bmatrix}.$$

Verify the validity of the associative law for matrix multiplication and the distributive law for matrix multiplication for this special case.

19. Let

$$A = \begin{bmatrix} 1 & 2 \\ 3 & 4 \end{bmatrix} \quad \text{and} \quad B = \begin{bmatrix} 2 & 1 \\ 4 & 3 \end{bmatrix}.$$

Compute AB and BA and hence deduce that matrix multiplication is, in general, not commutative.

20. Let

$$A = \begin{bmatrix} 0 & 3 & 0 \\ 1 & 0 & 1 \\ 0 & 2 & 0 \end{bmatrix}, \quad B = \begin{bmatrix} 2 & 4 & 5 \\ 3 & -1 & -6 \\ 4 & 3 & 4 \end{bmatrix}, \quad \text{and} \quad C = \begin{bmatrix} 4 & 5 & 6 \\ 3 & -1 & -6 \\ 2 & 2 & 3 \end{bmatrix}.$$

a. Compute AB.

b. Compute AC.

c. Using the results of parts (a) and (b), conclude that $AB = AC$ does not imply that $B = C$.

21. Let

$$A = \begin{bmatrix} 3 & 0 \\ 8 & 0 \end{bmatrix} \quad \text{and} \quad B = \begin{bmatrix} 0 & 0 \\ 4 & 5 \end{bmatrix}.$$

Show that $AB = O$, thereby demonstrating that for matrix multiplication the equation $AB = O$ does not imply that one or both of the matrices A and B must be the zero matrix.

22. Let

$$A = \begin{bmatrix} 2 & 2 \\ -2 & -2 \end{bmatrix}.$$

Show that $A^2 = O$. Compare this with the equation $a^2 = 0$, where a is a real number.

23. Find the matrix A such that

$$A \begin{bmatrix} 1 & 0 \\ -1 & 3 \end{bmatrix} = \begin{bmatrix} -1 & -3 \\ 3 & 6 \end{bmatrix}.$$

Hint: Let

$$A = \begin{bmatrix} a & b \\ c & d \end{bmatrix}.$$

24. Let

$$A = \begin{bmatrix} 3 & 1 \\ 0 & 2 \end{bmatrix} \quad \text{and} \quad B = \begin{bmatrix} 4 & -2 \\ 2 & 1 \end{bmatrix}.$$

a. Compute $(A + B)^2$.

b. Compute $A^2 + 2AB + B^2$.

c. From the results of parts (a) and (b) show that in general $(A + B)^2 \neq A^2 + 2AB + B^2$.

25. If

$$A = \begin{bmatrix} 2 & 4 \\ 5 & -6 \end{bmatrix} \quad \text{and} \quad B = \begin{bmatrix} 4 & 8 \\ -7 & 3 \end{bmatrix}.$$

a. Find A^T and show that $(A^T)^T = A$.

b. Show that $(A + B)^T = A^T + B^T$.

c. Show that $(AB)^T = B^T A^T$.

In Exercises 26–29, write the given system of linear equations in matrix form.

26. $2x - 3y = 7$
$3x - 4y = 8$

27. $2x \qquad = 7$
$3x - 2y = 12$

28. $2x - 3y + 4z = 6$
$2y - 3z = 7$
$x - y + 2z = 4$

29. $x - 2y + 3z = -1$
$3x + 4y - 2z = 1$
$2x - 3y + 7z = 6$

30. The Cinema Center comprises four theaters: Cinemas I, II, III, and IV. The admission price for one feature at the Center is \$2 for children, \$3 for students, and \$4 for adults. The attendance for the Sunday matinee is given by the matrix

$$
\begin{array}{c}
\begin{array}{ccc} \text{Children} & \text{Students} & \text{Adults} \end{array} \\
A = \begin{array}{c} \text{Cinema I} \\ \text{Cinema II} \\ \text{Cinema III} \\ \text{Cinema IV} \end{array}
\begin{bmatrix}
225 & 110 & 50 \\
75 & 180 & 225 \\
280 & 85 & 110 \\
0 & 250 & 225
\end{bmatrix}
\end{array}
$$

Write a column vector B representing the admission prices. Then compute AB, the column vector showing the gross receipts for each theater. Finally, find the total revenue collected at the Cinema Center for admission that Sunday afternoon.

31. The admissions committee of a university anticipates an enrollment of 8000 students in its freshman class next year. To satisfy admission quotas, the incoming students have been categorized according to their sex and place of residence. The number of students in each category is given by the matrix

$$
\begin{array}{c}
\begin{array}{cc} \text{Male} & \text{Female} \end{array} \\
A = \begin{array}{c} \text{In-state} \\ \text{Out-of-state} \\ \text{Foreign} \end{array}
\begin{bmatrix}
2700 & 3000 \\
800 & 700 \\
500 & 300
\end{bmatrix}
\end{array}
$$

By using data accumulated in previous years, the admissions committee has determined that these students will elect to enter the College of Letters and Science, the College of Fine Arts, the School of Business Administration, and the School of Engineering according to the percentages that appear in the following matrix:

$$
\begin{array}{c}
\begin{array}{cccc} \text{L. \& S.} & \text{Fine Arts} & \text{Bus. Ad.} & \text{Eng.} \end{array} \\
B = \begin{array}{c} \text{Male} \\ \text{Female} \end{array}
\begin{bmatrix}
.25 & .20 & .30 & .25 \\
.30 & .35 & .25 & .10
\end{bmatrix}
\end{array}
$$

Find the matrix AB that shows the number of in-state, out-of-state, and foreign students expected to enter each discipline.

32. Refer to Example 19. Suppose that Ace Novelty received an order from another amusement park for 1200 Pink Panthers, 1800 Giant Pandas, and 1400 Big Birds. The quantity of each type of stuffed animal to be produced at each plant is shown in the following production matrix:

$$
\begin{array}{c}
\begin{array}{ccc} \text{Panthers} & \text{Pandas} & \text{Birds} \end{array} \\
P = \begin{array}{c} \text{L.A.} \\ \text{Seattle} \end{array}
\begin{bmatrix}
700 & 1000 & 800 \\
500 & 800 & 600
\end{bmatrix}
\end{array}
$$

Each Panther requires 1.3 square yards of plush, 20 cubic feet of stuffing, and 12 pieces of trim. If the materials required to produce the other two stuffed animals and the unit cost for each type of material are the same as those given in Example 19

   a. Find how much of each type of material must be purchased for each plant.

   b. Find the total cost of materials that will be incurred at each plant.

   c. Find the total cost of materials incurred by Ace Novelty in filling the order.

33. The following figure shows the routes of an international airline connecting five cities. A line joining two cities indicates that there is a direct flight between them.

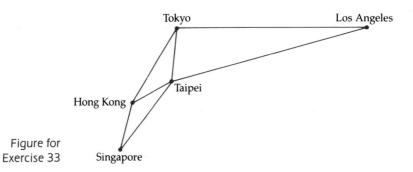

Figure for
Exercise 33

The same information can be represented by the matrix

	Singapore	Hong Kong	Taipei	Tokyo	Los Angeles
Singapore	0	1	1	0	0
Hong Kong	1	0	1	1	0
$A =$ Taipei	1	1	0	1	1
Tokyo	0	1	1	0	1
L.A.	0	0	1	1	0

where $a_{ij} = 1$ indicates that there is a direct flight between the ith city and the jth city, and $a_{ij} = 0$ indicates that there is no direct flight between the cities.

a. Compute A^2.

b. By referring to the figure, verify that the entry a_{ij} of A^2 represents the number of one-stop routes between the ith city and the jth city.

34. The following figure shows the routes of a domestic airline connecting five cities. (Refer to Exercise 33.)

a. Represent this information by a matrix A.

b. Write the matrix that shows the number of one-stop routes between any two cities.

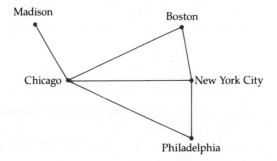

Figure for
Exercise 34

Solutions to Self-Check Exercises

2.4

1. We compute

$$\begin{bmatrix} 1 & 3 & 0 \\ 2 & 4 & -1 \end{bmatrix} \begin{bmatrix} 3 & 1 & 4 \\ 2 & 0 & 3 \\ 1 & 2 & -1 \end{bmatrix}$$

$$= \begin{bmatrix} 1(3) + 3(2) + 0(1) & 1(1) + 3(0) + 0(2) & 1(4) + 3(3) + 0(-1) \\ 2(3) + 4(2) - 1(1) & 2(1) + 4(0) - 1(2) & 2(4) + 4(3) - 1(-1) \end{bmatrix}$$

$$= \begin{bmatrix} 9 & 1 & 13 \\ 13 & 0 & 21 \end{bmatrix}.$$

2. Let

$$A = \begin{bmatrix} 0 & 1 & -2 \\ 2 & -1 & 3 \\ 1 & 0 & 4 \end{bmatrix}, \quad X = \begin{bmatrix} x \\ y \\ z \end{bmatrix}, \quad \text{and} \quad B = \begin{bmatrix} 1 \\ 0 \\ 7 \end{bmatrix}.$$

Then the given system may be written as the matrix equation

$$AX = B.$$

3. Write

$$B = \begin{bmatrix} 24 \\ 47 \\ 150 \\ 14 \end{bmatrix} \begin{matrix} \text{AT\&T} \\ \text{GTE} \\ \text{IBM} \\ \text{TWA} \end{matrix}$$

and compute

$$AB = \begin{matrix} \text{Ash} \\ \text{Joan} \end{matrix} \begin{bmatrix} 2000 & 1000 & 500 & 5000 \\ 1000 & 500 & 2000 & 0 \end{bmatrix} \begin{bmatrix} 24 \\ 47 \\ 150 \\ 14 \end{bmatrix}$$

$$= \begin{bmatrix} 240{,}000 \\ 347{,}500 \end{bmatrix} \begin{matrix} \text{Ash} \\ \text{Joan} \end{matrix}$$

We conclude that Ash's stock holdings were worth $240,000 and Joan's stock holdings were worth $347,500 on July 1.

▶ ## 2.5

Inverse of a Square Matrix

▶ Inverse of a Square Matrix

▶ Solving Systems of Equations with Inverses

▶ Applications

▶ Inverse of a Square Matrix

In this section, we will discuss a procedure for finding the inverse of a matrix and show how the inverse can be used to help us solve a system of linear equations. The inverse of a matrix also plays a central role in the Leontief input-output model (discussed in Section 2.6).

A square matrix A of order n is said to have an inverse if there exists a matrix A^{-1} (read "A, *inverse*") of order n with the property that $AA^{-1} = A^{-1} A = I$. Thus, the **inverse of a matrix** plays the same role under the multiplication of matrices that the inverse (reciprocal) of a nonzero real number plays under the multiplication of real numbers. (Compare this with $aa^{-1} = a^{-1} a = 1$, where a is a nonzero real number). By way of example, the matrix

$$A = \begin{bmatrix} 1 & 2 \\ 3 & 4 \end{bmatrix}$$

has as its inverse

$$A^{-1} = \begin{bmatrix} -2 & 1 \\ \frac{3}{2} & -\frac{1}{2} \end{bmatrix}$$

since

$$AA^{-1} = \begin{bmatrix} 1 & 2 \\ 3 & 4 \end{bmatrix} \begin{bmatrix} -2 & 1 \\ \frac{3}{2} & -\frac{1}{2} \end{bmatrix} = \begin{bmatrix} 1 & 0 \\ 0 & 1 \end{bmatrix} = \begin{bmatrix} -2 & 1 \\ \frac{3}{2} & -\frac{1}{2} \end{bmatrix} \begin{bmatrix} 1 & 2 \\ 3 & 4 \end{bmatrix} = A^{-1} A.$$

Not every square matrix has an inverse. A matrix that does not have an inverse is called **singular.** An example of a singular matrix is given by

$$B = \begin{bmatrix} 0 & 1 \\ 0 & 0 \end{bmatrix}.$$

If B has an inverse given by

$$B^{-1} = \begin{bmatrix} a & b \\ c & d \end{bmatrix}$$

where a, b, c, and d are some appropriate numbers, then, by the definition of an inverse, we must have $BB^{-1} = I$, that is,

$$\begin{bmatrix} 0 & 1 \\ 0 & 0 \end{bmatrix} \begin{bmatrix} a & b \\ c & d \end{bmatrix} = \begin{bmatrix} 1 & 0 \\ 0 & 1 \end{bmatrix}$$

or

$$\begin{bmatrix} c & d \\ 0 & 0 \end{bmatrix} = \begin{bmatrix} 1 & 0 \\ 0 & 1 \end{bmatrix}$$

which implies that $0 = 1$!

The methods of Section 2.4 can be used to find the inverse of a nonsingular matrix. In order to discover such an algorithm let us find the inverse of the matrix A given by

$$A = \begin{bmatrix} 1 & 2 \\ -1 & 3 \end{bmatrix}.$$

Suppose A^{-1} exists and is given by

$$A^{-1} = \begin{bmatrix} a & b \\ c & d \end{bmatrix}$$

where a, b, c, and d are to be determined. By the definition of an inverse, we have $AA^{-1} = I$,

or

$$\begin{bmatrix} 1 & 2 \\ -1 & 3 \end{bmatrix} \begin{bmatrix} a & b \\ c & d \end{bmatrix} = \begin{bmatrix} 1 & 0 \\ 0 & 1 \end{bmatrix},$$

which simplifies to

$$\begin{bmatrix} a + 2c & b + 2d \\ -a + 3c & -b + 3d \end{bmatrix} = \begin{bmatrix} 1 & 0 \\ 0 & 1 \end{bmatrix}.$$

But this matrix equation is equivalent to the following two systems of linear equations:

$$\left. \begin{array}{r} a + 2c = 1 \\ -a + 3c = 0 \end{array} \right\} \qquad \left. \begin{array}{r} b + 2d = 0 \\ -b + 3d = 1 \end{array} \right\}$$

with augmented matrices given by

$$\begin{bmatrix} 1 & 2 & | & 1 \\ -1 & 3 & | & 0 \end{bmatrix} \quad \text{and} \quad \begin{bmatrix} 1 & 2 & | & 0 \\ -1 & 3 & | & 1 \end{bmatrix}.$$

Note that the matrices of coefficients of the two systems are identical. This suggests that we solve the two systems of linear equations simultaneously by writing the following augmented matrix, which we obtain by joining the coefficient matrix and the two columns of constants

$$\begin{bmatrix} 1 & 2 & | & 1 & 0 \\ -1 & 3 & | & 0 & 1 \end{bmatrix}.$$

Using the Gauss-Jordan elimination method, we obtain the following sequence of equivalent matrices:

$$\begin{bmatrix} 1 & 2 & | & 1 & 0 \\ -1 & 3 & | & 0 & 1 \end{bmatrix} \xrightarrow{R_2+R_1} \begin{bmatrix} 1 & 2 & | & 1 & 0 \\ 0 & 5 & | & 1 & 1 \end{bmatrix} \xrightarrow{\frac{1}{5}R_2} \begin{bmatrix} 1 & 2 & | & 1 & 0 \\ 0 & 1 & | & \frac{1}{5} & \frac{1}{5} \end{bmatrix}$$

$$\xrightarrow{R_1-2R_2} \begin{bmatrix} 1 & 0 & | & \frac{3}{5} & -\frac{2}{5} \\ 0 & 1 & | & \frac{1}{5} & \frac{1}{5} \end{bmatrix}.$$

Thus, $a = 3/5$, $c = 1/5$, $b = -2/5$, and $d = 1/5$, giving

$$A^{-1} = \begin{bmatrix} \frac{3}{5} & -\frac{2}{5} \\ \frac{1}{5} & \frac{1}{5} \end{bmatrix}.$$

The following computations verify that A^{-1} is indeed the inverse of A:

$$\begin{bmatrix} 1 & 2 \\ -1 & 3 \end{bmatrix}\begin{bmatrix} \frac{3}{5} & -\frac{2}{5} \\ \frac{1}{5} & \frac{1}{5} \end{bmatrix} = \begin{bmatrix} 1 & 0 \\ 0 & 1 \end{bmatrix} = \begin{bmatrix} \frac{3}{5} & -\frac{2}{5} \\ \frac{1}{5} & \frac{1}{5} \end{bmatrix}\begin{bmatrix} 1 & 2 \\ -1 & 3 \end{bmatrix}$$

The preceding example reveals the general algorithm for computing the inverse of a square matrix of order n when it exists.

Finding the Inverse of a Matrix

Given the $n \times n$ matrix A:

1. Join the $n \times n$ matrix I to it to obtain the augmented matrix

$$[A \mid I].$$

2. Use a sequence of row operations to reduce $[A \mid I]$ to the form

$$[I \mid B]$$

if possible.
The matrix B is the inverse of A.

EXAMPLE

21 Find the inverse of the matrix

$$A = \begin{bmatrix} 2 & 1 & 1 \\ 3 & 2 & 1 \\ 2 & 1 & 2 \end{bmatrix}.$$

SOLUTION We form the augmented matrix

$$\begin{bmatrix} 2 & 1 & 1 & | & 1 & 0 & 0 \\ 3 & 2 & 1 & | & 0 & 1 & 0 \\ 2 & 1 & 2 & | & 0 & 0 & 1 \end{bmatrix}$$

and use the Gauss-Jordan elimination method to reduce it to the form $[I \mid B]$ as follows:

$$\begin{bmatrix} 2 & 1 & 1 & | & 1 & 0 & 0 \\ 3 & 2 & 1 & | & 0 & 1 & 0 \\ 2 & 1 & 2 & | & 0 & 0 & 1 \end{bmatrix} \xrightarrow{-R_1+R_2} \begin{bmatrix} 1 & 1 & 0 & | & -1 & 1 & 0 \\ 3 & 2 & 1 & | & 0 & 1 & 0 \\ 2 & 1 & 2 & | & 0 & 0 & 1 \end{bmatrix} \begin{array}{c} -R_2+3R_1 \\ \overrightarrow{-R_3+2R_1} \end{array}$$

$$\begin{bmatrix} 1 & 1 & 0 & | & -1 & 1 & 0 \\ 0 & 1 & -1 & | & -3 & 2 & 0 \\ 0 & 1 & -2 & | & -2 & 2 & -1 \end{bmatrix} \begin{array}{c} R_1-R_2 \\ \overrightarrow{-R_3+R_2} \end{array} \begin{bmatrix} 1 & 0 & 1 & | & 2 & -1 & 0 \\ 0 & 1 & -1 & | & -3 & 2 & 0 \\ 0 & 0 & 1 & | & -1 & 0 & 1 \end{bmatrix}$$

$$\begin{array}{c} R_1-R_3 \\ \overrightarrow{R_2+R_3} \end{array} \begin{bmatrix} 1 & 0 & 0 & | & 3 & -1 & -1 \\ 0 & 1 & 0 & | & -4 & 2 & 1 \\ 0 & 0 & 1 & | & -1 & 0 & 1 \end{bmatrix}$$

The inverse of A is the matrix

$$A^{-1} = \begin{bmatrix} 3 & -1 & -1 \\ -4 & 2 & 1 \\ -1 & 0 & 1 \end{bmatrix}.$$

We leave it to you to verify the results. ◀

Example 22 illustrates what happens to the reduction process when a matrix A does not have an inverse.

EXAMPLE

22 Find the inverse of the matrix

$$A = \begin{bmatrix} 1 & 2 & 3 \\ 2 & 1 & 2 \\ 3 & 3 & 5 \end{bmatrix}.$$

SOLUTION

We form the augmented matrix

$$\begin{bmatrix} 1 & 2 & 3 & | & 1 & 0 & 0 \\ 2 & 1 & 2 & | & 0 & 1 & 0 \\ 3 & 3 & 5 & | & 0 & 0 & 1 \end{bmatrix}$$

and use the Gauss-Jordan elimination method as follows:

$$\begin{bmatrix} 1 & 2 & 3 & | & 1 & 0 & 0 \\ 2 & 1 & 2 & | & 0 & 1 & 0 \\ 3 & 3 & 5 & | & 0 & 0 & 1 \end{bmatrix} \begin{array}{c} -R_2+2R_1 \\ \overrightarrow{-R_3+3R_1} \end{array} \begin{bmatrix} 1 & 2 & 3 & | & 1 & 0 & 0 \\ 0 & 3 & 4 & | & 2 & -1 & 0 \\ 0 & 3 & 4 & | & 3 & 0 & -1 \end{bmatrix}$$

$$\xrightarrow{R_3-R_2} \begin{bmatrix} 1 & 2 & 3 & | & 1 & 0 & 0 \\ 0 & 3 & 4 & | & 2 & -1 & 0 \\ 0 & 0 & 0 & | & 1 & 1 & -1 \end{bmatrix}$$

Since all entries in the last row of the 3×3 submatrix that comprises the left-hand side of the augmented matrix just obtained are all equal to zero, the latter cannot be reduced to the form $[I \mid B]$. Accordingly, we draw the conclusion that A is singular, that is, does not have an inverse. ◀

▶ Solving Systems of Equations with Inverses

We now show how the inverse of a matrix may be used to solve certain systems of linear equations in which the number of equations in the system is equal to the number of variables. For simplicity, let us illustrate the process for a system of three linear equations in three variables.

$$a_{11}x_1 + a_{12}x_2 + a_{13}x_3 = c_1$$
$$a_{21}x_1 + a_{22}x_2 + a_{23}x_3 = c_2$$
$$a_{31}x_1 + a_{32}x_2 + a_{33}x_3 = c_3 \tag{17}$$

Let us write

$$A = \begin{bmatrix} a_{11} & a_{12} & a_{13} \\ a_{21} & a_{22} & a_{23} \\ a_{31} & a_{32} & a_{33} \end{bmatrix}, \quad X = \begin{bmatrix} x_1 \\ x_2 \\ x_3 \end{bmatrix}, \quad \text{and} \quad B = \begin{bmatrix} b_1 \\ b_2 \\ b_3 \end{bmatrix}.$$

You should verify that the system (17) of linear equations may be written in the form of the matrix equation

$$AX = B. \tag{18}$$

If A is nonsingular, then the method of this section may be used to compute A^{-1}. Next, multiplying both sides of (18) by A^{-1} (on the left), we obtain

$$A^{-1}AX = A^{-1}B \quad \text{or} \quad IX = A^{-1}B \quad \text{or} \quad X = A^{-1}B,$$

the desired solution to the problem.

In the case of a system of n equations with n unknowns, we have the following more general result.

Using Inverses to Solve Systems of Equations

If $AX = B$ is a linear system of n equations in n unknowns, and if A^{-1} exists, then

$$X = A^{-1}B$$

is the unique solution of the system.

Examples 23 and 24 show situations in which using the inverse of a matrix to solve systems of linear equations is particularly advantageous.

▶ Applications

EXAMPLE

23 Solve the following systems of linear equations

$$a.\ \begin{aligned} 2x + y + z &= 1 \\ 3x + 2y + z &= 2 \\ 2x + y + 2z &= -1 \end{aligned} \qquad b.\ \begin{aligned} 2x + y + z &= 2 \\ 3x + 2y + z &= -3 \\ 2x + y + 2z &= 1 \end{aligned}$$

SOLUTION We may write the given system of equations in the form

$$AX = B \quad \text{and} \quad AX = C$$

respectively, where

$$A = \begin{bmatrix} 2 & 1 & 1 \\ 3 & 2 & 1 \\ 2 & 1 & 2 \end{bmatrix}, \quad X = \begin{bmatrix} x \\ y \\ z \end{bmatrix}, \quad B = \begin{bmatrix} 1 \\ 2 \\ -1 \end{bmatrix}, \quad \text{and} \quad C = \begin{bmatrix} 2 \\ -3 \\ 1 \end{bmatrix}.$$

The matrix A is precisely that given in Example 21. Using the results obtained there, we find that the solution of the first system is

$$X = A^{-1}B = \begin{bmatrix} 3 & -1 & -1 \\ -4 & 2 & 1 \\ -1 & 0 & 1 \end{bmatrix}\begin{bmatrix} 1 \\ 2 \\ -1 \end{bmatrix}$$

$$= \begin{bmatrix} 3 \cdot 1 & + (-1) \cdot 2 + (-1) \cdot (-1) \\ (-4) \cdot 1 + 2 \cdot 2 & + 1 \cdot (-1) \\ (-1) \cdot 1 + 0 \cdot 2 & + 1 \cdot (-1) \end{bmatrix} = \begin{bmatrix} 2 \\ -1 \\ -2 \end{bmatrix},$$

or $x = 2$, $y = -1$, and $z = -2$.
The solution of the second system is

$$X = A^{-1}C = \begin{bmatrix} 3 & -1 & -1 \\ -4 & 2 & 1 \\ -1 & 0 & 1 \end{bmatrix}\begin{bmatrix} 2 \\ -3 \\ 1 \end{bmatrix} = \begin{bmatrix} 8 \\ -13 \\ -1 \end{bmatrix},$$

or $x = 8$, $y = -13$, $z = -1$. ◀

EXAMPLE

24 The management of Checkers Rent-A-Car plans to expand its fleet of rental cars for the next quarter by purchasing compact and full-size cars. The average cost of a compact is $7,000, and the average cost of a full-size car is $15,000.

a. If a total of 800 cars are to be purchased with a budget of $9,000,000, how many cars of each size will be acquired?

b. If the predicted demand calls for a total purchase of 1,000 cars with a budget of $10,000,000, how many cars of each type will be acquired?

SOLUTION
Let x and y denote the number of compact and full-size cars to be purchased. Furthermore, let n denote the total number of cars to be acquired and b the amount of money budgeted for the purchase of these cars. Then

$$x + y = n$$
$$7,000x + 15,000y = b.$$

This system of two equations in two variables may be written in the matrix form

$$AX = B,$$

where

$$A = \begin{bmatrix} 1 & 1 \\ 7,000 & 15,000 \end{bmatrix}, \quad X = \begin{bmatrix} x \\ y \end{bmatrix}, \quad \text{and} \quad B = \begin{bmatrix} n \\ b \end{bmatrix}.$$

Therefore,

$$X = A^{-1}B.$$

But

$$A^{-1} = \begin{bmatrix} \frac{15000}{8000} & -\frac{1}{8000} \\ -\frac{7000}{8000} & \frac{1}{8000} \end{bmatrix} \quad \text{(See Exercises 2.5, Problem 30.)}$$

so

$$X = \begin{bmatrix} \frac{15000}{8000} & -\frac{1}{8000} \\ -\frac{7000}{8000} & \frac{1}{8000} \end{bmatrix} \begin{bmatrix} n \\ b \end{bmatrix}.$$

a. Here $n = 800$ and $b = 9,000,000$, so

$$X = A^{-1}B = \begin{bmatrix} \frac{15000}{8000} & -\frac{1}{8000} \\ -\frac{7000}{8000} & \frac{1}{8000} \end{bmatrix} \begin{bmatrix} 800 \\ 9,000,000 \end{bmatrix} = \begin{bmatrix} 375 \\ 425 \end{bmatrix}.$$

Therefore, 375 small cars and 425 full-size cars will be acquired in this case.

b. Here $n = 1,000$ and $b = 10,000,000$, so

$$X = A^{-1}B = \begin{bmatrix} \frac{15000}{8000} & -\frac{1}{8000} \\ -\frac{7000}{8000} & \frac{1}{8000} \end{bmatrix} \begin{bmatrix} 1,000 \\ 10,000,000 \end{bmatrix} = \begin{bmatrix} 625 \\ 375 \end{bmatrix}.$$

Therefore, 625 small cars and 375 full-size cars will be purchased in this case. ◀

Self-Check Exercises

2.5

1. Find the inverse of the matrix

$$\begin{bmatrix} 2 & 1 & -1 \\ 1 & 1 & -1 \\ -1 & -2 & 3 \end{bmatrix}$$

if it exists.

2. Solve the system of linear equations

$$\begin{aligned} 2x + y - z &= b_1 \\ x + y - z &= b_2 \\ -x - 2y + 3z &= b_3 \end{aligned}$$

where (a) $b_1 = 5$, $b_2 = 4$, $b_3 = -8$ and (b) $b_1 = 2$, $b_2 = 0$, $b_3 = 5$ by finding the inverse of the coefficient matrix.

3. Grand Canyon Tours, Inc. offers air and ground scenic tours of the Grand Canyon. Tickets for the 7-1/2 hour tour cost $169 for an adult and $129 for a child, and each tour group is limited to 19 people. On three recent fully booked tours, total receipts were $2931 for the first tour, $3011 for the second tour, and $2771 for the third tour. Determine how many adults and how many children were in each tour.

Solutions to Self-Check Exercises 2.5 can be found on page 134.

▶

Exercises 2.5

In Exercises 1–4, show that the given matrices are inverses of each other by showing that their product is the identity matrix I.

1. $\begin{bmatrix} 1 & -3 \\ 1 & -2 \end{bmatrix}$ and $\begin{bmatrix} -2 & 3 \\ -1 & 1 \end{bmatrix}$

2. $\begin{bmatrix} 4 & 5 \\ 2 & 3 \end{bmatrix}$ and $\begin{bmatrix} \frac{3}{2} & -\frac{5}{2} \\ -1 & 2 \end{bmatrix}$

3. $\begin{bmatrix} 3 & 2 & 3 \\ 2 & 2 & 1 \\ 2 & 1 & 1 \end{bmatrix}$ and $\begin{bmatrix} -\frac{1}{3} & -\frac{1}{3} & \frac{4}{3} \\ 0 & 1 & -1 \\ \frac{2}{3} & -\frac{1}{3} & -\frac{2}{3} \end{bmatrix}$

4. $\begin{bmatrix} 2 & 4 & -2 \\ -4 & -6 & 1 \\ 3 & 5 & -1 \end{bmatrix}$ and $\begin{bmatrix} \frac{1}{2} & -3 & -4 \\ -\frac{1}{2} & 2 & 3 \\ -1 & 1 & 2 \end{bmatrix}$

In Exercises 5–14, find the inverse (if it exists) of the given matrix. Verify your answer.

5. $\begin{bmatrix} 3 & 4 \\ -2 & 2 \end{bmatrix}$

6. $\begin{bmatrix} 4 & 2 \\ 6 & 3 \end{bmatrix}$

7. $\begin{bmatrix} 2 & -3 & -4 \\ 0 & 0 & -1 \\ 1 & -2 & 1 \end{bmatrix}$

8. $\begin{bmatrix} 1 & -1 & 3 \\ 2 & 1 & 2 \\ -2 & -2 & 1 \end{bmatrix}$

9. $\begin{bmatrix} 4 & 2 & 2 \\ -1 & -3 & 4 \\ 3 & -1 & 6 \end{bmatrix}$

10. $\begin{bmatrix} 1 & 2 & 0 \\ -3 & 4 & -2 \\ -5 & 0 & -2 \end{bmatrix}$

11. $\begin{bmatrix} 1 & 4 & -1 \\ 2 & 3 & -2 \\ -1 & 2 & 3 \end{bmatrix}$

12. $\begin{bmatrix} 3 & -2 & 7 \\ -2 & 1 & 4 \\ 6 & -5 & 8 \end{bmatrix}$

13. $\begin{bmatrix} 1 & 1 & -1 & 1 \\ 2 & 1 & 1 & 0 \\ 2 & 1 & 0 & 1 \\ 2 & -1 & -1 & 3 \end{bmatrix}$

14. $\begin{bmatrix} 1 & 1 & 2 & 3 \\ 2 & 3 & 0 & -1 \\ 0 & 2 & -1 & 1 \\ 1 & 2 & 1 & 1 \end{bmatrix}$

In Exercises 15–22, write each system of linear equations as a matrix equation and solve it by finding the inverse of the coefficient matrix.

15. $x + 2y = b_1$
    $2x - y = b_2$

    where *a.* $b_1 = 14, b_2 = 5$ and *b.* $b_1 = 4, b_2 = -1$

16. $3x - 2y = b_1$
    $4x + 3y = b_2$

    where *a.* $b_1 = -6, b_2 = 10$ and *b.* $b_1 = 3, b_2 = -2$

17. $x + 2y + z = b_1$
    $x + y + z = b_2$
    $3x + y + z = b_3$

    where *a.* $b_1 = 7, b_2 = 4, b_3 = 2$
    and *b.* $b_1 = 5, b_2 = -3, b_3 = -1$

18. $x_1 + x_2 + x_3 = b_1$
    $x_1 - x_2 + x_3 = b_2$
    $x_1 - 2x_2 - x_3 = b_3$

    where *a.* $b_1 = 5, b_2 = -3, b_3 = -1$
    and *b.* $b_1 = 1, b_2 = 4, b_3 = -2$

19. $3x + 2y - z = b_1$
    $2x - 3y + z = b_2$
    $x - y - z = b_3$

    where *a.* $b_1 = 2, b_2 = -2, b_3 = 4$
    and *b.* $b_1 = 8, b_2 = -3, b_3 = 6$

20. $2x_1 + x_2 + x_3 = b_1$
    $x_1 - 3x_2 + 4x_3 = b_2$
    $-x_1 + x_3 = b_3$

    where *a.* $b_1 = 1, b_2 = 4, b_3 = -3$
    and *b.* $b_1 = 2, b_2 = -5, b_3 = 0$

21. $x_1 + x_2 + x_3 + x_4 = b_1$
    $x_1 - x_2 - x_3 + x_4 = b_2$
    $x_2 + 2x_3 + 2x_4 = b_3$
    $x_1 + 2x_2 + x_3 - 2x_4 = b_4$

    where *a.* $b_1 = 1, b_2 = -1, b_3 = 4, b_4 = 0$
    and *b.* $b_1 = 2, b_2 = 8, b_3 = 4, b_4 = -1$

22. $x_1 + x_2 + 2x_3 + x_4 = b_1$
    $4x_1 + 5x_2 + 9x_3 + x_4 = b_2$
    $3x_1 + 4x_2 + 7x_3 + x_4 = b_3$
    $2x_1 + 3x_2 + 4x_3 + 2x_4 = b_4$

    where *a.* $b_1 = 3, b_2 = 6, b_3 = 5, b_4 = 7$
    and *b.* $b_1 = 1, b_2 = -1, b_3 = 0, b_4 = -4$

23. Let

$$A = \begin{bmatrix} 2 & 3 \\ -4 & -5 \end{bmatrix}$$

a. Find A^{-1}.

b. Show that $(A^{-1})^{-1} = A$.

24. Let

$$A = \begin{bmatrix} 6 & -4 \\ -4 & 3 \end{bmatrix} \quad \text{and} \quad B = \begin{bmatrix} 3 & -5 \\ 4 & -7 \end{bmatrix}$$

a. Find AB, A^{-1}, B^{-1}.

b. Show that $(AB)^{-1} = B^{-1}A^{-1}$.

25. Let

$$A = \begin{bmatrix} 2 & -5 \\ 1 & -3 \end{bmatrix}, \quad B = \begin{bmatrix} 4 & 3 \\ 1 & 1 \end{bmatrix}, \quad \text{and} \quad C = \begin{bmatrix} 2 & 3 \\ -2 & 1 \end{bmatrix}$$

a. Find ABC, A^{-1}, B^{-1}, C^{-1}.

b. Show that $(ABC)^{-1} = C^{-1}B^{-1}A^{-1}$.

26. Let

$$A = \begin{bmatrix} a & b \\ c & d \end{bmatrix}$$

a. Find A^{-1}.

b. Find the necessary condition for A to be nonsingular.

c. Verify that $AA^{-1} = A^{-1}A = I$.

27. Rainbow Harbor Cruises charges $8 per adult and $4 per child for a round-trip ticket. The records show that on a certain weekend 1000 people took the cruise on Saturday and 800 people took the cruise on Sunday. The total receipts for Saturday were $6400, and the total receipts for Sunday were $4800. Determine how many adults and children took the cruise on Saturday and on Sunday.

28. Bel Air Publishing Company publishes a deluxe leather edition and a standard edition of its Daily Organizer. The company's marketing department estimates that x copies of the deluxe edition and y copies of the standard edition will be demanded per month when the unit prices are p dollars and q dollars, respectively, where x, y, p, and q are related by the following system of linear equations:

$$5x + y = 1000(70 - p)$$
$$x + 3y = 1000(40 - q).$$

Find the monthly demand for the deluxe edition and the standard edition when the unit prices are set according to the following schedules:

    a. $p = 50$ and $q = 25$ b. $p = 45$ and $q = 25$ c. $p = 45$ and $q = 20$

29. Bob, a nutritionist attached to the University Medical Center, has been asked to prepare special diets for two patients, Susan and Tom. Bob has decided that Susan's meals should contain at least 400 mg of calcium, 20 mg of iron, and 50 mg of vitamin C, whereas Tom's meals should contain at least 350 mg of calcium, 15 mg of iron, and 40 mg of vitamin C. Bob has also decided that the meals are to be prepared from three basic foods: food A, food B, and food C. The special nutritional contents of these foods are summarized in the following table. Find how many ounces of each type of food should be used in a meal so that the minimum requirements of calcium, iron, and vitamin C are met for each patient's meals.

	Contents in mg/oz		
	Calcium	*Iron*	*Vitamin C*
Food A	30	1	2
Food B	25	1	5
Food C	20	2	4

30. Refer to Example 24. Verify that the matrix A^{-1} given on page 130 is the inverse of the matrix

$$A = \begin{bmatrix} 1 & 1 \\ 7{,}000 & 15{,}000 \end{bmatrix}.$$

Solutions to Self-Check Exercises

2.5

1. We form the augmented matrix

$$\left[\begin{array}{rrr|rrr} 2 & 1 & -1 & 1 & 0 & 0 \\ 1 & 1 & -1 & 0 & 1 & 0 \\ -1 & -2 & 3 & 0 & 0 & 1 \end{array}\right]$$

and row-reduce as follows:

$$\left[\begin{array}{rrr|rrr} 2 & 1 & -1 & 1 & 0 & 0 \\ 1 & 1 & -1 & 0 & 1 & 0 \\ -1 & -2 & 3 & 0 & 0 & 1 \end{array}\right] \xrightarrow{R_1 \leftrightarrow R_2} \left[\begin{array}{rrr|rrr} 1 & 1 & -1 & 0 & 1 & 0 \\ 2 & 1 & -1 & 1 & 0 & 0 \\ -1 & -2 & 3 & 0 & 0 & 1 \end{array}\right] \xrightarrow[R_3 + R_1]{-R_2 + 2R_1}.$$

$$\left[\begin{array}{rrr|rrr} 1 & 1 & -1 & 0 & 1 & 0 \\ 0 & 1 & -1 & -1 & 2 & 0 \\ 0 & -1 & 2 & 0 & 1 & 1 \end{array}\right] \xrightarrow[R_3 + R_2]{R_1 - R_2} \left[\begin{array}{rrr|rrr} 1 & 0 & 0 & 1 & -1 & 0 \\ 0 & 1 & -1 & -1 & 2 & 0 \\ 0 & 0 & 1 & -1 & 3 & 1 \end{array}\right] \xrightarrow{R_2 + R_3}$$

$$\left[\begin{array}{rrr|rrr} 1 & 0 & 0 & 1 & -1 & 0 \\ 0 & 1 & 0 & -2 & 5 & 1 \\ 0 & 0 & 1 & -1 & 3 & 1 \end{array}\right]$$

From the preceding results, we see that

$$A^{-1} = \begin{bmatrix} 1 & -1 & 0 \\ -2 & 5 & 1 \\ -1 & 3 & 1 \end{bmatrix}.$$

2. a. We write the systems of linear equations in the matrix form

$$AX = B_1$$

where

$$A = \begin{bmatrix} 2 & 1 & -1 \\ 1 & 1 & -1 \\ -1 & -2 & 3 \end{bmatrix}, \quad X = \begin{bmatrix} x \\ y \\ z \end{bmatrix}, \quad \text{and} \quad B_1 = \begin{bmatrix} 5 \\ 4 \\ -8 \end{bmatrix}.$$

Now, using the results of Exercise 1, we have

$$X = \begin{bmatrix} x \\ y \\ z \end{bmatrix} = A^{-1}B_1 = \begin{bmatrix} 1 & -1 & 0 \\ -2 & 5 & 1 \\ -1 & 3 & 1 \end{bmatrix} \begin{bmatrix} 5 \\ 4 \\ -8 \end{bmatrix} = \begin{bmatrix} 1 \\ 2 \\ -1 \end{bmatrix}.$$

Therefore $x = 1$, $y = 2$, and $z = -1$.

b. Here A and X are as in part (a) but

$$B_2 = \begin{bmatrix} 2 \\ 0 \\ 5 \end{bmatrix}.$$

Therefore,

$$X = \begin{bmatrix} x \\ y \\ z \end{bmatrix} = A^{-1}B_2 = \begin{bmatrix} 1 & -1 & 0 \\ -2 & 5 & 1 \\ -1 & 3 & 1 \end{bmatrix} \begin{bmatrix} 2 \\ 0 \\ 5 \end{bmatrix} = \begin{bmatrix} 2 \\ 1 \\ 3 \end{bmatrix},$$

or $x = 2$, $y = 1$, and $z = 3$.

3. Let x denote the number of adults and y the number of children in a tour. Since the tours are filled to capacity, we have

$$x + y = 19.$$

Next, using the fact that the total receipts for the first tour were $2931 leads to the equation

$$169x + 129y = 2931.$$

Therefore the number of adults and the number of children in the first tour is found by solving the system of linear equations

$$\begin{aligned} x + \quad y &= \quad 19 \\ 169x + 129y &= 2931. \end{aligned} \tag{a}$$

Similarly, we see that the number of adults and the number of children in the second and third tours are found by solving the systems

$$\begin{aligned} x + \quad y &= \quad 19 \\ 169x + 129y &= 3011 \end{aligned} \tag{b}$$

and

$$\begin{aligned} x + \quad y &= \quad 19 \\ 169x + 129y &= 2771. \end{aligned} \tag{c}$$

These systems may be written in the form

$$AX = B_1, \quad AX = B_2, \quad \text{and} \quad AX = B_3,$$

where

$$A = \begin{bmatrix} 1 & 1 \\ 169 & 129 \end{bmatrix}, \quad X = \begin{bmatrix} x \\ y \end{bmatrix}, \quad B_1 = \begin{bmatrix} 19 \\ 2931 \end{bmatrix}, \quad B_2 = \begin{bmatrix} 19 \\ 3011 \end{bmatrix}, \quad \text{and} \quad B_3 = \begin{bmatrix} 19 \\ 2771 \end{bmatrix}.$$

To solve these systems, we first find A^{-1}. Using the Gauss-Jordan procedure, we obtain

$$\begin{bmatrix} 1 & 1 & | & 1 & 0 \\ 169 & 129 & | & 0 & 1 \end{bmatrix} \xrightarrow{R_2 - 169R_1} \begin{bmatrix} 1 & 1 & | & 1 & 0 \\ 0 & -40 & | & -169 & 1 \end{bmatrix} \xrightarrow{-\frac{1}{40}R_2}$$

$$\begin{bmatrix} 1 & 1 & | & 1 & 0 \\ 0 & 1 & | & \frac{169}{40} & -\frac{1}{40} \end{bmatrix} \xrightarrow{R_1 - R_2} \begin{bmatrix} 1 & 0 & | & -\frac{129}{40} & \frac{1}{40} \\ 0 & 1 & | & \frac{169}{40} & -\frac{1}{40} \end{bmatrix}$$

and conclude that

$$A^{-1} = \begin{bmatrix} -\frac{129}{40} & \frac{1}{40} \\ \frac{169}{40} & -\frac{1}{40} \end{bmatrix}.$$

Therefore, solving each system, we find

$$X = \begin{bmatrix} x \\ y \end{bmatrix} = A^{-1}B_1 = \begin{bmatrix} -\frac{129}{40} & \frac{1}{40} \\ \frac{169}{40} & -\frac{1}{40} \end{bmatrix} \begin{bmatrix} 19 \\ 2931 \end{bmatrix} = \begin{bmatrix} 12 \\ 7 \end{bmatrix} \qquad \text{(a)}$$

$$X = \begin{bmatrix} x \\ y \end{bmatrix} = A^{-1}B_2 = \begin{bmatrix} -\frac{129}{40} & \frac{1}{40} \\ \frac{169}{40} & -\frac{1}{40} \end{bmatrix} \begin{bmatrix} 19 \\ 3011 \end{bmatrix} = \begin{bmatrix} 14 \\ 5 \end{bmatrix} \qquad \text{(b)}$$

$$X = \begin{bmatrix} x \\ y \end{bmatrix} = A^{-1}B_3 = \begin{bmatrix} -\frac{129}{40} & \frac{1}{40} \\ \frac{169}{40} & -\frac{1}{40} \end{bmatrix} \begin{bmatrix} 19 \\ 2771 \end{bmatrix} = \begin{bmatrix} 8 \\ 11 \end{bmatrix}. \qquad \text{(c)}$$

We conclude that:

a. There were 12 adults and 7 children on the first tour.

b. There were 14 adults and 5 children on the second tour.

c. There were 8 adults and 11 children on the third tour.

2.6

Leontief Input-Output Model (Optional)

▶ Input-Output Analysis

▶ Applications

▶ Input-Output Analysis

One of the many important applications of matrix theory to the field of economics is in the study of the relationship between industrial production and consumer demand. At the heart of this analysis is the Leontief input-output model, pioneered by Wassily Leontief, who was awarded a Nobel Prize in economics in 1973 for his contributions to the field.

In order to illustrate this concept, let us consider an oversimplified economy consisting of three sectors: agriculture (A), manufacturing (M), and service (S). In general, part of the output of one sector is absorbed by another sector through inter-industry purchases, with the excess available to fulfill consumer demands. The relationship governing both intra-industrial and inter-industrial sales (and purchases) is conveniently represented by means of an **input-output matrix:**

$$\begin{array}{c} \text{Output (amount produced)} \\[4pt] \begin{array}{ccc} & \text{A} \quad \text{M} \quad \text{S} \end{array} \end{array}$$

$$\begin{array}{cc} & \begin{array}{c} \text{A} \\ \text{M} \\ \text{S} \end{array} \end{array} \begin{bmatrix} 0.2 & 0.2 & 0.1 \\ 0.2 & 0.4 & 0.1 \\ 0.1 & 0.2 & 0.3 \end{bmatrix} \qquad (19)$$

Input (amount used in production)

The first column (read from top to bottom) tells us that the production of 1 unit of agricultural products requires the consumption of 0.2 units of agricultural products, 0.2 units of manufactured goods, and 0.1 units of services. The second column tells us that the production of 1 unit of manufactured products requires the consumption of 0.2 units of agricultural products, 0.4 units of manufactured products, and 0.2 units of services. Finally, the third column tells us that the production of 1 unit of services requires the consumption of 0.1 units each of agricultural goods and manufactured products, and 0.3 units of services.

EXAMPLE

25 Refer to the input-output matrix (19).

a. If the units are measured in millions of dollars, determine the amount of agricultural products consumed in the production of $100 million worth of manufactured goods.

b. Determine the dollar amount of manufactured products required to produce $200 million worth of all goods and services in the economy.

SOLUTION

a. The production of 1 unit, that is, $1 million worth of manufactured goods, requires the consumption of 0.2 units of agricultural products. Thus, the amount of agricultural products consumed in the production of $100 million worth of manufactured goods is given by (100)(0.2), or $20 million.

b. The amount of manufactured goods required to produce 1 unit of all goods and services in the economy is given by adding the numbers of the second row of the input-output matrix; that is, $0.2 + 0.4 + 0.1$, or 0.7 units. Therefore, the production of $200 million worth of all goods and services in the economy requires 200(0.7), or $140 million worth, of manufactured products. ◀

Next, suppose that the total output of goods of the agriculture and manufacturing sectors and the total output from the service sector of the economy are given by x, y, and z units, respectively. What is the value of agricultural products consumed in the internal process of producing this total output of various goods and services?

To answer this question, we first note, by examining the input-output matrix

$$
\begin{array}{c}
\phantom{\text{Input} \quad} \text{Output} \\
\phantom{\text{Input} \quad} \begin{array}{ccc} A & M & S \end{array} \\
\text{Input} \quad \begin{array}{c} A \\ M \\ S \end{array}
\begin{bmatrix}
0.2 & 0.2 & 0.1 \\
0.2 & 0.4 & 0.1 \\
0.1 & 0.2 & 0.3
\end{bmatrix}
\end{array}
$$

that 0.2 units of agricultural products are required to produce 1 unit of agricultural products, so the amount of agricultural goods required to produce x units of agricultural products is given by $0.2x$ units. Next, again referring to the input-output matrix, we see that 0.2 units of agricultural products are required to produce 1 unit of manufactured products, so the requirement for producing y units of the latter is $0.2y$ units of agricultural products. Finally, we see that 0.1 units of agricultural goods are required to produce 1 unit of services, so the value of agricultural products required to produce z units of services is $0.1z$ units. Thus, the total amount of agricultural products required to produce the total output of goods and services in the economy is

$$0.2x + 0.2y + 0.1z$$

units. In a similar manner, we see that the total amount of manufactured products and the total value of services to produce the total output of goods and services in the economy are given by

$$0.2x + 0.4y + 0.1z$$

and

$$0.1x + 0.2y + 0.3z,$$

respectively.

These results could also be obtained using matrix multiplication. To see this, write the total output of goods and services x, y, and z as a 3×1 matrix

$$X = \begin{bmatrix} x \\ y \\ z \end{bmatrix} \quad \text{(Gross production matrix)}$$

The matrix X is called the **gross production matrix.** Letting A denote the input-output matrix, we have

$$A = \begin{bmatrix} 0.2 & 0.2 & 0.1 \\ 0.2 & 0.4 & 0.1 \\ 0.1 & 0.2 & 0.3 \end{bmatrix} \quad \text{(Input-output matrix)}$$

Then the product

$$AX = \begin{bmatrix} 0.2 & 0.2 & 0.1 \\ 0.2 & 0.4 & 0.1 \\ 0.1 & 0.2 & 0.3 \end{bmatrix} \begin{bmatrix} x \\ y \\ z \end{bmatrix}$$

$$= \begin{bmatrix} 0.2x + 0.2y + 0.1z \\ 0.2x + 0.4y + 0.1z \\ 0.1x + 0.2y + 0.3x \end{bmatrix} \quad \text{(Internal consumption matrix)}$$

is a 3×1 matrix whose entries represent the respective values of the agricultural products, manufactured products, and services consumed in the internal process of production. The matrix AX is referred to as the **internal consumption matrix.**

Now, since X gives the total production of goods and services in the economy, and AX, as we have just seen, gives the amount of products and services consumed in the production of these goods and services, the 3×1 matrix $X - AX$ gives the net output of goods and services that is exactly enough to satisfy consumer demands. Letting matrix D represent these consumer demands, we are led to the following matrix equation:

$$X - AX = D$$

or

$$(I - A)X = D,$$

where I is the 3×3 identity matrix.

Assuming that the inverse of $(I - A)$ exists, multiplying both sides of the last equation by $(I - A)^{-1}$ yields

$$X = (I - A)^{-1}D. \tag{20}$$

I

Leontief Input-Output Model

In a Leontief input-output model, the matrix equation giving the net output of goods and services needed to satisfy consumer demand is

$$
\begin{array}{cccc}
\text{Total} & \text{Internal} & \text{Consumer} \\
\text{output} & \text{consumption} & \text{demand}
\end{array}
$$

$$
X \quad - \quad AX \quad = \quad D
$$

where X is the total output matrix, A is the input-output matrix, and D is the matrix representing consumer demand.
The solution to this equation is

$$
X = (I - A)^{-1}D \quad \text{(Assuming that } (I - A)^{-1} \text{ exists)}
$$

which gives the amount of goods and services that must be produced to satisfy consumer demand.

► Applications

Equation (20) gives us a means of finding the amount of goods and services to be produced in order to satisfy a given level of consumer demand, as illustrated by the following example.

EXAMPLE

26 For the three-sector economy with input-output matrix given by (19), which is reproduced here

$$
\begin{bmatrix}
0.2 & 0.2 & 0.1 \\
0.2 & 0.4 & 0.1 \\
0.1 & 0.2 & 0.3
\end{bmatrix} \quad \text{(Each unit equals \$1 million.)}
$$

a. Find the gross output of goods and services needed to satisfy a consumer demand of $100 million worth of agricultural products, $80 million worth of manufactured products, and $50 million worth of services.

b. Find the value of the goods and services consumed in the internal process of production in order to meet this gross output.

SOLUTION

a. We are required to determine the gross production matrix

$$
X = \begin{bmatrix} x \\ y \\ z \end{bmatrix},
$$

where x, y, and z denote the value of the agricultural products, the manufactured products, and services. The matrix representing the consumer demand is given by

$$D = \begin{bmatrix} 100 \\ 80 \\ 50 \end{bmatrix}.$$

Next, we compute

$$I - A = \begin{bmatrix} 1 & 0 & 0 \\ 0 & 1 & 0 \\ 0 & 0 & 1 \end{bmatrix} - \begin{bmatrix} 0.2 & 0.2 & 0.1 \\ 0.2 & 0.4 & 0.1 \\ 0.1 & 0.2 & 0.3 \end{bmatrix} = \begin{bmatrix} 0.8 & -0.2 & -0.1 \\ -0.2 & 0.6 & -0.1 \\ -0.1 & -0.2 & 0.7 \end{bmatrix}.$$

Using the method of Section 2.5, we find (to two decimal places)

$$(I - A)^{-1} = \begin{bmatrix} 1.43 & 0.57 & 0.29 \\ 0.54 & 1.96 & 0.36 \\ 0.36 & 0.64 & 1.57 \end{bmatrix}.$$

Finally, using Equation (20), we find

$$X = (I - A)^{-1}D = \begin{bmatrix} 1.43 & 0.57 & 0.29 \\ 0.54 & 1.96 & 0.36 \\ 0.36 & 0.64 & 1.57 \end{bmatrix} \begin{bmatrix} 100 \\ 80 \\ 50 \end{bmatrix} = \begin{bmatrix} 203.1 \\ 228.8 \\ 165.7 \end{bmatrix};$$

that is, to fulfill consumer demands, \$203.1 million worth of agricultural products, \$228.8 million worth of manufactured products, and \$165.7 million worth of services should be produced.

b. The amount of goods and services consumed in the internal process of production is given by AX, or equivalently by $X - D$. In this case, it is more convenient to use the latter, which gives the required result of

$$\begin{bmatrix} 203.1 \\ 228.8 \\ 165.7 \end{bmatrix} - \begin{bmatrix} 100 \\ 80 \\ 50 \end{bmatrix} = \begin{bmatrix} 103.1 \\ 148.8 \\ 115.7 \end{bmatrix},$$

or \$103.1 million worth of agricultural products, \$148.8 million worth of manufactured products, and \$115.7 million worth of services. ◀

EXAMPLE

27 The TKK Corporation, a large conglomerate, has three subsidiaries engaged in producing raw rubber, manufacturing tires, and manufacturing other rubber-based goods. The production of 1 unit of raw rubber requires the consumption of 0.08 units of rubber, 0.04 units of tires, and 0.02 units of other rubber-based goods. To produce 1 unit of tires requires 0.6 units of raw rubber, 0.02 units of tires, and 0 units of other rubber-based goods. To produce 1 unit of other rubber-based goods requires 0.3 units of raw rubber, 0.01 units of tires, and 0.06 units of other rubber-based goods. Market research indicates that the demand for the following year will be $200 million for raw rubber, $800 million for tires, and $120 million for rubber-based products. Find the level of production for each subsidiary in order to satisfy this demand.

SOLUTION   View the corporation as an economy having three sectors with an input-output matrix given by

$$
\begin{array}{c}
\\
\\
\text{Raw rubber} \\
A = \text{Tires} \\
\text{Goods}
\end{array}
\begin{array}{c}
\text{Raw} \\
\text{rubber} \quad \text{Tires} \quad \text{Goods} \\
\begin{bmatrix}
0.08 & 0.60 & 0.30 \\
0.04 & 0.02 & 0.01 \\
0.02 & 0 & 0.06
\end{bmatrix}.
\end{array}
$$

Using (20), the required level of production is given by

$$
X = \begin{bmatrix} x \\ y \\ z \end{bmatrix} = (I - A)^{-1}D,
$$

where x, y, and z denote the outputs of raw rubber, tires, and other rubber-based goods, and

$$
D = \begin{bmatrix} 200 \\ 800 \\ 120 \end{bmatrix}.
$$

Now,

$$
I - A = \begin{bmatrix}
0.92 & -0.60 & -0.30 \\
-0.04 & 0.98 & -0.01 \\
-0.02 & 0 & 0.94
\end{bmatrix}.
$$

You are asked to verify that

$$
(I - A)^{-1} = \begin{bmatrix}
1.13 & 0.69 & 0.37 \\
0.05 & 1.05 & 0.03 \\
0.02 & 0.02 & 1.07
\end{bmatrix} \quad \text{(See Exercises 2.6, Problem 7.)}
$$

Therefore,

$$
X = (I - A)^{-1}D = \begin{bmatrix}
1.13 & 0.69 & 0.37 \\
0.05 & 1.05 & 0.03 \\
0.02 & 0.02 & 1.07
\end{bmatrix}\begin{bmatrix} 200 \\ 800 \\ 120 \end{bmatrix} = \begin{bmatrix} 822.4 \\ 853.6 \\ 148.4 \end{bmatrix};
$$

that is, to fulfill the predicted demand, $822.4 million worth of raw rubber, $853.6 million worth of tires, and $148.4 million worth of other rubber-based goods should be produced.   ◀

Self-Check Exercises

2.6

1. Solve the matrix equation $(I - A)X = D$ for x and y given that

$$A = \begin{bmatrix} 0.4 & 0.1 \\ 0.2 & 0.2 \end{bmatrix}, \quad X = \begin{bmatrix} x \\ y \end{bmatrix}, \quad \text{and} \quad D = \begin{bmatrix} 50 \\ 10 \end{bmatrix}.$$

2. A simple economy consists of two sectors: agriculture (A) and transportation (T). The input-output matrix for this economy is given by

$$\begin{array}{cc} & \begin{array}{cc} A & T \end{array} \\ A = \begin{array}{c} A \\ T \end{array} & \begin{bmatrix} 0.4 & 0.1 \\ 0.2 & 0.2 \end{bmatrix}. \end{array}$$

   a. Find the gross output of agricultural products needed to satisfy a consumer demand for $50 million worth of agricultural products and $10 million worth of transportation.

   b. Find the value of agricultural products and transportation consumed in the internal process of production in order to meet the gross output.

Solutions to Self-Check Exercises 2.6 can be found on page 145.

▶

Exercises 2.6

1. A simple economy consists of three sectors: agriculture (A), manufacturing (M), and transportation (T). The input-output matrix for this economy is given by

$$\begin{array}{cc} & \begin{array}{ccc} A & M & T \end{array} \\ \begin{array}{c} A \\ M \\ T \end{array} & \begin{bmatrix} 0.4 & 0.1 & 0.1 \\ 0.1 & 0.4 & 0.3 \\ 0.2 & 0.2 & 0.2 \end{bmatrix}. \end{array}$$

   a. If the units are measured in millions of dollars, determine the amount of agricultural products consumed in the production of $100 million worth of manufactured goods.

   b. Determine the dollar amount of manufactured products required to produce $200 million worth of all goods in the economy.

   c. Which sector consumes the greatest amount of agricultural products in the production of a unit of goods in that sector? the least?

2. The relationship governing the intra-industrial and inter-industrial sales and purchases of four basic industries—agriculture (A), manufacturing (M), transportation

(T), and energy (E)—of a certain economy is given by the following input-output matrix:

$$
\begin{array}{c@{}c}
 & \begin{array}{cccc} A & M & T & E \end{array} \\
\begin{array}{c} A \\ M \\ T \\ E \end{array} &
\begin{bmatrix}
0.3 & 0.2 & 0 & 0.1 \\
0.2 & 0.3 & 0.2 & 0.1 \\
0.2 & 0.2 & 0.1 & 0.3 \\
0.1 & 0.2 & 0.3 & 0.2
\end{bmatrix}
\end{array}
$$

a. How many units of energy are required to produce 1 unit of manufactured goods?

b. How many units of energy are required to produce 3 units of all goods in the economy?

c. Which sector of the economy is least dependent on the cost of energy?

d. Which sector of the economy has the smallest intra-industry purchases (sales)?

In Exercises 3–6, solve the matrix equation $(I - A)X = D$ for the given matrices A and D.

3. $A = \begin{bmatrix} 0.4 & 0.2 \\ 0.3 & 0.1 \end{bmatrix}$, $D = \begin{bmatrix} 10 \\ 12 \end{bmatrix}$

4. $A = \begin{bmatrix} 0.2 & 0.3 \\ 0.5 & 0.2 \end{bmatrix}$, $D = \begin{bmatrix} 4 \\ 8 \end{bmatrix}$

5. $A = \begin{bmatrix} 0.5 & 0.2 \\ 0.2 & 0.5 \end{bmatrix}$, $D = \begin{bmatrix} 10 \\ 20 \end{bmatrix}$

6. $A = \begin{bmatrix} 0.6 & 0.2 \\ 0.1 & 0.4 \end{bmatrix}$, $D = \begin{bmatrix} 8 \\ 12 \end{bmatrix}$

7. Let

$$
A = \begin{bmatrix}
0.08 & 0.60 & 0.30 \\
0.04 & 0.02 & 0.01 \\
0.02 & 0 & 0.06
\end{bmatrix}.
$$

Show that

$$
(I - A)^{-1} = \begin{bmatrix}
1.13 & 0.69 & 0.37 \\
0.05 & 1.05 & 0.03 \\
0.02 & 0.02 & 1.07
\end{bmatrix}.
$$

8. Consider the economy of Exercise 1, consisting of three sectors: agriculture (A), manufacturing (M), and transportation (T), with an input-output matrix given by

$$
\begin{array}{c@{}c}
 & \begin{array}{ccc} A & M & T \end{array} \\
\begin{array}{c} A \\ M \\ T \end{array} &
\begin{bmatrix}
0.4 & 0.1 & 0.1 \\
0.1 & 0.4 & 0.3 \\
0.2 & 0.2 & 0.2
\end{bmatrix}.
\end{array}
$$

a. Find the gross output of goods needed to satisfy a consumer demand for $200 million worth of agricultural products, $100 million worth of manufactured products, and $60 million worth of transportation.

b. Find the value of goods and transportation consumed in the internal process of production in order to meet this gross output.

In Exercises 9–12, matrix A is an input-output matrix associated with an economy, and matrix D (units in millions of dollars) is a demand vector. In each problem find the *final inputs of each industry so that the demands of both industry and the open sector are met.*

9. $A = \begin{bmatrix} 0.4 & 0.2 \\ 0.3 & 0.5 \end{bmatrix}$, $D = \begin{bmatrix} 12 \\ 24 \end{bmatrix}$

10. $A = \begin{bmatrix} 0.1 & 0.4 \\ 0.3 & 0.2 \end{bmatrix}$, $D = \begin{bmatrix} 5 \\ 10 \end{bmatrix}$

11. $A = \begin{bmatrix} \frac{1}{5} & \frac{2}{5} & \frac{1}{5} \\ \frac{1}{2} & 0 & \frac{1}{2} \\ 0 & \frac{1}{5} & 0 \end{bmatrix}$, $D = \begin{bmatrix} 10 \\ 5 \\ 15 \end{bmatrix}$

12. $A = \begin{bmatrix} 0.2 & 0.4 & 0.1 \\ 0.3 & 0.2 & 0.1 \\ 0.1 & 0.2 & 0.2 \end{bmatrix}$, $D = \begin{bmatrix} 6 \\ 8 \\ 10 \end{bmatrix}$

Solutions to Self-Check Exercises

2.6

1. Multiplying both sides of the given equation on the left by $(I - A)^{-1}$, we see that

$$X = (I - A)^{-1}D.$$

Now

$$I - A = \begin{bmatrix} 1 & 0 \\ 0 & 1 \end{bmatrix} - \begin{bmatrix} 0.4 & 0.1 \\ 0.2 & 0.2 \end{bmatrix} = \begin{bmatrix} 0.6 & -0.1 \\ -0.2 & 0.8 \end{bmatrix}.$$

Next, we use the Gauss-Jordan procedure to compute $(I - A)^{-1}$ (to two decimal places) as follows:

$$\begin{bmatrix} 0.6 & -0.1 & | & 1 & 0 \\ -0.2 & 0.8 & | & 0 & 1 \end{bmatrix} \xrightarrow{\frac{1}{0.6}R_1} \begin{bmatrix} 1 & -0.17 & | & 1.67 & 0 \\ -0.2 & 0.8 & | & 0 & 1 \end{bmatrix} \xrightarrow{R_2 + 0.2R_1}$$

$$\begin{bmatrix} 1 & -0.17 & | & 1.67 & 0 \\ 0 & 0.77 & | & 0.33 & 1 \end{bmatrix} \xrightarrow{\frac{1}{0.77}R_2} \begin{bmatrix} 1 & -0.17 & | & 1.67 & 0 \\ 0 & 1 & | & 0.43 & 1.30 \end{bmatrix} \xrightarrow{R_1 + 0.17R_2}$$

$$\begin{bmatrix} 1 & 0 & | & 1.74 & 0.22 \\ 0 & 1 & | & 0.43 & 1.30 \end{bmatrix}$$

giving

$$(I - A)^{-1} = \begin{bmatrix} 1.74 & 0.22 \\ 0.43 & 1.30 \end{bmatrix}.$$

Therefore,

$$X = \begin{bmatrix} x \\ y \end{bmatrix} = (I - A)^{-1}D = \begin{bmatrix} 1.74 & 0.22 \\ 0.43 & 1.30 \end{bmatrix} \begin{bmatrix} 50 \\ 10 \end{bmatrix} = \begin{bmatrix} 89.2 \\ 34.5 \end{bmatrix},$$

or $x = 89.2$ and $y = 34.5$.

2. *a.* Let

$$X = \begin{bmatrix} x \\ y \end{bmatrix}$$

denote the gross production matrix where x denotes the value of the agricultural products and y the value of transportation. Also, let

$$D = \begin{bmatrix} 50 \\ 10 \end{bmatrix}$$

denote the consumer demand. Then

$$(I - A)X = D$$

or equivalently,

$$X = (I - A)^{-1}D.$$

Using the results of Exercise 1, we find that $x = 89.2$ and $y = 34.5$. That is, to fulfill consumer demands, $89.2 million worth of agricultural products must be produced and $34.5 million worth of transportation services must be used.

b. The amount of agricultural products consumed and transportation services used is given by

$$X - D = \begin{bmatrix} 89.2 \\ 34.5 \end{bmatrix} - \begin{bmatrix} 50 \\ 10 \end{bmatrix} = \begin{bmatrix} 39.2 \\ 24.5 \end{bmatrix},$$

or $39.2 million worth of agricultural products and $24.5 million worth of transportation services.

▶

Chapter 2 Review Exercises

In Exercises 1–4, perform the given operations, if possible.

1. $\begin{bmatrix} 1 & 2 \\ -1 & 3 \\ 2 & 1 \end{bmatrix} + \begin{bmatrix} 1 & 0 \\ 0 & 1 \\ 1 & 2 \end{bmatrix}$

2. $\begin{bmatrix} -1 & 2 \\ 3 & 4 \end{bmatrix} - \begin{bmatrix} 1 & 2 \\ 5 & -2 \end{bmatrix}$

3. $[-3 \quad 2 \quad 1] \begin{bmatrix} 2 & 1 \\ -1 & 0 \\ 2 & 1 \end{bmatrix}$

4. $\begin{bmatrix} 1 & 3 & 2 \\ -1 & 2 & 3 \end{bmatrix} \begin{bmatrix} 1 \\ 4 \\ 2 \end{bmatrix}$

In Exercises 5–12, compute the given expressions, if possible, given that

$$A = \begin{bmatrix} 1 & 3 & 1 \\ -2 & 1 & 3 \\ 4 & 0 & 2 \end{bmatrix}, \quad B = \begin{bmatrix} 2 & 1 & 3 \\ -2 & -1 & -1 \\ 1 & 4 & 2 \end{bmatrix}, \quad \text{and} \quad C = \begin{bmatrix} 3 & -1 & 2 \\ 1 & 6 & 4 \\ 2 & 1 & 3 \end{bmatrix}.$$

5. $2A + 3B$

6. $3A - 2B$

7. $2(3A)$

8. $2(3A - 4B)$

9. $A(B - C)$

10. $AB + AC$

11. $A(BC)$

12. $(1/2)(CA - CB)$

13. Mr. Spaulding bought 10,000 shares of stock X, 20,000 shares of stock Y, and 30,000 shares of stock Z at a unit price of $20, $30, and $50 per share, respectively. Six months later, the closing prices of stocks X, Y, and Z were $22, $35, and $51 per share, respectively. Spaulding made no other stock transactions during the period in question. Compare the value of Mr. Spaulding's stock holdings at the time of purchase and six months later.

14. Ms. Newburg operates three self-service gasoline stations in different parts of town. On a certain day, Station A sold 600 gallons of premium, 1000 gallons of regular, 800 gallons of super-unleaded, and 1400 gallons of regular-unleaded gasoline; Station B sold 700 gallons of premium, 800 gallons of regular, 600 gallons of super-unleaded,

and 1200 gallons of regular-unleaded gasoline; Station C sold 1200 gallons of premium, 800 gallons of regular, 1000 gallons of super-unleaded, and 900 gallons of regular-unleaded gasoline. If the price of gasoline on that day was $1.60 per gallon for premium, $1.20 per gallon for regular, $1.50 per gallon for super-unleaded, and $1.30 per gallon for regular-unleaded gasoline, use matrix algebra to find the total revenue at each station.

In Exercises 15–20, solve the system of linear equations using the Gauss-Jordan elimination method.

15. $2x - 3y = 5$
    $3x + 4y = -1$

16. $x - y + 2z = 5$
    $3x + 2y + z = 10$
    $2x - 3y - 2z = -10$

17. $3x - 2y + 4z = 16$
    $2x + y - 2z = -1$
    $x + 4y - 8z = -18$

18. $3x - 2y + 4z = 11$
    $2x - 4y + 5z = 4$
    $x + 2y - z = 10$

19. $x - 2y + 3z + 4w = 17$
    $2x + y - 2z - 3w = -9$
    $3x - y + 2z - 4w = 0$
    $4x + 2y - 3z + w = -2$

20. $3x - 2y + z = 4$
    $x + 3y - 4z = -3$
    $2x - 3y + 5z = 7$
    $x - 8y + 9z = 10$

In Exercises 21–23, find the inverse of the given matrix (if it exists).

21. $A = \begin{bmatrix} 2 & 4 \\ 1 & 6 \end{bmatrix}$

22. $A = \begin{bmatrix} 1 & 2 & 4 \\ 3 & 1 & 2 \\ 1 & 0 & -6 \end{bmatrix}$

23. $A = \begin{bmatrix} 2 & 1 & -3 \\ 1 & 2 & -4 \\ 3 & 1 & -2 \end{bmatrix}$

In Exercises 24–26, write each system of linear equations in the form $AX = C$. Find A^{-1} and use the result to solve the system.

24. $x - 3y = -1$
    $2x + 4y = 8$

25. $2x - 3y + 4z = 17$
    $x + 2y - 4z = -7$
    $3x - y + 2z = 14$

26. $x - 2y + 4z = 13$
    $2x + 3y - 2z = 0$
    $x + 4y - 6z = -15$

27. Desmond Jewelry, Inc. wishes to produce three types of pendants: type A, type B, and type C. To manufacture a type-A pendant requires 2 minutes on machines I and II and 3 minutes on machine III. A type-B pendant requires 2 minutes on machine I, 3 minutes on machine II, and 4 minutes on machine III. A type-C pendant requires 3 minutes on machine I, 4 minutes on machine II, and 3 minutes on machine III. There are 3-1/2 hours available on machine I, 4-1/2 hours available on machine II, and 5 hours available on machine III. How many pendants of each type should the company make in order to use all the available time?

Linear Programming:
A Geometric
Approach

How are the aircraft engines to be shipped? Curtis-Roe Aviation Industries manufactures jet engines in two different locations. These engines are to be shipped to the company's two main assembly plants. In Example 9, page 162, we will show how many engines should be produced and shipped from each manufacturing plant to each assembly plant in order to minimize shipping costs.

▶ CHAPTER

THREE

3.1

Graphing Systems of Linear Inequalities in Two Variables

▶ Graphing Linear Inequalities

▶ Graphing Systems of Linear Inequalities

▶ Graphing Linear Inequalities

In Chapter 1, we saw that a linear equation in two variables x and y

$$ax + by + c = 0 \quad (a, b \text{ not both equal to zero})$$

has a *solution set* that may be exhibited graphically as points on a straight line in the xy-plane. We will now show that there is also a simple graphical representation for **linear inequalities** in two variables

$$ax + by + c < 0, \quad ax + by + c \leq 0$$
$$ax + by + c > 0, \quad ax + by + c \geq 0.$$

Before turning to a general procedure for graphing such inequalities, let us consider a specific example. Suppose we wish to graph

$$2x + 3y < 6. \tag{1}$$

We first graph the equation $2x + 3y = 6$, which is obtained from the given inequality by replacing the inequality "<" with an equality (Figure 3.1).

Observe that this line divides the xy-plane into two half-planes—an upper half-plane and a lower half-plane. Let us show that the upper half-plane is the graph of the linear inequality

$$2x + 3y > 6, \tag{2}$$

whereas the lower half-plane is the graph of the linear inequality

$$2x + 3y < 6. \tag{3}$$

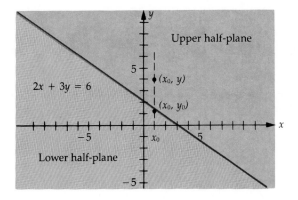

Figure 3.1

To see this, let us write (2) and (3) in the equivalent forms

$$y > -\frac{2}{3}x + 2 \tag{4}$$

and

$$y < -\frac{2}{3}x + 2. \tag{5}$$

The equation of the line itself is, of course,

$$y = -\frac{2}{3}x + 2. \tag{6}$$

Now, fix $x = x_0$ and observe that if (x_0, y_0) is any point satisfying (6), it lies on the line L (Figure 3.1). On the other hand, a point (x_0, y) satisfying (2) and therefore (4) lies above the point (x_0, y_0); that is, it lies in the upper half-plane, since $y > y_0$ in this case. Similarly, a point (x_0, y) satisfying (3) or (5) lies in the lower half-plane, since $y < y_0$. Since this is true for every real number x_0, our assertion is verified.

This analysis shows that the lower half-plane provides a solution to our problem (Figure 3.2). (The dotted line shows that the points on L do not belong

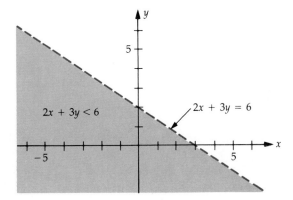

Figure 3.2

to the solution set.) Observe that the two half-planes in question are clearly mutually exclusive (that is, they do not have any points in common). Because of this, there is, from the practical point of view, an alternative and easier method of determining the solution to the problem.

To determine the required half-plane, let us pick *any* point lying in one of the half-planes. For simplicity, let us pick the origin (0, 0), which lies in the lower-half plane. Substituting $x = 0$ and $y = 0$ (the coordinates of this point) into the given inequality (1), we find

$$2(0) + 3(0) < 6,$$

or $0 < 6$, which is certainly true. This tells us that the required half-plane is the half-plane containing the test point, namely the lower half-plane. Next, let us see what happens if we choose the point (2, 3), which lies in the upper half-plane. Substituting $x = 2$ and $y = 3$ into the given inequality, we find

$$2(2) + 3(3) < 6,$$

or $10 < 6$, which is false. This tells us that the upper half-plane is *not* the required half-plane, as expected. Note too that no point (x, y) lying on the line constitutes a solution to our problem, because of the *strict* inequality "<".

This discussion suggests the following procedure for graphing a linear inequality in two variables.

Procedure for Graphing Linear Inequalities

1. Draw the graph of the equation obtained for the given inequality by replacing the inequality sign with an equals sign. Use a dotted line if the problem involves a strict inequality, "<" or ">". Otherwise, use a solid line to indicate that the line itself constitutes part of the solution.

2. Pick a test point lying in one of the half-planes determined by the line sketched in (1) and substitute the values of x and y into the given inequality. Use the origin whenever possible.

3. If the inequality is satisfied, the graph of the inequality includes the half-plane containing the test point. Otherwise, the solution includes the half-plane not containing the test point.

EXAMPLE

1 Determine the solution set for the inequality $2x + 3y \geq 6$.

SOLUTION

Replacing the inequality "$\geq$" with an equality, we obtain the equation $2x + 3y = 6$, whose graph is the straight line shown in Figure 3.3. Instead of a dotted line as before, we use a solid line to show that all points on the line are also solutions to the problem. Picking the origin as our test point, we find

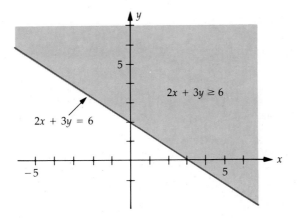

Figure 3.3

$2(0) + 3(0) \geq 6$, or $0 \geq 6$, which is impossible. So we conclude that the solution set comprises the half-plane not containing the origin, including in this case the line given by $2x + 3y = 6$. ◀

EXAMPLE

2 Graph $x \leq -1$.

SOLUTION The graph of $x = -1$ is the vertical line shown in Figure 3.4. Picking the origin $(0, 0)$ as a test point, we find $0 \leq -1$, which is false. Therefore, the required solution is the *left* half-plane, which does not contain the origin.

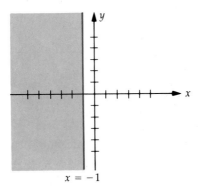

Figure 3.4

$x = -1$

◀

EXAMPLE

3 Graph $x - 2y > 0$.

SOLUTION We first graph the equation $x - 2y = 0$, or $y = (1/2)x$ (Figure 3.5). Since the origin lies on the line, we may not use it as a test point. (Why?) Let us pick $(1, 2)$ as

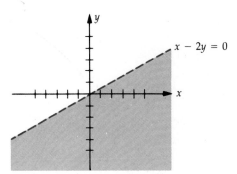

Figure 3.5

a test point. Substituting $x = 1$ and $y = 2$ into the given inequality, we find $1 - 2(2) > 0$, or $-3 > 0$, which is false. Therefore, the required solution is the half-plane that does not contain the test point, namely, the lower half-plane. ◀

▶ Graphing Systems of Linear Inequalities

By the *solution set of a system of linear inequalities* in the two variables x and y, we mean the set of all points (x, y) satisfying each inequality of the system. The graphical solution of such a system may be obtained by graphing the solution set for each inequality independently and then determining the region in common with each solution set.

EXAMPLE

4 Determine the solution set for the system

$$4x + 3y \geq 12$$
$$x - y \leq 0.$$

SOLUTION

Proceeding as the previous examples, you should have no difficulty locating the half-planes determined by each of the linear inequalities that make up the system. These half-planes are shown in Figure 3.6. The intersection of the two half-planes is the shaded region. A point in this region is an element of the solution set for the given system. The point P, the intersection of the two straight lines determined by the equations

$$4x + 3y = 12$$
$$x - y = 0,$$

is found by solving the equations simultaneously.

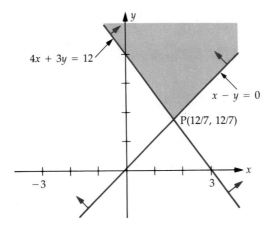

Figure 3.6

EXAMPLE

5 Sketch the solution set for the system

$$x \geq 0$$
$$y \geq 0$$
$$x + y - 6 \leq 0$$
$$2x + y - 8 \leq 0.$$

SOLUTION The first inequality in the system defines the right half-plane—all points to the right of the y-axis plus all points lying on the y-axis itself. The second inequality in the system defines the upper half-plane, including the x-axis. The half-planes defined by the third and fourth inequalities are indicated by arrows in Figure 3.7. Thus, the required region, the intersection of the four half-planes defined by the four inequalities in the given system of linear inequalities, is the shaded region. The point P is found by solving the equations $x + y - 6 = 0$ and $2x + y - 8 = 0$ simultaneously.

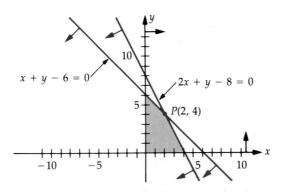

Figure 3.7

EXAMPLE

6 Determine the graphical solution set for the following system of linear inequalities:

$$2x + y \geq 50$$
$$x + 2y \geq 40$$
$$x \geq 0$$
$$y \geq 0$$

SOLUTION The required solution set is the unbounded region shown in Figure 3.8.

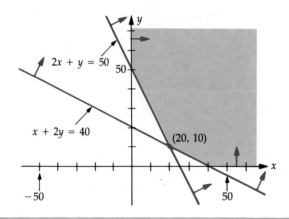

Figure 3.8

Self-Check Exercises

3.1

1. Determine graphically the solution set for the following system of inequalities:

$$x + 2y \leq 10$$
$$5x + 3y \leq 30$$
$$x \geq 0, y \geq 0$$

2. Determine graphically the solution set for the following system of inequalities:

$$5x + 3y \geq 30$$
$$x - 3y \leq 0$$
$$x \geq 2$$

Solutions to Self-Check Exercises 3.1 can be found on page 157.

Exercises 3.1

In Exercises 1–8, find the graphical solution of each inequality:

1. $4x - 8 < 0$

2. $3y + 2 > 0$

3. $x - y \leq 0$

4. $3x + 4y \leq -2$

5. $2x + y \leq 4$

6. $-3x + 6y \geq 12$

7. $4x - 3y \leq -24$

8. $5x - 3y \geq 15$

In Exercises 9–25, determine graphically the solution set for each system of inequalities:

9. $2x + 4y > 16$
   $-x + 3y \geq 7$

10. $3x - 2y > -13$
    $-x + 2y > 5$

11. $x + 2y \geq 3$
    $2x + 4y \leq -2$

12. $x + y \geq -2$
    $3x - y \leq 6$

13. $x - y \leq 0$
    $2x + 3y \geq 10$

14. $x + y \leq 6$
    $0 \leq x \leq 3$
    $y \geq 0$

15. $3x - 6y \leq 12$
    $-x + 2y \leq 4$
    $x \geq 0 \; y \geq 0$

16. $x + y \geq 20$
    $x + 2y \geq 40$
    $x \geq 0 \; y \geq 0$

17. $4x - 3y \leq 12$
    $5x + 2y \leq 10$
    $x \geq 0 \; y \geq 0$

18. $3x - 7y \geq -24$
    $x + 3y \geq 8$
    $x \geq 0, \; y \geq 0$

19. $x + 2y \geq 3$
    $5x - 4y \leq 16$
    $0 \leq y \leq 2$
    $x \geq 0$

20. $3x + 4y \geq 12$
    $2x - y \geq -2$
    $0 \leq y \leq 3$
    $x \geq 0$

21. $x + y \leq 4$
    $2x + y \leq 6$
    $2x - y \geq -1$
    $x \geq 0, \; y \geq 0$

22. $6x + 5y \geq 30$
    $3x + y \leq 6$
    $x + y \leq 4$
    $x \geq 0, \; y \geq 0$

23. $6x + 7y \leq 84$
    $12x - 11y \leq 18$
    $6x - 7y \leq 28$
    $x \geq 0, \; y \geq 0$

24. $x - y \geq -6$
    $x - 2y \leq -2$
    $x + 2y \geq 6$
    $x - 2y \geq -14$
    $x \geq 0, \; y \geq 0$

25. $x - 3y \geq -18$
    $3x - 2y \geq 2$
    $x - 3y \leq -4$
    $3x - 2y \leq 16$
    $x \geq 0, \; y \geq 0$

Solutions to Self-Check Exercises

3.1

1. The required solution set is shown in the figure on page 158.

The point P is found by solving the system of equations

$$x + 2y = 10$$
$$5x + 3y = 30.$$

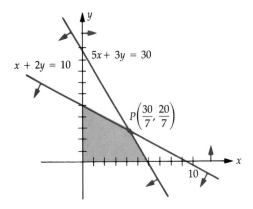

Solving the first equation for x in terms of y gives

$$x = 10 - 2y.$$

Substituting this value of x into the second equation of the system gives

$$5(10 - 2y) + 3y = 30$$
$$50 - 10y + 3y = 30$$
$$-7y = -20,$$

so $y = 20/7$. Substituting this value of y into the expression for x found earlier, we obtain

$$x = 10 - 2\left(\frac{20}{7}\right) = \frac{30}{7},$$

giving the point of intersection as (30/7, 20/7).

2. The required solution set is shown in the following figure.

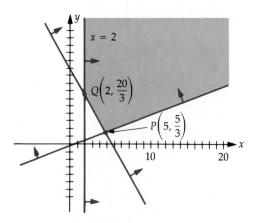

To find the coordinates of P, we solve the system

$$5x + 3y = 30$$
$$x - 3y = 0.$$

Solving the second equation for x in terms of y and substituting this value of x in the first equation gives

$$5(3y) + 3y = 30,$$

or $y = 5/3$. Substituting this value of y into the second equation gives $x = 5$. Next, the coordinates of Q are found by solving the system

$$5x + 3y = 30$$
$$x = 2,$$

obtaining $x = 2$ and $y = 20/3$.

3.2

Linear Programming Problems

▶ Maximization Problems

▶ Minimization Problems

▶ A Transportation Problem

▶ A Warehouse Problem

Many business and economics problems are concerned with optimizing (maximizing or minimizing) a function subject to a system of equalities or inequalities. The function to be optimized is called the **objective function.** (Examples of objective functions are profit functions and cost functions.) The system of equalities or inequalities to which the objective function is subjected reflects the constraints (for example, limitations on resources such as materials and labor) imposed on the solution(s) to the problem. Problems of this nature are called **mathematical programming problems.** In particular, problems in which both the objective function and the constraints are expressed as linear equations or inequalities are called **linear programming problems.**

A Linear Programming Problem

A linear programming problem consists of a linear objective function to be maximized or minimized subject to certain constraints in the form of linear equalities or inequalities.

▶ Maximization Problems

As an example of a linear programming problem in which the objective function is to be maximized, let us consider the following simplified version of a production problem involving two variables.

EXAMPLE

7 The Ace Novelty Company wishes to produce two types of souvenirs: type A and type B. Each type-A souvenir will result in a profit of $1.00, and each type-B souvenir will result in a profit of $1.20. To manufacture a type-A souvenir requires 2 minutes on machine I and 1 minute on machine II. A type-B souvenir requires 1 minute on machine I and 3 minutes on machine II. There are 3 hours available on machine I and 5 hours available on machine II for processing the order. How many souvenirs of each type should the company make in order to maximize profit?

SOLUTION

As a first step toward the mathematical formulation of this problem, we tabulate the given information in Table 3.1.

	Type A	Type B	Time Available
Machine I	2 min.	1 min.	180 min.
Machine II	1 min.	3 min.	300 min.
Profit per unit	$1.00	$1.20	

Table 3.1

Let x be the number of type-A souvenirs and y be the number of type-B souvenirs to be made. Then the total profit P, in dollars, is given by

$$P = x + 1.2y,$$

which is the objective function to be maximized.

The total amount of time that machine I is used is given by $2x + y$ minutes and must not exceed 180 minutes. Thus we have the inequality

$$2x + y \leq 180.$$

Similarly, the total amount of time that machine II is used is $x + 3y$ minutes, which cannot exceed 300 minutes, so we are led to the inequality

$$x + 3y \leq 300.$$

Finally, neither x nor y can be negative, so

$$x \geq 0$$
$$y \geq 0.$$

To summarize, the problem at hand is one of maximizing the objective function $P = x + 1.2y$ subject to the system of inequalities

$$2x + y \leq 180$$
$$x + 3y \leq 300$$
$$x \geq 0$$
$$y \geq 0.$$

The solution to this problem will be completed in Example 11, Section 3.3. ◄

▶ Minimization Problems

In the following example of a linear programming problem, the objective function is to be minimized.

EXAMPLE

8 A nutritionist advises that an individual who is suffering from iron- and vitamin-B deficiency take at least 2400 mg of iron, 2100 mg of vitamin B-1 (Thiamine), and 1500 mg of vitamin B-2 (Riboflavin) over a period of time. Two vitamin pills are suitable, brand A and brand B. Each brand A pill contains 40 mg of iron, 10 mg of vitamin B-1, 5 mg of vitamin B-2, and costs 6¢. Each brand B pill contains 10 mg of iron, 15 mg each of vitamins B-1 and B-2, and costs 8¢ (see Table 3.2). What combination of pills should the individual purchase in order to meet the minimum iron and vitamin requirements at the lowest cost?

	Brand A	Brand B	Minimum Requirement
Iron	40 mg	10 mg	2400 mg
Vitamin B-1	10 mg	15 mg	2100 mg
Vitamin B-2	5 mg	15 mg	1500 mg
Cost per pill	6¢	8¢	

Table 3.2

SOLUTION Let x be the number of brand A pills and y be the number of brand B pills to be purchased. The cost C, measured in cents, is given by

$$C = 6x + 8y$$

and is the objective function to be minimized.

The amount of iron contained in x brand A pills and y brand B pills is given by $40x + 10y$ mg, and this must be greater than or equal to 2400 mg. This translates into the inequality

$$40x + 10y \geq 2400.$$

Similar considerations involving the minimum requirements of vitamins B-1 and B-2 lead to the inequalities

$$10x + 15y \geq 2100$$

and
$$5x + 15y \geq 1500,$$

respectively. Thus, the problem here is to minimize $C = 6x + 8y$ subject to

$$40x + 10y \geq 2400$$
$$10x + 15y \geq 2100$$
$$5x + 15y \geq 1500$$
$$x \geq 0, y \geq 0.$$

The solution to this problem will be completed in Example 12, Section 3.3. ◀

▶ A Transportation Problem

EXAMPLE

9 Curtis–Roe Aviation Industries has two plants, I and II, that produce the "Zephyr" jet engines used in their light commercial airplanes. The maximum production capacities of these two plants are 100 units and 110 units per month, respectively. The engines are shipped to two of Curtis–Roe's main assembly plants, A and B. The shipping costs (in dollars) per engine from plants I and II to the main assembly plants A and B are as follows:

	To Assembly Plant	
From	*A*	*B*
Plant I	100	60
Plant II	120	70

In a certain month, assembly plant A needs at most 80 engines, whereas assembly plant B needs at most 70 engines. Find the number of engines to be shipped from each plant to each main assembly plant if shipping costs are to be kept to a minimum.

SOLUTION

Let x denote the number of engines shipped from plant I to assembly plant A and let y denote the number of engines shipped from plant I to assembly plant B. Since the requirements of assembly plants A and B are 80 and 70 engines, respectively, the number of engines shipped from plant II to assembly plants A and B are $(80-x)$ and $(70-y)$, respectively. These numbers may be displayed in a schematic. With the aid of the following schematic and the shipping cost schedule, we find that the total shipping costs incurred by the company are given by

$$C = 100x + 60y + 120(80 - x) + 70(70 - y)$$
$$= 14,500 - 20x - 10y.$$

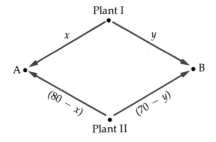

Next, the production constraints on plants I and II lead to the inequalities

$$x + y \leq 100$$

and
$$(80 - x) + (70 - y) \leq 110.$$

The last inequality simplifies to

$$x + y \geq 40.$$

Also, the requirements of the two main assembly plants lead to the inequalities

$$x \geq 0, \ y \geq 0, \ 80 - x \geq 0, \quad \text{and} \quad 70 - y \geq 0.$$

The last two may be written as $x \leq 80$ and $y \leq 70$.

Summarizing, we have the following linear programming problem: Minimize the objective (cost) function $C = 14,500 - 20x - 10y$ subject to the constraints

$$x + y \geq 40$$
$$x + y \leq 100$$
$$x \leq 80$$
$$y \leq 70$$

where $x \geq 0$ and $y \geq 0$.

You will be asked to complete the solution to this problem in Problem 22, Exercises 3.3. ◀

▶ A Warehouse Problem

EXAMPLE

10 The Acrosonic Company manufactures its model F loudspeaker systems in two separate locations, plant I and plant II. The output at plant I is at most 400 per month, whereas the output at plant II is at most 600 per month. These loudspeaker systems are shipped to three warehouses that serve as distribution centers for the company. In order for the warehouses to meet their orders, the minimum monthly requirements of warehouses A, B, and C are 200, 300, and 400, respectively. Shipping costs from plant I to warehouses A, B, and C are $20, $8, and $10 per loudspeaker system, respectively, and shipping costs from plant II to each of these warehouses are $12, $22, and $18, respectively. What should the shipping schedule be if Acrosonic wishes to meet the requirements of the distribution centers and at the same time keep its shipping costs to a minimum?

SOLUTION   The respective shipping costs per loudspeaker system in dollars may be tabulated as in Table 3.3:

	Warehouse		
Plant	A	B	C
I	20	8	10
II	12	22	18

Table 3.3

Letting x_1 denote the number of loudspeaker systems shipped from plant I to warehouse A, x_2 the number shipped from plant I to warehouse B, and so on, leads to Table 3.4.

	Warehouse			
Plant	A	B	C	Max. Prod.
I	x_1	x_2	x_3	400
II	x_4	x_5	x_6	600
Min. req.	200	300	400	

Table 3.4

From Tables 3.3 and 3.4 we see that the cost of shipping x_1 loudspeaker systems from plant I to warehouse A is $20x_1$, the cost of shipping x_2 loudspeaker systems from plant I to warehouse B is $8x_2$, and so on. Thus the total monthly shipping cost incurred by Acrosonic is given by

$$C = 20x_1 + 8x_2 + 10x_3 + 12x_4 + 22x_5 + 18x_6.$$

Next, the production constraints on plants I and II lead to the inequalities

$$x_1 + x_2 + x_3 \leq 400$$

and

$$x_4 + x_5 + x_6 \leq 600$$

(see Table 3.4). Also, the minimum requirements of each of the three warehouses lead to the three inequalities

$$x_1 + x_4 \geq 200$$
$$x_2 + x_5 \geq 300$$

and

$$x_3 + x_6 \geq 400.$$

Summarizing, we have the following linear programming problem: Minimize $C = 20x_1 + 8x_2 + 10x_3 + 12x_4 + 22x_5 + 18x_6$ subject to

$$x_1 + x_2 + x_3 \leq 400$$
$$x_4 + x_5 + x_6 \leq 600$$
$$x_1 + x_4 \geq 200$$
$$x_2 + x_5 \geq 300$$
$$x_3 + x_6 \geq 400$$
$$x_1 \geq 0, x_2 \geq 0, \ldots, x_6 \geq 0.$$

The solution to this problem will be completed in Section 4.2, Example 9. ◀

Self-Check Exercises

3.2

1. Mr. Balduzzi, the proprietor of Luigi's Pizza Palace, allocates $9000 a month for advertising in two newspapers, the *City Tribune* and the *Daily News*. The *City Tribune* charges $300 for a certain advertisement, whereas the *Daily News* charges $100 for the same ad. Mr. Balduzzi has stipulated that the ad is to appear in at least 15 but no more than 30 editions of the *Daily News* per month. The *City Tribune* has a daily circulation of 50,000, and the *Daily News* has a circulation of 20,000. Under these conditions, determine how many ads Balduzzi should place in each newspaper in order to reach the largest number of readers. Formulate but do not solve the problem. (The solution to this problem can be found in Exercise 3 of Solutions to Self-Check Exercises 3.3.)

2. The United Motors Corporation manufactures compact, mid-size, and full-size cars. For next year the company plans to build at most 500,000 cars, of which a maximum of 80,000 will be full size. To comply with government fuel economy standards for passenger cars, the fleet must average at least 27 mpg (miles per gallon) to avoid extensive fines. United's engineering division has estimated that the compact model will average 35 mpg, the mid-size model will average 24 mpg, and the full-size model will average 20 mpg. Each compact model can be expected to bring in a profit of $2000, each mid-size model a profit of $3000, and each full-size model a profit of $5000. Determine how many cars of each size United should build for the upcoming model year if its profits are to be maximized. *Hint:* Let x, y, and z denote the number of compact, mid-size, and full-size cars to be built. Then, the condition that the fleet average be at least 27 mpg is

$$\frac{35x + 24y + 20z}{x + y + z} \geq 27.$$

Show that this may be written as $-8x + 3y + 7z \leq 0$. (The solution to this problem can be found in Exercise 2, Solutions to Self-Check Exercises 4.1.)

Solutions to Self-Check Exercises 3.2 can be found on page 169.

▶

Exercises 3.2

Formulate, but do not solve, each of the following exercises as a linear programming problem.

1. A company manufactures two products, A and B, on two machines I and II. It has been determined that the company will realize a profit of $3 on each unit of product A and a profit of $4 on each unit of product B. To manufacture a unit of product A requires 6 minutes on machine I and 5 minutes on machine II. To manufacture a unit of product B requires 9 minutes on machine I and 4 minutes on machine II. There are 5 hours of machine time available on machine I and 3 hours of machine time available on machine II in each work shift. How many units of each product should be produced in each shift to maximize the company's profit?

2. Kane Manufacturing has a division that produces two models of hibachis, model A and model B. To produce each model A hibachi requires 3 pounds of cast iron and 6 minutes of labor. To produce each model B hibachi requires 4 pounds of cast iron and 3 minutes of labor. The profit for each model A hibachi is $2.00, and the profit for each model B hibachi is $1.50. If 1000 pounds of cast iron and 20 hours of labor are available for the production of hibachis per day, how many hibachis of each model should the division produce in order to help maximize the company's profits?

3. Refer to Exercise 2. Because of a backlog of orders in model A hibachis, the manager of Kane Manufacturing has decided to produce at least 150 of these models a day. Operating under this additional constraint, how many hibachis of each model should the company produce to maximize profit?

4. Madison Finance has a total of $20 million earmarked for homeowner and auto loans. On the average, homeowner loans will have a 10 percent rate of return on the loans given out, whereas auto loans will yield a 12 percent rate of return. Management has also stipulated that the total amount of homeowner loans should be greater than or equal to four times the total amount of automobile loans. Determine the total amount of loans of each type the company should extend to each category in order to maximize its returns.

5. A nutritionist at the Medical Center has been asked to prepare a special diet for certain patients. She has decided that the meals should contain a minimum of 400 mg of calcium, 10 mg of iron, and 40 mg of vitamin C. She has further decided that the meals are to be prepared from foods A and B. Each ounce of food A contains 30 mg of calcium, 1 mg of iron, 2 mg of vitamin C, and 2 mg of cholesterol. Each ounce of food B contains 25 mg of calcium, 0.5 mg of iron, 5 mg of vitamin C, and 5 mg of cholesterol. Find how many ounces of each type of food should be used in a meal so that the cholesterol content is minimized and the minimum requirements of calcium, iron, and vitamin C are met.

6. A farmer has 150 acres of land suitable for cultivating crops A and B. The cost of cultivating crop A is $40 per acre, whereas that of crop B is $60 per acre. The farmer has a maximum of $7400 available for land cultivation. Each acre of crop A requires 20 hours of labor and each acre of crop B requires 25 hours of labor. The farmer has a maximum of 3300 hours of labor available. If he expects to make a profit of $150 per acre on crop A and $200 per acre on crop B, how many acres of each crop should he plant in order to maximize his profit?

7. The TMA Company manufactures 19-inch color television picture tubes in two separate locations, location I and location II. The output at location I is at most 6000 tubes per month, whereas the output at location II is at most 5000 per month. TMA is the main supplier of picture tubes to the Pulsar Corporation, its holding company, which has priority in having all its requirements met. In a certain month, Pulsar placed orders for 3000 and 4000 picture tubes to be shipped to two of its factories located in city A and city B, respectively. The shipping costs (in dollars) per picture tube from the two plants of TMA to the two Pulsar factories are as follows:

Shipping Costs per Picture Tube

| | To Pulsar Factories | |
From	*City A*	*City B*
TMA (Loc. I)	$3	$2
TMA (Loc. II)	$4	$5

Find a shipping schedule that meets the requirements of both companies while keeping costs at a minimum.

8. A company manufactures products A, B, and C. Each product is processed in three departments: I, II, and III. The total available labor hours per week for departments I, II, and III are 900, 1080, and 840, respectively. The time requirements (in hours per unit) and profit per unit for each product are as follows:

	Product A	*Product B*	*Product C*
Dept. I	2	1	2
Dept. II	3	1	2
Dept. III	2	2	1
Profit	$18	$12	$15

How many units of each product should the company produce in order to maximize its profit?

9. As part of a campaign to promote its Annual Clearance Sale, the Excelsior Company decided to buy television advertising time on Station KAOS. Excelsior's advertising budget was $102,000. Morning time costs $3,000 per minute, afternoon time costs $1,000 per minute, and evening (prime) time costs $12,000 per minute. Because of previous commitments, KAOS could not offer Excelsior more than 6 minutes of prime time or more than a total of 25 minutes of advertising time over the two weeks in which the commercials were to be run. KAOS estimated that morning commercials would be seen by 200,000 people, afternoon commercials would be seen by 100,000 people, and evening commercials would be seen by 600,000 people. How much morning, afternoon, and evening advertising time should Excelsior buy to maximize exposure of its commercials?

10. The Custom Office Furniture Company is introducing a new line of executive desks made from a specially selected grade of walnut. Initially, three different models—A, B, and C—are to be marketed. Each model A desk requires $1\frac{1}{4}$ hours for fabrication, 1 hour for assembly, and 1 hour for finishing; each model B desk requires $1\frac{1}{2}$ hours for fabrication, 1 hour for assembly, and 1 hour for finishing; each model C desk requires $1\frac{1}{2}$ hours, $\frac{3}{4}$ hour, and $\frac{1}{2}$ hour for fabrication, assembly, and finishing, respectively. The profit on each model A desk is $26, the profit on each model B desk

is $28, and the profit on each model C desk is $24. The total time available in the fabrication department, the assembly department, and the finishing department in the first month of production is 310 hours, 205 hours, and 190 hours, respectively. To maximize the company's profit, how many desks of each model should be made in the month?

11. The Acrosonic Company of Example 10 also manufactures a model G loudspeaker system in plant I and plant II. The output at plant I is at most 800 systems per month, whereas the output at plant II is at most 600 per month. These loudspeaker systems are also shipped to the three warehouses—A, B, and C—whose minimum monthly requirements are 500, 400, and 400, respectively. Shipping costs from plant I to warehouse A, warehouse B, and warehouse C are $16, $20, and $22 per loudspeaker system, respectively, and shipping costs from plant II to each of these warehouses are $18, $16, and $14, respectively. What shipping schedule will enable Acrosonic to meet the warehouses' requirements and at the same time keep its shipping costs to a minimum?

12. Steinwelt Piano manufactures uprights and consoles in two plants, plant I and plant II. The output of plant I is at most 300 per month, whereas the output of plant II is at most 250 per month. These pianos are shipped to three warehouses that serve as distribution centers for the company. In order to fill current and projected future orders, warehouse A requires a minimum of 200 pianos per month, warehouse B requires at least 150 pianos per month, and warehouse C requires at least 200 pianos per month. The shipping cost of each piano from plant I to warehouse A, warehouse B, and warehouse C is $60, $60, and $80, respectively, and the shipping cost of each piano from plant II to warehouse A, warehouse B, and warehouse C is $80, $70, and $50, respectively. What shipping schedule will enable Steinwelt to meet the warehouses' requirements while keeping the shipping costs to a minimum?

13. Boise Lumber has decided to enter the lucrative prefabricated housing business. Initially, it plans to offer three models: a standard model, a deluxe model, and a luxury model. Each house is prefabricated and partially assembled in the factory, and the final assembly is completed on site. Each standard model requires $6,000 worth of building material, each deluxe model requires $8,000 worth of building material, and each luxury model requires $10,000 worth of building material. The amount of labor required to prefabricate and partially assemble each standard, deluxe, and luxury model in the factory is estimated to be 240, 220, and 200 labor hours, respectively. The amount of on-site labor required to erect and finish each standard, deluxe, and luxury model is estimated to be 180, 210, and 300 labor hours, respectively. The profit from the sale of each standard, deluxe, and luxury model is expected to be $3,400, $4,000, and $5,000, respectively. For the first year's production, a sum of $8,200,000 is budgeted for the building material; the number of labor hours available for work in the factory (for prefabrication and partial assembly) is not to exceed 218,000 hours; and the amount of labor for on-site work is to be less than or equal to 237,000 labor hours. Determine how many houses of each type Boise should produce (market research has confirmed that there should be no problems with sales) to maximize its profit from this new venture.

14. Bayer Pharmaceutical manufactures three kinds of cold formulas: formula I, formula II, and formula III. It takes 2.5 hours to produce 1,000 bottles of formula I, 3 hours to produce 1,000 bottles of formula II, and 4 hours to produce 1,000 bottles of formula III. The profits for each 1,000 bottles of formula I, formula II, and formula III are $180, $200, and $300, respectively. Suppose that for a certain production run there are enough ingredients on hand to make at most 9,000 bottles of formula I, 12,000 bottles of formula II, and 6,000 bottles of formula III. Furthermore, suppose that the time for the production run is limited to a maximum of 70 hours. Find how many bottles of each formula should be produced in this production run so that the profit is maximized.

Solutions to Self-Check Exercises

3.2

1. Let x denote the number of ads to be placed in the *City Tribune* and y the number to be placed in the *Daily News*. The total cost for placing x ads in the *City Tribune* and y ads in the *Daily News* is $300x + 100y$ dollars, and since the monthly budget is $9000, we must have

$$300x + 100y \le 9000.$$

Next, the condition that the ad must appear in at least 15 but no more than 30 editions of the *Daily News* translates into the inequalities

$$y \ge 15$$
and
$$y \le 30.$$

Finally, the objective function to be maximized is

$$P = 50{,}000x + 20{,}000y.$$

To summarize, we have the following linear programming problem: Maximize $P = 50{,}000x + 20{,}000y$ subject to

$$
\begin{aligned}
300x + 100y &\le 9000 \\
y &\ge 15 \\
y &\le 30 \\
x \ge 0, \; y &\ge 0.
\end{aligned}
$$

2. The condition that the number of cars to be built be at most 500,000 gives

$$x + y + z \le 500{,}000.$$

Next, the condition that the number of full-size cars to be built not exceed 80,000 is described by the inequality

$$z \le 80{,}000.$$

The average fuel economy of the fleet is given by

$$\frac{\text{total fuel consumption of the fleet}}{\text{total number of vehicles produced}}$$
$$= \frac{35x + 24y + 20z}{x + y + z}.$$

Since this is to be at least 27 mpg, we have

$$\frac{35x + 24y + 20z}{x + y + z} \ge 27.$$

Multiplying both sides of the inequality by $x + y + z$ (which is a positive number), we find

$$35x + 24y + 20z \geq 27x + 27y + 27z,$$

or $\qquad 8x - 3y - 7z \geq 0.$

Multiplying both sides of the inequality by -1 and recalling that this reverses the sense of the inequality, we have

$$-8x + 3y + 7z \leq 0.$$

Finally, the profit function is

$$P = 2000x + 3000y + 5000z.$$

To summarize, we have the following linear programming problem: Maximize $P = 2000x + 3000y + 5000z$ subject to

$$
\begin{aligned}
x + y + z &\leq 500{,}000 \\
z &\leq 80{,}000 \\
-8x + 3y + 7z &\leq 0 \\
x \geq 0,\ y \geq 0,\ z &\geq 0.
\end{aligned}
$$

3.3

Graphical Solution of Linear Programming Problems

▶ The Graphical Method

▶ Applications

▶ The Graphical Method

Linear programming problems in two variables have relatively simple geometrical interpretations. For example, the system of linear constraints associated with a two-dimensional linear programming problem, unless it is inconsistent, defines a planar region whose boundary is composed of straight line segments and/or half lines. Such problems are therefore amenable to graphical analysis.

Consider the following two-dimensional linear programming problem: Maximize $P = 3x + 2y$ subject to

$$
\begin{aligned}
2x + 3y &\leq 12 \\
2x + y &\leq 8 \\
x \geq 0,\ y &\geq 0
\end{aligned}
\tag{7}
$$

The system of linear inequalities (7) defines the planar region S shown in Figure 3.9. Each point in S is a candidate for the solution of the problem at hand and is referred to as a **feasible solution.** The set S itself is referred to as a **feasible set.** Our goal is to find from among all the points in the set S the point(s) that optimizes (in this case, maximizes) the objective function P. Such a feasible solution is called an **optimal solution** and constitutes the solution to the linear programming problem under consideration.

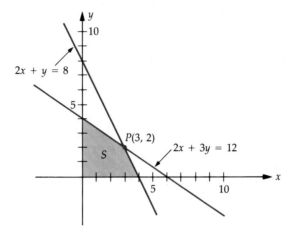

Figure 3.9

As noted earlier, each point $P(x, y)$ in S is a candidate for the (optimal) solution to the problem at hand. For example, the point $(1, 3)$ is easily seen to lie in S and is therefore in the running. The value of the objective function P at the point $(1, 3)$ is given by $P = 3(1) + 2(3) = 9$. Now, if we could compute the value of P corresponding to each point in S, then the point(s) in S that gave the largest value to P would constitute the solution set sought. Unfortunately, in most problems the number of candidates is either too large or, as in this problem, the number is infinite. Thus this method is at best unwieldy and at worst impractical.

Let us turn the question around. Instead of asking for the value of the objective function P at a feasible point, let us assign a value to the objective function P and ask whether there are feasible points that would correspond to the given value of P. To this end, suppose we assign a value of 6 to P. Then the objective function P becomes $3x + 2y = 6$, a linear equation in x and y, and is therefore representable as the graph of a straight line L_1 in the plane. In Figure 3.10 we have drawn the graph of this straight line superimposed on the feasible set S.

It is clear that each point on the straight line segment given by the intersection of the straight line L_1 and the feasible set S corresponds to the given value, 6, of P. Let us repeat the process, this time assigning a value of 10 to P. We obtain the equation $3x + 2y = 10$ and the line L_2 (see Figure 3.10), which suggests that there are feasible points that correspond to a larger value of P. Observe that the

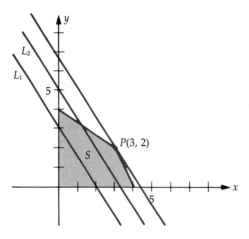

Figure 3.10

line L_2 is parallel to the line L_1, each with slope equal to $-3/2$, which is easily seen by casting the corresponding equations in the slope-intercept form.

In general, by assigning different values to the objective function, we obtain a family of parallel lines, each with slope equal to $-3/2$. Furthermore, a line corresponding to a larger value of P lies farther away from the origin than one with a smaller value of P. The implication is clear. To obtain the optimal solution(s) to the problem at hand, find the straight line, from this family of straight lines, that is farthest away from the origin and still intersects the feasible set S. The required line is the one that passes through the point $P(3, 2)$ [see Figure 3.10], so the solution to the problem is given by $x = 3$, $y = 2$, resulting in a maximum value of $P = 3(3) + 2(2) = 13$.

That the optimal solution to this problem was found to occur at a vertex of the feasible set S is no accident. In fact, the result is a consequence of the following basic theorem on linear programming, which we state without proof.

Theorem

If a unique (optimal) solution exists to a linear programming problem, then it must occur at a vertex, or corner point, of the feasible set S associated with the problem. Furthermore, if the objective function P is optimized at two adjacent vertices of S, then it is optimized at every point on the line segment joining these vertices, in which case, there are infinitely many solutions to the problem.

This theorem tells us that our search for the solution(s) to a linear programming problem may be restricted to the examination of the set of vertices of the feasible set S associated with the problem. Since a feasible set S has finitely many vertices, the theorem suggests that the solution(s) to the linear programming problem may be found by inspecting the values of the objective function P at these vertices.

The **method of corners,** a simple procedure for solving linear programming problems based on this theorem, follows.

The Graphical Solution of a Linear Programming Problem

1. Graph the feasible set of solutions.
2. Find all the corner points (vertices) of the feasible set.
3. Evaluate the objective function at each corner point.
4. Find the vertex that renders the objective function a maximum (minimum). If there is only one such vertex, then this vertex constitutes a unique solution to the problem. If the objective function is maximized (minimized) at two adjacent corner points of S, there are infinitely many optimal solutions given by the points on the line segment determined by these two vertices.

▶ Applications

EXAMPLE

11 We are now in a position to complete the solution to the production problem posed in Example 7 of Section 3.2. Recall that the mathematical formulation led to the following linear programming problem: Maximize $P = x + 1.2y$ subject to

$$2x + y \le 180$$
$$x + 3y \le 300$$
$$x \ge 0, y \ge 0.$$

SOLUTION

The feasible set S for the problem is shown in Figure 3.11. The vertices of the feasible set are $A(0, 0)$, $B(90, 0)$, $C(48, 84)$, and $D(0, 100)$. The values of P at these vertices may be tabulated as follows:

Vertex	$P = x + 1.2y$
$A(0, 0)$	0
$B(90, 0)$	90
$C(48, 84)$	148.8
$D(0, 100)$	120

From the table, one can see that the maximum of $P = x + 1.2y$ occurs at the vertex $(48, 84)$ and has a value of 148.8. Recalling what the symbols x, y, and P represent, we conclude that Ace Novelty would maximize its profit (a figure of $148.80) by producing 48 type-A souvenirs and 84 type-B souvenirs.

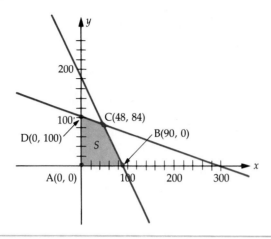

Figure 3.11

EXAMPLE

12 Complete the solution of the nutrition problem posed in Example 8.

SOLUTION Recall that the mathematical formulation of the problem led to the following linear programming problem in two variables: Minimize $C = 6x + 8y$ subject to the constraints

$$40x + 10y \geq 2400$$
$$10x + 15y \geq 2100$$
$$5x + 15y \geq 1500$$
$$x \geq 0, y \geq 0.$$

The feasible set S defined by the system of constraints is shown in Figure 3.12. The vertices of the feasible set S are $A(0, 240)$, $B(30, 120)$, $C(120, 60)$, and $D(300, 0)$. The values of the objective function C at these vertices are given by the following table:

Vertex	$C = 6x + 8y$
$A(0, 240)$	1920
$B(30, 120)$	1140
$C(120, 60)$	1200
$D(300, 0)$	1800

From the table, we can see that the minimum for the objective function $C = 6x + 8y$ occurs at the vertex $B(30, 120)$ and has a value of 1140. Thus, the individual should purchase 30 Brand A pills and 120 Brand B pills at a (minimum) cost of $11.40.

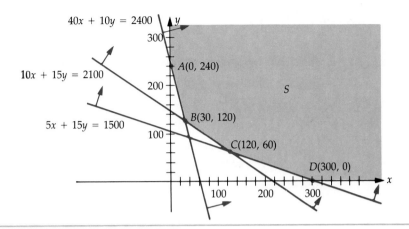

Figure 3.12

EXAMPLE

13 Find the maximum and minimum of $P = 2x + 3y$ subject to the following system of linear inequalities:

$$2x + 3y \leq 30$$
$$y - x \leq 5$$
$$x + y \geq 5$$
$$x \leq 10$$
$$x \geq 0, y \geq 0$$

SOLUTION The feasible set S is shown in Figure 3.13. The vertices of the feasible set S are $A(5, 0)$, $B(10, 0)$, $C(10, 10/3)$, $D(3, 8)$, and $E(0, 5)$. The values of the objective function P at these vertices are given by the following table.

Vertex	$P = 2x + 3y$
$A(5, 0)$	10
$B(10, 0)$	20
$C(10, 10/3)$	30
$D(3, 8)$	30
$E(0, 5)$	15

From the table, we can see that the maximum for the objective function $P = 2x + 3y$ occurs at the vertices $C(10, 10/3)$ and $D(3, 8)$. This tells us that every point on the line segment joining the points $C(10, 10/3)$ and $D(3, 8)$ maximizes P, giving it a value of 30 at each of these points. From the table it is also clear that P is minimized at the point $(5, 0)$, where it attains a value of 10.

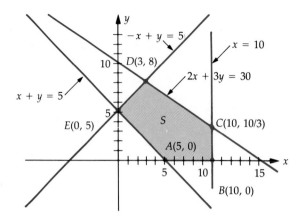

Figure 3.13

We close this section by examining two situations in which a linear programming problem may have no solution.

EXAMPLE

14 Solve the following linear programming problem: Maximize $P = x + 2y$ subject to

$$-2x + y \leq 4$$
$$x - 3y \leq 3$$
$$x \geq 0, y \geq 0.$$

SOLUTION The feasible set S for this problem is shown in Figure 3.14.

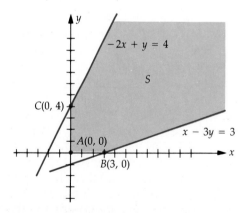

Figure 3.14

Since the set S is clearly unbounded (x and y can take on *any* positive value), we see that we can make P as large as we please by making x and y large enough. Therefore, the problem has no solution. In this situation we say that the solution is **unbounded.**

EXAMPLE

15 Solve the following linear programming problem: Maximize $P = x + 2y$ subject to

$$x + 2y \le 4$$
$$2x + 3y \ge 12$$
$$x \ge 0, y \ge 0$$

SOLUTION   The half-planes described by the constraints (inequalities) have no points in common (Figure 3.15).

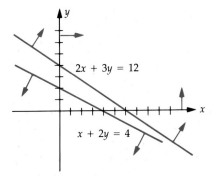

Figure 3.15

Therefore, there are no feasible points and the problem has no solution. In this situation, we say that this problem is **infeasible** or **inconsistent.** (These situations are unlikely to occur in well-posed problems arising from practical applications of linear programming.)   ◀

The method of corners is particularly effective in solving two-variable linear programming problems with a small number of constraints, as has been amply demonstrated in the preceding examples. Its effectiveness, however, decreases rapidly as the number of variables and/or constraints increases. For example, it may be shown that a linear programming problem in 3 variables and 5 constraints may have up to 56 feasible corner points. The determination of the feasible corner points calls for the solution of 56 3 × 3 systems of linear equations and then the verification, by the substitution of each of these solutions into the system of constraints, that it is in fact a feasible point. When the number of variables and constraints goes up to 5 and 10, respectively (still a very small system from the standpoint of applications in economics), the number of vertices to be found and checked for feasible corner points increases dramatically to 3003, and each of these vertices is found by solving a 5 × 5 linear system! For this reason, the method of corners is seldom used to solve linear programming problems; its redeeming value lies in the fact that much insight is gained into the nature of the solutions of linear programming problems through its use in solving two-variable problems.

Self-Check Exercises

3.3

1. Use the method of corners to solve the following linear programming problem: Maximize $P = 4x + 5y$ subject to

$$x + 2y \leq 10$$
$$5x + 3y \leq 30$$
$$x \geq 0, y \geq 0.$$

2. Use the method of corners to solve the following linear programming problem: Minimize $C = 5x + 3y$ subject to

$$5x + 3y \geq 0$$
$$x - 3y \leq 0$$
$$x \geq 2.$$

3. Mr. Balduzzi, the proprietor of Luigi's Pizza Palace, allocates $9000 a month for advertising in two newspapers, the *City Tribune* and the *Daily News*. The *City Tribune* charges $300 for a certain advertisement, whereas the *Daily News* charges $100 for the same ad. Mr. Balduzzi has stipulated that the ad is to appear in at least 15 but no more than 30 editions of the *Daily News* per month. The *City Tribune* has a daily circulation of 50,000, and the *Daily News* has a circulation of 20,000. Under these conditions, determine how many ads Balduzzi should place in each newspaper in order to reach the largest number of readers.

Solutions to Self-Check Exercises 3.3 can be found on page 182.

▶

Exercises 3.3

In Exercises 1–15, solve each linear programming problem by the method of corners.

1. Maximize $P = 2x + 3y$ subject to

$$x + y \leq 6$$
$$x \leq 3$$
$$x \geq 0, y \geq 0$$

2. Maximize $P = 3x + 4y$ subject to

$$x + 3y \leq 15$$
$$4x + y \leq 16$$
$$x \geq 0, y \geq 0$$

3. Minimize $C = 2x + 10y$ subject to

$$5x + 2y \geq 40$$
$$x + 2y \geq 20$$
$$y \geq 3, x \geq 0$$

4. Minimize $C = 2x + 5y$ subject to

$$4x + y \geq 40$$
$$2x + y \geq 30$$
$$x + 3y \geq 30$$
$$x \geq 0, y \geq 0$$

5. Minimize $C = 6x + 3y$ subject to the constraints of Exercise 4.

6. Maximize $P = 2x + 5y$ subject to

$$2x + y \leq 16$$
$$2x + 3y \leq 24$$
$$y \leq 6$$
$$x \geq 0, y \geq 0$$

7. Minimize $C = 10x + 15y$ subject to

$$x + y \leq 10$$
$$3x + y \geq 12$$
$$-2x + 3y \geq 3$$
$$x \geq 0, y \geq 0$$

8. Maximize $P = 2x + 5y$ subject to the constraints of Exercise 7.

9. Maximize $P = 3x + 4y$ subject to

$$x + 2y \leq 50$$
$$5x + 4y \leq 145$$
$$2x + y \geq 25$$
$$y \geq 5, x \geq 0$$

10. Maximize $P = 4x + 3y$ subject to the constraints of Exercise 9.

11. Maximize $P = 2x + 3y$ subject to

$$x + y \leq 48$$
$$x + 3y \geq 60$$
$$9x + 5y \leq 320$$
$$x \geq 10, y \geq 0$$

12. Minimize $C = 5x + 3y$ subject to the constraints of Exercise 11.

13. Find the maximum and minimum of $P = 10x + 12y$ subject to

$$5x + 2y \geq 63$$
$$x + y \geq 18$$
$$3x + 2y \leq 51$$
$$x \geq 0, y \geq 0$$

14. Find the maximum and minimum of $P = 4x + 3y$ subject to

$$3x + 5y \geq 20$$
$$3x + y \leq 16$$
$$-2x + y \leq 1$$
$$x \geq 0, y \geq 0$$

15. Find the maximum and minimum of $P = 2x + 4y$ subject to

$$x + y \le 20$$
$$-x + y \le 10$$
$$x \le 10$$
$$x + y \ge 5$$
$$y \ge 5, x \ge 0$$

16. A company manufactures two products, A and B, on two machines, I and II. It has been determined that the company will realize a profit of $3 on each unit of product A and a profit of $4 on each unit of product B. To manufacture a unit of product A requires 6 minutes on machine I and 5 minutes on machine II. To manufacture a unit of product B requires 9 minutes on machine I and 4 minutes on machine II. There are 5 hours of machine time available on machine I and 3 hours of machine time available on machine II in each work shift. How many units of each product should be produced in each shift to maximize the company's profit?

17. Kane Manufacturing has a division that produces two models of hibachis, model A and model B. To produce each model A hibachi requires 3 pounds of cast iron and 6 minutes of labor. To produce each model B hibachi requires 4 pounds of cast iron and 3 minutes of labor. The profit for each model A hibachi is $2.00, and the profit for each model B hibachi is $1.50. If 1000 pounds of cast iron and 20 hours of labor are available for the production of hibachis per day, how many hibachis of each model should the division produce in order to help maximize the company's profits?

18. Refer to Exercise 17. Because of a backlog of orders in model A hibachis, the manager of Kane Manufacturing has decided to produce at least 150 of these models per day. Operating under this additional constraint, how many hibachis of each model should the company produce to maximize profit?

19. Madison Finance has a total of $20 million earmarked for homeowner and auto loans. On the average, homeowner loans will have a 10 percent rate of return on the loans given out, whereas auto loans will yield a 12 percent rate of return. Management has also stipulated that the total amount of homeowner loans should be greater than or equal to four times the total amount of auto loans. Determine how much money the company should extend to each category in order to maximize its returns.

20. A nutritionist at the Medical Center has been asked to prepare a special diet for certain patients. She has decided that the meals should contain a minimum of 400 mg of calcium, 10 mg of iron, and 40 mg of vitamin C. She has further decided that the meals are to be prepared from foods A and B. Each ounce of food A contains 30 mg of calcium, 1 mg of iron, 2 mg of vitamin C, and 2 mg of cholesterol. Each ounce of food B contains 25 mg of calcium, 0.5 mg of iron, 5 mg of vitamin C, and 5 mg of cholesterol. Find how many ounces of each type of food should be used in a meal so that the cholesterol content is minimized and the minimum requirements of calcium, iron, and vitamin C are met.

21. A farmer has 150 acres of land suitable for cultivating crops A and B. The cost of cultivating crop A is $40 per acre, whereas that of crop B is $60 per acre. The farmer has a maximum of $7400 available for land cultivation. Each acre of crop A requires 20 hours of labor and each acre of crop B requires 25 hours of labor. The farmer has a maximum of 3300 hours of labor available. If he expects to make a profit of $150 per acre on crop A and $200 per acre on crop B, how many acres of each crop should he plant in order to maximize his profit?

22. Complete the solution to Example 9, Section 3.2.

23. The TMA Company manufactures 19-inch color television picture tubes in two separate locations, location I and location II. The output at location I is at most 6000 tubes per month whereas the output at location II is at most 5000 per month. TMA is the main supplier of picture tubes to the Pulsar Corporation, its holding company, which has priority in having all its requirements met. In a certain month, Pulsar placed orders for 3000 and 4000 picture tubes to be shipped to two of its factories located in city A and city B, respectively. The shipping costs (in dollars) per picture tube from the two plants of TMA to the two Pulsar factories are as follows:

Shipping Costs per Picture Tube

	To Pulsar Factories	
From	*City A*	*City B*
TMA (Loc. I)	$3	$2
TMA (Loc. II)	$4	$5

Find a shipping schedule that meets the requirements of both companies while keeping costs at a minimum.

24. Patricia has $30,000 to invest in securities in the form of corporate stocks. She has narrowed her choices to two groups of stocks: growth stocks that she assumes will yield a 15 percent return (dividends and capital appreciation) within a year, and speculative stocks that she assumes will yield a 25 percent return (mainly in capital appreciation) within a year. Determine how much she should invest in each group of stocks in order to maximize the return on her investments within a year if she has decided to invest three times as much in growth stocks as in speculative stocks.

25. A veterinarian has been asked to prepare a diet for a group of dogs to be used in a nutrition study at the School of Animal Science. It has been stipulated that each serving should be no larger than 8 ounces and must contain at least 29 units of nutrient I and 20 units of nutrient II. The vet has decided that the diet may be prepared from two brands of dog food: brand A and brand B. Each ounce of brand A contains 3 units of nutrient I and 4 units of nutrient II. Each ounce of brand B contains 5 units of nutrient I and 2 units of nutrient II. The cost of brand A is 3¢ per ounce and that of brand B is 4¢ per ounce. Determine how many ounces of each brand of dog food should be used to meet the given requirements at a minimum cost.

26. The Trendex Corporation, a telephone survey company, has been hired to conduct a television viewing poll among urban and suburban families in the Los Angeles area. The client has stipulated that a maximum of 1500 families are to be interviewed. At least 500 urban families must be interviewed, and at least half of the total number of families interviewed must be from the suburban area. For this service, Trendex will be paid $1500 plus $2.00 for each completed interview. From previous experience Trendex has determined that it will incur an expense of $1.10 for each successful interview with an urban family and $1.25 for each successful interview with a suburban family. Find the number of urban and suburban families that Trendex should interview in order to maximize its profit.

Solutions to Self-Check Exercises

3.3

1. The feasible set S for the problem was graphed in the solution to Exercise 1 of Self-Check Exercises 3.1. It is reproduced in the following figure.

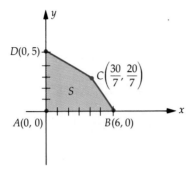

The values of the objective function P at the vertices of S are summarized in the following table.

Vertex	$P = 4x + 5y$
$A(0, 0)$	0
$B(6, 0)$	24
$C\left(\dfrac{30}{7}, \dfrac{20}{7}\right)$	$\dfrac{220}{7} = 31\dfrac{3}{7}$
$D(0, 5)$	25

From the table we see that the maximum for the objective function P is attained at the vertex $C(30/7, 20/7)$. Therefore, the solution to the problem is $x = 30/7$, $y = 20/7$ and $P = 31\frac{3}{7}$.

2. The feasible set S for the problem was graphed in the solution to Exercise 2 of Self-Check Exercises 3.1. It is reproduced in the following figure.

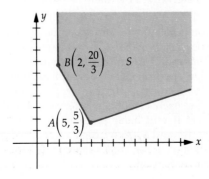

Evaluating the objective function $C = 5x + 3y$ at each corner point, we obtain the table

Vertex	$C = 5x + 3y$
$A\left(5, \dfrac{5}{3}\right)$	30
$B\left(2, \dfrac{20}{3}\right)$	30

We conclude that the objective function is minimized at every point on the line segment joining the points (5, 5/3) and (2, 20/3) and $C = 30$.

3. Refer to Exercise 1 of Self-Check Exercises 3.2. The problem is to maximize $P = 50,000x + 20,000y$ subject to

$$300x + 100y \leq 9000$$
$$y \geq 15$$
$$y \leq 30$$
$$x \geq 0, y \geq 0.$$

The feasible set S for the problem is shown in the following figure.

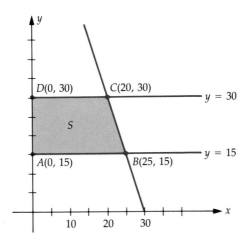

Evaluating the objective function $P = 50,000x + 20,000y$ at each vertex of S, we obtain

Vertex	$P = 50,000x + 20,000y$
$A(0, 15)$	300,000
$B(25, 15)$	1,550,000
$C(20, 30)$	1,600,000
$D(0, 30)$	600,000

From the table we see that P is maximized when $x = 20$ and $y = 30$. Therefore, Balduzzi should place 20 ads in the *City Tribune* and 30 with the *Daily News*.

▶

Chapter 3 Review Exercises

In Exercises 1–8, use the method of corners to solve the given linear programming problems.

1. Maximize $P = 3x + 5y$ subject to

$$2x + 3y \leq 12$$
$$x + y \leq 5$$
$$x \geq 0, y \geq 0$$

2. Maximize $P = 2x + 3y$ subject to

$$2x + y \leq 12$$
$$x - 2y \leq 1$$
$$x \geq 0, y \geq 0$$

3. Maximize $P = 3x + 2y$ subject to

$$2x + y \leq 16$$
$$2x + 3y \leq 36$$
$$4x + 5y \geq 28$$
$$x \geq 0, y \geq 0$$

4. Maximize $P = 6x + 2y$ subject to

$$x + 2y \leq 12$$
$$x + y \leq 8$$
$$2x - 3y \geq 6$$
$$x \geq 0, y \geq 0$$

5. Minimize $C = 2x + 7y$ subject to

$$3x + 5y \geq 45$$
$$3x + 10y \geq 60$$
$$x \geq 0, y \geq 0$$

6. Minimize $C = 4x + y$ subject to

$$6x + y \geq 18$$
$$2x + y \geq 10$$
$$x + 4y \geq 12$$
$$x \geq 0, y \geq 0$$

7. Find the maximum and minimum of $Q = x + y$ subject to

$$5x + 2y \geq 20$$
$$x + 2y \geq 8$$
$$x + 4y \leq 22$$
$$x \geq 0, y \geq 0$$

8. Find the maximum and minimum of $Q = 2x + 5y$ subject to

$$x + y \geq 4$$
$$-x + y \leq 6$$
$$x + 3y \leq 30$$
$$x \leq 12$$
$$x \geq 0, y \geq 0$$

9. An investor has decided to commit no more than $80,000 to the purchase of the common stocks of two companies, Company A and Company B. He has also estimated that there is a chance of a 1 percent capital loss on his investment in Company A and a chance of a 4 percent loss on his investment in Company B, and he has decided that these losses should not exceed $2000. On the other hand, he expects to make a 14 percent profit from his investment in Company A and a 20 percent profit from his investment in Company B. Determine how much he should invest in the stock of each company in order to maximize his investment returns.

10. The Soundex Company produces two models of clock/radios. Model A requires 15 minutes of work on assembly line I and 10 minutes of work on assembly line II. Model B requires 10 minutes of work on assembly line I and 12 minutes of work on assembly line II. At most, 25 hours of assembly time on line I and 22 hours of assembly time on line II are available per day. It is anticipated that Soundex will realize a profit of $12 on model A and $10 on model B. How many clock/radios of each model should be produced per day in order to maximize the company's profit?

Linear Programming: An Algebraic Approach

How much profit? The Ace Novelty Company produces three types of souvenirs. Each type requires a certain amount of time on each of three different machines. Each machine may be operated for a certain amount of time per day. In Example 5, page 201, we will determine how many souvenirs of each type Ace Novelty should make per day in order to maximize its daily profits.

▶CHAPTER

FOUR

4.1

The Simplex Method: Standard Problems and Maximization

▶ The Simplex Method

▶ Application

▶ The Simplex Method

As we mentioned toward the end of Chapter 3, the method of corners is not suitable for solving linear programming problems when the number of variables and/or constraints is large. Its major shortcoming is that a knowledge of all the corner points of the feasible set S associated with the problem is required. What we need is a method of solution that is based on a judicious selection of the corner points of the feasible set S, thereby reducing the number of points to be inspected. One such technique, called the **simplex method,** was developed in the late 1940s by George Dantzig and is based on the Gauss-Jordan elimination method. The simplex method is readily adaptable to the electronic computer, which makes it ideally suitable for solving linear programming problems involving large numbers of variables and constraints.

Basically, the simplex method is an iterative procedure; that is, it is repeated over and over again. Beginning at some initial feasible solution (a corner point of the feasible set S, usually the origin) each iteration brings us to another corner point of S with an improved (but certainly no worse) value of the objective function. The iteration is terminated when the optimal solution is reached (if it exists).

In this section, we will describe the simplex method for solving a large class of problems that are referred to as **standard linear programming problems** (nonstandard linear programming problems will be dealt with in Section 4.2).

Before stating a formal procedure for solving standard linear programming problems based on the simplex method, let us consider the following analysis of a two-variable problem. The ensuing discussions will clarify the general procedure and at the same time enhance our understanding of the simplex method by examining the motivation that led to the steps of the procedure.

A Standard Linear Programming Problem

A standard linear programming problem is one in which

1. The objective function is to be maximized.
2. All the variables involved in the problem are nonnegative.
3. Each linear constraint may be written so that the expression involving the variables is less than or equal to a nonnegative constant.

Consider the standard linear programming problem presented at the beginning of Section 3.3.

Maximize

$$P = 3x + 2y \tag{1}$$

subject to

$$2x + 3y \leq 12$$
$$2x + y \leq 8$$
$$x \geq 0, y \geq 0 \tag{2}$$

The feasible set S associated with the above problem is reproduced in Figure 4.1, where we have labeled the four feasible corner points $A(0, 0)$, $B(4, 0)$, $C(3, 2)$, and $D(0, 4)$. Recall that the optimal solution to the problem occurs at the corner point $C(3, 2)$.

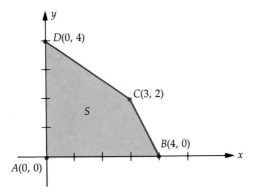

Figure 4.1

As a first step in the solution, using the simplex method, we replace the system of inequality constraints (2) with a system of equality constraints. This may be accomplished by using nonnegative variables called **slack variables** (the reason for the name will soon become clear). By adding the slack variable u to the left-hand side of the first inequality $2x + 3y \leq 12$, we obtain the equation $2x + 3y + u = 12$. Note that u takes up the slack between the left-hand side and the right-hand side of the inequality $2x + 3y \leq 12$. Similarly, the inequality

$2x + y \le 8$ is converted into the equation $2x + y + v = 8$ through the introduction of the slack variable v. The system of linear inequalities (2) may now be viewed as the system of linear equations

$$2x + 3y + u \qquad = 12$$
$$2x + y \qquad + v = 8$$

where x, y, u, and v are all nonnegative.

Finally, rewriting the objective function (1) in the form $-3x - 2y + P = 0$, where the coefficient of P is $+1$, we are led to the following system of linear equations:

$$2x + 3y + u \qquad \qquad = 12$$
$$2x + y \qquad + v \qquad = 8$$
$$-3x - 2y \qquad \qquad + P = 0 \qquad (3)$$

Since the system (3) comprises three linear equations in the five variables x, y, u, v, and P, we may solve for three of the variables in terms of the other two. Thus, there are infinitely many solutions to this system expressible in terms of two parameters. Our linear programming problem is now seen to be equivalent to the following: From among all the solutions of the system (3) for which x, y, u, and v are nonnegative (such solutions are called **feasible solutions**), determine the solution(s) that renders P a maximum.

The augmented matrix associated with the system (3) is

$$
\begin{array}{ccccc}
x & y & u & v & P \\
\end{array}
$$
$$
\left[
\begin{array}{ccccc|c}
2 & 3 & 1 & 0 & 0 & 12 \\
2 & 1 & 0 & 1 & 0 & 8 \\
-3 & -2 & 0 & 0 & 1 & 0 \\
\end{array}
\right] \qquad (4)
$$

Observe that each of the u-, v-, and P-columns of the augmented matrix (4) is in **unit form**; that is, one of the entries in the column is a 1, whereas the rest of the entries in the column are equal to zero.

A Unit Column

A column in an augmented matrix is in unit form if one of the entries in the column is a 1 and the other entries are zeros.

The variables associated with columns in unit form are called **basic variables**; all other variables are called **nonbasic variables**.

Now, the configuration of the augmented matrix (4) suggests that we solve for the basic variables u, v, and P in terms of the nonbasic variables x and y, obtaining

$$u = 12 - 2x - 3y$$
$$v = 8 - 2x - y$$
$$P = 3x + 2y \qquad (5)$$

Of the infinitely many feasible solutions obtainable by assigning arbitrary nonnegative values to the parameters x and y, a particular solution is obtained by letting $x = 0$ and $y = 0$. In fact, this solution is given by

$$x = 0, \qquad y = 0, \qquad u = 12, \qquad v = 8, \quad \text{and} \quad P = 0$$

Such a solution, obtained by setting the nonbasic variables equal to zero, is called a **basic solution** of the system. This particular solution corresponds to the corner point $A(0, 0)$ of the feasible set associated with the linear programming problem (see Figure 4.1). Observe that $P = 0$ at this point.

Now, if the value of P could not be increased we would have found the optimal solution to the problem at hand. To determine whether the value of P could, in fact, be improved, let us turn our attention to the objective function in (1). Since both the coefficients of x and y are positive, the value of P could be improved by increasing x and/or y; that is, by moving away from the origin. Note that we arrive at the same conclusion by observing that the last row of the augmented matrix (4) contains entries that are *negative*. (Compare the original objective function, $P = 3x + 2y$, with the rewritten objective function, $-3x - 2y + P = 0$.)

Continuing our quest for an optimal solution, our next task is to determine whether it is more profitable to increase the value of x or that of y (increasing x and y simultaneously is more difficult). Since the coefficient of x is greater than that of y, a unit increase in the x-direction would result in a greater increase in the value of the objective function P than a unit increase in the y-direction. Thus, we should increase the value of x while holding y constant. How much can x be increased while holding $y = 0$? From system (5) we see that, upon setting $y = 0$,

$$u = 12 - 2x$$
$$v = 8 - 2x \qquad (6)$$

Since u must be nonnegative, the first equation of (6) implies that x cannot exceed 12/2, or 6. The second equation of (6) and the nonnegativity of v implies that x cannot exceed 8/2, or 4. Thus, we conclude that x can be increased by at most 4.

Now, if we set $y = 0$ and $x = 4$ in system (5) we obtain the solution

$$x = 4, \qquad y = 0, \qquad u = 4, \qquad v = 0, \quad \text{and} \quad P = 12$$

which is a basic solution to (3), this time with y and v as nonbasic variables. (Recall that the nonbasic variables are precisely the variables that are set equal to zero.)

Let us see how this basic solution may be found by working with the augmented matrix of the system. Since x is to replace v as a basic variable, our aim is to find an augmented matrix that is equivalent to the matrix (4) and has a configuration in which the x-column is in the unit form

$$\begin{bmatrix} 0 \\ 1 \\ 0 \end{bmatrix}$$

replacing what is presently the form of the v-column in (4). This may be accomplished using elementary row operations as follows:

$$
\begin{array}{ccccc}
x & y & u & v & P
\end{array}
\begin{bmatrix}
2 & 3 & 1 & 0 & 0 & 12 \\
② & 1 & 0 & 1 & 0 & 8 \\
-3 & -2 & 0 & 0 & 1 & 0
\end{bmatrix}
\xrightarrow{\frac{1}{2}R_2}
\begin{array}{ccccc}
x & y & u & v & P
\end{array}
\begin{bmatrix}
2 & 3 & 1 & 0 & 0 & 12 \\
① & ½ & 0 & ½ & 0 & 4 \\
-3 & -2 & 0 & 0 & 1 & 0
\end{bmatrix}
\tag{7}
$$

$$
\xrightarrow[R_3 + 3R_2]{R_1 - 2R_2}
\begin{array}{ccccc}
x & y & u & v & P
\end{array}
\begin{bmatrix}
0 & 2 & 1 & -1 & 0 & 4 \\
1 & ½ & 0 & ½ & 0 & 4 \\
0 & -½ & 0 & 3/2 & 1 & 12
\end{bmatrix}
\tag{8}
$$

Using (8), we now easily solve for the basic variables x, u, and P in terms of the nonbasic variables y and v, obtaining

$$
\begin{aligned}
x &= 4 - (1/2)y - (1/2)v \\
u &= 4 - 2y + v \\
P &= 12 + (1/2)y - (3/2)v.
\end{aligned}
$$

Setting the nonbasic variables y and v equal to zero gives

$$x = 4, \quad y = 0, \quad u = 4, \quad v = 0, \quad \text{and} \quad P = 12$$

as before.

We have completed one iteration of the simplex procedure and our search has brought us from the feasible corner point $A(0, 0)$ where $P = 0$ to the feasible corner point $B(4, 0)$ where P attained a value of 12, which is certainly an improvement!

Before going on, let us introduce the following terminology. The circled element 2 in the first augmented matrix of (7), which was to be converted into a 1, is called a **pivot element**. The column containing the pivot element is called the **pivot column**. The pivot column is associated with the nonbasic variable, which is to be converted to a basic variable. Note that *the last entry in the pivot column is the negative number with the largest absolute value in the last row*—precisely the criterion for choosing the direction of maximum increase in P.

The row containing the pivot element is called the **pivot row**. The pivot row can also be found by dividing each positive number in the pivot column into the corresponding number in the last column (the column of constants). *The pivot row is the one with the smallest ratio.* In the augmented matrix (7), the pivot row is the second row, since the ratio 8/2, or 4, is less than the ratio 12/2, or 6. You should compare this with the earlier analysis pertaining to the determination of the largest permissible increase in the value of x.

The following is a summary of the procedure for selecting the pivot element:

Selecting the Pivot Element

1. Select the pivot column: Locate the most negative entry in the bottom row. The column containing this entry is the **pivot column.** (If there is more than one such column, choose any one.)

2. Select the pivot row: Divide each positive entry in the pivot column into its corresponding entry in the column of constants. The **pivot row** is the row corresponding to the smallest ratio thus obtained. (If there is more than one such entry, choose any one.)

3. The **pivot element** is the element common to both the pivot column and the pivot row.

Continuing with the solution to our problem, we observe that the last row of the augmented matrix (8) contains a negative number, namely, $-1/2$. This indicates that P is not maximized at the feasible corner point $B(4, 0)$ and another iteration is required. Without once again going into a detailed analysis, we proceed immediately to the selection of a pivot element. In accordance with the rules, we perform the necessary row operations as follows:

$$
\begin{array}{c}
\\
\text{pivot} \to \\
\text{row} \\
\\
\end{array}
\begin{array}{ccccc}
x & y & u & v & P \\
\end{array}
\left[
\begin{array}{ccccc|c}
0 & ② & 1 & -1 & 0 & 4 \\
1 & \frac{1}{2} & 0 & \frac{1}{2} & 0 & 4 \\
0 & -\frac{1}{2} & 0 & \frac{3}{2} & 1 & 12 \\
\end{array}
\right]
\begin{array}{c}
\text{ratio} \\
\frac{4}{2} = 2 \\
\frac{4}{1/2} = 8 \\
\end{array}
$$

pivot column

$$
\xrightarrow{\frac{1}{2}R_1}
\begin{array}{ccccc}
x & y & u & v & P \\
\end{array}
\left[
\begin{array}{ccccc|c}
0 & ① & \frac{1}{2} & -\frac{1}{2} & 0 & 2 \\
1 & \frac{1}{2} & 0 & \frac{1}{2} & 0 & 4 \\
0 & -\frac{1}{2} & 0 & \frac{3}{2} & 1 & 12 \\
\end{array}
\right]
$$

$$
\begin{array}{c}
R_2 - \frac{1}{2}R_1 \\
\xrightarrow{\hspace{1cm}} \\
R_3 + \frac{1}{2}R_1 \\
\end{array}
\begin{array}{ccccc}
x & y & u & v & P \\
\end{array}
\left[
\begin{array}{ccccc|c}
0 & 1 & \frac{1}{2} & -\frac{1}{2} & 0 & 2 \\
1 & 0 & -\frac{1}{4} & \frac{3}{4} & 0 & 3 \\
0 & 0 & \frac{1}{4} & \frac{5}{4} & 1 & 13 \\
\end{array}
\right]
$$

Interpreting the last augmented matrix in the usual fashion we find the basic solution $x = 3$, $y = 2$, and $P = 13$. Since there are no negative entries in the last row and the solution is optimal, P cannot be increased further. The optimal solution is the feasible corner point $C(3, 2)$, which agrees with the solution to the problem obtained in Section 3.3.

Having seen how the simplex method works, let us list the steps involved in the procedure. The first step is to set up the initial simplex tableau.

Setting Up the Initial Simplex Tableau

1. Transform the system of linear inequalities into a system of linear equations by introducing slack variables.

2. Rewrite the objective function

$$P = c_1x_1 + c_2x_2 + \cdots + c_nx_n$$

in the form

$$-c_1x_1 - c_2x_2 - \cdots - c_nx_n + P = 0$$

where all the variables are on the left and the coefficient of P is $+1$. Write this equation below the equation of part 1.

3. Write the augmented matrix associated with this system of linear equations.

EXAMPLE

1 Set up the initial simplex tableau for the linear programming problem posed in Chapter 3, Example 7 (page 160).

SOLUTION The problem at hand is to maximize

$$P = x + 1.2y$$

or, equivalently,

$$P = x + \frac{6}{5}y$$

subject to the following constraints:

$$2x + y \leq 180$$
$$x + 3y \leq 300$$
$$x \geq 0, y \geq 0 \tag{9}$$

This is a standard linear programming problem and may be solved by the simplex method. Since the system (9) comprises two linear inequalities, we introduce the two slack variables u and v to convert it to the system of linear equations

$$\begin{aligned} 2x + y + u \quad\quad &= 180 \\ x + 3y \quad\quad + v &= 300. \end{aligned}$$

Next, by rewriting the objective function in the form

$$-x - \frac{6}{5}y + P = 0$$

where the coefficient of P is $+1$ and placing it below the above system, we obtain the following system of linear equations:

$$\begin{aligned} 2x + y + u \quad\quad\quad\quad &= 180 \\ x + 3y \quad\quad + v \quad\quad &= 300 \\ -x - \frac{6}{5}y \quad\quad\quad\quad + P &= \quad 0 \end{aligned}$$

The initial simplex tableau associated with the above system is

x	y	u	v	P	Constant
2	1	1	0	0	180
1	3	0	1	0	300
-1	$-6/5$	0	0	1	0

◀

Before completing the solution to the problem posed in Example 1, let us summarize the main steps of the simplex method.

The Simplex Method

1. Set up the initial simplex tableau.

2. Determine whether the optimal solution has been reached.

   a. If all the entries in the last row are nonnegative, the optimal solution has been reached. Proceed to Step 4.

   b. If there is at least one negative entry in the last row of the simplex tableau, the optimal solution has not been reached. Proceed to Step 3.

3. Perform the following iteration:
Locate the pivot element and convert it to a 1 by dividing all the elements in the pivot row by the pivot element. Using row operations, convert the pivot column into a unit column by adding suitable multiples of the pivot row to each of the other rows as required. Return to Step 2.

4. Determine the optimal solution(s).

   a. If no unit columns are identical, then there is only one solution to the problem. The value of the variable heading each unit column is given by the entry lying in the column of constants in the row containing the 1. The variables heading columns not in unit form are assigned the value zero.

   b. (Interpretation Rule) If two or more unit columns are identical, then more than one solution exists. Choose any one of these columns and proceed as in (a), with the exception that the variables heading the other identical unit column(s) are assigned the value zero.

EXAMPLE

2 Complete the solution to the problem discussed in Example 1.

SOLUTION The first step in our procedure, setting up the initial simplex tableau, was completed in Example 1. We continue with Step 2.

Step 2 Determine whether the optimal solution has been reached.
First, refer to the initial simplex tableau:

x	y	u	v	P	Constant
2	1	1	0	0	180
1	3	0	1	0	300
-1	$-6/5$	0	0	1	0

(10)

Since there are negative entries in the last row of the initial simplex tableau, the initial solution is not optimal. Thus we proceed to Step 3.

Step 3 Perform the following iterations:
First, locate the pivot element:

a. Since the entry $-6/5$ in the last row of the initial simplex tableau is the negative entry with the larger absolute value, the second column in the tableau is the pivot column.

b. Divide each positive number of the pivot column into the corresponding entry in the column of constants and compare the ratios thus obtained. We see that the ratio 300/3 is less than the ratio 180/1, so the second row in the tableau is the pivot row.

c. The entry 3 lying in the pivot column and the pivot row is the pivot element.

Next, we convert this pivot element into a 1 by multiplying all the entries in the pivot row by 1/3. Then, using elementary operations, we complete the conversion of the pivot column into a unit column.

The details of the iteration are recorded as follows:

	x	y	u	v	P	Constant	ratio
	2	1	1	0	0	180	$^{180}/_1 = 180$
pivot → row	1	③	0	1	0	300	$^{300}/_3 = 100$
	-1	$-^6/_5$	0	0	1	0	

↑
pivot
column

	x	y	u	v	P	Constant
$\frac{1}{3}R_2$ →	2	1	1	0	0	180
	$^1/_3$	①	0	$^1/_3$	0	100
	-1	$-^6/_5$	0	0	1	0

	x	y	u	v	P	Constant
$R_1 - R_2$ → $R_3 + \frac{6}{5}R_2$	$^5/_3$	0	1	$-^1/_3$	0	80
	$^1/_3$	1	0	$^1/_3$	0	100
	$-^3/_5$	0	0	$^2/_5$	1	120

(11)

This completes one iteration. The last row of the simplex tableau contains a negative number, so an optimal solution has not been reached. Therefore, we repeat the iterative step once again as follows:

	x	y	u	v	P	Constant	ratio
pivot → row	⑤⁄₃	0	1	$-^1/_3$	0	80	$\frac{80}{5/3} = 48$
	$^1/_3$	1	0	$^1/_3$	0	100	$\frac{100}{1/3} = 300$
	$-^3/_5$	0	0	$^2/_5$	1	120	

↑
pivot
column

	x	y	u	v	P	Constant
$\frac{3}{5}R_1$ →	①	0	$^3/_5$	$-^1/_5$	0	48
	$^1/_3$	1	0	$^1/_3$	0	100
	$-^3/_5$	0	0	$^2/_5$	1	120

	x	y	u	v	P	Constant
$R_2 - \frac{1}{3}R_1$ → $R_3 + \frac{3}{5}R_1$	1	0	$^3/_5$	$-^1/_5$	0	48
	0	1	$-^1/_5$	$^2/_5$	0	84
	0	0	$^9/_{25}$	$^7/_{25}$	1	$148^4/_5$

(12)

The last row of the simplex tableau (12) contains no negative numbers, and we therefore conclude that the optimal solution has been reached.

Step 4 Determine the optimal solution.

Locate the basic variables in the final tableau. In this case, the basic variables (those heading unit columns) are x, y, and P. The value assigned to the basic variable x is the number 48, which is the entry lying in the column of constants and in the first row (the row that contains the 1).

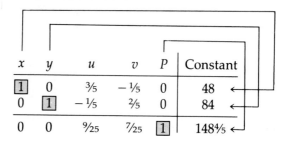

x	y	u	v	P	Constant
[1]	0	$\frac{3}{5}$	$-\frac{1}{5}$	0	48
0	[1]	$-\frac{1}{5}$	$\frac{2}{5}$	0	84
0	0	$\frac{9}{25}$	$\frac{7}{25}$	[1]	$148\frac{4}{5}$

Similarly, we conclude that $y = 84$ and $P = 148.8$. Next, we note that the variables u and v are nonbasic and are accordingly assigned the values $u = 0$ and $v = 0$. These results agree with those obtained in Chapter 3, Example 11 (page 173). ◀

EXAMPLE

3 Maximize $P = 2x + 2y + z$ subject to

$$2x + y + 2z \le 14$$
$$2x + 4y + z \le 26$$
$$x + 2y + 3z \le 28$$
$$x \ge 0, y \ge 0, z \ge 0.$$

SOLUTION Introducing the slack variables u, v, and w and rewriting the objective function in the standard form gives the following system of linear equations:

$$2x + y + 2z + u \qquad\qquad = 14$$
$$2x + 4y + z \qquad + v \qquad\quad = 26$$
$$x + 2y + 3z \qquad\quad + w \quad = 28$$
$$-2x - 2y - z \qquad\qquad\quad + P = 0$$

The initial simplex tableau is given by

x	y	z	u	v	w	P	Constant
2	1	2	1	0	0	0	14
2	4	1	0	1	0	0	26
1	2	3	0	0	1	0	28
−2	−2	−1	0	0	0	1	0

Since the most negative entry in the last row (-2) occurs twice, we may choose either the x- or the y-column as the pivot column. Choosing the x-column as the pivot column and proceeding with the first iteration, we obtain the following sequence of tableaus:

	x	y	z	u	v	w	P	Constant	ratio
pivot row →	②	1	2	1	0	0	0	14	$^{14}\!/_2 = 7$
	2	4	1	0	1	0	0	26	$^{26}\!/_2 = 13$
	1	2	3	0	0	1	0	28	$^{28}\!/_1 = 28$
	-2	-2	-1	0	0	0	1	0	

↑
pivot
column

	x	y	z	u	v	w	P	Constant
$\tfrac{1}{2}R_1 \longrightarrow$	①	½	1	½	0	0	0	7
	2	4	1	0	1	0	0	26
	1	2	3	0	0	1	0	28
	-2	-2	-1	0	0	0	1	0

	x	y	z	u	v	w	P	Constant
$R_2 - 2R_1 \atop R_3 - R_1 \atop R_4 + 2R_1 \longrightarrow$	1	½	1	½	0	0	0	7
	0	3	-1	-1	1	0	0	12
	0	³⁄₂	2	$-½$	0	1	0	21
	0	-1	1	1	0	0	1	14

Since there is a negative number in the last row of the simplex tableau, we perform another iteration as follows:

	x	y	z	u	v	w	P	Constant	ratio
pivot row →	1	½	1	½	0	0	0	7	$\frac{7}{1/2} = 14$
	0	③	-1	-1	1	0	0	12	$\frac{12}{3} = 4$
	0	³⁄₂	2	$-½$	0	1	0	21	$\frac{21}{3/2} = 14$
	0	-1	1	1	0	0	1	14	

↑
pivot
column

	x	y	z	u	v	w	P	Constant
$\frac{1}{3}R_2$	1	½	1	½	0	0	0	7
$\longrightarrow$	0	①	$-\frac{1}{3}$	$-\frac{1}{3}$	⅓	0	0	4
	0	3/2	2	$-\frac{1}{2}$	0	1	0	21
	0	−1	1	1	0	0	1	14

	x	y	z	u	v	w	P	Constant
$R_1-\frac{1}{2}R_2$ $\longrightarrow$	1	0	$\frac{7}{6}$	$\frac{2}{3}$	$-\frac{1}{6}$	0	0	5
$R_3-\frac{3}{2}R_2$	0	1	$-\frac{1}{3}$	$-\frac{1}{3}$	⅓	0	0	4
R_4+R_2	0	0	$\frac{5}{2}$	0	$-\frac{1}{2}$	1	0	15
	0	0	$\frac{2}{3}$	$\frac{2}{3}$	⅓	0	1	18

All entries in the last row are nonnegative, so we have reached the optimal solution. By the interpretation rule, the solution is given by $x = 5$, $y = 4$, $z = 0$, $u = 0$, $v = 0$, $w = 15$, and $P = 18$. ◀

EXAMPLE

4   Maximize the function $P = 2x + 2y - 4z$ subject to

$$3x + 3y - 2z \leq 100$$
$$5x + 5y + 3z \leq 150$$
$$x \geq 0, y \geq 0, z \geq 0$$

SOLUTION   Introducing the slack variables u and v, we obtain the following system of linear equations:

$$3x + 3y - 2z + u \qquad\qquad = 100$$
$$5x + 5y + 3z \qquad + v \qquad = 150$$
$$-2x - 2y + 4z \qquad\qquad + P = 0$$

The initial simplex tableau and the tableaus resulting from the first iteration follow.

	x	y	z	u	v	P	Constant	ratio
	3	3	−2	1	0	0	100	$^{100}\!/_3 = 33\frac{1}{3}$
pivot row →	⑤	5	3	0	1	0	150	$^{150}\!/_5 = 30$
	−2	−2	4	0	0	1	0	

↑
pivot column

(We could, of course, have chosen the second column as the pivot column.)

	x	y	z	u	v	P	Constant
	3	3	-2	1	0	0	100
$\frac{1}{5}R_2 \longrightarrow$	①	1	$\frac{3}{5}$	0	$\frac{1}{5}$	0	30
	-2	-2	4	0	0	1	0

	x	y	z	u	v	P	Constant
$R_1 - 3R_2 \longrightarrow$	0	0	$-\frac{19}{5}$	1	$-\frac{3}{5}$	0	10
$R_3 + 2R_2 \longrightarrow$	1	1	$\frac{3}{5}$	0	$\frac{1}{5}$	0	30
	0	0	$\frac{26}{5}$	0	$\frac{2}{5}$	1	60

All entries in the last row are nonnegative and the tableau is final. By the interpretation rule (Step 4b in this instance) we conclude that

$$x = 30, y = 0, z = 0, u = 10, v = 0, P = 60$$

and
$$x = 0, y = 30, z = 0, u = 10, v = 0, P = 60$$

are optimal solutions. ◀

▶ Application

EXAMPLE

5 The Ace Novelty Company has determined that the profits are $6, $5, and $4 for each type-A, type-B, and type-C souvenir that it plans to produce. To manufacture a type-A souvenir requires 2 minutes on machine I, 1 minute on machine II, and 2 minutes on machine III. A type-B souvenir requires 1 minute on machine I, 3 minutes on machine II, and 1 minute on machine III. A type-C souvenir requires 1 minute on machine I and 2 minutes on each of machines II and III. There are 3 hours available on machine I, 5 hours available on machine II, and 4 hours available on machine III for manufacturing these souvenirs each day. How many souvenirs of each type should Ace Novelty make per day in order to maximize its profit? (Compare with Example 1, Chapter 2.)

SOLUTION

The given information may be tabulated as follows:

	Machine			
Souvenir	I	II	III	Profit per Unit ($)
A	2	1	2	6
B	1	3	1	5
C	1	2	2	4
Max. machine-time available (minutes)	180	300	240	

Let x, y, and z denote the respective numbers of type-A, type-B, and type-C souvenirs to be made. The total amount of time that machine I is used is given by $2x + y + z$ minutes and must not exceed 180 minutes. Thus, we have the inequality

$$2x + y + z \le 180.$$

Similar considerations on the use of machines II and III lead to the following inequalities:

$$x + 3y + 2z \le 300$$
$$2x + y + 2z \le 240$$

The profit resulting from the sale of the souvenirs produced is given by

$$P = 6x + 5y + 4z.$$

The mathematical formulation of the product at hand has led to the following standard linear programming problem:

Maximize the objective (profit) function $P = 6x + 5y + 4z$ subject to the constraints

$$2x + y + z \le 180$$
$$x + 3y + 2z \le 300$$
$$2x + y + 2z \le 240$$
$$x \ge 0, y \ge 0, z \ge 0$$

Introducing the slack variables u, v, and w gives the following system of linear equations:

$$2x + y + z + u \qquad\qquad = 180$$
$$x + 3y + 2z \qquad + v \qquad\quad = 300$$
$$2x + y + 2z \qquad\qquad + w \quad = 240$$
$$-6x - 5y - 4z \qquad\qquad\qquad + P = 0$$

The tableaus resulting from the use of the simplex algorithm are as follows:

	x	y	z	u	v	w	P	Constant	ratio
pivot row →	②	1	1	1	0	0	0	180	$180/2 = 90$
	1	3	2	0	1	0	0	300	$300/1 = 300$
	2	1	2	0	0	1	0	240	$240/2 = 120$
	−6	−5	−4	0	0	0	1	0	

↑
pivot
column

	x	y	z	u	v	w	P	Constant
$\frac{1}{2}R_1$	①	½	½	½	0	0	0	90
$\longrightarrow$	1	3	2	0	1	0	0	300
	2	1	2	0	0	1	0	240
	-6	-5	-4	0	0	0	1	0

	x	y	z	u	v	w	P	Constant	ratio
$R_2 - R_1$	1	½	½	½	0	0	0	90	$\frac{90}{1/2} = 180$
$R_3 - 2R_1$	0	⑤⁄₂	3⁄2	$-½$	1	0	0	210	$\frac{210}{5/2} = 84$
$R_4 + 6R_1$	0	0	1	-1	0	1	0	60	—
pivot row	0	-2	-1	3	0	0	1	540	

pivot column

	x	y	z	u	v	w	P	Constant
$\frac{2}{5}R_2$	1	½	½	½	0	0	0	90
$\longrightarrow$	0	①	3⁄5	$-⅕$	2⁄5	0	0	84
	0	0	1	-1	0	1	0	60
	0	-2	-1	3	0	0	1	540

	x	y	z	u	v	w	P	Constant
$R_1 - \frac{1}{2}R_2$	1	0	⅕	3⁄5	$-⅕$	0	0	48
$\longrightarrow$	0	1	3⁄5	$-⅕$	2⁄5	0	0	84
$R_4 + 2R_2$	0	0	1	-1	0	1	0	60
	0	0	⅕	13⁄5	4⁄5	0	1	708

From the final simplex tableau we read off the solution

$$x = 48, \quad y = 84, \quad z = 0, \quad u = 0, \quad v = 0, \quad w = 60, \quad \text{and} \quad P = 708$$

Thus, in order to maximize its profit, Ace Novelty should produce 48 type-A souvenirs, 84 type-B souvenirs, and no type-C souvenirs. The resulting profit is $708 per day. The value of the slack variable $w = 60$ tells us that 1 hour of the available time on machine III is left unused.   ◀

It is instructive to compare the results obtained here with those obtained in Example 1, Chapter 2. Recall that in order to use all of the machine time available on each of the three machines, Ace Novelty had to produce 36 type-A, 48 type-B, and 60 type-C souvenirs. This would have resulted in a profit of $696. This example shows how, through the optimal use of equipment, a company may boost its profit and at the same time minimize machine wear!

Self-Check Exercises

4.1

1. Solve the following linear programming problem by the simplex method: Maximize $P = 2x + 3y + 6z$ subject to

$$\begin{aligned}
2x + 3y + z &\leq 10 \\
x + y + 2z &\leq 8 \\
2y + 3z &\leq 6 \\
x \geq 0, \quad y \geq 0, \quad z &\geq 0
\end{aligned}$$

2. The United Motors Corporation manufactures compact, mid-size, and full-size automobiles. For the next year the company plans to build at most 500,000 cars, of which a maximum of 80,000 will be full size. To comply with government fuel economy standards for passenger cars, the fleet must average at least 27 mpg (miles per gallon) to avoid extensive fines. United's engineering division has estimated that the compacts will average 35 mpg, mid-size models will average 24 mpg, and full-size models will average 20 mpg. Each compact can be expected to bring in a profit of $2000, each mid-size model a profit of $3000, and each full-size model a profit of $5000. Determine how many models of each size United should build for the upcoming year if its profits are to be maximized. *Hint:* Let x, y, and z denote the number of compact, mid-size, and full-size cars to be built. Then, the condition that the fleet average be at least 27 mpg is

$$\frac{35x + 24y + 20z}{x + y + z} \geq 27.$$

Show that this may be written as $-8x + 3y + 7z \leq 0$.

Solutions to Self-Check Exercises 4.1 can be found on page 209.

▶

Exercises 4.1

In Exercises 1–6, determine whether the given simplex tableau is in final form. If so, find the solution to the associated regular linear programming problem. If not, find the pivot element to be used in the next iteration of the simplex method.

1.

x	y	u	v	P	Constant
0	1	$5/7$	$-1/7$	0	$20/7$
1	0	$-3/7$	$2/7$	0	$30/7$
0	0	$13/7$	$3/7$	1	$220/7$

2.

x	y	u	v	P	Constant
1	1	1	0	0	6
1	0	-1	1	0	2
3	0	5	0	1	30

3.

x	y	u	v	P	Constant
0	$\frac{1}{2}$	1	$-\frac{1}{2}$	0	2
1	$\frac{1}{2}$	0	$\frac{1}{2}$	0	4
0	$-\frac{1}{2}$	0	$\frac{3}{2}$	1	12

4.

x	y	z	u	v	w	P	Constant
3	0	5	1	1	0	0	28
2	1	3	0	1	0	0	16
2	0	8	0	3	0	1	48

5.

x	y	z	u	v	w	P	Constant
1	$-\frac{1}{3}$	0	$\frac{1}{3}$	0	$-\frac{2}{3}$	0	$\frac{1}{3}$
0	2	0	0	1	1	0	6
0	$\frac{2}{3}$	1	$\frac{1}{3}$	0	$\frac{1}{3}$	0	$\frac{13}{3}$
0	4	0	1	0	2	1	17

6.

x	y	z	u	v	w	P	Constant
$\frac{1}{2}$	0	$\frac{1}{4}$	1	$-\frac{1}{4}$	0	0	$\frac{19}{2}$
$\frac{1}{2}$	0	$\frac{3}{4}$	0	$\frac{1}{4}$	0	0	$\frac{21}{2}$
2	0	3	0	0	1	0	30
-1	0	$-\frac{1}{2}$	6	$\frac{3}{2}$	0	1	63

In Exercises 7–20, solve each linear programming problem by the simplex method.

7. Maximize $P = 5x + 3y$ subject to

$$\begin{aligned} x + y &\le 80 \\ 3x &\le 90 \\ x \ge 0, y &\ge 0 \end{aligned}$$

8. Maximize $P = 10x + 12y$ subject to

$$\begin{aligned} x + 2y &\le 12 \\ 3x + 2y &\le 24 \\ x \ge 0, y &\ge 0 \end{aligned}$$

9. Maximize $P = 5x + 4y$ subject to

$$\begin{aligned} 3x + 5y &\le 78 \\ 4x + y &\le 36 \\ x \ge 0, y &\ge 0 \end{aligned}$$

10. Maximize $P = 4x + 6y$ subject to

$$3x + y \leq 24$$
$$2x + y \leq 18$$
$$x + 3y \leq 24$$
$$x \geq 0, y \geq 0$$

11. Maximize $P = 15x + 12y$ subject to

$$x + y \leq 12$$
$$3x + y \leq 30$$
$$10x + 7y \leq 70$$
$$x \geq 0, y \geq 0$$

12. Maximize $P = 3x + 4y + 5z$ subject to

$$x + y + z \leq 8$$
$$3x + 2y + 4z \leq 24$$
$$x \geq 0, y \geq 0, z \geq 0$$

13. Maximize $P = 3x + 3y + 4z$ subject to

$$x + y + 3z \leq 15$$
$$4x + 4y + 3z \leq 65$$
$$x \geq 0, y \geq 0, z \geq 0$$

14. Maximize $P = 3x + 4y + z$ subject to

$$3x + 10y + 5z \leq 120$$
$$5x + 2y + 8z \leq 6$$
$$8x + 10y + 3z \leq 105$$
$$x \geq 0, y \geq 0, z \geq 0$$

15. Maximize $P = x + 2y - z$ subject to

$$2x + y + z \leq 14$$
$$4x + 2y + 3z \leq 28$$
$$2x + 5y + 5z \leq 30$$
$$x \geq 0, y \geq 0, z \geq 0$$

16. Maximize $P = 4x + 6y + 5z$ subject to

$$x + y + z \leq 20$$
$$2x + 4y + 3z \leq 42$$
$$2x + 3z \leq 30$$
$$x \geq 0, y \geq 0, z \geq 0$$

17. Maximize $P = x + 4y - 2z$ subject to

$$3x + y - z \leq 80$$
$$2x + y - z \leq 40$$
$$-x + y + z \leq 80$$
$$x \geq 0, y \geq 0, z \geq 0$$

18. Maximize $P = 12x + 10y + 5z$ subject to

$$2x + y + z \le 10$$
$$3x + 5y + z \le 45$$
$$2x + 5y + z \le 40$$
$$x \ge 0, y \ge 0, z \ge 0$$

19. Maximize $P = 2x + 6y + 6z$ subject to

$$2x + y + 3z \le 10$$
$$4x + y + 2z \le 56$$
$$6x + 4y + 3z \le 126$$
$$2x + y + z \le 32$$
$$x \ge 0, y \ge 0, z \ge 0$$

20. Maximize $P = 4x + y + 5z$ subject to

$$x + 2y + 5z \le 30$$
$$2x + y + z \le 10$$
$$3x + 2y + z \le 12$$
$$x + y + z \le 8$$
$$x \ge 0, y \ge 0, z \ge 0$$

21. A company manufactures two products, A and B, on two machines, I and II. It has been determined that the company will realize a profit of $3 on each unit of product A and a profit of $4 on each unit of product B. To manufacture a unit of product A requires 6 minutes on machine I and 5 minutes on machine II. To manufacture a unit of product B requires 9 minutes on machine I and 4 minutes on machine II. There are 5 hours of machine time available on machine I and 3 hours of machine time available on machine II in each work shift. How many units of each product should the company produce in each shift to maximize its profit?

22. The Kane Manufacturing Company has a division that produces two models of hibachis, model A and model B. To produce each model-A hibachi requires 3 pounds of cast iron and 6 minutes of labor. To produce each model-B hibachi requires 4 pounds of cast iron and 3 minutes of labor. The profit for each model-A hibachi is $2.00, and the profit for each model-B hibachi is $1.50. If 1000 pounds of cast iron and 20 hours of labor are available for the production of hibachis in a day, how many hibachis of each model should the division produce in order to maximize the company's profits?

23. A farmer has 150 acres of land suitable for the cultivation of crop A and crop B. The cost of cultivating crop A is $40 per acre, and that of crop B is $60 per acre. The farmer has a maximum of $7400 available for land cultivation. Each acre of crop A requires 20 hours of labor, and each acre of crop B requires 25 hours of labor. The farmer has a maximum of 3300 hours of labor available. If he expects to make a profit of $150 per acre on crop A and $200 per acre on crop B, how many acres of each crop should he plant in order to maximize his profit?

24. A company manufactures products A, B, and C. Each product is processed in three departments: I, II, and III. The total available labor hours per week for departments I, II, and III are 900, 1080, and 840, respectively. The time requirements (in hours per unit) and profit per unit for each product are as follows:

	Product A	Product B	Product C
Dept. I	2	1	2
Dept. II	3	1	2
Dept. III	2	2	1
Profit	$18	$12	$15

Table for
Exercise 24

How many units of each product should the company produce in order to maximize its profit?

25. As part of a campaign to promote its Annual Clearance Sale, the Excelsior Company decided to buy television advertising time on Station KAOS. Excelsior's television advertising budget was $102,000. Morning time costs $3,000 per minute, afternoon time costs $1,000 per minute, and evening (prime) time costs $12,000 per minute. Because of previous commitments, KAOS could not offer Excelsior more than 6 minutes of prime time or more than a total of 25 minutes of advertising time over the two weeks in which the commercials were to be run. KAOS estimated that morning commercials would be seen by 200,000 people, afternoon commercials would be seen by 100,000 people, and evening commercials would be seen by 600,000 people. How much morning, afternoon, and evening advertising time should Excelsior buy to maximize exposure of its commercials?

26. The Custom Office Furniture Company is introducing a new line of executive desks made from a specially selected grade of walnut. Initially, three models—A, B, and C—are to be marketed. Each model-A desk requires 1¼ hours for fabrication, 1 hour for assembly, and 1 hour for finishing; each model-B desk requires 1½ hours for fabrication, 1 hour for assembly, and 1 hour for finishing; each model-C desk requires 1½ hours, ¾ hour, and ½ hour for fabrication, assembly, and finishing, respectively. The profit on each model-A desk is $26, the profit on each model-B desk is $28, and the profit on each model-C desk is $24. The total time available in the fabrication department, the assembly department, and the finishing department in the first month of production is 310 hours, 205 hours, and 190 hours, respectively. To maximize the company's profit, how many desks of each model should be made in the month?

27. The Boise Lumber Company has decided to enter the lucrative prefabricated housing business. Initially, it plans to offer three models of houses: a standard model, a deluxe model, and a luxury model. Each house is prefabricated and partially assembled in the factory, and the final assembly is completed on site. Each standard model requires $6,000 worth of building material, each deluxe model requires $8,000 worth of building material, and each luxury model requires $10,000 worth of building material. The amount of labor required to prefabricate and partially assemble each standard, deluxe, and luxury model in the factory is estimated to be 240, 220, and 200 labor hours, respectively. The amount of on-site labor required to erect and finish each standard, deluxe, and luxury model is estimated to be 180, 210, and 300 labor hours, respectively. The profit from the sale of each standard, deluxe, and luxury model house is expected to be $3,400, $4,000, and $5,000, respectively. For the first year's production, a sum of $8,200,000 is budgeted for building material; the number of labor hours available for work in the factory (for prefabrication and partial assembly) is not to exceed 218,000 hours, and the amount of labor for on-site work is to be less than or equal to 237,000 labor hours. Determine how many houses of each type Boise Lumber should produce to maximize its profit from this new venture. (Market research has confirmed that there should be no problems with sales.)

28. Bayer Pharmaceutical manufactures three kinds of cold formulas: formulas I, II, and III. It takes 2.5 hours to produce 1,000 bottles of formula I, 3 hours to produce 1,000 bottles of formula II, and 4 hours to produce 1,000 bottles of formula III. The profits for each 1,000 bottles of formulas I, II, and III are $180, $200, and $300, respectively. Suppose that for a certain production run there are enough ingredients on hand to make at most 9,000 bottles of formula I, 12,000 bottles of formula II, and 6,000 bottles of formula III. Furthermore, suppose that the time for the production run is limited to a maximum of 70 hours. Find how many bottles of each formula should be produced in this production run so that the profit is maximized.

*29. Solve the following regular linear programming problem using the simplex method: Maximize $Q = x + 2y$ subject to

$$
\begin{aligned}
-2x + \quad y &\le 4 \\
x - 3y &\le 3 \\
x \ge 0, \ y &\ge 0
\end{aligned}
$$

Also, solve the problem graphically to show that it has an unbounded solution. [*Hint:* What part of the simplex procedure breaks down?]

Solutions to Self-Check Exercises

4.1

1. Introducing the slack variables u, v, and w, we obtain the following system of linear equations:

$$
\begin{aligned}
2x + 3y + \quad z + u &\quad\quad\quad\quad = 10 \\
x + \quad y + 2z \quad\quad + v &\quad\quad\quad = 8 \\
2y + 3z \quad\quad\quad + w \quad &= 6 \\
-2x - 3y - 6z \quad\quad\quad\quad\quad + P &= 0
\end{aligned}
$$

The initial simplex tableau and the successive tableaus resulting from the use of the simplex procedure follow:

	x	y	z	u	v	w	P	Constant	ratio
	2	3	1	1	0	0	0	10	$^{10}/_{13} = 10$
	1	1	2	0	1	0	0	8	$^{8}/_{2} = 4$
pivot → row	0	2	③	0	0	1	0	6	$^{6}/_{3} = 2$
	-2	-3	-6	0	0	0	1	0	

$\frac{1}{3}R_3$ →

↑
pivot
column

	x	y	z	u	v	w	P	Constant	
	2	3	1	1	0	0	0	10	$R_1 - R_3$
	1	1	2	0	1	0	0	8	$R_2 - 2R_3$
	0	$^2/_3$	①	0	0	$^1/_3$	0	2	$R_4 + 6R_3$
	-2	-3	-6	0	0	0	1	0	

	x	y	z	u	v	w	P	Constant	ratio	
	2	$\tfrac{7}{3}$	0	1	0	$-\tfrac{1}{3}$	0	8	$\tfrac{8}{2}=4$	R_1-2R_2
pivot row →	①	$-\tfrac{1}{3}$	0	0	1	$-\tfrac{2}{3}$	0	4	$\tfrac{4}{1}=4$	$\overrightarrow{R_4+2R_2}$
	0	$\tfrac{2}{3}$	1	0	0	$\tfrac{1}{3}$	0	2	—	
	-2	1	0	0	0	2	1	12		

pivot column (↑ under x)

x	y	z	u	v	w	P	Constant
0	3	0	1	-2	1	0	0
1	$-\tfrac{1}{3}$	0	0	1	$-\tfrac{2}{3}$	0	4
0	$\tfrac{2}{3}$	1	0	0	$\tfrac{1}{3}$	0	2
0	$\tfrac{1}{3}$	0	0	2	$\tfrac{2}{3}$	1	20

All entries in the last row are nonnegative and the tableau is final. We conclude that $x = 4$, $y = 0$, $z = 2$, and $P = 20$.

2. Refer to Exercise 2 of Self-Check Exercises 3.2. The problem is to maximize $P = 2,000x + 3,000y + 5,000z$ subject to

$$x + y + z \le 500,000$$
$$z \le 80,000$$
$$-8z + 3y + 7z \le 0$$
$$x \ge 0,\ y \ge 0,\ z \ge 0$$

This is a regular linear programming problem and may be solved using the simplex method. We obtain the following tableaus, the first being, naturally, the initial simplex tableau.

	x	y	z	u	v	w	P	Constant	ratio	
	1	1	1	1	0	0	0	500,000	500,000	
	0	0	1	0	1	0	0	80,000	80,000	$\tfrac{1}{7}R_3$
pivot row →	-8	3	⑦	0	0	1	0	0	0	→
	$-2,000$	$-3,000$	$-5,000$	0	0	0	1	0		

pivot column (↑ under z)

	x	y	z	u	v	w	P	Constant	
	1	1	1	1	0	0	0	500,000	R_1-R_3
	0	0	1	0	1	0	0	80,000	$\overrightarrow{R_2-R_3}$
	$-\tfrac{8}{7}$	$\tfrac{3}{7}$	①	0	0	$\tfrac{1}{7}$	0	0	$R_4+5,000R_3$
	$-2,000$	$-3,000$	$-5,000$	0	0	0	1	0	

	x	y	z	u	v	w	P	Constant	ratio	
	$15/7$	$4/7$	0	1	0	$-1/7$	0	500,000	$233{,}333\frac{1}{3}$	$\frac{7}{8}R_2$
pivot row →	⑧/₇	$-3/7$	0	0	1	$-1/7$	0	80,000	70,000	→
	$-8/7$	$3/7$	1	0	0	$1/7$	0	0	0	
	$-\dfrac{54{,}000}{7}$	$-\dfrac{6{,}000}{7}$	0	0	0	$\dfrac{5{,}000}{7}$	1	0		

↑ pivot column

	x	y	z	u	v	w	P	Constant	
	$15/7$	$4/7$	0	1	0	$-1/7$	0	500,000	$R_1 - \frac{15}{7}R_2$
	①	$-3/8$	0	0	$7/8$	$-1/8$	0	70,000	$R_3 + \frac{8}{7}R_2$
	$-8/7$	$3/7$	1	0	0	$1/7$	0	0	$R_4 + \frac{54{,}000}{7}R_2$
	$-\dfrac{54{,}000}{7}$	$-\dfrac{6{,}000}{7}$	0	0	0	$\dfrac{5{,}000}{7}$	1	0	

	x	y	z	u	v	w	P	Constant	ratio	
pivot row →	0	⑪/₈	0	1	$-15/8$	$1/8$	0	350,000	$254{,}545\,5/11$	$\frac{8}{11}R_1$
	1	$-3/8$	0	0	$7/8$	$-1/8$	0	70,000	—	→
	0	0	1	0	1	0	0	80,000	—	
	0	$-3{,}750$	0	0	6,750	-250	1	540,000,000		

↑ pivot column

	x	y	z	u	v	w	P	Constant	
	0	①	0	$8/11$	$-15/11$	$1/11$	0	$254{,}545\,5/11$	$R_2 + \frac{3}{8}R_1$
	1	$-3/8$	0	0	$7/8$	$-1/8$	0	70,000	$R_4 + 3{,}750R_1$
	0	0	1	0	1	0	0	80,000	
	0	$-3{,}750$	0	0	6,750	-250	1	540,000,000	

x	y	z	u	v	w	P	Constant
0	1	0	$8/11$	$-15/11$	$1/11$	0	$254{,}544\,5/11$
1	0	0	$3/11$	$4/11$	$-1/11$	0	$165{,}454\,6/11$
0	0	1	0	1	0	0	80,000
0	0	0	$2{,}727\,3/11$	$1{,}636\,4/11$	$90\,10/11$	1	1,494,545,455

The last tableau is in final form, and we see that $x = 165{,}455$, $y = 254{,}545$, and $z = 80{,}000$. Therefore, the company should build 165,455 compact, 254,545 mid-size, and 80,000 full-size cars. The maximum profit realizable is \$1,494,545,455.

4.2

The Simplex Method: Nonstandard Problems and Minimization

▶ Minimization with ≤ Constraints

▶ The Dual Problem

▶ Application

▶ Minimization with ≤ Constraints

In the last section we developed a procedure, called the simplex method, for solving standard linear programming problems. Recall that a standard linear programming problem is one in which the following three conditions are satisfied:

1. The objective function is to be maximized.

2. All the variables involved are nonnegative.

3. Each linear constraint may be written so that the expression involving the variables is less than or equal to a nonnegative constant.

Linear programming problems in which at least one of the above conditions does not hold true are called **nonstandard linear programming problems.**

In this section we will see how, through slight modifications, the simplex method may be adapted to solve certain classes of nonstandard problems. In particular, we will see how the modified procedure may be used to help solve problems involving the minimization of objective functions.

We begin by considering the class of nonstandard linear programming problems that call for the minimization of the objective functions but otherwise satisfy conditions (2) and (3) for standard problems. The method of solving this type of problem is illustrated in the following example.

EXAMPLE

6 Minimize $C = -2x - 3y$ subject to

$$5x + 4y \leq 32$$
$$x + 2y \leq 10$$
$$x \geq 0, y \geq 0$$

SOLUTION This problem involves the minimization of the objective function and is accordingly a nonstandard linear programming problem. Note, however, that all other conditions for a regular linear programming problem hold true. To solve a problem of this type we observe that minimizing the objective function C is equivalent to maximizing the objective function $P = -C$. Thus, the solution to this problem may be found by solving the following associated standard linear programming problem: Maximize $P = 2x + 3y$ subject to the given constraints. Using the simplex method with u and v as slack variables, we obtain the following sequence of simplex tableaus:

	x	y	u	v	P	Constant	ratio
	5	4	1	0	0	32	$32/4 = 8$
pivot row →	1	②	0	1	0	10	$10/2 = 5$
	-2	-3	0	0	1	0	

$$\underset{\substack{\uparrow \\ \text{pivot} \\ \text{column}}}{}$$

	x	y	u	v	P	Constant
$\frac{1}{2}R_2$ →	5	4	1	0	0	32
	½	①	0	½	0	5
	-2	-3	0	0	1	0

	x	y	u	v	P	Constant	ratio
pivot row →	③	0	1	-2	0	12	$\frac{12}{3} = 4$
$\xrightarrow{\substack{R_1 - 4R_2 \\ R_3 + 3R_2}}$	½	1	0	½	0	5	$\frac{5}{1/2} = 10$
	$-\frac{1}{2}$	0	0	³⁄₂	1	15	

$$\underset{\substack{\uparrow \\ \text{pivot} \\ \text{column}}}{}$$

	x	y	u	v	P	Constant
$\frac{1}{3}R_1$ →	①	0	⅓	$-\frac{2}{3}$	0	4
	½	1	0	½	0	5
	$-\frac{1}{2}$	0	0	³⁄₂	1	15

	x	y	u	v	P	Constant
$\xrightarrow{\substack{R_2 - \frac{1}{2}R_1 \\ R_3 + \frac{1}{2}R_1}}$	1	0	⅓	$-\frac{2}{3}$	0	4
	0	1	$-\frac{1}{6}$	⅚	0	3
	0	0	⅙	⅞	1	17

The last tableau is in final form. The solution to the standard linear programming problem associated with the given linear programming problem is $x = 4$, $y = 3$, and $P = 17$, so the required solution is given by $x = 4$, $y = 3$, and $C = -17$. You may verify that the solution is correct by using the method of corners. ◀

▶ The Dual Problem

Another special class of nonstandard linear programming problems we encounter in practical applications is characterized by the following conditions:

1. The objective function is to be *minimized.*

2. All the variables involved are nonnegative.

3. Each linear constraint may be written so that the expression involving the variables is *greater than* or equal to a constant.

A convenient method for solving this type of problem is based on the following observation. Each maximization linear programming problem is associated with a minimization problem, and vice-versa. For the purpose of identification, the given problem is called the **primal problem,** whereas the problem related to it is called the **dual problem.** The following example illustrates the technique for constructing the dual of a given linear programming problem.

EXAMPLE

7 Write the dual problem associated with the following problem:
Minimize the objective function $C = 6x + 8y$ subject to the constraints

$$40x + 10y \geq 2400$$
$$10x + 15y \geq 2100$$
$$5x + 15y \geq 1500$$
$$x \geq 0, y \geq 0$$

SOLUTION We first write down the following tableau for the given (primal) problem:

x	y	Constant
40	10	2400
10	15	2100
5	15	1500
6	8	

Next, we interchange the columns and rows of the above tableau and head the three columns of the resulting array with the three variables u, v, and w, obtaining the tableau

u	v	w	Constant
40	10	5	6
10	15	15	8
2400	2100	1500	

Interpreting the last tableau as if it were part of the initial simplex tableau for a standard linear programming problem, with the exception that the signs of the coefficients pertaining to the objective function are not reversed, we construct the required (dual) problem as follows:

Maximize the objective function $P = 2400u + 2100v + 1500w$ subject to the constraints

$$40u + 10v + 5w \leq 6$$
$$10u + 15v + 15w \leq 8$$

where $u \geq 0$, $v \geq 0$, and $w \geq 0$. ◀

The connection between the solution of the primal problem and that of the dual problem is given by the following theorem. The theorem, attributed to John von Neumann (1903–1957), is stated without proof.

Theorem 1

A primal problem has a solution if and only if the corresponding dual problem has a solution. Furthermore, if a solution exists, then

1. The objective functions of both the primal and the dual problem attain the same optimal value.

2. The optimal solution to the primal problem appears under the slack variables in the last row of the final simplex tableau associated with the dual problem.

Armed with this theorem, we will solve the problem posed in Example 7.

EXAMPLE

8 Complete the solution to the problem posed in Example 7.

SOLUTION

Observe that the dual problem associated with the given (primal) problem is a standard linear programming problem. The solution may thus be found using the simplex algorithm. Introducing the slack variables x and y, we obtain the following system of linear equations:

$$
\begin{aligned}
40u + 10v + 5w + x &= 6 \\
10u + 15v + 15w + y &= 8 \\
-2400u - 2100v - 1500w + P &= 0
\end{aligned}
$$

Continuing with the simplex algorithm, we obtain the following sequence of simplex tableaus:

	u	v	w	x	y	P	Constant	ratio
pivot row $\rightarrow$	(40)	10	5	1	0	0	6	$6/40 = 3/20$
	10	15	15	0	1	0	8	$8/10 = 4/5$
	-2400	-2100	-1500	0	0	1	0	

$\uparrow$ pivot column

$\frac{1}{40}R_1$	u	v	w	x	y	P	Constant
$\longrightarrow$	(1)	$1/4$	$1/8$	$1/40$	0	0	$3/20$
	10	15	15	0	1	0	8
	-2400	-2100	-1500	0	0	1	0

	u	v	w	x	y	P	Constant	ratio
$R_2 - 10R_1$	1	$1/4$	$1/8$	$1/40$	0	0	$3/20$	$\frac{3/20}{1/4} = \frac{3}{5}$
$R_3 + 2400R_1$	0	$(25/2)$	$55/4$	$-1/4$	1	0	$13/2$	$\frac{13/2}{25/2} = \frac{13}{25}$
	0	-1500	-1200	60	0	1	360	

$\frac{2}{25}R_2$	u	v	w	x	y	P	Constant
$\longrightarrow$	1	$1/4$	$1/8$	$1/40$	0	0	$3/20$
	0	(1)	$11/10$	$-1/50$	$2/25$	0	$13/25$
	0	-1500	-1200	60	0	1	360

	u	v	w	x	y	P	Constant
$R_1 - \frac{1}{4}R_2$	1	0	$-3/20$	$3/100$	$-1/50$	0	$1/50$
$R_3 + 1500R_2$	0	1	$11/10$	$-1/50$	$2/25$	0	$13/25$
	0	0	450	30	120	1	1140

Solution for the primal problem

The last tableau is final. Theorem 1 tells us that the solution to the primal problem is $x = 30$ and $y = 120$ with a minimum value for C of 1140. Observe that the solution to the dual (maximization) problem may be read off from the simplex tableau in the usual manner: $u = 1/50$, $v = 13/25$, $w = 0$, and $P = 1140$. Note that the maximum value of P is equal to the minimum value of C as guaranteed by Theorem 1. The solution to the primal problem agrees with the solution of the same problem solved in Section 3.3, Example 12 (page 174) using the method of corners.   ◀

▶ Application

Our last example illustrates how the warehouse problem posed in Section 3.2 may be solved by duality.

EXAMPLE

9 Complete the solution to the warehouse problem given in Section 3.2, Example 10 (page 163).

$$\text{Minimize } C = 20x_1 + 8x_2 + 10x_3 + 12x_4 + 22x_5 + 18x_6 \text{ subject to} \qquad (13)$$

$$
\begin{aligned}
x_1 + x_2 + x_3 & & & \le 400 \\
& x_4 + x_5 + x_6 & & \le 600 \\
x_1 & \quad + x_4 & & \ge 200 \\
x_2 & \quad + x_5 & & \ge 300 \\
x_3 & \quad + x_6 & & \ge 400 \\
\end{aligned}
$$

$$x_1 \ge 0,\ x_2 \ge 0,\ \ldots,\ x_6 \ge 0 \qquad (14)$$

SOLUTION Upon multiplying each of the first two inequalities of (14) by -1, we obtain the following equivalent system of constraints in which each of the expressions involving the variables is greater than or equal to a constant:

$$
\begin{aligned}
-x_1 - x_2 - x_3 & & & \ge -400 \\
& -x_4 - x_5 - x_6 & & \ge -600 \\
x_1 & \quad + x_4 & & \ge \quad 200 \\
x_2 & \quad + x_5 & & \ge \quad 300 \\
x_3 & \quad + x_6 & & \ge \quad 400 \\
\end{aligned}
$$

The problem may now be solved by duality. First we write the array of numbers

x_1	x_2	x_3	x_4	x_5	x_6	Constant
-1	-1	-1	0	0	0	-400
0	0	0	-1	-1	-1	-600
1	0	0	1	0	0	200
0	1	0	0	1	0	300
0	0	1	0	0	1	400
20	8	10	12	22	18	

Interchanging the rows and columns of this array of numbers and heading the five columns of the resulting array of numbers by the variables u_1, u_2, u_3, u_4, and u_5 leads to the following:

u_1	u_2	u_3	u_4	u_5	Constant
-1	0	1	0	0	20
-1	0	0	1	0	8
-1	0	0	0	1	10
0	-1	1	0	0	12
0	-1	0	1	0	22
0	-1	0	0	1	18
-400	-600	200	300	400	

from which we construct the associated dual problem: Maximize $P = -400u_1 - 600u_2 + 200u_3 + 300u_4 + 400u_5$ subject to

$$
\begin{aligned}
-u_1 \qquad\quad + u_3 \qquad\qquad &\le 20 \\
-u_1 \qquad\qquad\quad + u_4 \qquad &\le 8 \\
-u_1 \qquad\qquad\qquad + u_5 &\le 10 \\
-u_2 + u_3 \qquad\qquad &\le 12 \\
-u_2 \qquad + u_4 \qquad &\le 22 \\
-u_2 \qquad\qquad + u_5 &\le 18 \\
u_1 \ge 0, u_2 \ge 0, \ldots, u_5 \ge 0
\end{aligned}
$$

Solving the regular linear programming problem by the simplex algorithm, we obtain the following sequence of tableaus ($x_1, x_2, \ldots, x_6$ are slack variables):

	u_1	u_2	u_3	u_4	u_5	x_1	x_2	x_3	x_4	x_5	x_6	P	Constant	ratio
	-1	0	1	0	0	1	0	0	0	0	0	0	20	—
pivot	-1	0	0	1	0	0	1	0	0	0	0	0	8	—
row →	-1	0	0	0	①	0	0	1	0	0	0	0	10	10
	0	-1	1	0	0	0	0	0	1	0	0	0	12	—
	0	-1	0	1	0	0	0	0	0	1	0	0	22	—
	0	-1	0	0	1	0	0	0	0	0	1	0	18	18
	400	600	-200	-300	-400	0	0	0	0	0	0	1	0	

↑
pivot
column

	u_1	u_2	u_3	u_4	u_5	x_1	x_2	x_3	x_4	x_5	x_6	P	Constant	ratio
	-1	0	1	0	0	1	0	0	0	0	0	0	20	—
pivot row →	-1	0	0	①	0	0	1	0	0	0	0	0	8	8
	-1	0	0	0	1	0	0	1	0	0	0	0	10	—
$R_6 - R_3$	0	-1	1	0	0	0	0	0	1	0	0	0	12	—
$R_7 + 400R_3$	0	-1	0	1	0	0	0	0	0	1	0	0	22	22
	1	-1	0	0	0	0	0	-1	0	0	1	0	8	—
	0	600	-200	-300	0	0	0	400	0	0	0	1	4,000	

↑
pivot
column

	u_1	u_2	u_3	u_4	u_5	x_1	x_2	x_3	x_4	x_5	x_6	P	Constant	ratio
	-1	0	1	0	0	1	0	0	0	0	0	0	20	—
	-1	0	0	1	0	0	1	0	0	0	0	0	8	—
$R_5 - R_2$	-1	0	0	0	1	0	0	1	0	0	0	0	10	—
$R_7 + 300R_2$	0	-1	1	0	0	0	0	0	1	0	0	0	12	—
	1	-1	0	0	0	0	-1	0	0	1	0	0	14	14
pivot row →	①	-1	0	0	0	0	0	-1	0	0	1	0	8	8
	-300	600	-200	0	0	0	300	400	0	0	0	1	6,400	

↑
pivot
column

	u_1	u_2	u_3	u_4	u_5	x_1	x_2	x_3	x_4	x_5	x_6	P	Constant	ratio
	0	-1	1	0	0	1	0	-1	0	0	1	0	28	28
$R_1 + R_6$	0	-1	0	1	0	0	1	-1	0	0	1	0	16	—
$R_2 + R_6$	0	-1	0	0	1	0	0	0	0	0	1	0	18	—
$R_3 + R_6$ →	0	-1	①	0	0	0	0	0	0	1	0	0	12	12
$R_5 - R_6$	0	0	0	0	0	0	-1	1	0	1	-1	0	6	—
$R_7 + 300R_6$ pivot row	1	-1	0	0	0	0	0	-1	0	0	1	0	8	—
	0	300	-200	0	0	0	300	100	0	0	300	1	8,800	

↑
pivot
column

	u_1	u_2	u_3	u_4	u_5	x_1	x_2	x_3	x_4	x_5	x_6	P	Constant
	0	0	0	0	0	1	0	-1	-1	0	1	0	16
	0	-1	0	1	0	0	1	-1	0	0	1	0	16
$R_1 - R_4$	0	-1	0	0	1	0	0	0	0	0	1	0	18
$R_7 + 200R_4$	0	-1	1	0	0	0	0	0	1	0	0	0	12
	0	0	0	0	0	0	-1	1	0	1	-1	0	6
	1	-1	0	0	0	0	0	-1	0	0	1	0	8
	0	100	0	0	0	0	300	100	200	0	300	1	11,200

The last tableau is final and we find that

$$x_1 = 0, \quad x_2 = 300, \quad x_3 = 100, \quad x_4 = 200, \quad x_5 = 0, \quad x_6 = 300,$$

and
$$P = 11{,}200$$

Thus, to minimize shipping costs, Acrosonic should ship 300 loudspeaker systems from plant I to warehouse B, 100 systems from plant I to warehouse C, 200 systems from plant II to warehouse A, and 300 systems from plant II to warehouse C at a total cost of $11,200.   ◀

In this section we have touched on two important classes of nonstandard linear programming problems. The solution of other classes of nonstandard linear programming problems may be found in more advanced texts on linear programming.

Self-Check Exercises
4.2

1. Write the dual problem associated with the following problem: Minimize $C = 2x + 5y$ subject to

$$
\begin{aligned}
4x + y &\geq 40 \\
2x + y &\geq 30 \\
x + 3y &\geq 30 \\
x \geq 0, \, y &\geq 0
\end{aligned}
$$

2. Solve the (primal) problem posed in Exercise 1.

Solutions to Self-Check Exercises 4.2 can be found on page 223.

▶
Exercises 4.2

In Exercises 1–6, use the technique developed in this section to solve each of the following nonstandard linear programming problems.

1. Minimize $C = -2x + y$ subject to

$$
\begin{aligned}
x + 2y &\leq 6 \\
3x + 2y &\leq 12 \\
x \geq 0, \, y &\geq 0
\end{aligned}
$$

2. Minimize $C = -2x - 3y$ subject to

$$3x + 4y \leq 24$$
$$7x - 4y \leq 16$$
$$x \geq 0, y \geq 0$$

3. Minimize $C = -3x - 2y$ subject to the constraints of Exercise 2.

4. Minimize $C = x - 2y + z$ subject to

$$x - 2y + 3z \leq 10$$
$$2x + y - 2z \leq 15$$
$$2x + y + 3z \leq 20$$
$$x \geq 0, y \geq 0, z \geq 0$$

5. Minimize $C = 2x - 3y - 4z$ subject to

$$-x + 2y - z \leq 8$$
$$x - 2y + 2z \leq 10$$
$$2x + 4y - 3z \leq 12$$
$$x \geq 0, y \geq 0, z \geq 0$$

6. Minimize $C = -3x - 2y - z$ subject to the constraints of Exercise 5.

In Exercises 7–16, construct the dual problem associated with the given (primal) problem. Solve the (primal) problem.

7. Minimize $C = 2x + 5y$ subject to

$$x + 2y \geq 4$$
$$3x + 2y \geq 6$$
$$x \geq 0, y \geq 0$$

8. Minimize $C = 3x + 2y$ subject to

$$2x + 3y \geq 90$$
$$3x + 2y \geq 120$$
$$x \geq 0, y \geq 0$$

9. Minimize $C = 6x + 4y$ subject to

$$6x + y \geq 60$$
$$2x + y \geq 40$$
$$x + y \geq 30$$
$$x \geq 0, y \geq 0$$

10. Minimize $C = 10x + y$ subject to

$$4x + y \geq 16$$
$$x + 2y \geq 12$$
$$x \geq 2$$
$$x \geq 0, y \geq 0$$

11. Minimize $C = 200x + 150y + 120z$ subject to

$$20x + 10y + z \geq 10$$
$$x + y + 2z \geq 20$$
$$x \geq 0, y \geq 0, z \geq 0$$

12. Minimize $C = 40x + 30y + 11z$ subject to

$$2x + y + z \geq 8$$
$$x + y - z \geq 6$$
$$x \geq 0, y \geq 0, z \geq 0$$

13. Minimize $C = 6x + 8y + 4z$ subject to

$$x + 2y + 2z \geq 10$$
$$2x + y + z \geq 24$$
$$x + y + z \geq 16$$
$$x \geq 0, y \geq 0, z \geq 0$$

14. Minimize $C = 12z + 4y + 8z$ subject to

$$2x + 4y + z \geq 6$$
$$3x + 2y + 2z \geq 2$$
$$4x + y + z \geq 2$$
$$x \geq 0, y \geq 0, z \geq 0$$

15. Minimize $C = 30x + 12y + 20z$ subject to

$$2x + 4y + 3z \geq 6$$
$$6x + z \geq 2$$
$$6y + 2z \geq 4$$
$$x \geq 0, y \geq 0, z \geq 0$$

16. Minimize $C = 8x + 6y + 4z$ subject to

$$2x + 3y + z \geq 6$$
$$x + 2y - 2z \geq 4$$
$$x + y + 2z \geq 2$$
$$x \geq 0, y \geq 0, z \geq 0$$

17. The Acrosonic Company also manufactures a model G loudspeaker system in plant I and plant II. The output at plant I is at most 800 per month, and the output at plant II is at most 600 per month. Model G loudspeaker systems are also shipped to the three warehouses—A, B, and C—whose minimum monthly requirements are 500, 400, and 400, respectively. Shipping costs from plant I to warehouse A, warehouse B, and warehouse C are $16, $20, and $22 per loudspeaker system, respectively, and shipping costs from plant II to each of these warehouses are $18, $16, and $14, respectively. What shipping schedule will enable Acrosonic to meet the requirements of the warehouses while keeping its shipping costs to a minimum?

18. The Steinwelt Piano Company manufactures uprights and consoles in two plants, plant I and plant II. The output of plant I is at most 300 per month, and the output of plant II is at most 250 per month. These pianos are shipped to three warehouses that

serve as distribution centers for Steinwelt. In order to fill current and projected future orders, warehouse A requires a minimum of 200 pianos per month, warehouse B requires at least 150 pianos per month, and warehouse C requires at least 200 pianos per month. The shipping cost of each piano from plant I to warehouse A, warehouse B, and warehouse C is $60, $60, and $80, respectively, and the shipping cost of each piano from plant II to warehouse A, warehouse B, and warehouse C is $80, $70, and $50, respectively. What shipping schedule will enable Steinwelt to meet the requirements of the warehouses while keeping the shipping costs to a minimum?

19. The owner of the Health Juice-Bar wishes to prepare a low-calorie fruit juice with a high vitamin A and C content by blending orange juice and pink-grapefruit juice. Each glass of the blended juice is to contain at least 1200 International Units (I.U.) of vitamin A and 200 I.U. of vitamin C. One ounce of orange juice contains 60 I.U. of vitamin A, 16 I.U. of vitamin C, and 14 calories; each ounce of pink-grapefruit juice contains 120 I.U. of vitamin A, 12 I.U. of vitamin C, and 11 calories. How many ounces of each juice should a glass of the blend contain if it is to meet the minimum vitamin requirements and at the same time contain a minimum number of calories?

20. An oil company operates two refineries in a certain city. Refinery I has an output of 200, 100, and 100 barrels of low-, medium-, and high-grade oil per day, respectively. Refinery II has an output of 100, 200, and 600 barrels of low-, medium-, and high-grade oil per day, respectively. The company wishes to produce at least 1000, 1400, and 3000 barrels of low-, medium-, and high-grade oil to fill an order. If it costs $200 per day to operate refinery I and $300 per day to operate refinery II, determine how many days each refinery should be operated to meet the requirements of the order at minimum cost to the company.

Solutions to Self-Check Exercises

4.2

1. We first write down the following tableau for the given (primal) problem:

x	y	Constant
4	1	40
2	1	30
1	3	30
2	5	0

Next, we interchange the columns and rows of the tableau and head the three columns of the resulting array with the three variables u, v, and w, obtaining the tableau

u	v	w	Constant
4	2	1	2
1	1	3	5
40	30	30	0

Interpreting the last tableau as if it were the initial tableau for a standard linear programming problem with the exception that the signs of the coefficients pertaining to the objective function are not reversed, we construct the required (dual) problem as follows: Maximize $P = 40u + 30v + 30w$ subject to

$$4u + 2v + w \le 2$$
$$u + v + 3w \le 5$$
$$u \ge 0, v \ge 0, w \ge 0$$

2. We introduce slack variables x and y to obtain the following system of linear equations

$$4u + 2v + w + x = 2$$
$$u + v + 3w + y = 5$$
$$-40u - 30v - 30w + P = 0$$

Using the simplex algorithm, we obtain the following sequence of simplex tableaus:

	u	v	w	x	y	P	Constant	ratio	
pivot row →	④	2	1	1	0	0	2	$2/4 = 1/2$	$\frac{1}{4}R_1$
	1	1	3	0	1	0	5	$5/1 = 5$	→
	-40	-30	-30	0	0	1	0		

↑ pivot column

	u	v	w	x	y	P	Constant	
	①	$1/2$	$1/4$	$1/4$	0	0	$1/2$	$R_2 - R_1$
	1	1	3	0	1	0	5	$R_3 + 40R_1$
	-40	-30	-30	0	0	1	0	→

	u	v	w	x	y	P	Constant	ratio	
	1	$1/2$	$1/4$	$1/4$	0	0	$1/2$	$\frac{1/2}{1/4} = 2$	$\frac{4}{11}R_2$
pivot row →	0	$1/2$	⑪⁄₄	$-1/4$	1	0	$9/2$	$\frac{9/2}{11/4} = \frac{18}{11}$	→
	0	-10	-20	10	0	1	20		

↑ pivot column

	u	v	w	x	y	P	Constant	
	1	$1/2$	$1/4$	$1/4$	0	0	$1/2$	$R_1 - \frac{1}{4}R_2$
	0	$2/11$	①	$-1/11$	$4/11$	0	$18/11$	$R_3 + 20R_2$
	0	-10	-20	10	0	1	20	→

	u	v	w	x	y	P	Constant	ratio	
pivot row →	1	⑤⁄₁₁	0	³⁄₁₁	$-\frac{1}{11}$	0	¹⁄₁₁	$\frac{1/11}{5/11} = \frac{1}{5}$	$\frac{11}{5}R_1 \longrightarrow$
	0	²⁄₁₁	1	$-\frac{1}{11}$	⁴⁄₁₁	0	¹⁸⁄₁₁	$\frac{18/11}{2/11} = 9$	
	0	$-\frac{70}{11}$	0	⁹⁰⁄₁₁	⁸⁰⁄₁₁	1	⁵⁸⁰⁄₁₁		

↑
pivot
column

	u	v	w	x	y	P	Constant	
	¹¹⁄₅	①	0	³⁄₅	$-\frac{1}{5}$	0	¹⁄₅	$R_2 - \frac{2}{11}R_1$
	0	²⁄₁₁	1	$-\frac{1}{11}$	⁴⁄₁₁	0	¹⁸⁄₁₁	$\overrightarrow{}$
	0	$-\frac{70}{11}$	0	⁹⁰⁄₁₁	⁸⁰⁄₁₁	1	⁵⁸⁰⁄₁₁	$R_3 + \frac{70}{11}R_1$

	u	v	w	x	y	P	Constant
	¹¹⁄₅	1	0	³⁄₅	$-\frac{1}{5}$	0	¹⁄₅
	$-\frac{2}{5}$	0	1	$-\frac{1}{5}$	²⁄₅	0	⁸⁄₅
	14	0	0	12	6	1	54

solution for the
primal problem

The last tableau is final, and the solution to the primal problem is $x = 12$ and $y = 6$ with a minimum value for C of 54.

Chapter 4 Review Exercises

In Exercises 1–6, use the simplex method to solve the given linear programming problem.

1. Maximize $P = 3x + 4y$ subject to

$$x + 3y \leq 15$$
$$4x + y \leq 16$$
$$x \geq 0, y \geq 0$$

2. Maximize $P = 2x + 5y$ subject to

$$2x + y \leq 16$$
$$2x + 3y \leq 24$$
$$y \leq 6$$
$$x \geq 0, y \geq 0$$

3. Maximize $P = 2x + 3y + 5z$ subject to

$$x + 2y + 3z \leq 12$$
$$x - 3y + 2z \leq 10$$
$$x \geq 0, y \geq 0, z \geq 0$$

4. Maximize $P = x + 2y + 3z$ subject to

$$2x + y + z \leq 14$$
$$3x + 2y + 4z \leq 24$$
$$2x + 5y - 2z \leq 10$$
$$x \geq 0, y \geq 0, z \geq 0$$

5. Minimize $C = 24x + 18y + 24z$ subject to

$$3x + 2y + z \geq 4$$
$$x + y + 3z \geq 6$$
$$x \geq 0, y \geq 0, z \geq 0$$

6. Minimize $C = 4x + 2y + 6z$ subject to

$$x + 2y + z \geq 4$$
$$2x + y + 2z \geq 2$$
$$3x + 2y + z \geq 3$$
$$x \geq 0, y \geq 0, z \geq 0$$

7. A company manufactures three products, A, B, and C, on two machines, I and II. It has been determined that the company will realize a profit of $4 on each unit of product A, $6 on each unit of product B, and $8 on each unit of product C. To manufacture a unit of product A requires 9 minutes on machine I and 6 minutes on machine II; to manufacture a unit of product B requires 12 minutes on machine I and 6 minutes on machine II; and to manufacture a unit of product C requires 18 minutes on machine I and 10 minutes on machine II. There are 6 hours of machine time available on machine I and 4 hours of machine time available on machine II in each work shift. How many units of each product should be produced in each shift in order to maximize the company's profit?

8. Mr. Moody has decided to invest at most $100,000 in securities in the form of corporate stocks. He has classified his options into three groups of stocks: "blue-chip" stocks that he assumes will yield a 10 percent return (dividends and capital appreciation) within a year, growth stocks that he assumes will yield a 15 percent return within a year, and speculative stocks that he assumes will yield a 20 percent return (mainly due to capital appreciation) within a year. Because of the relative risks involved in his investment, Mr. Moody has further decided that no more than 30 percent of his investment should be in growth and speculative stocks and at least 50 percent of his investment should be in "blue-chip" and speculative stocks. Determine how much Mr. Moody should invest in each group of stocks in the hope of maximizing the return on his investments.

Mathematics of Finance

How much is the home mortgage payment? The Blakelys received a bank loan to help finance the purchase of a house. They have agreed to repay the loan in equal monthly installments over a certain period of time. In Example 16, page 253, we will show how to determine the size of the monthly installment so that the loan is fully amortized at the end of the term.

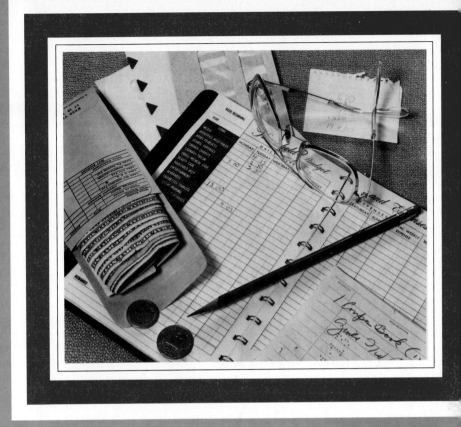

▶CHAPTER

FIVE

5.1

Compound Interest

▶ Compound Interest

▶ Effective Rate of Interest

▶ Present Value

▶ Applications

▶ Compound Interest

We discussed simple interest in Section 1.3. Recall that in calculating simple interest, the amount of interest is computed on the original principal. Frequently, however, the interest earned is periodically added to the principal and thereafter earns interest itself at the same rate. This is called **compound interest.**

If the original principal is P dollars and if the interest is compounded annually at the rate of r per year, then at the end of the first year the new principal, or *accumulated value*, is given by

$$A_1 = P + Pr = P(1 + r).$$

The interest earned in the second year is computed on A_1, so at the end of the second year, the accumulated amount is

$$A_2 = A_1(1 + r) = P(1 + r)^2,$$

and at the end of the third year, the accumulated amount is

$$A_3 = A_2(1 + r) = P(1 + r)^3.$$

Letting A_0 denote the original principal P, we see that the accumulated amount at the end of the nth year (n, a positive integer) is given by

$$A_n = A_0(1 + r)^n = P(1 + r)^n \tag{1}$$

EXAMPLE

1   An amount of $1000 is deposited in a bank that pays 8 percent interest per year, compounded annually. What is the total amount on deposit at the end of three years, assuming that no deposits or withdrawals were made during those three years? What is the interest earned in that period of time?

SOLUTION   Using Equation (1) with $P = 1000$, $r = 0.08$, and $n = 3$, we have

$$A_3 = 1000(1.08)^3$$
$$= 1259.71,$$

or $1259.71. The interest earned over the three-year period is given by

$$1259.71 - 1000 = 259.71,$$

or $259.71. Compare this result with Example 15, Chapter 1.   ◀

Using a similar argument, we can easily derive a formula for the case in which interest is compounded more than once a year. To this end, suppose that interest at the rate of r per year (called the **stated rate,** or **nominal rate**) is compounded m times a year on a principal of P dollars. The interval of time between successive interest calculations is called the **conversion period,** and the (simple) interest rate per conversion period is given by $i = r/m$ per period. Therefore, the accumulated amount at the end of the first period is given by

$$A_1 = P + Pi = P(1 + i).$$

At the end of the second period the accumulated amount is

$$A_2 = A_1(1 + i) = P(1 + i)^2,$$

at the end of the third period the accumulated amount is

$$A_3 = A_2(1 + i) = P(1 + i)^3,$$

and at the end of the nth period the accumulated amount is

$$A_n = A_{n-1}(1 + i) = P(1 + i)^n.$$

Thus, we are led to the **compound interest formula:**

Compound Interest Formula

If P dollars are invested at a nominal interest rate of r per year, compounded m times a year, then at the end of each n conversion periods the accumulated amount is given by

$$A_n = P(1 + i)^n \qquad (2)$$

where $i = r/m$. Normally, the life of a loan is stated in terms of years (say, t years), in which case the number of conversion periods n is given by $n = tm$.

EXAMPLE

2 Find the accumulated amount after three years if $1000 is invested at 8 percent per year compounded (a) annually, (b) semiannually, (c) quarterly, and (d) monthly.

SOLUTION

a. Here $P = 1000$, $r = 0.08$, and $m = 1$. Thus, $i = r = 0.08$ and $n = 3$, so (2) gives

$$A_3 = 1000(1.08)^3$$
$$= 1259.71,$$

or $1259.71.

b. Here $P = 1000$, $r = 0.08$, and $m = 2$. Thus, $i = 0.08/2 = 0.04$ and $n = (3)(2) = 6$, so that (2) gives

$$A_6 = 1000(1.04)^6$$
$$= 1265.32,$$

or $1265.32.

c. In this case, $P = 1000$, $r = 0.08$, and $m = 4$. Thus, $i = 0.08/4 = 0.02$ and $n = (3)(4) = 12$, so (2) gives

$$A_{12} = 1000(1.02)^{12}$$
$$= 1268.24,$$

or $1268.24.

d. Here $P = 1000$, $r = 0.08$, and $m = 12$. Thus, $i = 0.08/12 = 0.0067$ and $n = (3)(12) = 36$, so (2) gives

$$A_{36} = 1000(1.0067)^{36}$$
$$= 1271.75,$$

or $1271.75. These results are summarized in Table 5.1.

Nominal Rate (r)	Conversion Period	Interest Rate/ Conversion Period	Initial Investment	Accumulated Amount
8%	Annual ($m = 1$)	8%	$1000	$1259.71
8%	Semiannual ($m = 2$)	4%	$1000	$1265.32
8%	Quarterly ($m = 4$)	2%	$1000	$1268.24
8%	Monthly ($m = 12$)	2/3%	$1000	$1271.75

Table 5.1

◀

▶ Effective Rate of Interest

In the last example, we saw that the interest actually earned on the investment depended on the frequency with which the interest was compounded. Thus, the stated, or nominal, rate of 8 percent per year did not reflect the actual rate at which interest was earned. Now, the **effective rate** (also called the annual percentage rate, abbreviated APR) is the rate that when compounded annually yields the same accumulated amount as the nominal rate compounded m times a year.

To derive a relationship between the nominal interest rate r per year compounded m times a year and its corresponding effective rate R per year, let us suppose that a principal of P dollars is invested for t years and earns interest at the stated rate. Then by definition, the accumulated amount obtained by compounding the interest annually at the effective rate R should be the same as the accumulated amount obtained by compounding the interest m times a year. Thus,

$$P(1 + R)^t = P\left(1 + \frac{r}{m}\right)^{tm}.$$

Dividing both sides of the equation by P and taking the tth root of both sides of the resulting equation yields

$$1 + R = \left(1 + \frac{r}{m}\right)^m.$$

Solving for R, we obtain the formula for computing the effective rate of interest:

Effective Rate of Interest Formula

$$R = \left(1 + \frac{r}{m}\right)^m - 1 \tag{3}$$

where
R = effective rate of interest
r = nominal interest rate per year
m = number of conversion periods per year

EXAMPLE

3 Find the effective rate of interest corresponding to a nominal rate of 8 percent per year compounded (a) annually, (b) semiannually, (c) quarterly, and (d) monthly.

SOLUTION *a.* The effective rate of interest corresponding to a nominal rate of 8 percent per year compounded annually is, of course, given by 8 percent per year. This result is also confirmed by using (3) with $r = 0.08$ and $m = 1$.

Thus, $\qquad\qquad\qquad R = (1 + 0.08) - 1 = 0.08,$

so the effective rate required is 8 percent per year.

b. Let $r = 0.08$ and $m = 2$. Then (3) yields

$$R = \left(1 + \frac{0.08}{2}\right)^2 - 1$$
$$= (1.04)^2 - 1$$
$$= 0.0816,$$

so the required effective rate is 8.16 percent per year.

c. Let $r = 0.08$ and $m = 4$. Then (3) yields

$$R = \left(1 + \frac{0.08}{4}\right)^4 - 1$$
$$= (1.02)^4 - 1$$
$$= 0.08243,$$

so the corresponding effective rate in this case is 8.243 percent per year.

d. Let $r = 0.08$ and $m = 12$. Then (3) yields

$$R = \left(1 + \frac{0.08}{12}\right)^{12} - 1$$
$$= (1.0067)^{12} - 1$$
$$= 0.08343,$$

so the corresponding effective rate in this case is 8.343 percent per year. ◀

Now, if the effective rate of interest R is known, the accumulated amount after n years on an investment of P dollars may be more readily computed by using the formula

$$A_n = P(1 + R)^n.$$

The Truth in Lending Act passed by Congress in 1968 requires that the effective rate of interest be disclosed in all contracts involving interest charges. The passage of this act has benefited consumers because they now have a common basis for comparing the various nominal rates quoted by different financial institutions. Furthermore, knowing the effective rate enables consumers to compute the actual charges involved in a transaction. Thus, if the effective rates of interest found in the last example were known, the accumulated values of Example 2, shown in Table 5.2, could have been readily found.

Nominal Rate	Frequency of Interest Payment	Effective Rate	Initial Investment	Accumulated Amount After 3 Years	
8%	Annually	8%	$1000	$1000(1 + 0.08)^3$	= $1259.71
8%	Semiannually	8.16%	$1000	$1000(1 + 0.0816)^3$	= $1265.32
8%	Quarterly	8.243%	$1000	$1000(1 + 0.08243)^3$	= $1268.23
8%	Monthly	8.343%	$1000	$1000(1 + 0.08343)^3$	= $1271.75

Table 5.2

▶ Present Value

Let us return to Equation (2), which expresses the accumulated amount at the end of n periods when interest at the rate of r is compounded m times a year. The principal P in (2) is often referred to as the **present value,** and the accumulated value A_n is called the **future value,** since it is realized at a future date. In certain instances, an investor may wish to determine how much money he should invest now, at a fixed rate of interest, so that he will realize a certain sum at some future date. This problem may be solved by expressing P in terms of A_n. Thus, from (2) we find

$$P = A_n(1 + i)^{-n}.$$

Here, as before, $i = r/m$ where m is the number of conversion periods per year.

Present Value Formula for Compound Interest

$$P = A_n(1 + i)^{-n} \tag{4}$$

EXAMPLE

4 Find how much money should be deposited in a bank paying interest at the rate of 6 percent per year compounded monthly so that at the end of three years the accumulated amount will be $20,000.

SOLUTION Here, $r = 0.06$ and $m = 12$, so $i = 0.06/12 = 0.005$ and $n = (3)(12) = 36$. Thus, the problem is to determine P given that $A_{36} = 20,000$. Using (4), we obtain

$$P = 20,000(1.005)^{-36}$$
$$= 16,713,$$

or $16,713. ◀

EXAMPLE

5 Find the present value of $49,158.60 due in five years at an interest rate of 10 percent per year compounded quarterly.

SOLUTION Using (4) with $r = 0.1$ and $m = 4$, so that $i = 0.1/4 = 0.025$, $n = (4)(5) = 20$, and $A_{20} = 49,158.6$, we obtain

$$P = (49,158.6)(1.025)^{-20}$$
$$= 30,000,$$

or $30,000. ◀

▶ Applications

The returns on certain investments such as zero coupon certificates of deposit (CDs) and zero coupon bonds are compared by quoting the time it takes for each investment to triple, or even quadruple. These calculations make use of the compound interest formula (2).

EXAMPLE

6 Jane has narrowed her investment options down to two:

a. to purchase a CD that matures in 12 years and pays interest upon maturity at the rate of 12 percent per year compounded daily (assume 360 days in a year)

b. to purchase a zero coupon CD that will quadruple her investment in the same period

Which option will optimize her investment?

SOLUTION   Let us compute the accumulated amount under option (a). Here

$$r = 0.12, m = 360, \text{ and } t = 12,$$

so, $n = 12(360) = 4320$ and $i = 0.12/360 \approx 0.0003333$. The accumulated amount at the end of 12 years (after 4320 conversion periods) is

$$A_{4320} = P(1.0003333)^{4320}$$
$$\approx 4.22P,$$

or $4.22P$ dollars. Had Jane invested in option (b), the accumulated amount of her investment would have been $4P$ dollars at the end of 12 years. Therefore, she should choose option (a).   ◀

EXAMPLE

7 An investment earns interest at the rate of 8 percent per year compounded monthly. Find how long it will take for the investment to double in value (called the **doubling time** for the investment).

SOLUTION   Suppose P dollars are invested initially. Then, our immediate objective is to find n given that

$$A_n = 2P, r = 0.08, \text{ and } m = 12.$$

Now, $i = 0.08/12 = 0.0067$, so (2) gives

$$2P = P(1.0067)^n$$
$$2 = (1.0067)^n$$
$$\log 2 = n \log 1.0067$$
$$n = \frac{\log 2}{\log 1.0067} = \frac{0.30103}{0.0029001} \approx 103.8,$$

or 103.8 conversion periods. Therefore, bearing in mind that $m = 12$, we conclude that the doubling time is 103.8/12, or approximately 8.7 years. ◄

EXAMPLE

8 Diane has an Individual Retirement Account (IRA) with a brokerage firm. Her money is invested in a money market mutual fund that pays interest on a daily basis. Over a nine-month period in which no deposits or withdrawals were made, her account grew from \$4654.74 to \$4934.02. Find the effective rate at which Diane's account is earning interest over this period (assume 360 days in a year).

SOLUTION First we determine the nominal rate of interest r percent (compounded daily). Using (2) with $A = 4934.02$, $P = 4654.74$, $m = 360$, and $t = 3/4$ so that $n = (360)(3/4) = 270$, we have

$$4934.02 = 4654.74(1 + i)^{270}$$
$$(1 + i)^{270} = \frac{4934.02}{4654.74}$$
$$1 + i = \left(\frac{4934.02}{4654.74}\right)^{1/270}$$

or

$$i = 1.00021583 - 1$$
$$= 0.00021583.$$

Therefore,

$$r = mi \quad \text{(Formula 2)}$$
$$= (360)(0.00021583)$$
$$= 0.0777,$$

or 7.77 percent per year. To find the corresponding effective rate of interest, we use (3), which gives

$$R = \left(1 + \frac{.0777}{360}\right)^{360} - 1$$
$$= 0.0808,$$

so the required effective rate is 8.08 percent per year. ◄

Self-Check Exercises

5.1

1. Find the present value of $20,000 due in three years at an interest rate of 12 percent per year compounded monthly.

2. Mr. Baker is a retiree living on social security and the income from his investment. Currently, his $100,000 investment in a one-year CD is yielding 11.6 percent interest compounded daily. If he reinvests the principal ($100,000) on the due date of the CD in another one-year CD paying 9.2 percent interest compounded daily, find the net decrease in his yearly income from his investment.

Solutions to Self-Check Exercises 5.1 can be found on page 240.

▶

Exercises 5.1

In Exercises 1–10, find the accumulated amount A if the principal P is invested at the interest rate of r per year for n years.

1. $P = \$1000$, $r = 7$ percent, $n = 8$, compounded annually

2. $P = \$1000$, $r = 8\frac{1}{2}$ percent, $n = 6$, compounded annually

3. $P = \$2500$, $r = 7$ percent, $n = 10$, compounded semiannually

4. $P = \$2500$, $r = 9$ percent, $n = 10.5$, compounded semiannually

5. $P = \$12,000$, $r = 8$ percent, $n = 10.5$, compounded quarterly

6. $P = \$42,000$, $r = 7\frac{3}{4}$ percent, $n = 8$, compounded quarterly

7. $P = \$150,000$, $r = 14$ percent, $n = 4$, compounded monthly

8. $P = \$180,000$, $r = 9$ percent, $n = 6\frac{1}{4}$, compounded monthly

9. $P = \$150,000$, $r = 12$ percent, $n = 3$, compounded daily

10. $P = \$200,000$, $r = 8$ percent, $n = 4$, compounded daily

In Exercises 11–14, find the effective rate corresponding to the given nominal rate.

11. 10 percent compounded semiannually

12. 9 percent compounded quarterly

13. 8 percent compounded monthly

14. 8 percent compounded daily

In Exercises 15–18, find the present value of $40,000 due in four years at the given rate of interest.

15. 8 percent compounded semiannually

16. 8 percent compounded quarterly

17. 7 percent compounded monthly

18. 9 percent compounded daily

In Exercises 19 and 20, find the time t that it will take for an investment to triple in value at the given rate.

19. 8 percent compounded quarterly   *20.* 9 percent compounded monthly

21. An amount of $25,000 is deposited in a bank that pays interest at the rate of 7 percent per year, compounded annually. What is the total amount on deposit at the end of six years, assuming that there are no deposits or withdrawals during those six years? What is the interest earned in that period of time?

22. If the cost of a semiprivate room in a hospital was $120 per day five years ago, and hospital costs have risen at the rate of 12 percent per year since that time, what rate would you expect to pay for a semiprivate room today?

23. Today a typical family of four spends $400 monthly for food. If inflation occurs at the rate of 9 percent per year over the next six years, how much should the typical family of four expect to spend for food six years from now?

24. The Brennans are planning to buy a house four years from now. Housing experts in their area have estimated that the cost of a home will increase at a rate of 11 percent per year during that period. If this economic prediction holds true, how much can the Brennans expect to pay for a house that currently costs $80,000?

25. A utility company in a western city of the United States expects the consumption of electricity to increase by 8 percent per year during the next decade, due mainly to the expected increase in population. If consumption does increase at this rate, find the amount by which the utility company will have to increase its generating capacity to meet the needs of the area at the end of the decade.

26. Find the accumulated amount after ten years if $10,000 is invested at 9 percent per year compounded (a) annually, (b) semiannually, (c) quarterly, (d) monthly.

27. The managers of a pension fund have invested $1.5 million in U.S. government certificates of deposit that pay interest at the rate of 9.5 percent per year compounded semiannually over a period of ten years. At the end of this period, how much will the investment be worth?

28. A young man is the beneficiary of a trust fund established for him 21 years ago at his birth. If the original amount placed in trust was $10,000, how much will he receive if the money has earned interest at the rate of (a) 8 percent compounded annually? (b) 8 percent compounded quarterly? (c) 8 percent compounded monthly?

29. At what annual rate of interest compounded quarterly should an individual invest his money if he wishes to double his investment within six years?

30. What rate of interest compounded quarterly is needed to triple an investment in ten years?

31. Find how much money should be deposited in a bank paying interest at the rate of 8.5 percent per year compounded quarterly so that at the end of five years the accumulated amount will be $40,000.

32. Find the present value of $80,000 due in four years at an interest rate of 9 percent per year compounded semiannually.

33. An individual purchased a four-year, $10,000 promissory note with an interest rate of 8.5 percent compounded semiannually. How much did she purchase the note for?

34. The parents of a child have just come into a large inheritance and wish to establish a trust fund for the child's college education. If they estimate that they will need $60,000 in 13 years, how much should they set aside in the trust now if they can invest the money at 8½ percent (a) compounded annually? (b) compounded semiannually? (c) compounded quarterly?

35. Mr. Kaplan invested a sum of money five years ago in a savings account that has since paid interest at the rate of 8 percent compounded quarterly. His investment is now worth $22,289.22. How much did he originally invest?

36. The proprietors of The Coachmen Inn secured two loans from the Union Bank: one for $8,000 due in three years and one for $15,000 due in six years, both at an interest rate of 10 percent compounded semiannually. The bank has agreed to allow the two loans to be consolidated into one loan payable in five years at the same interest rate. What amount will the proprietors of the inn be required to pay the bank at the end of five years?

37. Find the effective rate of interest corresponding to a nominal rate of 9 percent per year compounded (a) annually, (b) semiannually, (c) quarterly, (d) monthly.

38. At an annual inflation rate of 7.5 percent, how long will it take the Consumer Price Index (CPI) to double?

39. Ms. Collins purchased a house in 1980 for $60,000. In 1986 she sold the house and made a net profit of $28,000. Find the effective annual rate of return on her investment over the six-year period.

40. Mr. Stevens purchased 1,000 shares of a certain stock for $25,250 (including commissions). He sold the shares two years later and received $32,100 after deducting commissions. Find the effective annual rate of return on his investment over the two-year period.

Solutions to Self-Check Exercises

5.1

1. Using (4) with $r = 0.12$ and $m = 12$, so that

$$i = \frac{0.12}{12} = 0.01, n = (12)(3) = 36, \text{ and } A_{36} = 20,000,$$

we find the required present value to be

$$P = 20,000(1.01)^{-36}$$
$$= 13,978.50,$$

or $13,978.50.

2. The accumulated amount of Mr. Baker's current investment is found by using (2) with $P = 100,000$, $r = 0.116$, and $m = 360$. Thus,

$$i = \frac{0.116}{360} = 0.0003222 \text{ and } n = 360,$$

so the required accumulated amount is

$$A_{360} = 100,000(1.0003222)^{360}$$
$$= 112,296.59,$$

or $112,296.59.

Next, we compute the accumulated amount of Mr. Baker's reinvestment. Once again, using (2) with $P = 100{,}000$, $r = 0.092$, and $m = 360$ so that

$$i = \frac{0.092}{360} = 0.0002556 \text{ and } n = 360,$$

we find the required accumulated amount in this case to be

$$\overline{A}_{360} = 100{,}000(1.0002556)^{360},$$

or $109,636.95. Therefore, Mr. Baker can expect to experience a net decrease in yearly income of

$$112{,}296.59 - 109{,}636.95$$

dollars, or $2,659.64.

5.2

Annuities

▶ Future Value of an Annuity

▶ Present Value of an Annuity

▶ Applications

▶ Future Value of an Annuity

An **annuity** is a sequence of payments made at regular time intervals. The time period in which these payments are made is called the **term** of the annuity. Depending on whether the term is given by a fixed time interval, a time interval that begins at a definite date but extends indefinitely, or one that is not fixed in advance, an annuity is called an **annuity certain,** a **perpetuity,** or a **contingent annuity,** respectively. In general, the payments in an annuity need not be equal, but in many important applications they are equal. In this section, we will assume that annuity payments are equal. Examples of annuities are regular deposits to a savings account, monthly home mortgage payments, and monthly insurance payments.

Annuities are also classified by payment dates. An annuity in which the payments are made at the end of each payment period is called an **ordinary annuity,** whereas an annuity in which the payments are made at the beginning of each period is called an **annuity due.** Furthermore, an annuity in which the payment period coincides with the interest conversion period is called a **simple annuity,** whereas an annuity in which the payment period differs from the interest conversion period is called a **complex annuity.**

In this section we will consider ordinary annuities, which are certain and simple. In other words, we will study annuities that are subject to the following conditions:

1. The terms are given by fixed time intervals.
2. The periodic payments are equal in size.
3. The payments are made at the end of the payment periods.
4. The payment periods coincide with the interest conversion periods.

To find the amount A of an annuity (ordinary, certain, and simple), let

$$P = \text{size of each payment in the annuity}$$
$$r = \text{interest rate compounded } m \text{ times per year}$$
$$i = \text{interest rate per period} = r/m$$
$$n = \text{number of payment periods}$$

The first payment of P dollars made at the end of the first period earns compound interest at the rate of i over the remaining $(n - 1)$ periods and therefore has an accumulated amount of $P(1 + i)^{n-1}$ dollars at the end of the term of the annuity (see Figure 5.1).

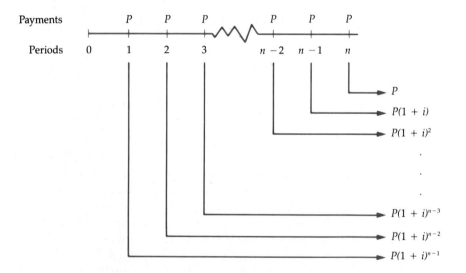

Figure 5.1

The second payment of P dollars made at the end of the second period earns compound interest at the rate of i over the remaining $(n - 2)$ periods and has an accumulated amount of $P(1 + i)^{n-2}$ dollars at the end of the term, and so on. The last payment earns no interest, since it is due at the end of the term. The amount of the annuity is obtained by adding all the terms in Figure 5.1. Thus,

$$S = P + P(1 + i) + P(1 + i)^2 + \cdots + P(1 + i)^{n-1}.$$

The sum on the right is the sum of the first n terms of a *geometric progression* with first term P and common ratio $(1 + i)$. We will show in Section 5.4 that S can be written in the more compact form

$$S = P\left[\frac{(1 + i)^n - 1}{i}\right] \tag{5}$$

The expression inside the brackets is commonly denoted by $s_{\overline{n}|i}$ (read "s angle n at i") and is called the **compound-amount factor.** Extensive tables have been constructed that give values of $s_{\overline{n}|i}$ for different values of i and n (such as Table I, Appendix C). In terms of the compound-amount factor,

$$S = Ps_{\overline{n}|i} \tag{6}$$

The quantity S in Formulas (5) and (6) is realizable at some future date and is accordingly called the *future value of an annuity.*

Future Value of an Annuity

The **future value S of an annuity** of n payments of P dollars each, paid at the end of each investment period into an account that earns interest at the rate of i per period, is

$$S = P\left[\frac{(1 + i)^n - 1}{i}\right].$$

EXAMPLE

9 Find the amount of an ordinary annuity of 12 monthly payments of $100 that earn interest at 12 percent per year compounded monthly.

SOLUTION

Here $P = 100$, $r = 0.12$, and $m = 12$, so $i = 0.12/12 = 0.01$ and $n = 12$. Thus, by Formula (5), the required amount of the annuity is given by

$$S = \frac{100[(1.01)^{12} - 1]}{0.01}$$

$$= 1268.25 \quad \text{(Using a hand calculator)}$$

or $1268.25. The same result is obtained by observing that

$$S = 100s_{\overline{12}|0.01}$$

$$= 100(12.6825)$$

$$= 1268.25 \quad \text{(Using Table I, Appendix C)} \quad \blacktriangleleft$$

▶ Present Value of an Annuity

In certain instances, you may want to determine the current value of a sequence of equal periodic payments that will be made over a certain period of time. After each payment is made, the new balance continues to earn interest at some nominal rate. Such an amount is referred to as the *present value of an annuity*. To derive a formula for determining the present value A of an annuity, we appeal to (4). Since the future value of the annuity is, by (5),

$$S = P\left[\frac{(1 + i)^n - 1}{i}\right],$$

we see, by (4), that the present value of the annuity A is given by

$$A = S(1 + i)^{-n}$$

$$= (1 + i)^{-n} P\left[\frac{(1 + i)^n - 1}{i}\right]$$

or

$$A = P\left[\frac{1 - (1 + i)^{-n}}{i}\right]$$

$$= Pa_{\overline{n}|i},$$

where the factor $a_{\overline{n}|i}$ (read "*a* angle *n* at *i*") represents the expression inside the brackets. Extensive tables have also been constructed giving values of $a_{\overline{n}|i}$ for different values of i and n (see Table I, Appendix C).

Present Value of an Annuity

The **present value A of an annuity** of n payments of P dollars each, paid at the end of each investment period into an account that earns interest at the rate of i per period, is

$$A = P\left[\frac{1 - (1 + i)^{-n}}{i}\right] \qquad (7)$$

EXAMPLE

10 Find the present value of an ordinary annuity of 24 payments of $100 each made monthly and earning interest at 12 percent per year compounded monthly.

SOLUTION   Here $P = 100$, $i = r/m = 0.12/12 = 0.01$, and $n = 24$, so by (7),

$$A = \frac{100[1 - (1.01)^{-24}]}{0.01}$$

$$= 2124.34,$$

or $2124.34. The same result may be obtained by using Table I, Appendix C. Thus,

$$A = 100a_{\overline{24}|0.01}$$
$$= 100(21.2434)$$
$$= 2124.34.$$
◄

▶ Applications

EXAMPLE

11 As a savings program toward their child's college education, the parents decide to deposit $100 at the end of every month into a time-deposit bank account paying interest at the rate of 6 percent per year compounded monthly. If the savings program began when the child was 6 years old, how much money would have accumulated by the time the child turns 18?

SOLUTION By the time the child turns 18, the parents would have made 144 deposits into the account. Thus, $n = 144$. Furthermore, we have $P = 100$, $r = 0.06$, and $m = 12$, so $i = 0.06/12 = 0.005$. Using Formula (5), we find that the amount of money that would have accumulated is given by

$$S = \frac{100[(1.005)^{144} - 1]}{0.005}$$
$$= 21,015,$$

or $21,015.
◄

EXAMPLE

12 After making a down payment of $2000 for an automobile, Mr. McMurphy paid $200 per month for 36 months with interest charged at 12 percent per year compounded monthly on the unpaid balance. What was the original cost of the car? What portion of Mr. McMurphy's total car payments went toward interest charges?

SOLUTION The loan taken up by Mr. McMurphy is given by the present value of the annuity

$$A = \frac{200[1 - (1.01)^{-36}]}{0.01} = 200a_{\overline{36}|0.01}$$
$$= 6021.50,$$

or $6021.50. Therefore, the original cost of the automobile is $8021.50 ($6021.50 plus the $2000 down payment). The interest charges paid by Mr. McMurphy are given by $(36)(200) - 6021.50$, or $1178.50.
◄

One important application of annuities is in the area of tax planning. During the past decade, Congress has created many special tax-deferred retirement vehicles such as Individual Retirement Accounts (IRAs) for the employed individual, and Keogh plans and Simplified Employer Pension (SEP) plans for those who are self-employed. All these plans are examples of annuities in which the individual is allowed to make tax-deductible contributions to an investment account. The amount of the contribution is limited by congressional legislation. The taxes on *both* the contributions and the interest and/or capital gains accumulated in the account are deferred until the money is withdrawn, ideally during retirement, when tax brackets should be lower. In the interim period, the individual has the benefit of tax-free growth on his or her investment.

Suppose, for example, that you are in a marginal tax bracket of 28 percent and that you receive a year-end bonus of $2000 from your employer on which you have not yet paid taxes but wish to deposit into either an IRA or a regular savings account. Additionally, suppose that both accounts earn interest at an effective annual rate of 8 percent per year. If you choose to invest your bonus in a regular savings account, you will first have to pay taxes on the $2000, leaving $1400 to invest. At the end of one year, you will also have to pay taxes on the interest earned, leaving you with

Accumulated amount	Tax on interest	Net amount
1555.20	− 32.26	= 1522.94,

or $1478.40.

On the other hand, if you put the money into an IRA account, the entire sum will earn interest, and at the end of one year you will have (1.08)($2000), or $2160, in your account. Of course, you will still have to pay taxes on this money when you withdraw it, but you will have gained the advantage of tax-free growth of the larger principal over the years. The disadvantage of this option is that if you withdraw the money before you reach the age of 59½, you will be liable for taxes on both your contributions and the interest earned and you will also have to pay a 10 percent penalty.

In practice, the size of the contributions the individual might make to the various retirement plans might vary from year to year. Also, he or she might make the contributions at different payment periods. To simplify our discussion, we will consider an example in which an individual makes a fixed payment at regular intervals.

EXAMPLE

13 Ms. Coughlin is planning to make the maximum contribution of $2000 on January 31 of each year into an IRA earning interest at the rate of 9 percent per year. Twenty-five years from now, on January 31 of the year following her retirement, how much will she have in her IRA?

SOLUTION   The amount of money Ms. Coughlin will have in her account when she retires is found by using (5) with $P = 2000$, $r = 0.09$, and $m = 1$, so $i = r/m = 0.09$ and $n = 25$. The required amount is given by

$$S = \frac{2000[(1.09)^{25} - 1]}{0.09}$$
$$= 169{,}401.79,$$

or $169,401.79.   ◀

Tax-deferred annuities are another type of investment vehicle that allow an individual to build assets for retirement, college funds, or other future needs. The advantage gained in this type of investment is that the tax on the accumulated interest is deferred to a later date. Note that in this type of investment, as opposed to those mentioned earlier, the contributions themselves are not tax deductible. At first glance the advantage thus gained may seem to be relatively inconsequential, but its true effect is illustrated by the next example.

EXAMPLE

14   Both Mr. Clark and Mr. Colby are salaried individuals, 45 years of age, who are saving for their retirement 20 years from now. Both Clark and Colby are also in the 28 percent marginal tax bracket. Mr. Clark makes a $1000 contribution annually on December 31 into a savings account earning an effective rate of 8 percent per year. At the same time, Mr. Colby makes a $1000 annual payment to an insurance company for a tax-deferred annuity. The annuity also earns interest at an effective rate of 8 percent per year. (Assume that both men remain in the same tax bracket throughout this period, and disregard state income taxes.)

a. Calculate how much each man will have in his investment account at the end of 20 years.

b. Compute the interest earned in each account.

c. Show that even if the interest on Mr. Colby's investment were subjected to a tax of 28 percent upon withdrawal of his investment at the end of 20 years, the net accumulated amount of his investment would still be greater than that of Mr. Clark's.

SOLUTION   *a.* Because Mr. Clark is in the 28 percent marginal tax bracket, the net yield for his investment is $(0.72)(8)$, or 5.76 percent, per year.

Using (5) with $P = 1000$, $r = 0.0576$, and $m = 1$, so that $i = 0.0576$ and $n = 20$, we see that Clark's investment will be worth

$$S = \frac{1000[(1 + 0.0576)^{20} - 1]}{0.0576}$$
$$= 35{,}850.49,$$

or \$35,850.49 at his retirement.

Mr. Colby has a tax-sheltered investment with an effective yield of 8 percent per year. Using (5) with $P = 1000$, $r = 0.08$, and $m = 1$, so that $i = 0.08$ and $n = 20$, we see that Colby's investment will be worth

$$S = \frac{1000[(1 + 0.08)^{20} - 1]}{0.08}$$
$$= 45{,}761.96,$$

or \$45,761.96 at his retirement.

b. Each man will have paid 20(1,000), or 20,000 dollars, into his account. Therefore, the total interest earned in Mr. Clark's account will be (35,850.49 − 20,000), or \$15,850.49, whereas the total interest earned in Mr. Colby's account would be (45,761.96 − 20,000), or \$25,761.96.

c. From (b) we see that the total interest earned in Mr. Colby's account would be \$25,761.96. If it were taxed at 28 percent, he would still end up with (0.72)(25,761.96), or \$18,548.61. This is larger than the total interest of \$15,850.49 earned by Mr. Clark. ◀

Self-Check Exercises

5.2

1. Mr. and Mrs. Fletcher opened a spousal IRA on January 31, 1985 with the maximum permissible contribution of \$2250. They plan to make a contribution of \$2250 thereafter on January 31 of each year until their retirement in the year 2005. If the account earns interest at the rate of 8 percent per year compounded yearly, how much will the Fletchers have in their account when they retire?

2. The Denver Wildcatting Company has an immediate need for a loan. In an agreement worked out with its banker, Denver assigns its royalty income of \$4800 per month for the next three years from certain oil properties to the bank, with the first payment due at the end of the first month. If the bank charges interest at the rate of 9 percent per year compounded monthly, what is the amount of the loan negotiated between the parties?

Solutions to Self-Check Exercises 5.2 can be found on page 250.

▶

Exercises 5.2

In Exercises 1–6, find the amount (future value) of each ordinary annuity.

1. $1000 a year for ten years at 10 percent per year compounded annually

2. $1500 per semiannual period for eight years at 9 percent per year compounded semiannually

3. $1800 per quarter for six years at 8 percent per year compounded quarterly

4. $500 per semiannual period for 12 years at 11 percent per year compounded semiannually

5. $600 per quarter for nine years at 12 percent per year compounded quarterly

6. $150 per month for 15 years at 10 percent per year compounded monthly

In Exercises 7–12, find the present value of each ordinary annuity.

7. $5000 a year for eight years at 8 percent per year compounded annually

8. $1200 per semiannual period for six years at 10 percent per year compounded semiannually

9. $4000 a year for five years at 9 percent per year compounded yearly

10. $3000 per semiannual period for six years at 11 percent per year compounded semiannually

11. $800 per quarter for seven years at 12 percent per year compounded quarterly

12. $150 per month for ten years at 8 percent per year compounded monthly

13. If a merchant deposits $1500 annually at the end of each tax year in an IRA account paying interest at the rate of 8 percent per year compounded annually, how much will she have in her account at the end of 25 years?

14. If Mr. Pruitt deposits $100 a month in a savings account earning interest at the rate of 8 percent per year compounded monthly, how much will he have on deposit in his savings account at the end of six years, assuming that he makes no withdrawals during that period?

15. Mrs. Lynde has joined a "Christmas Fund Club" at her bank. At the end of every month, December through October inclusive, she will make a deposit of $40 in her fund. If the money earns interest at the rate of 7 percent per year compounded monthly, how much will she have in her account on December 1 of the following year?

16. Mr. Robinson, who is self-employed, contributes $5000 a year into a Keogh Account. How much will he have in the account after 25 years if the account earns interest at the rate of 10 percent per year compounded yearly?

17. As a fringe benefit for the past 12 years, Mr. Collin's employer has contributed $100 a month into an employee retirement account for Collins that pays interest at the rate of 7 percent per year compounded monthly. Mr. Collins has also contributed $2000 per year for the past eight years into an IRA that pays interest at the rate of 9 percent per year compounded yearly. How much does Mr. Collins have in his retirement fund at this time?

18. The Pirerra's are planning to go to Europe three years from now and have agreed to set aside $150 each month for their trip. If they deposit this money at the end of each month into a savings account paying interest at the rate of 8 percent per year compounded monthly, how much money will be in their travel fund at the end of the third year?

19. Mr. and Mrs. Betz have leased an auto for two years at $450 per month. If money is worth 9 percent per year compounded monthly, what is the equivalent cash payment (present value) of this annuity?

20. Ms. Newburg made a down payment of $2000 toward the purchase of a new car. In order to pay the balance of the purchase price, she has secured a loan from her bank at the rate of 12 percent per year compounded monthly. Under the terms of her finance agreement, she is required to make payments of $210 per month for 36 months. What is the cash price of the car?

21. R. G. Pierce Publishing Company sells encyclopedias under two payment plans: cash or installment. Under the installment plan, the customer pays $22 per month over three years with interest charged on the balance at a rate of 18 percent per year compounded monthly. Find the cash price for a set of encyclopedias if it is equivalent to the price paid by a customer using the installment plan.

22. A state lottery commission pays the winner of the "Million Dollar" lottery 20 installments of $50,000 per year. Determine how much money the commission should have in the bank initially to guarantee the payments assuming that the balance on deposit with the bank earns interest at the rate of 8 percent per year compounded yearly. [*Hint:* Find the present value of an annuity.]

23. The Johnsons have accumulated a nest egg of $25,000 that they intend to use as a down payment toward the purchase of a new house. Because their present gross income has placed them in a relatively high tax bracket, they have decided to invest a minimum of $800 per month in monthly payments (to take advantage of their tax deductions) toward the purchase of their house. However, because of other financial obligations their monthly payments should not exceed $1000. If local mortgage rates are 11 percent per year compounded monthly for a conventional 25-year mortgage, what is the price range of houses that they should consider?

24. Refer to Exercise 23. If local mortgage rates were lowered to 10 percent, how would this affect the price range of houses the Johnsons should consider?

25. Refer to Exercise 23. If the Johnsons decide to secure a 15-year mortgage instead of a 25-year mortgage, what is the price range of houses they should consider when the local mortgage rate for this type of loan is 10.5 percent?

Solutions to Self-Check Exercises

5.2

1. The amount the Fletchers will have in their account when they retire may be found by using (5) with $P = 2250$, $r = 0.08$, $m = 1$, so that $i = r = 0.08$ and $n = 20$. Thus

$$S = \frac{2250[(1.08)^{20} - 1]}{0.08}$$

$$= 102,964.42,$$

or $102,964.42.

2. We want to find the present value of an ordinary annuity of 36 payments of $4800 each made monthly and earning interest at 9 percent per year compounded monthly.

Using (7) with $P = 4800$, $m = 12$, so that $i = r/m = 0.09/12 = 0.0075$ and $n = (12)(3) = 36$, we find

$$A = \frac{4800[1 - (1.0075)^{-36}]}{0.0075}$$

$$= 150,944.67,$$

or $150,944.67, the amount of the loan negotiated.

5.3

Amortization and Sinking Funds

▶ Amortization of Loans

▶ Financing a Home

▶ Sinking Funds

▶ Amortization of Loans

The annuity formulas derived in Section 5.2 may be used to answer questions involving the amortization of certain types of installment loans. For example, in a typical housing loan, the mortgagor makes periodic payments toward reducing his indebtedness to the lender, who charges interest at a fixed rate on the unpaid portion of the debt. In practice, the borrower is required to repay the lender in periodic installments, usually of the same size and over a fixed term, so that the loan (principal plus interest charges) is amortized at the end of the term.

By thinking of the monthly loan repayments P as the payments in an annuity, we see that the original amount of the loan is given by A, the present value of the annuity. From (7), we have

$$A = P\left[\frac{1 - (1 + i)^{-n}}{i}\right]$$

$$= Pa_{\overline{n}|i}.$$

A question a financier might ask is: How much should the monthly installment be so that the loan will be amortized at the end of the term of the loan? To answer this question, we simply solve Formula (7) for P in terms of A, obtaining

$$P = \frac{Ai}{1 - (1 + i)^{-n}}$$

$$= \frac{A}{a_{\overline{n}|i}}$$

Amortization Formula

The periodic payment P on a loan of A dollars to be amortized over n periods with interest charged at the rate of i per period is

$$P = \frac{Ai}{1 - (1 + i)^{-n}}. \tag{8}$$

EXAMPLE

15 A sum of $50,000 is to be repaid over a five-year period through equal install-ments made at the end of each year. If an interest rate of 8 percent per year is charged on the unpaid balance, and interest calculations are made at the end of each year, determine the size of each installment so that the loan (principal plus interest charges) is amortized at the end of five years. Verify the result by displaying the amortization schedule.

SOLUTION Substituting $A = 50,000$, $i = r = 0.08$ (here $m = 1$), and $n = 5$ into (8), we obtain

$$P = \frac{(50,000)(0.08)}{1 - (1.08)^{-5}}$$
$$= 12,522.82,$$

giving the required yearly installment as $12,522.82.
The amortization schedule is presented in Table 5.3.

End of Period	Interest Charged	Repayment Made	Payment Toward Principal	Outstanding Principal
0				50,000.00
1	4000.00	12,522.82	8522.82	41,477.18
2	3318.17	12,522.82	9204.65	32,272.53
3	2581.80	12,522.82	9941.02	22,331.51
4	1786.52	12,522.82	10,736.30	11,595.21
5	927.62	12,522.82	11,595.20	0.01

Table 5.3

The outstanding principal at the end of five years is, of course, zero. (The figure of $.01 in Table 5.3 is the result of round-off errors.) Observe that initially the larger portion of the repayment goes toward payment of interest charges, but as time goes by the larger portion of the installment goes toward repayment of the principal. ◀

▶ Financing a Home

EXAMPLE

16 The Blakelys borrowed $80,000 from a bank to help finance the purchase of a house. The bank charges interest at a rate of 9 percent per year on the unpaid balance, with interest computations made at the end of each month. The Blakelys have agreed to repay the loan in equal monthly installments over 30 years. How much should each payment be if the loan is to be amortized at the end of the term?

SOLUTION Here $A = 80,000$, $i = r/m = .09/12 = 0.0075$, and $n = (30)(12) = 360$. Using Formula (8), we find that the size of each monthly installment required is given by

$$P = \frac{(80,000)(0.0075)}{1 - (1.0075)^{-360}}$$

$$= 643.70 \quad \text{(Using a pocket calculator)}$$

or $643.70.   ◀

EXAMPLE

17 Refer to Example 16 and find the amount of equity the Blakelys will have in their home (disregarding the down payment and appreciation) after the

a. 60th payment (5 years)

b. 240th payment (20 years)

SOLUTION a. After 60 payments have been made, the outstanding principal is given by the sum of the present values of the remaining installments (that is, $360 - 60 = 300$ installments). But this sum is just the present value of an annuity with $n = 300$, $P = 643.70$, and $i = 0.0075$ (see Example 16). Using Formula (7), we find that

$$A = 643.70 \left[\frac{1 - (1 + 0.0075)^{-300}}{0.0075} \right]$$

$$= 76,704.32,$$

or $76,704.32. Therefore, we may find how much equity the Blakelys have in their home by subtracting the outstanding principal from the original principal:

$$80,000 - 76,704.32 = 3,295.68,$$

or $3,295.68.

b. As in part (a), we compute the present value of the annuity with $n = 360 - 240 = 120$. Thus,

$$A = 643.70 \left[\frac{1 - (1 + 0.0075)^{-120}}{0.0075} \right]$$

$$= 50,814.79,$$

or $50,814.79. Therefore, the Blakelys' equity after the 240th payment is

$$80,000 - 50,814.79 = 29,185.21,$$

or $29,185.21. ◄

EXAMPLE

18 The Jacksons have determined that after making a down payment they could afford at most $1000 for a monthly house payment. The bank charges interest at the rate of 10.5 percent per year on the unpaid balance, with interest computations made at the end of each month. If the loan is to be amortized in equal monthly installments over 30 years, what is the maximum amount that the Jacksons could borrow from the bank?

SOLUTION

Here $i = r/m = 0.105/12 = 0.00875$, $n = (30)(12) = 360$, $P = 1000$, and we are required to find A. From (8), we have

$$P = \frac{Ai}{1 - (1 + i)^{-n}},$$

or, solving for A,

$$A = \frac{P[1 - (1 + i)^{-n}]}{i}.$$

Substituting the numerical values for P, n, and i into this expression for A, we obtain

$$A = \frac{1000[1 - (1.00875)^{-360}]}{0.00875}$$

$$\approx 109,321.$$

Therefore, the Jacksons could borrow at most $109,321. ◄

▶ Sinking Funds

Another important application of the annuity formulas is in the computation of sinking funds. Briefly, a **sinking fund** is a fund that is set up by a party to use for a specific purpose at some future date. For example, an individual might establish a sinking fund for the purpose of discharging a debt at a future date. A corporation might establish a sinking fund in order to accumulate sufficient capital to replace equipment that is expected to be obsolete at some future date.

By thinking of the amount to be accumulated by a specific date in the future as the future value of an annuity, we can answer questions about a large class of sinking fund problems by invoking Formula (5).

EXAMPLE

19 The proprietor of Carson Hardware has decided to set up a sinking fund for the purpose of purchasing a computer in two years' time. It is expected that the computer will cost $30,000. If the fund earns 10 percent per year compounded quarterly, determine the size of each (equal) quarterly installment the proprietor should pay into the fund. Verify the result by displaying the schedule.

SOLUTION   The problem at hand is to find the size of each (quarterly) payment P of an annuity given that its future value is $S = 30,000$, the interest earned per conversion period is $i = r/m = 0.1/4 = 0.025$, and the number of payments is $n = (2)(4) = 8$. The formula for the annuity, from (5), is

$$S = P\left[\frac{(1 + i)^n - 1}{i}\right],$$

which when solved for P yields

$$P = \frac{iS}{(1 + i)^n - 1} \tag{9}$$

or equivalently,

$$P = \frac{S}{s_{\overline{n}|i}}.$$

Substituting the appropriate numerical values for i, S, and n into (9), we obtain the desired quarterly payment

$$P = \frac{(0.025)(30,000)}{(1.025)^8 - 1}$$

$$= 3434.07,$$

or $3434.07. Table 5.4 shows the required schedule. Once again, the $.44 in excess of the accumulated amount of $30,000 is the result of round-off errors.

End of Period	Deposit Made	Interest Earned	Addition to Fund	Accumulated Amount in Fund
1	3434.07	0	3434.07	3434.07
2	3434.07	85.85	3519.92	6953.99
3	3434.07	173.85	3607.92	10,561.91
4	3434.07	264.05	3698.12	14,260.03
5	3434.07	356.50	3790.57	18,050.60
6	3434.07	451.27	3885.34	21,935.94
7	3434.07	548.40	3982.47	25,918.41
8	3434.07	647.96	4082.03	30,000.44

Table 5.4

◀

The formula derived in this last example is restated below.

Sinking Fund Payment

The periodic payment P that is required to accumulate a sum of S dollars over n periods with interest charged at the rate of i per period is

$$P = \frac{iS}{(1 + i)^n - 1}.$$

Self-Check Exercises

5.3

1. The Moores wish to borrow $100,000 from a bank to help finance the purchase of a house. Their banker has offered the following plans for their consideration. In plan I, the Moores have 30 years to repay the loan in monthly installments with interest on the unpaid balance charged at 10.5 percent per year compounded monthly. In plan II, the loan is to be repaid in monthly installments over 15 years with interest on the unpaid balance charged at 9.75 percent per year compounded monthly.

   a. Find how much the monthly repayments would be in plan I and plan II.

   b. How much would the Moores save in choosing plan II over plan I?

2. Mr. Harris, a self-employed individual who is 45 years old, is setting up a Defined-Benefit Keogh Plan for his retirement. If he wishes to have $250,000 in this Keogh account by age 65, what is the size of each yearly installment he will be required to make into a savings account earning interest at 8¼ percent per year? (Assume that Mr. Harris is eligible to make the required contributions.)

Solutions to Self-Check Exercises 5.3 can be found on page 259.

the past 20 years. His investment has earned interest at the rate of 8 percent per year compounded quarterly over this period. Now, at age 60, he is considering retirement. What quarterly payment will he receive over the next 15 years (assuming that the money is earning interest at the same rate)? If he continues working and makes quarterly payments of the same amount in his IRA account until age 65, what quarterly payment will he receive from his fund upon retirement over the following 10 years?

*28. Ms. Dwyer purchased a new car during a special sales promotion by the manufacturer. She secured a loan from the manufacturer in the amount of $10,000 at a rate of 7.9 percent per year compounded monthly. Her bank is now charging 11.5 percent per year compounded monthly for new car loans. Assuming that each loan would be amortized by 36 equal monthly installments, determine the amount of interest she will have saved at the end of three years by borrowing from the manufacturer rather than the bank. [*Hint:* Let $A = \$10,000$, $i = .079/12$, and $i = .115/12$, $n = 36$, and find P. Then calculate the future value of an annuity using these values.]

*29. The Sandersons are planning to refinance their home. The outstanding principal on their original loan is $100,000 and was to be amortized in 240 equal monthly installments at an interest rate of 14.2 percent per year compounded monthly. The new loan they expect to secure is to be amortized over the same period at an interest rate of 10.6 percent per year compounded monthly. How much can they expect to save over the life of the loan in interest payments by refinancing the loan at this time?

*30. After making a down payment of $25,000, the Meyers need to secure a loan of $90,000 to purchase a certain house. Their bank's current rate for 25-year home loans is 11 percent compounded monthly. The owner has offered to finance the loan at 9.8 percent compounded monthly. Assuming that both loans would be amortized over a 25-year period by 300 equal monthly installments, determine the amount of interest the Meyers would save by choosing the seller's financing.

Solutions to Self-Check Exercises

5.3

1. *a.* We use (8) in each instance. Under plan I,

$$A = 100,000, \quad i = \frac{r}{m} = \frac{0.105}{12} = 0.00875, \text{ and } n = (30)(12) = 360.$$

Therefore, the size of each monthly repayment under plan I is

$$P = \frac{100,000(0.00875)}{1 - (1.00875)^{-360}}$$
$$= 914.74,$$

or $914.74.
   Under plan II,

$$A = 100,000, \quad i = \frac{r}{m} = \frac{.0975}{12} = 0.008125, \text{ and } n = (15)(12) = 180.$$

Therefore, the size of each monthly payment under plan II is

$$P = \frac{100{,}000(0.008125)}{1 - (1.008125)^{-180}}$$
$$= 1059.36,$$

or $1,059.36.

b. Under plan I, the total amount of repayments would have been

$$(360)(914.74) \quad \text{(Number of payments times the size of each installment)}$$
$$= 329{,}306.40,$$

or $329,306.40. Under plan II, the total amount of repayments would have been

$$(180)(1059.36)$$
$$= 190{,}684.80,$$

or $190,684.80. Therefore, the Moores would save

$$329{,}306.40 - 190{,}684.80 = 138{,}621.60,$$

or $138,621.60.

2. We use (9) with

$$S = 200{,}000, \; i = r = 0.0825 \text{ (since } m = 1\text{), and } n = 20,$$

giving the required size of each installment as

$$P = \frac{(0.0825)(250{,}000)}{(1.0825)^{20} - 1}$$
$$= 5313.59,$$

or $5313.59.

▶ 5.4

Arithmetic and Geometric Progressions (Optional)

▶ Arithmetic Progressions

▶ Geometric Progressions

▶ Double Declining-Balance Method of Depreciation

▶ Arithmetic Progressions

An **arithmetic progression** is a sequence of numbers in which each term after the first is obtained by adding a constant d to the preceding term. The constant d is called the **common difference.** For example, the sequence

$$3, 6, 9, 12, \ldots$$

is an arithmetic progression with the common difference equal to 3.

Observe that an arithmetic progression is completely determined if the first term and the common difference are known. In fact, if

$$a_1, a_2, a_3, \ldots, a_n, \ldots$$

is an arithmetic progression with the first term given by a and common difference given by d, then by definition,

$$
\begin{aligned}
a_1 &= a \\
a_2 &= a_1 + d = a + d \\
a_3 &= a_2 + d = (a + d) + d = a + 2d \\
a_4 &= a_3 + d = (a + 2d) + d = a + 3d \\
&\vdots \\
a_n &= a_{n-1} + d = a + (n - 2)d + d = a + (n - 1)d.
\end{aligned}
$$

Thus we see that the nth term of an arithmetic progression with first term a and common difference d is given by

$$a_n = a + (n - 1)d \tag{10}$$

EXAMPLE

20 Find the twelfth term of the arithmetic progression

$$2, 7, 12, 17, 22, \ldots$$

SOLUTION

The first term of the arithmetic progression is $a_1 = a = 2$, and the common difference is $d = 5$, so upon setting $n = 12$ in Equation (10), we find

$$a_{12} = 2 + (12 - 1)5 = 57. \qquad \blacktriangleleft$$

EXAMPLE

21 Write the first five terms of an arithmetic progression whose third and eleventh terms are 21 and 85, respectively.

SOLUTION   Using Equation (10) we obtain

$$a_3 = a + 2d = 21$$
$$a_{11} = a + 10d = 85.$$

Subtracting the first equation from the second gives $8d = 64$, or $d = 8$. Substituting this value of d into the first equation yields $a + 16 = 21$, or $a = 5$. Thus, the required arithmetic progression is given by the sequence

$$5, 13, 21, 29, 37, \ldots .$$ ◀

Let S_n denote the sum of the first n terms of an arithmetic progression with first term $a_1 = a$ and common difference d. Then

$$S_n = a + (a + d) + (a + 2d) + \cdots + [a + (n - 1)d] \tag{11}$$

Rewriting the expression for S_n with the terms in reverse order gives

$$S_n = [a + (n - 1)d] + [a + (n - 2)d] + \cdots + (a + d) + a \tag{12}$$

Adding (11) and (12), we obtain

$$2S_n = [2a + (n - 1)d] + [2a + (n - 1)d] + \cdots + [2a + (n - 1)d]$$
$$= n[2a + (n - 1)d],$$

or $$S_n = \frac{n}{2}[2a + (n - 1)d] \tag{13}$$

EXAMPLE

22   Find the sum of the first 20 terms of the arithmetic progression of Example 20.

SOLUTION   Letting $a = 2$, $d = 5$, and $n = 20$ in Equation (13), we obtain

$$S_{20} = \frac{20}{2}[2 \cdot 2 + 19 \cdot 5]$$
$$= 990.$$ ◀

EXAMPLE

23   The Madison Electric Company had sales of \$200,000 in its first year of operation. If the sales increased by \$30,000 per year thereafter, find Madison's sales in the fifth year and its total sales over the first five years of operation.

SOLUTION   Madison's yearly sales follow an arithmetic progression with the first term given by $a = 200{,}000$ and the common difference given by $d = 30{,}000$. The sales in the fifth year are found by using (10) with $n = 5$. Thus,

$$a_5 = 200{,}000 + (5 - 1)30{,}000$$
$$= 320{,}000{,}$$

or $320,000.

The total sales of the company over the first five years of operation are found by using (13) with $n = 5$. Thus,

$$S_5 = \frac{5}{2}[2(200{,}000) + (5 - 1)30{,}000]$$

$$= 1{,}300{,}000{,}$$

or $1,300,000. ◀

▶ Geometric Progressions

A **geometric progression** is a sequence of numbers in which each term after the first is obtained by multiplying the preceding term by a constant r. The constant r is called the **common ratio.**

A geometric progression is completely determined if the first term and the common ratio are known. Thus, if

$$a_1, a_2, a_3, \ldots, a_n, \ldots$$

is a geometric progression with the first term given by a and common ratio given by r, then by definition,

$$a_1 = a$$
$$a_2 = a_1 r = ar$$
$$a_3 = a_2 r = ar^2$$
$$a_4 = a_3 r = ar^3$$
$$\vdots$$
$$a_n = a_{n-1} r = ar^{n-1}.$$

Thus we see that the nth term of a geometric progression with first term a and common ratio r is given by

$$a_n = ar^{n-1} \tag{14}$$

EXAMPLE

24 Find the eighth term of a geometric progression whose first five terms are 162, 54, 18, 6, 2.

SOLUTION The common ratio is found by taking the ratio of any term other than the first to the preceding term. Taking the ratio of the fourth term to the third term, for example, gives $r = 6/18 = 1/3$. To find the eighth term of the geometric progression, use (14) with $a = 162$, $r = 1/3$, and $n = 8$, obtaining

$$a_8 = 162\left(\frac{1}{3}\right)^7$$

$$= \frac{2}{27}.$$

◀

EXAMPLE

25 Find the tenth term of a geometric progression whose third term is 16 and whose seventh term is 1.

SOLUTION

Using (14) with $n = 3$ and $n = 7$, respectively, yields

$$a_3 = ar^2 = 16$$

and

$$a_7 = ar^6 = 1.$$

Dividing a_7 by a_3 gives

$$\frac{ar^6}{ar^2} = \frac{1}{16},$$

from which we obtain $r^4 = 1/16$, or $r = 1/2$. Substituting this value of r into the expression for a_3, we obtain

$$a\left(\frac{1}{2}\right)^2 = 16, \text{ or } a = 64.$$

Finally, using (14) once again with $a = 64$, $r = 1/2$, and $n = 10$ gives

$$a_{10} = 64\left(\frac{1}{2}\right)^9 = \frac{1}{8}.$$

◀

To find the sum of the first n terms of a geometric progression with the first term $a_1 = a$ and common ratio r, denote the required sum by S_n. Then

$$S_n = a + ar + ar^2 + \cdots + ar^{n-2} + ar^{n-1} \tag{15}$$

Upon multiplying (15) by r, we obtain

$$rS_n = ar + ar^2 + ar^3 + \cdots + ar^{n-1} + ar^n \tag{16}$$

Subtracting (16) from (15) gives

$$S_n - rS_n = a - ar^n$$

or

$$(1 - r)S_n = a(1 - r^n).$$

If $r \neq 1$, we may divide both sides of the last equation by $(1 - r)$, obtaining

$$S_n = \frac{a(1 - r^n)}{(1 - r)}.$$

If $r = 1$, then (15) gives

$$S_n = a + a + a + \cdots + a \quad (n \text{ terms})$$
$$= na.$$

Thus,

$$S_n = \begin{cases} \dfrac{a(1 - r^n)}{1 - r} & \text{if } r \neq 1 \\ na & \text{if } r = 1 \end{cases} \tag{17}$$

EXAMPLE

26 Find the sum of the first six terms of the geometric series

$$3, 6, 12, 24, \ldots.$$

SOLUTION Here $a = 3$ and $r = 6/3 = 2$, so (17) gives

$$S_6 = \frac{3(1 - 2^6)}{1 - 2} = 189.$$ ◀

EXAMPLE

27 The Michaelson Land Development Company had sales of $1 million in its first year of operation. If the sales increased by 10 percent per year thereafter, find Michaelson's sales in the fifth year and its total sales over the first five years of operation.

SOLUTION Michaelson's yearly sales follow a geometric progression with the first term given by $a = 1{,}000{,}000$ and the common ratio given by $r = 1.1$. The sales in the fifth year are found by using (14) with $n = 5$. Thus,

$$a_5 = 1{,}000{,}000(1.1)^4$$
$$= 1{,}464{,}100,$$

or $1,464,100.

Michaelson's total sales over the first five years of operation are found by using (17) with $n = 5$. Thus,

$$S_5 = \frac{1{,}000{,}000[1 - (1.1)^5]}{1 - 1.1}$$

$$= 6{,}105{,}100,$$

or $6,105,100.   ◄

► Double Declining-Balance Method of Depreciation

In Section 1.3 we discussed the straight-line, or linear, method of depreciating an asset. Linear depreciation assumes that the asset depreciates at a constant rate. For certain assets such as machines, whose market values drop rapidly in the early years of usage and thereafter less rapidly, another method of depreciation called the **double declining-balance method** is often used. In practice, a business firm normally employs the double declining-balance method for depreciating such assets for a certain number of years and then switches over to the linear method.

To derive an expression for the book value of an asset being depreciated by the double declining-balance method, let C (in dollars) denote the original cost of the asset and suppose it is to be depreciated over N years. Using this method, the amount depreciated each year is $2/N$ times the value of the asset at the beginning of that year. Thus, the amount by which the asset is depreciated in its first year of use is given by $2C/N$, so if $V(1)$ denotes the book value of the asset at the end of the first year, then

$$V(1) = C - \frac{2C}{N} = C\left(1 - \frac{2}{N}\right).$$

Next, if $V(2)$ denotes the book value of the asset at the end of the second year, then a similar argument leads to

$$V(2) = C\left(1 - \frac{2}{N}\right) - C\left(1 - \frac{2}{N}\right)\frac{2}{N}$$

$$= C\left(1 - \frac{2}{N}\right)\left(1 - \frac{2}{N}\right)$$

$$= C\left(1 - \frac{2}{N}\right)^2.$$

Continuing, we find that if $V(n)$ denotes the book value of the asset at the end of n years, then the terms $C, V(1), V(2), \ldots, V(n)$ form a geometric progression with first term C and common ratio $(1 - 2/N)$. Consequently, the nth term, $V(n)$, is given by

$$V(n) = C\left(1 - \frac{2}{N}\right)^n \quad (1 \le n \le N) \tag{18}$$

Also, if $D(n)$ denotes the amount by which the asset has been depreciated by the end of the nth year, then

$$D(n) = C - C\left(1 - \frac{2}{N}\right)^n$$

$$= C\left[1 - \left(1 - \frac{2}{N}\right)^n\right] \tag{19}$$

EXAMPLE

28 A tractor purchased at a cost of $60,000 is to be depreciated by the double declining-balance method over ten years. What is the book value of the tractor at the end of five years? By what amount has the tractor been depreciated by the end of the fifth year?

SOLUTION

We have $C = 60,000$ and $N = 10$. Thus, using (18) with $n = 5$ gives the book value of the tractor at the end of five years as

$$V(5) = 60,000\left(1 - \frac{2}{10}\right)^5$$

$$= 60,000\left(\frac{4}{5}\right)^5$$

$$= 19,660.80,$$

or $19,660.80.

The amount by which the tractor has been depreciated by the end of the fifth year is given by

$$60,000 - 19,660.80 = 40,339.20,$$

or $40,339.20. You may verify the last result by using (19) directly.   ◀

Self-Check Exercises

5.4

1. Find the sum of the first five terms of the geometric sequence with first term −24 and common ratio −1/2.

2. Office equipment purchased for $75,000 is to be depreciated by the double declining-balance method over five years. Find the book value at the end of three years.

3. Derive Formula 5 of Section 5.2 (page 243).

Solutions to Self-Check Exercises 5.4 can be found on page 271.

▶

Exercises 5.4

In Exercises 1–4, find the nth term of the arithmetic progression that has the given values of a, d, and n.

1. $a = 6, d = 3, n = 9$

2. $a = -5, d = 3, n = 7$

3. $a = -15, d = 3/2, n = 8$

4. $a = 1.2, d = 0.4, n = 98$

5. Find the first five terms of the arithmetic progression whose fourth and eleventh terms are 30 and 107, respectively.

6. Find the first five terms of the arithmetic progression whose seventh and twenty-third terms are -5 and -29, respectively.

7. Find the seventh term of the arithmetic progression: $x, x + y, x + 2y, \ldots$.

8. Find the eleventh term of the arithmetic progression: $a + b, 2a, 3a - b, \ldots$.

9. Find the sum of the first 15 terms of the arithmetic progression: $4, 11, 18, \ldots$.

10. Find the sum of the first 20 terms of the arithmetic progression: $5, -1, -7, \ldots$.

11. Find the sum of all the odd integers between 14 and 58.

12. Find the sum of the even integers between 21 and 99.

13. Find $f(1) + f(2) + f(3) + \cdots + f(20)$, given that $f(x) = 3x - 4$.

14. Find $g(1) + g(2) + g(3) + \cdots + g(50)$, given that $g(x) = 12 - 4x$.

15. Show that Equation (13) can be written as $S_n = \frac{n}{2}(a + a_n)$ where a_n represents the last term of an arithmetic progression. Use this formula to find:

   a. the sum of the first 11 terms of the arithmetic progression whose first and last terms are 3 and 47, respectively.

   b. the sum of the first 20 terms of the arithmetic progression whose first and last terms are 5 and -33, respectively.

16. The Moderne Furniture Company had sales of $150,000 during its first year of operation. If the sales increased by $16,000 per year thereafter, find the sales of the company in the fifth year and its total sales over the first five years of operation.

17. As part of her physical-fitness program, Karen has taken up jogging. If she jogs 1 mile the first day and increases her daily run by 1/2 mile every four days, how long will it take her to reach her goal of 10 miles per day?

18. A 100-foot well is to be drilled. The cost of drilling the first foot is $6.00 and the cost of drilling each additional foot is $2.50 more than that of the preceding foot. Find the cost of drilling the entire 100 feet.

19. A tourist wishes to go from the airport to his hotel, which is 25 miles away. The taxi rate is $1.00 for the first mile and $.60 for each additional mile. The airport limousine also goes to his hotel and charges a flat rate of $7.50. How much money will the tourist save by taking the airport limousine?

20. A recent college graduate received two job offers. Company A offered her an initial salary of $16,800 with guaranteed annual increases of $1,250 per year for the first five years. Company B offered an initial salary of $18,000 with guaranteed annual increases of $900 per year for the first five years.

a. Which company is offering a higher salary for the fifth year of employment?

b. Which company is offering more money for the first five years of employment?

21. One of the methods that the Internal Revenue Service allows for computing depreciation of certain business property is the sum-of-the-years-digits method. If a property valued at C dollars has an estimated useful life of N years and a salvage value of S dollars, then the amount of depreciation D_n allowed during the nth year is given by

$$D_n = (C - S)\frac{N - (n - 1)}{S_N} \quad (0 \le n \le N)$$

where S_N is the sum of the first N positive integers representing the estimated useful life of the property. Thus,

$$S_N = 1 + 2 + \cdots + N = \frac{N(N + 1)}{2}.$$

a. Verify that the sum of the arithmetic progression $S_n = 1 + 2 + \cdots + N$ is given by

$$\frac{N(N + 1)}{2}.$$

b. If office furniture worth $6000 is to be depreciated by this method over $N = 10$ years, and the salvage value of the furniture is $500, find the depreciation for the third year by computing D_3.

22. Referring to Example 19 in Section 1.3, the amount of depreciation allowed for a printing machine, which has an estimated useful life of five years and an initial value of $100,000 (with no salvage value) was $20,000 per year using the straight-line method of depreciation. Determine the amount of depreciation that would be allowed for the first year if the printing machine were depreciated using the sum-of-the-years-digits method described in Exercise 21. Which method would result in a larger depreciation of the asset in its first year of use?

In Exercises 23–28, determine which of the given sequences are geometric progressions. For each geometric progression, find the seventh term and the sum of the first seven terms.

23. 4, 8, 16, 32, . . .

24. 1, $-1/2$, 1/4, $-1/8$, . . .

25. 1/2, $-3/8$, 1/4, $-9/64$, . . .

26. 0.004, 0.04, 0.4, 4, . . .

27. 243, 81, 27, 9, . . .

28. -1, 1, 3, 5, . . .

29. Find the twentieth term and sum of the first 20 terms of the geometric progression: $-3, 3, -3, 3, \ldots$.

30. Find the twenty-third term in a geometric progression having the first term $a = 0.1$ and ratio $r = 2$.

31. It has been projected that the population of a certain city in the Southwest will increase by 8 percent during each of the next five years. If the current population is 200,000, what is the expected population in five years?

32. The Metro Cable T.V. Company had sales of $2,500,000 in its first year of operation. If thereafter the sales increased by 12 percent of the previous year, find the sales of the company in the fifth year and the total sales over the first five years of operation.

33. Suppose that the cost-of-living index had increased by 9 percent during each of the past six years and that a member of the E.U.W. Union had been guaranteed an annual increase equal to 2 percent above the cost-of-living index over that period. What would the present salary of a union member be whose salary was $18,000 six years ago?

34. The parents of a nine-year-old boy have agreed to deposit $10 in their son's bank account on his tenth birthday and to double the size of their deposit every year thereafter until his eighteenth birthday.

   a. How much will they have to deposit on his eighteenth birthday?

   b. How much will they have deposited by his eighteenth birthday?

35. Suppose that an employee of the Stenton Printing Company whose current annual salary is $18,000 has the option of taking an annual raise of 8 percent per year for the next four years or a fixed annual raise of $1,500 per year. Which option would be more profitable to him considering his total earnings over the four-year period?

36. A culture of a certain bacteria is known to double in number every three hours. If the culture has an initial count of 20, what will the population of the culture be at the end of 24 hours?

37. Ms. Simpson is the recipient of a trust fund that she will receive over a period of six years. Under the terms of the trust, she is to receive $10,000 the first year and each succeeding annual payment is to be increased by 15 percent.

   a. How much will she receive during the sixth year?

   b. What is the total amount of the six payments that she will receive?

In Exercises 38–40, find the book value of office equipment purchased at a cost C at the end of the nth year if it is to be depreciated by the double declining-balance method over ten years. Assume a salvage value of $0.

38. $C = \$20,000, n = 4$

39. $C = \$150,000, n = 8$

40. $C = \$80,000, n = 7$

41. Restaurant equipment purchased at a cost of $150,000 is to be depreciated by the double declining-balance method over ten years. What is the book value of the equipment at the end of six years? By what amount has the equipment been depreciated at the end of the sixth year?

42. Refer to Exercise 22. Recall that a printing machine that had an estimated useful life of five years and an initial value of $100,000 (with no salvage value) was to be depreciated. At the end of the first year using the straight-line method of depreciation, the amount of depreciation allowed was $20,000, and when the sum-of-the-years-digits method was used the depreciation was $33,333. Determine the amount of depreciation that would be allowed for the first year if the printing machine were depreciated by the double declining-balance method. Which of these three methods would result in the largest depreciation of the printing machine at the end of its first year of use?

Solutions to Self-Check Exercises

5.4

1. Use (17) with $a = -24$ and $r = -1/2$, obtaining

$$S_5 = \frac{-24\left[1 - \left(-\dfrac{1}{2}\right)^5\right]}{1 - \left(-\dfrac{1}{2}\right)}$$

$$= \frac{-24\left(1 + \dfrac{1}{32}\right)}{\dfrac{3}{2}}$$

$$= -\frac{33}{2}.$$

2. Use (18) with $C = 75{,}000$, $N = 5$, and $n = 3$, giving the book value of the office equipment at the end of three years as

$$V(3) = 75{,}000\left(1 - \frac{2}{5}\right)^3$$

$$= 16{,}200,$$

or $16,200.

3. We have

$$S = P + P(1 + i) + P(1 + i)^2 + \cdots + P(1 + i)^{n-1}$$

Now, the sum on the right is easily seen to be the sum of the first n terms of a geometric progression with first term P and common ratio $(1 + i)$, so by virtue of Formula 17

$$S = \frac{P[1 - (1 + i)^n]}{1 - (1 + i)}$$

$$= P\left[\frac{(1 + i)^n - 1}{i}\right]$$

Chapter 5 Review Exercises

1. Find the accumulated amount after four years if $5000 is invested at 10 percent per year compounded (a) annually, (b) semiannually, (c) quarterly, and (d) monthly.

2. Find the effective rate of interest corresponding to a nominal rate of 12 percent per year compounded (a) annually, (b) semiannually, (c) quarterly, and (d) monthly.

3. Find the present value of $64,540 due in six years at an interest rate of 8 percent per year compounded monthly.

4. What rate of interest compounded quarterly is needed to double an investment in eight years?

5. Find the amount (future value) of an ordinary annuity of $150 per quarter for seven years at 8 percent per year compounded quarterly.

6. Find the future value of an ordinary annuity of $120 per month for ten years at 9 percent compounded monthly.

7. Find the present value of an ordinary annuity of 36 payments of $250 each made monthly and earning interest at 9 percent per year compounded monthly.

8. Find the present value of an ordinary annuity of 60 payments of $5000 each made quarterly and earning interest at 8 percent per year compounded quarterly.

9. Find the payment P that is needed to amortize a loan of $10,000 at 9.2 percent per year compounded monthly with 36 monthly installments over a period of three years.

10. Find the payment P that is needed to accumulate $15,000 with 60 monthly installments over a period of five years at an interest rate of 7.2 percent compounded monthly.

11. Find the rate of interest per year compounded on a daily basis that is equivalent to 7.2 percent per year compounded monthly.

12. The J.C.N. Media Corporation had sales of $1,750,000 in the first year of operation. If the sales increased by 14 percent per year thereafter, find the sales of the company in the fourth year and the total sales over the first four years of operation.

13. The Blakes have decided to start a monthly savings program in order to provide for their son's college education. How much should they deposit each month in a time-deposit bank account earning interest at the rate of 8 percent per year compounded monthly so that at the end of the tenth year the accumulated amount will be $40,000?

14. Ms. Cantwell made a down payment of $400 toward the purchase of new furniture. In order to pay the balance of the purchase price she has secured a loan from her bank at 12 percent per year compounded monthly. Under the terms of her finance agreement she is required to make payments of $75.32 for 24 months. What was the purchase price of the furniture?

15. The Turners have purchased a house for $120,000. They made an initial down payment of $24,000 and have secured a mortgage with interest charged at the rate of 12 percent per year on the unpaid balance. Interest computations are made at the end of each month. If the loan is to be amortized over 25 years,

    a. what monthly payment will the Turners be required to make?

    b. what is the total interest payment?

    c. what is their equity (disregarding appreciation) after ten years?

16. The management of a corporation anticipates a capital expenditure of $500,000 in five years' time for the purpose of purchasing replacement machinery. In order to finance this purchase, a sinking fund that earns interest at the rate of 10 percent per year compounded quarterly will be set up. Determine the amount of each (equal) quarterly installment that should be deposited in the fund.

17. The outstanding balance on Mr. Baker's credit card account is $3200. The bank

issuing the credit card is charging 18.6 percent per year compounded monthly. If Mr. Baker decides to pay off this balance in 18 equal monthly installments, how much will his monthly payment be?

18. Refer to Exercise 17. What is the effective rate of interest the bank is charging Mr. Baker?

Sets and

Counting

What are the investment options? An investor has decided to purchase shares of stocks from a recommended list of aerospace companies, companies engaged in energy development, and electronics companies. In Example 30, page 305, we will determine how many ways the investor may select the group of three companies from the list.

►CHAPTER

SIX

6.1

Sets and Set Operations

▶ Set Terminology and Notation

▶ Set Operations

▶ Application

▶ Set Terminology and Notation

The word *set* permeates much of modern mathematics. For our purpose, a **set** is a well-defined aggregate or collection of objects. Thus, a set is not just any collection of objects, but it must be well defined in the sense that, given any object, we should be able to determine whether or not it belongs to the collection.

EXAMPLE

1 *a.* By our interpretation of the term *set*, each of the following are sets:

    (i) The set of days in the year 1981.

    (ii) The set of members of the New York Philharmonic Orchestra.

    (iii) The set of Ford Pintos produced in the year 1976.

    (iv) The set of negative integers.[†]
    (v) The set of real numbers.[†]

b. The following collections of objects are not well defined and are therefore not sets:

    (i) The collection of fine champagne.

    (ii) The collection of intelligent people.

    (iii) The collection of hot days this month. ◀

[†]See Appendix B.

The objects of a set are called the **elements,** or *members,* of the set and are usually denoted by lowercase letters a, b, c, . . . , but the sets themselves are usually denoted by uppercase letters A, B, C, ... The elements of a set may be displayed by listing each element between braces. For example, using the **roster notation,** the set A consisting of the first three letters of the English alphabet is written

$$A = \{a, b, c\}.$$

The set B of all letters of the alphabet may be written as

$$B = \{a, b, c, \ldots, z\}.$$

Another kind of set notation commonly used is **set-builder notation.** Here, a rule is given that describes the definite property or properties to be satisfied by an object x to qualify for membership in the set. Using this notation, the set B is written as

$$B = \{x \mid x \text{ is a letter of the English alphabet}\}$$

and is read "B is the set of all elements x such that x is a letter of the English alphabet."

EXAMPLE

2   List the elements of the set

$$A = \{x \mid x \text{ is a day of the week}\}.$$

SOLUTION

$A = \{$Monday, Tuesday, Wednesday, Thursday, Friday, Saturday, Sunday$\}$. ◀

EXAMPLE

3   Write the set

$$B = \{5, 6, 7, 8, 9\}$$

in set-builder notation.

SOLUTION

$B = \{x \mid x \text{ is an integer greater than 4 but less than 10}\}$.   ◀

EXAMPLE

4   List the elements of the set

$$C = \{x \mid x + 3 = 0; x \text{ an integer}\}.$$

SOLUTION Only one integer satisfies the condition $x + 3 = 0$, namely, the integer -3. Thus, the set C consists of only the number -3, and $C = \{-3\}$. ◀

EXAMPLE

5 List the elements of the set of letters in the word *MISSISSIPPI*.

SOLUTION The required set is $\{I, M, P, S\}$. Note that, by convention, an element belonging to a set is listed only once. ◀

If a is an element of a set A, we write $a \in A$, read "a belongs to A" or "a is an element of A." If, on the other hand, the element a does not belong to the set A, then we write $a \notin A$, read "a does not belong to A." For example, if $A = \{1, 2, 3, 4, 5\}$, then $3 \in A$ but $6 \notin A$.

Equality of Sets

Two sets A and B are **equal,** written $A = B$, if and only if they have exactly the same elements.

EXAMPLE

6 Let A, B, and C be the sets

$$A = \{a, e, i, o, u\}$$
$$B = \{a, i, o, e, u\}$$
$$C = \{a, e, i, o\}$$

Then $A = B$, since they both contain exactly the same elements. Note that the order in which the elements are displayed is immaterial. Also, $A \neq C$, since $u \in A$ but $u \notin C$. Similarly, we conclude that $B \neq C$. ◀

Subset

If every element of a set A is also an element of a set B, then we say that A is a **subset** of B and write $A \subseteq B$.

By this definition, two sets A and B are equal if and only if (1) $A \subseteq B$ and (2) $B \subseteq A$. You may verify this (see Exercises 6.1, Problem 22).

EXAMPLE

7 Referring to Example 6, we find that $C \subseteq B$, since every element of C is also an element of B. Also, if D is the set

$$D = \{a, e, i, o, x\},$$

then D is not a subset of A, written $D \not\subseteq A$, since $x \in D$ but $x \notin A$. Observe that $A \not\subseteq D$ as well, since $u \in A$ but $u \notin D$. ◄

If A and B are sets such that $A \subseteq B$ but $A \neq B$, then we say that A is a **proper subset** of B. In other words, a set A is a proper subset of a set B, written $A \subset B$, if (1) $A \subseteq B$ and (2) there exists at least one element in B that is not in A. The latter condition states that the set A is properly "smaller" than the set B.

EXAMPLE

8 Let $A = \{1, 2, 3, 4, 5, 6\}$ and $B = \{2, 4, 6\}$. Then B is a proper subset of A, since (1) $B \subseteq A$, which is easily verified, and (2) there exists at least one element in A that is not in B—for example, the element 1. ◄

Empty Set

The set that contains no elements is called the **empty set** and is denoted ϕ.

EXAMPLE

9 The following are examples of the empty set.

a. $\{x \mid x$ is a person who stands over twelve feet tall$\}$
b. $\{x \mid x$ is a horse that flies$\}$
c. $\{x \mid x + 2 = 0; x$ is a positive integer$\}$ ◄

Observe that the empty set, ϕ, is a subset of every set. To see this, let A be any set. If $\phi \not\subseteq A$, then there must exist at least one element in ϕ that is not in A. But ϕ is empty, so this contradiction establishes the assertion.

EXAMPLE

10 List all subsets of the set $A = \{a, b, c\}$.

SOLUTION

The subsets are

$$\phi, \{a\}, \{b\}, \{c\}, \{a, b\}, \{a, c\}, \{b, c\}, \{a, b, c\}. \qquad \blacktriangleleft$$

In contrast with the empty set we have on the other extreme, the notion of a largest, or *universal*, set. A **universal set** is the set of all elements of interest in a particular discussion. It is the largest in the sense that all sets considered in the discussion of the problem are subsets of the universal set. Of course, different universal sets are associated with different problems.

EXAMPLE

11 *a.* If the problem at hand is to determine the ratio of female to male students in a college, then a logical choice of a universal set is the set comprising the whole student body of the college.

b. If the problem is to determine the ratio of female to male students in the business department of the college in (a), then the set of all students in the business department may be chosen as the universal set. $\qquad \blacktriangleleft$

Observe that the universal set in Example 11(a) will serve as a universal set for the problem posed in Example 11(b), but not vice-versa. Note also that the set comprising the student body of another college will not serve as a universal set for the problem in Example 11(a), nor can the set of business students in a different college be taken as a universal set for the problem in Example 11(b). In practice, the choice of a universal set for a particular problem is usually clear from the context of the problem, and in this case the set need not be mentioned specifically.

A visual representation of sets is realized through the use of **Venn diagrams,** which are of considerable help in understanding the concepts introduced earlier as well as in solving problems involving sets. The universal set U is represented by a rectangle, and subsets of U are represented by regions lying inside the rectangle.

EXAMPLE

12 Use Venn diagrams to illustrate the following statements:

a. The sets A and B are equal.

b. The set A is a proper subset of the set B.

c. The sets A and B are not subsets of each other.

SOLUTION The respective Venn diagrams are shown in Figures 6.1(a), (b), and (c).

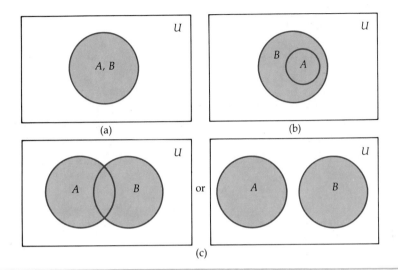

Figure 6.1

▶ Set Operations

Having introduced the concept of a set, our next task is to consider operations on sets; that is, to consider ways in which sets may be combined to yield other sets. These operations enable us to combine sets in much the same way the operations of addition and multiplication enable us to combine numbers to obtain other numbers. This leads to a mathematical system with a sufficiently rich structure to enable us to solve many problems of practical interest. In what follows, all sets are assumed to be subsets of a given universal set U.

Set Union

Let A and B be sets. The **union** of A and B, written $A \cup B$, is the set of all elements that belong to either A or B or both.

$$A \cup B = \{x \mid x \in A \text{ or } x \in B \text{ or both}\}$$

The shaded portion of the Venn diagram (Figure 6.2) depicts the set $A \cup B$.

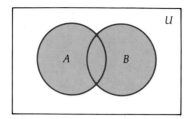

Figure 6.2

EXAMPLE

13   If $A = \{a, b, c\}$ and $B = \{a, c, d\}$ then $A \cup B = \{a, b, c, d\}$.   ◀

Set Intersection

Let A and B be sets. The set of elements in common with the sets A and B, written $A \cap B$, is called the **intersection** of A and B.

$$A \cap B = \{x \mid x \in A \text{ and } x \in B\}$$

The shaded portion of the Venn diagram (Figure 6.3) depicts the set $A \cap B$.

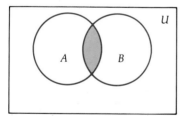

Figure 6.3

EXAMPLE

14   Let $A = \{a, b, c\}$ and $B = \{a, c, d\}$. Then $A \cap B = \{a, c\}$. (Compare this result with Example 13.)   ◀

EXAMPLE

15   Let $A = \{1, 3, 5, 7, 9\}$ and $B = \{2, 4, 6, 8, 10\}$. Then $A \cap B = \phi$.   ◀

The two sets of Example 15 have null intersection. In general, the sets A and B are said to be **disjoint** if they have no elements in common; that is, if $A \cap B = \phi$.

EXAMPLE

16 If U is the set of all students in the classroom and $M = \{x \in U \mid x \text{ is male}\}$ and $F = \{x \in U \mid x \text{ is female}\}$, then $F \cap M = \phi$, and F and M are disjoint. ◀

Complement of a Set

If U is a universal set and A is a subset of U, then the set of all elements in U that are not in A is called the **complement** of A and is denoted A^c.

$$A^c = \{x \mid x \in U, x \notin A\}$$

The shaded portion of the Venn diagram (Figure 6.4) shows the set A^c.

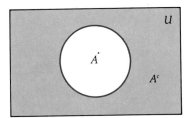

Figure 6.4

EXAMPLE

17 Let $U = \{1, 2, 3, 4, 5, 6, 7, 8, 9, 10\}$ and $A = \{2, 4, 6, 8, 10\}$. Then $A^c = \{1, 3, 5, 7, 9\}$. ◀

The following rules hold for the operation of complementation. See whether you can verify them.

Set Complementation

If U is a universal set and A is a subset of U, then

a. $U^c = \phi$ b. $\phi^c = U$ c. $(A^c)^c = A$

d. $A \cup A^c = U$ e. $A \cap A^c = \phi$

The following rules govern the operations on sets.

Set Operations

Let U be a universal set. If A, B, and C are arbitrary subsets of U, then

$A \cup B = B \cup A$	*Commutative law for union*
$A \cap B = B \cap A$	*Commutative law for intersection*
$A \cup (B \cup C) = (A \cup B) \cup C$	*Associative law for union*
$A \cap (B \cap C) = (A \cap B) \cap C$	*Associative law for intersection*
$A \cup (B \cap C) = (A \cup B) \cap (A \cup C)$	*Distributive law for union*
$A \cap (B \cup C) = (A \cap B) \cup (A \cap C)$	*Distributive law for intersection*

There are two additional rules, referred to as De Morgan's Laws, that govern the operations on sets.

De Morgan's Laws

Let A and B be sets. Then

$$(A \cup B)^c = A^c \cap B^c \tag{1}$$
$$(A \cap B)^c = A^c \cup B^c \tag{2}$$

Equation (1) states that the complement of the union of two sets is equal to the intersection of their complements. Equation (2) states that the complement of the intersection of two sets is equal to the union of their complements.

Although all of the above results may be proven rigorously, we will be content with demonstrating the plausibility of Equation (2) through the use of Venn diagrams, as the following example illustrates.

EXAMPLE

18 Using Venn diagrams, show that

$$(A \cap B)^c = A^c \cup B^c.$$

SOLUTION $(A \cap B)^c$ is the set of elements in U but not in $A \cap B$ and is thus the shaded region shown in Figure 6.5. Next, A^c and B^c are shown in Figures 6.6(a) and 6.6(b), respectively. Their union, $A^c \cup B^c$, is easily seen to be equivalent to $(A \cap B)^c$ by referring once again to Figure 6.5.

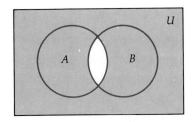

Figure 6.5

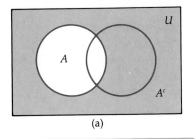

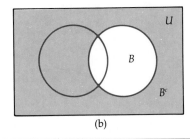

Figure 6.6   (a)   (b)   ◀

EXAMPLE

19 Let $U = \{1, 2, 3, 4, 5, 6, 7, 8, 9, 10\}$, $A = \{1, 2, 4, 8, 9\}$, and $B = \{3, 4, 5, 6, 8\}$. Verify by direct computation that $(A \cup B)^c = A^c \cap B^c$.

SOLUTION $A \cup B = \{1, 2, 3, 4, 5, 6, 8, 9\}$, so $(A \cup B)^c = \{7, 10\}$. On the other hand, $A^c = \{3, 5, 6, 7, 10\}$ and $B^c = \{1, 2, 7, 9, 10\}$, so $A^c \cap B^c = \{7, 10\}$. The required result follows. ◀

EXAMPLE

20 Let $A = \{1, 3, 4, 5, 8, 10\}$, $B = \{1, 2, 4, 6, 9\}$, and $C = \{2, 4, 6, 8\}$. Verify by direct computation that $A \cap (B \cup C) = (A \cap B) \cup (A \cap C)$.

SOLUTION $B \cup C = \{1, 2, 4, 6, 8, 9\}$, so $A \cap (B \cup C) = \{1, 4, 8\}$. On the other hand, we find $A \cap B = \{1, 4\}$ and $A \cap C = \{4, 8\}$, so $(A \cap B) \cup (A \cap C) = \{1, 4, 8\}$ and the desired result follows. ◀

▶ Application

EXAMPLE

21 Let U denote the set of all cars in a dealer's lot and

$A = \{x \in U \mid x \text{ is equipped with automatic transmission}\}$
$B = \{x \in U \mid x \text{ is equipped with air conditioning}\}$
$C = \{x \in U \mid x \text{ is equipped with power steering}\}$

Find an expression in terms of A, B, and C for each of the following sets:

a. The set of cars with at least one of the given options.

b. The set of cars with exactly one of the given options.

c. The set of cars with automatic transmission and power steering but no air conditioning.

SOLUTION

a. The set of cars with at least one of the given options is $A \cup B \cup C$ [Figure 6.7(a)].

b. The set of cars with automatic transmission only is given by $A \cap B^c \cap C^c$. Similarly, we find that the set of cars with air conditioning only is given by $B \cap C^c \cap A^c$, whereas the set of cars with power steering only is given by $C \cap A^c \cap B^c$. Thus, the set of cars with exactly one of the given options is $(A \cap B^c \cap C^c) \cup (B \cap C^c \cap A^c) \cup (C \cap A^c \cap B^c)$ [Figure 6.7(b)].

c. The set of cars with automatic transmission and power steering but no air conditioning is given by $A \cap C \cap B^c$ [Figure 6.7(c)].

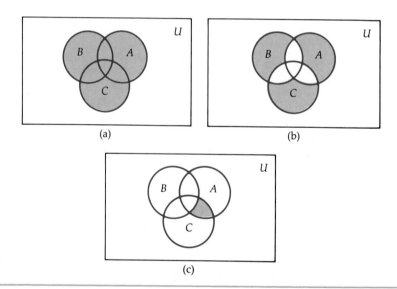

Figure 6.7

Self-Check Exercises

6.1

1. Let $U = \{1, 2, 3, 4, 5, 6, 7\}$, $A = \{1, 2, 3\}$, $B = \{3, 4, 5, 6\}$, and $C = \{2, 3, 4\}$. Find the following sets:

a. A^c *b.* $A \cup B$ *c.* $B \cap C$

d. $(A \cup B) \cap C$ *e.* $(A \cap B) \cup C$ *f.* $A^c \cap (B \cup C)^c$

2. Let U denote the set of all members of the House of Representatives. Let

$$D = \{x \in U \mid x \text{ is a Democrat}\}$$
$$R = \{x \in U \mid x \text{ is a Republican}\}$$
$$F = \{x \in U \mid x \text{ is a female}\}$$
$$L = \{x \in U \mid x \text{ is a lawyer by training}\}$$

Describe each of the following sets in words.

·*a.* $D \cap F$ *b.* $F \cap R$ *c.* $D \cap F \cap L^c$

Solutions to Self-Check Exercises 6.1 can be found on page 292.

▶

Exercises 6.1

1. Determine which of the following are sets:
   a. The past and present presidents of the United States
   b. The winners of the Indianapolis 500
   c. The warm days in the month of September 1981
   d. The streets of San Francisco

2. Determine which of the following are sets:
   a. The fast quarterbacks of the NFL teams
   b. The collection of the Library of Congress
   c. The days in the month of September 1980 in which the highest temperatures recorded exceeded 70°F
   d. The pictures at an exhibition

3. Write the following sets in set-builder notation.
   a. The set of gold medalists in the 1976 Olympic Games
   b. $\{3, 4, 5, 6, 7\}$
   c. The set of real numbers satisfying the equation $x^2 - 4x + 2 = 0$

4. Write the following sets in set-builder notation.
   a. $\{0, 2, 4, 6, 8, \ldots\}$
   b. $\{1, 3, 5, 7, 9, 11, \ldots, 39\}$
   c. The set of football teams in the NFL

In Exercises 5–8, list the elements of the given sets in roster notation.

5. *a.* $\{x \mid x \text{ is a digit in the number } 352{,}646\}$
   b. $\{x \mid x \text{ is a letter in the word } HIPPOPOTAMUS\}$

6. *a.* $\{x \mid 2 - x = 4; x, \text{ an integer}\}$
   b. $\{x \mid 2 - x = 4; x, \text{ a rational number}\}$

7. *a.* $\{x \mid (x + 2)(x - 3) = 0; x, \text{ a real number}\}$
   b. $\{x \mid (x + 1)(x - 2) = 0; x, \text{ a negative integer}\}$

8. a. $\{x \mid x^2 - 2 = 0; x,$ a rational number$\}$
   b. $\{x \mid x^2 - 2 = 0; x,$ a real number$\}$

In Exercises 9–12, state whether the given statements are true or false.

9. a. $\{a, b, c\} = \{c, a, b\}$ b. $\phi \in A$ c. $A \in A$

10. a. $A \subset A$ b. $0 \in \phi$ c. $0 = \phi$

11. a. $\{\phi\} = \phi$ b. $\{a, b\} \in \{a, b, c\}$

12. a. $\{$Chevrolet, Pontiac, Buick$\} \subset \{$General Motors$\}$
    b. $\{x \mid x$ is a silver medalist in the 1977 Olympic Games$\} = \phi$
    c. $\{x \mid x$ is a unicorn in the San Diego Zoo$\} = \phi$

13. Let $A = \{1, 2, 3, 4, 5\}$. Determine whether each of the following statements is true or false:
    a. $2 \in A$ b. $A \subseteq \{2, 4, 6\}$ c. $0 \in A$
    d. $\{1, 3, 5\} \in A$ e. $A \subseteq \{5, 4, 3, 2, 1\}$ f. $4 \subset A$

14. Let $A = \{1, 2, 3\}$. Which of the following sets are equal to A?
    a. $\{2, 1, 3\}$ b. $\{3, 2, 1\}$ c. $\{0, 1, 2, 3\}$
    d. $\{x \mid x$ is a whole number greater than 0 but less than 4$\}$
    e. $\{x \mid (x - 1)(x - 2)(x - 3) = 0; x,$ a real number$\}$

15. Let $A = \{a, e, l, t, r\}$. Which of the following sets are equal to A?
    a. $\{x \mid x$ is a letter of the word *later*$\}$
    b. $\{x \mid x$ is a letter of the word *latter*$\}$
    c. $\{x \mid x$ is a letter of the word *relate*$\}$

16. Let $A = \{a, b, c, d, e\}$. Determine whether each of the following statements is true or false:
    a. $\{a, c\} \subseteq A$ b. $\{a, g\} \subseteq A$ c. $\{e, d, c, b, a\} = A$
    d. $A \subseteq \{f, g, h, i, j\}$ e. $\phi \subset A$ f. $A \subseteq A$
    g. $\{a, c, e\}$ is a proper subset of A

17. List all subsets of the following sets:
    a. ϕ b. $\{1\}$ c. $\{1, 2\}$
    d. $\{1, 2, 3\}$ e. $\{1, 2, 3, 4\}$

18. List all subsets of the set $A = \{$Emerson, Hawthorne, Hemingway$\}$. Which of these are proper subsets of A?

19. List all subsets of the set $A = \{$IBM, U.S. Steel, Union Carbide, Boeing$\}$. Which of these are proper subsets of A?

20. Find the *smallest* possible set (that is, the set with the least number of elements) that contains the given sets as subsets:
    a. $\{1, 2\}, \{1, 3, 4\}, \{4, 6, 8, 10\}$
    b. $\{1, 2, 4\}, \{a, b\}$
    c. $\{$Jill, John, Jack$\}, \{$Susan, Sharon$\}$
    d. $\{$GM, Ford, Chrysler$\}, \{$Daimler-Benz, Volkswagen$\}, \{$Toyota, Datsun$\}$

21. Suppose that $A \subset B$ and $B \subset C$, where A and B are any two sets. What conclusion can be drawn regarding the sets A and C?

***22.** Verify the assertion that two sets A and B are equal if and only if (1) $A \subseteq B$ and (2) $B \subseteq A$.

23. Use Venn diagrams to represent the following relationships:

    a. $A \subset B$ and $B \subset C$.

    b. $A \subset U$ and $B \subset U$, where A and B have no elements in common.

    c. The sets A, B, and C are equal.

24. Let U denote the set of all students who applied for admission to the freshman class at Faber College for the upcoming academic year and let

$$A = \{x \in U \mid x \text{ is a successful applicant}\}$$
$$B = \{x \in U \mid x \text{ is a female student who enrolled in the freshman class}\}$$
$$C = \{x \in U \mid x \text{ is a male student who enrolled in the freshman class}\}$$

    a. Use Venn diagrams to represent the sets U, A, B, and C.

    b. Determine whether the following statements are true or false.

        (i) $A \subseteq B$ *(ii)* $B \subset A$ *(iii)* $C \subset B$

In Exercises 25–27, shade the portion of the accompanying figure that represents each of the given sets:

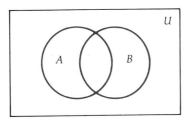

25. a. $A \cap B^c$ **b.** $A^c \cap B$

26. a. $A^c \cap B^c$ **b.** $(A \cup B)^c$

27. a. $A^c \cup B^c$ **b.** $(A \cap B)^c$

In Exercises 28–30, shade the portion of the accompanying figure that represents each of the given sets:

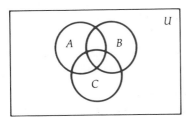

28. *a.* $A \cup B \cup C$ *b.* $A \cap B \cap C$ *c.* $A \cap B \cap C^c$

29. *a.* $A^c \cap B \cap C$ *b.* $A \cap B^c \cap C$ *c.* $A^c \cap B^c \cap C^c$

30. *a.* $(A \cup B)^c \cap C$ *b.* $A \cup (B \cap C)^c$ *c.* $(A \cup B \cup C)^c$

In Exercises 31–34, let $U = \{1, 2, 3, 4, 5, 6, 7, 8, 9, 10\}$, $A = \{1, 3, 5, 7, 9\}$, $B = \{2, 4, 6, 8, 10\}$, and $C = \{1, 2, 4, 5, 8, 9\}$. Find each of the given sets.

31. *a.* A^c *b.* $B \cup C$ *c.* $C \cup C^c$

32. *a.* $C \cap C^c$ *b.* $(A \cap C)^c$ *c.* $A \cup (B \cap C)$

33. *a.* $(A \cap B) \cup C$ *b.* $(A \cup B \cup C)^c$ *c.* $(A \cap B \cap C)^c$

34. *a.* $A^c \cap (B \cap C)$ *b.* $(A \cup B^c) \cup (B \cap C^c)$ *c.* $(A \cup B)^c \cap C^c$

35. Determine whether the following pairs of sets are disjoint:

a. $\{1, 2, 3, 4\}, \{4, 5, 6, 7\}$ *b.* $\{a, c, e, g\}, \{b, d, f\}$

c. $\phi, \{1, 3, 5\}$ *d.* $\{0, 1, 3, 4\}, \{0, 2, 5, 7\}$

e. $\{(\text{Michael, William}\}, \{\text{Jane, Michael, Mary}\}$

In Exercises 36–38, let U denote the set of all employees at the Universal Life Insurance Company. Let

$$T = \{x \in U \mid x \text{ drinks tea}\}$$
$$C = \{x \in U \mid x \text{ drinks coffee}\}$$

Describe each of the given sets in words:

36. *a.* T^c *b.* C^c *c.* $T \cup C$

37. *a.* $T \cap C$ *b.* $T \cap C^c$ *c.* $T^c \cap C$

38. *a.* $T^c \cap C^c$ *b.* $(T \cup C)^c$ *c.* $(T \cap C)^c$

39. Refer to Example 21. Find an expression in terms of A, B, and C for each of the following sets:

     a. The set of cars with all three of the given options

     b. The set of cars with exactly two of the given options

     c. The set of cars with at least two of the given options

     d. The set of cars with automatic transmission and power steering but no air conditioning

     e. The set of cars with none of the three given options

In Exercises 40–42, let U denote the set of all employees in a hospital. Let

$$N = \{x \in U \mid x \text{ is a nurse}\}$$
$$D = \{x \in U \mid x \text{ is a doctor}\}$$
$$A = \{x \in U \mid x \text{ is an administrator}\}$$
$$M = \{x \in U \mid x \text{ is a male}\}$$
$$F = \{x \in U \mid x \text{ is a female}\}$$

Describe each of the given sets in words:

40. *a.* D^c *b.* N^c *c.* $N \cup D$

41. *a.* $N \cap M$ *b.* $D \cap M^c$ *c.* $D \cap A$

42. *a.* $N \cap F$ *b.* $D^c \cap F$ *c.* $(D \cup N)^c$

In Exercises 43 and 44, let U denote the set of all senators in Congress. Let

$$D = \{x \in U \mid x \text{ is a Democrat}\}$$
$$R = \{x \in U \mid x \text{ is a Republican}\}$$
$$F = \{x \in U \mid x \text{ is a female}\}$$
$$L = \{x \in U \mid x \text{ is a lawyer by training}\}$$

Write the set that represents each of the given statements:

43. *a.* The set of all Democrats who are female

    b. The set of all Republicans who are male and are not lawyers by training

44. *a.* The set of all Democrats who are female or are lawyers by training

    b. The set of all senators who are not Democrats or are lawyers by training

In Exercises 45 and 46, let U denote the set of all students in the Business College of a certain university and let

$$A = \{x \in U \mid x \text{ had taken a course in accounting}\}$$
$$B = \{x \in U \mid x \text{ had taken a course in economics}\}$$
$$C = \{x \in U \mid x \text{ had taken a course in marketing}\}$$

Write the set that represents each of the given statements:

45. *a.* The set of students who have not had a course in economics

    b. The set of students who have had courses in accounting and economics

    c. The set of students who have had courses in accounting and economics but not marketing

46. *a.* The set of students who have had courses in economics but not courses in accounting or marketing

    b. The set of students who have had at least one of the three courses

    c. The set of students who have had all three courses

In Exercises 47–49, use Venn diagrams to illustrate each of the given statements:

47. *a.* $A \subset A \cup B$ and $B \subset A \cup B$ *b.* $A \cap B \subset A$ and $A \cap B \subset B$

48. *a.* $A \cup (B \cup C) = (A \cup B) \cup C$ *b.* $A \cap (B \cap C) = (A \cap B) \cap C$

49. *a.* $A \cap (B \cup C) = (A \cap B) \cup (A \cap C)$ *b.* $(A \cup B)^c = A^c \cap B^c$

In Exercises 50 and 51, let $U = \{1, 2, 3, 4, 5, 6, 7, 8, 9, 10\}$, $A = \{1, 3, 5, 7, 9\}$, $B = \{1, 2, 4, 7, 8\}$, *and* $C = \{2, 4, 6, 8\}$. *Verify by direct computation each of the given equations.*

50. *a.* $A \cup (B \cup C) = (A \cup B) \cup C$ *b.* $A \cap (B \cap C) = (A \cap B) \cap C$

51. *a.* $A \cap (B \cup C) = (A \cap B) \cup (A \cap C)$ *b.* $(A \cup B)^c = A^c \cap B^c$

In Exercises 52–54, refer to the accompanying figure and find the points that belong to each of the given sets.

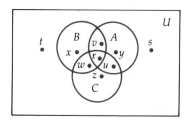

52. *a.* $A \cup B$ *b.* $A \cap B$ *c.* $A \cap (B \cup C)$

53. *a.* $(B \cup C)^c$ *b.* A^c *c.* $(B \cap C)^c$

54. *a.* $(A \cup B \cup C)^c$ *b.* $(A \cap B) \cap C^c$ *c.* $(A^c \cap C^c) \cup (B^c \cap C^c)$

Solutions to Self-Check Exercises

6.1

1. *a.* A^c is the set of all elements in U but not in A. Therefore,

$$A^c = \{4, 5, 6, 7\}.$$

b. $A \cup B$ consists of all elements in A and/or B. So,

$$A \cup B = \{1, 2, 3, 4, 5, 6\}.$$

c. $B \cap C$ is the set of all elements in B and C. Therefore,

$$B \cap C = \{3, 4\}.$$

d. Using the result from (b), we find

$$(A \cup B) \cap C = \{1, 2, 3, 4, 5, 6\} \cap \{2, 3, 4\}$$
$$= \{2, 3, 4\}.$$

e. First we compute

$$A \cap B = \{3\}.$$

Next, since $(A \cap B) \cup C$ is the set of all elements in $(A \cap B)$ and/or C, we conclude that

$$(A \cap B) \cup C = \{3\} \cup \{2, 3, 4\}$$
$$= \{2, 3, 4\}.$$

f. From (a), we have $A^c = \{4, 5, 6, 7\}$. Next, we compute

$$B \cup C = \{3, 4, 5, 6\} \cup \{2, 3, 4\}$$
$$= \{2, 3, 4, 5, 6\},$$

from which we deduce that

$$(B \cup C)^c = \{1, 7\} \quad \text{(The set of elements in } U \text{ but not in } B \cup C)$$

Finally, using these results, we obtain

$$A^c \cap (B \cup C)^c = \{4, 5, 6, 7\} \cap \{1, 7\}$$
$$= \{7\}.$$

2. *a.* $D \cap F$ denotes the set of all elements in D and F. Since an element in D is a Democrat and an element in F is a female representative, we see that $D \cap F$ is the set of all female Democrats in the House of Representatives.

b. Since F^c is the set of male representatives and R is the set of Republicans, we see that $F^c \cap R$ is the set of male Republicans in the House of Representatives.

c. L^c is the set of representatives who are not lawyers by training. Therefore, $D \cap F \cap L^c$ is the set of female Democratic representatives who are not lawyers by training.

6.2

The Number of Elements in a Finite Set

▶ Counting the Elements in a Set

▶ Applications

▶ Counting the Elements in a Set

The solution of many problems in mathematics often centers on determining the number of elements in certain sets. In the ensuing sections we will study some of the elementary techniques used to determine the number of elements in a set. A problem in which we find the number of elements in a set is often called a **counting problem** and constitutes a field of study known as **combinatorics.** Our study of combinatorics, however, is restricted to the results that will be required for our work in probability later on.

The number of elements in a finite set is determined by simply counting the elements in the set. If A is a set, then $n(A)$ denotes the number of elements in A. For example, if

$$A = \{1, 2, 3, \ldots , 20\}, \quad B = \{a, b\}, \quad C = \{8\},$$

then $n(A) = 20$, $n(B) = 2$, and $n(C) = 1$.

The empty set has no elements in it, so $n(\phi) = 0$. Another result that is easily seen to be true is the following: If A and B are disjoint sets, then

$$n(A \cup B) = n(A) + n(B) \tag{3}$$

EXAMPLE

22   If $A = \{a, c, d\}$ and $B = \{b, e, f, g\}$, then $n(A) = 3$ and $n(B) = 4$, so $n(A) + n(B) = 7$. On the other hand, $A \cup B = \{a, b, c, d, e, f, g\}$ and $n(A \cup B) = 7$. Thus, Equation (3) holds true in this case. Note that $A \cap B = \phi$.   ◀

In the general case, A and B need not be disjoint, which leads us to the general formula

$$n(A \cup B) = n(A) + n(B) - n(A \cap B)$$ (4)

To see this, we observe that the set $A \cup B$ may be viewed as the union of three mutually disjoint sets with x, y, and z elements, respectively (see Figure 6.8). The figure shows that

$$n(A \cup B) = x + y + z.$$

Also, $$n(A) = x + y$$
and $$n(B) = y + z,$$
so $$\begin{aligned} n(A) + n(B) &= (x + y) + (y + z) \\ &= (x + y + z) + y \\ &= n(A \cup B) + n(A \cap B) \quad [n(A \cap B) = y] \end{aligned}$$

Thus, solving for $n(A \cup B)$, we obtain

$$n(A \cup B) = n(A) + n(B) - n(A \cap B),$$

the desired result.

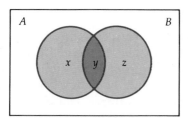

Figure 6.8

EXAMPLE

23   Let $A = \{a, b, c, d, e\}$ and $B = \{b, d, f, h\}$. Verify Equation (4) directly.

SOLUTION

$$A \cup B = \{a, b, c, d, e, f, h\}, \quad \text{so} \quad n(A \cup B) = 7.$$
$$A \cap B = \{b, d\}, \quad \text{so} \quad n(A \cap B) = 2.$$

Furthermore, $$n(A) = 5, \, n(B) = 4,$$
so $$n(A) + n(B) - n(A \cap B) = 5 + 4 - 2 = 7 = n(A \cup B).$$ ◀

▶ Applications

EXAMPLE

24   In a survey of 100 coffee drinkers it was found that 70 take sugar, 60 take cream, and 50 take both sugar and cream with their coffee. How many coffee drinkers take sugar or cream with their coffee?

SOLUTION   Let U denote the set of 100 coffee drinkers surveyed, and let

$$A = \{x \in U \mid x \text{ takes sugar}\}$$
$$B = \{x \in U \mid x \text{ takes cream}\}$$

Then $n(A) = 70$, $n(B) = 60$, and $n(A \cap B) = 50$. The set of coffee drinkers who take sugar or cream with their coffee is given by $A \cup B$. Using Equation (4), we find

$$n(A \cup B) = n(A) + n(B) - n(A \cap B)$$
$$= 70 + 60 - 50 = 80,$$

so 80 out of the 100 coffee drinkers surveyed take cream or sugar with their coffee.   ◀

An equation similar to Equation (4) may be derived for the case that involves any finite number of finite sets. For example, a relationship involving the number of elements in the sets A, B, and C is given by

$$n(A \cup B \cup C) = n(A) + n(B) + n(C) - n(A \cap B) - n(A \cap C) - n(B \cap C)$$
$$+ n(A \cap B \cap C) \qquad (5)$$

For a proof of this result, see Problem 30, Exercises 6.2.

As useful as equations such as (5) are, in practice it is often easier to attack a problem directly with the aid of Venn diagrams, as shown by the following example.

EXAMPLE

25   A leading cosmetics manufacturer advertised its products in three magazines: *Cosmopolitan*, *McCalls*, and the *Ladies Home Journal*. A survey of 500 customers by the manufacturer reveals the following information:

> 180 learned of its products from *Cosmopolitan*
>
> 200 learned of its products from *McCalls*
>
> 192 learned of its products from the *Ladies Home Journal*
>
> 84 learned of its products from *Cosmopolitan* and *McCalls*
>
> 52 learned of its products from *Cosmopolitan* and the *Ladies Home Journal*

64 learned of its products from *McCalls* and the *Ladies Home Journal*

38 learned of its products from all three magazines.

How many of the customers saw the manufacturer's advertisement in

a. at least one magazine?

b. exactly one magazine?

SOLUTION   Let U denote the set of all customers surveyed, and let

$$C = \{x \in U \mid x \text{ learned of the products from } Cosmopolitan\}$$
$$M = \{x \in U \mid x \text{ learned of the products from } McCalls\}$$
$$L = \{x \in U \mid x \text{ learned of the products from the } Ladies\ Home\ Journal\}$$

The result that 38 customers learned of the products from all three magazines translates into $n(C \cap M \cap L) = 38$ [see Figure 6.9(a)]. Next, the result that 64 learned of the products from *McCalls* and the *Ladies Home Journal* translates into $n(M \cap L) = 64$. This leaves

$$64 - 38 = 26$$

who learned of the products from *only McCalls* and the *Ladies Home Journal* [Figure 6.9(b)]. Similarly, $n(C \cap L) = 52$, so

$$52 - 38 = 14$$

learned of the product from *only Cosmopolitan* and the *Ladies Home Journal*, and $n(C \cap M) = 84$, so

$$84 - 38 = 46$$

learned of the product from *only Cosmopolitan* and *McCalls*. These numbers appear in the appropriate regions in Figure 6.9(b).

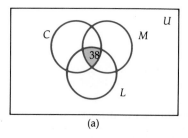 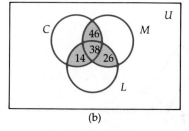

Figure 6.9   (a)   (b)

Continuing, we have $n(L) = 192$, so the number who learned of the products from the *Ladies Home Journal* only is given by

$$192 - 14 - 38 - 26 = 114 \quad \text{(See Figure 6.10.)}$$

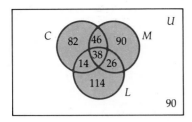

Figure 6.10

Similarly, $n(M) = 200$, so

$$200 - 46 - 38 - 26 = 90$$

learned of the products from *only McCalls,* and $n(C) = 180$, so

$$180 - 14 - 38 - 46 = 82$$

learned of the products from *only Cosmopolitan.* Finally,

$$500 - (90 + 26 + 114 + 14 + 82 + 46 + 38) = 90$$

learned of the products from other sources.

We are now in a position to answer questions (a) and (b).

a. Referring to Figure 6.10, we see that the number of customers who learned of the products from at least one magazine is given by

$$n(C \cup M \cup L) = 90 + 26 + 114 + 14 + 82 + 46 + 38 = 410.$$

b. The number of customers who learned of the products from exactly one magazine (see Figure 6.11) is given by

$$n(L \cap C^c \cap M^c) + n(M \cap C^c \cap L^c) + n(C \cap L^c \cap M^c) = 90 + 114 + 82 = 286.$$

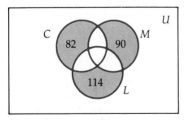

Figure 6.11

◀

Self-Check Exercises

6.2

1. Let A and B be subsets of a universal set U and suppose $n(U) = 100$, $n(A) = 60$, $n(B) = 40$, and $n(A \cap B) = 20$. Compute

    a. $n(A \cup B)$ *b.* $n(A \cap B^c)$ *c.* $n(A^c \cap B)$

2. In a recent survey of 1000 readers of *Video Magazine* it was found that 900 own at least one video cassette recorder (VCR) in the VHS format, 240 own at least one VCR in the Beta format, and 160 own VCRs in both the VHS and the Beta formats. How many of the readers surveyed own VCRs in the VHS format only? How many of the readers surveyed do not own a VCR in either the VHS or the Beta format?

Solutions to Self-Check Exercises 6.2 can be found on page 301.

▶

Exercises 6.2

In Exercises 1 and 2, verify the equation

$$n(A \cup B) = n(A) + n(B)$$

for the given disjoint sets.

1. $A = \{a, e, i, o, u\}$ and $B = \{g, h, k, l, m\}$

2. $A = \{x \mid x \text{ is a whole number between 0 and 4}\}$
   $B = \{x \mid x \text{ is a negative integer greater than } -4\}$

3. Let $A = \{2, 4, 6, 8\}$ and $B = \{6, 7, 8, 9, 10\}$. Compute

   a. $n(A)$ b. $n(B)$ c. $n(A \cup B)$ d. $n(A \cap B)$

4. Verify directly that $n(A \cup B) = n(A) + n(B) - n(A \cap B)$ for the sets in Exercise 3.

5. Let $A = \{a, e, i, o, u\}$ and $B = \{b, d, e, o, u\}$. Verify by direct computation that $n(A \cup B) = n(A) + n(B) - n(A \cap B)$.

6. If $n(A) = 15$, $n(A \cap B) = 5$, and $n(A \cup B) = 30$, what is $n(B)$?

7. If $n(A) = 10$, $n(A \cup B) = 15$, and $n(B) = 8$, what is $n(A \cap B)$?

In Exercises 8 and 9, let A and B be subsets of a universal set U and suppose $n(U) = 200$, $n(A) = 100$, $n(B) = 80$, and $n(A \cap B) = 40$. Compute each of the following:

8. a. $n(A \cup B)$ b. $n(A^c)$ c. $n(A \cap B^c)$

9. a. $n(A^c \cap B)$ b. $n(B^c)$ c. $n(A^c \cap B^c)$

10. Find $n(A \cup B)$ given that $n(A) = 6$, $n(B) = 10$, and $n(A \cap B) = 3$.

11. If $n(B) = 6$, $n(A \cup B) = 14$, and $n(A \cap B) = 3$, find $n(A)$.

12. If $n(A) = 4$, $n(B) = 5$, and $n(A \cup B) = 9$, find $n(A \cap B)$.

13. If $n(A) = 16$, $n(B) = 16$, $n(C) = 14$, $n(A \cap B) = 6$, $n(A \cap C) = 5$, $n(B \cap C) = 6$, and $n(A \cup B \cup C) = 31$, find $n(A \cap B \cap C)$.

14. If $n(A) = 12$, $n(B) = 12$, $n(A \cap B) = 5$, $n(A \cap C) = 5$, $n(B \cap C) = 4$, $n(A \cap B \cap C) = 2$, and $n(A \cup B \cup C) = 25$, find $n(C)$.

15. A survey of 1000 subscribers to the *Los Angeles Times* revealed that 900 people subscribe to the daily morning edition and 500 subscribe to both the daily morning and the Sunday editions. How many subscribe to the Sunday edition? How many subscribe to the Sunday edition only?

16. Of 100 clock/radios sold recently in a department store, 70 had FM circuitry and 90 had AM circuitry. How many radios had both FM and AM circuitry? How many could receive FM transmission only? How many could receive AM transmission only?

17. On a certain day the Wilton County Jail had 190 prisoners. Of these, 130 were accused of felonies and 121 were accused of misdemeanors. How many prisoners were accused of both a felony and a misdemeanor?

18. In a recent survey of 200 members of a local sports club, 100 members indicated that they plan to attend the next Summer Olympic Games, 60 indicated that they plan to attend the next Winter Olympic Games, and 40 indicated that they plan to attend both the Summer and the Winter Olympic Games. How many members of the club plan to attend

   a. at least one of the two games?

   b. exactly one of the games?

   c. the Summer Olympic Games only?

   d. none of the games?

19. In a survey of 120 consumers conducted in a shopping mall, 80 consumers indicated that they buy brand A of a certain product, 68 buy brand B, and 42 buy both brands. Determine the number of consumers participating in the survey who buy

   a. at least one of these brands.

   b. exactly one of these brands.

   c. only brand A.

   d. none of these brands.

20. Of 50 employees of a store located in downtown Boston, 18 people take the subway to work, 12 take the bus, and 7 take both the subway and the bus. Determine the number of employees who

   a. take the subway or the bus to work.

   b. take only the bus to work.

   c. take either the bus or the subway to work.

   d. get to work by some other means.

In Exercises 21–23, let A, B, and C be subsets of a universal set U and suppose $n(U) = 100$, $n(A) = 28$, $n(B) = 30$, $n(C) = 34$, $n(A \cap B) = 8$, $n(A \cap C) = 10$, $n(B \cap C) = 15$, and $n(A \cap B \cap C) = 5$. Compute each of the following:

21. *a.* $n(A \cup B \cup C)$ *b.* $n(A^c)$ *c.* $n(A^c \cap B \cap C)$

22. *a.* $n[A \cap (B \cup C)]$ *b.* $n[A \cap (B \cup C)^c]$ *c.* $n(A^c \cap B^c \cap C^c)$

23. *a.* $n[A^c \cap (B \cup C)]$ *b.* $n[A \cup (B \cap C)]$ *c.* $n(A^c \cap B^c \cap C^c)^c$

24. Data released by the Department of Education regarding the 1984 dropout rate (the percentage of ninth-grade students that don't graduate) showed that out of 50 states,

   12 states had an increase in the dropout rate since 1982

   15 states had a dropout rate of at least 30 percent since 1982

   21 states had an increase in the dropout rate and/or a dropout rate of at least 30 percent since 1982

How many states

   a. had both a dropout rate of at least 30 percent and an increase in the dropout rate over the two-year period?

b. had a drop-out rate that was less than 30 percent but that had increased over the two-year period?

25. A survey of the opinions of 10 leading economists in a certain country showed that because oil prices were expected to drop,

7 had lowered their estimate of the consumer inflation rate

8 had raised their estimate of the gross national product growth rate

2 had lowered their estimate of the consumer inflation rate but had not raised their estimate of the gross national product growth rate

in that country over the next 12 months. How many economists had both lowered their estimate of the consumer inflation rate and raised their estimate of the gross national product growth rate for that period?

26. Results of a Department of Education survey of 1985 SAT test scores in 22 states showed that since 1982

10 states had an average composite test score of at least 900

15 states had an increase of at least 10 points in the average composite score

8 states had both an average composite SAT score of at least 900 and an increase in the average composite score of at least 10 points

How many of the 22 states

a. had composite scores less than 900 and showed an increase of at least 10 points over the three-year period?

b. had composite scores of at least 900 and did not show an increase of at least 10 points over the three-year period?

27. A survey of 100 college students who frequent the reading lounge of a university revealed the following results:

40 read *Time* magazine

30 read *Newsweek*

25 read *U.S. News and World Report*

15 read *Time* magazine and *Newsweek*

12 read *Time* magazine and *U.S. News and World Report*

10 read *Newsweek* and *U.S. News and World Report*

4 read all three magazines

How many of the students surveyed read

a. at least one magazine?   *b.* exactly one magazine?

c. exactly two magazines?   *d.* none of these magazines?

28. To help plan the number of meals to be prepared in a college cafeteria, a survey was conducted and the following data were obtained:

130 students ate breakfast

180 students ate lunch

275 students ate dinner

68 students ate breakfast and lunch

112 students ate breakfast and dinner

90 students ate lunch and dinner

58 students ate all three meals

How many of the students

a. ate at least one meal in the cafeteria?

b. ate exactly one meal in the cafeteria?

c. ate only dinner in the cafeteria?

d. ate exactly two meals in the cafeteria?

29. The 120 consumers of Exercise 19 were also asked about their buying preferences concerning another product that is sold in the market under three labels. The results follow:

12 buy only those sold under label A

25 buy only those sold under label B

26 buy only those sold under label C

15 buy only those sold under labels A and B

10 buy only those sold under labels A and C

12 buy only those sold under labels B and C

8 buy the product sold under all three labels

How many of the consumers surveyed buy the product sold under

a. at least one of the three labels?   *b.* labels A and B but not C?

c. label A?   *d.* none of these labels?

*30. Derive Equation (5). [*Hint:* Equation (4) may be written as $n(D \cup E) = n(D) + n(E) - n(D \cap E)$. Now, put $D = A \cup B$ and $E = C$. Use Equation (4) again if necessary.]

Solutions to Self-Check Exercises

6.2

1. Refer to the following Venn diagram.

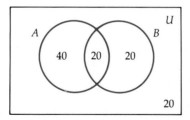

Using this result, we see that

    a. $n(A \cup B) = 40 + 20 + 20 = 80$

    b. $n(A \cap B^c) = 40$

    c. $n(A^c \cap B) = 20$

2. Let U denote the set of all readers surveyed, and let

$$A = \{x \in U \mid x \text{ owns at least one VCR in the VHS format}\}$$
$$B = \{x \in U \mid x \text{ owns at least one VCR in the Beta format}\}$$

Then the result that 160 of the readers own VCRs in both the VHS and the Beta formats gives $n(A \cap B) = 160$. Also, $n(A) = 900$ and $n(B) = 240$. Using this information, we obtain the following Venn diagram:

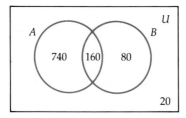

From the Venn diagram we see that the number of readers who own VCRs in the VHS format only is given by

$$n(A \cap B^c) = 740.$$

The number of readers who do not own a VCR in either the VHS or the Beta format is given by

$$n(A^c \cap B^c) = 20.$$

 ## 6.3

The Multiplication Principle

  ▶ The Fundamental Principle of Counting

  ▶ Applications

▶ The Fundamental Principle of Counting

The solution of certain problems requires more sophisticated counting techniques than those developed in the previous section. We will look at some such techniques in this and the following section. We begin by stating a fundamental principle of counting called the **multiplication principle.**

The Multiplication Principle

Suppose that there are m ways of performing a task T_1 and n ways of performing a task T_2. Then there are mn ways of performing the task T_1 followed by the task T_2.

EXAMPLE

26 Suppose that there are three trunk roads connecting town A and town B and two trunk roads connecting town B and town C.

a. Use the multiplication principle to find the number of ways a journey from town A to town C via town B may be completed.

b. Verify the result of (a) directly by exhibiting all possible routes.

SOLUTION   *a.* Since there are three ways of performing the first task (going from town A to town B) followed by two ways of performing the second task (going from town B to town C), the multiplication principle says that there are $3 \cdot 2$, or 6, ways to complete a journey from town A to town C via town B.

b. Label the trunk roads connecting town A and town B with the Roman numerals I, II, and III, and the trunk roads connecting town B and town C with the lowercase letters a and b. A schematic of this is shown in Figure 6.12.

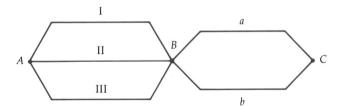

Figure 6.12

Then the routes from town A to town C via town B may be exhibited with the aid of a **tree diagram** (Figure 6.13).

The six routes are represented by the six ordered pairs

$$(I, a), \quad (I, b), \quad (II, a), \quad (II, b), \quad (III, a), \quad \text{and} \quad (III, b)$$

where (I, a) means that the journey from town A to town B is made on trunk road I with the rest of the journey from town B to town C to be completed on trunk road a, and so forth.

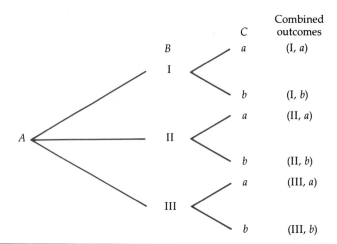

Figure 6.13

EXAMPLE

27 Diners at Angelo's Spaghetti Bar may select their entree from 6 varieties of pasta and 28 choices of sauce. How many such combinations are there that consist of one variety of pasta and one kind of sauce?

SOLUTION There are 6 ways of choosing a pasta followed by 28 ways of choosing a sauce, so by the multiplication principle, there are $6 \cdot 28$, or 168, combinations of this pasta dish. ◀

The multiplication principle may be easily extended, which leads to the **generalized multiplication principle**:

Generalized Multiplication Principle

Suppose a task T_1 can be performed in N_1 ways, a task T_2 can be performed in N_2 ways, . . . , and finally a task T_n can be performed in N_n ways. Then the number of ways of performing the tasks $T_1, T_2, \ldots, T_n$ in succession is given by the product

$$N_1 N_2 \cdots N_n$$

▶ Applications

We now illustrate the application of the generalized multiplication principle to several diverse situations.

EXAMPLE

28 A coin is tossed three times and the sequence of heads and tails is recorded.

a. Use the generalized multiplication principle to determine the number of outcomes of this activity.

b. Exhibit all the sequences by means of a tree diagram.

SOLUTION *a.* The coin may land in two ways. Therefore, in three tosses the number of outcomes (sequences) is given by $2 \cdot 2 \cdot 2$, or 8.

b. Let H and T denote the outcome "a head" and "a tail," respectively. Then the required sequences may be obtained as shown in Figure 6.14, giving the sequence as HHH, HHT, HTH, HTT, THH, THT, TTH, and TTT.

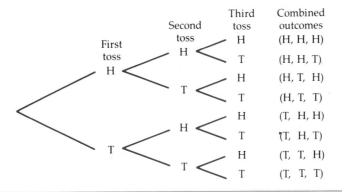

Figure 6.14

EXAMPLE

29 A combination lock is unlocked by dialing a sequence of numbers, first to the left, then to the right, and to the left again. If there are ten digits on the dial, determine the number of possible combinations.

SOLUTION There are ten choices for the first number, followed by ten for the second and ten for the third, so by the generalized multiplication principle, there are $10 \cdot 10 \cdot 10$, or 1000, possible combinations. ◄

EXAMPLE

30 An investor has decided to purchase shares in the stocks of three companies: one engaged in aerospace activities, one involved in energy development, and an electronics company. After some research the account executive of a brokerage firm has recommended that the investor consider stocks from five aerospace companies, three companies engaged in energy development, and four electronics companies. In how many ways may the investor select the group of three companies from the executive's list?

SOLUTION   The investor has five choices for selecting an aerospace company, three choices for selecting a company engaged in energy development, and four choices for selecting an electronics company. Therefore, by the generalized multiplication principle, there are $5 \cdot 3 \cdot 4$, or 60, ways in which she can select a group of three companies, one from each industry group.   ◄

Self-Check Exercises

6.3

1. Encore Travel, Inc. offers a "Theater Week in London" package originating from New York City. There is a choice of eight flights departing from New York City per week, a choice of five hotel accommodations, and a choice of one complimentary ticket to one of eight shows. How many such travel packages can one choose from?

2. The Cafe Napoleon offers a dinner special on Wednesdays consisting of a choice of two entrees (Beef Bourguignon and Chicken Basquaise); one dinner salad; one French roll; a choice of three vegetables; a choice of a carafe of Burgundy, Rosé, or Chablis wine; a choice of coffee or tea; and a choice of six French pastries for dessert. How many combinations of dinner specials are there?

Solutions to Self-Check Exercises 6.3 can be found on page 308.

► ───

Exercises 6.3

1. Lynbrook West, an apartment complex financed by the State Housing Finance Agency, consists of one-, two-, three-, and four-bedroom units. The rental rate for each type of unit—low, moderate, or market—is determined by the income of the tenant. How many different rates are there?

2. Five different types of monthly commuter passes are offered by a city's local transit authority for three different groups of passengers: youths, adults, and senior citizens. How many different kinds of passes must be printed each month?

3. In the game of Blackjack, a two-card hand consisting of an ace and a face card or a ten is called a "blackjack." If a standard 52-card deck is used, determine how many blackjack hands can be dealt.

4. A coin is tossed four times and the sequence of heads and tails is recorded.

   a. Use the generalized multiplication principle to determine the number of outcomes of this activity.

   b. Exhibit all the sequences by means of a tree diagram.

5. A female executive selecting her wardrobe purchased two blazers, four blouses, and three skirts in coordinating colors. How many ensembles consisting of a blazer, a blouse, and a skirt can she create from this collection?

6. There are four commuter trains and three express buses departing from city A to city B in the morning and three commuter trains and three express buses operating on the return trip in the evening. In how many ways can a commuter from city A to city B complete a daily round trip via bus and/or train?

7. A psychologist has constructed the following maze for use in an experiment. The maze is constructed so that a rat must pass through a series of one-way doors. How many different paths are there from start to finish?

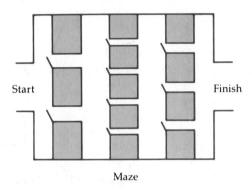

Maze

8. In a survey conducted by a union, members were asked to rate the importance of the following issues:

Reduced working hours
Increased fringe benefits
Improved working conditions

Five different responses were allowed for each issue. Among completed surveys, how many different responses to this survey were possible?

9. A new state employee is offered a choice of ten basic health plans, three dental plans, and two vision care plans. How many different health care plans are there to choose from if one plan is selected from each category?

10. How many three-letter code words can be constructed from the first ten letters of the Greek alphabet if no repetitions are allowed?

11. A computer dating service uses the results of its compatibility survey for arranging dates. The survey consists of 50 questions, each having 5 possible answers. How many different responses are possible if every question is answered?

12. Computers manufactured by a certain company have a serial number consisting of a letter of the alphabet followed by a four-digit number. If all the serial numbers of this type have been used, how many sets have already been manufactured?

13. Two soups, five entrees, and three desserts are listed on the "Special" menu at the Neptune Restaurant. How many different selections consisting of one soup, one entree, and one dessert can a customer choose from this menu?

14. An automobile manufacturer has three different subcompact cars in the line. Customers selecting one of these cars have a choice of three engine sizes, four body styles, and three color schemes. How many different selections can a customer make?

15. *a.* How many positive three-digit numbers can be formed that are greater than 700?

b. If no digit is repeated within the number, how many three-digit numbers greater than 700 can be formed?

16. An opinion poll is to be conducted among cable T.V. viewers. Six multiple-choice questions, each with four possible answers, will be asked. In how many different ways can a viewer complete the poll, if exactly one response is given to each question?

17. An opinion poll was conducted by the Morris Polling Group. Respondents were classified according to their sex (M or F), political affiliation (D, I, R), and the region of the country in which they reside (NW, W, C, S, E, NE).

    a. Use the generalized multiplication principle to determine the number of possible classifications.

    b. Construct a tree diagram to exhibit all possible classifications of females.

18. A certain state uses license plates consisting of three letters of the alphabet followed by three digits. How many different license plate numbers can be formed?

19. An exam consists of ten true-or-false questions. Assuming that every question is answered, in how many different ways can a student complete the exam? In how many ways may the exam be completed if a penalty is imposed for each incorrect answer, so that a student may leave some questions unanswered?

20. A warranty identification number for a certain product consists of a letter of the alphabet followed by a five-digit number. How many possible identification numbers are there if the first digit of the five-digit number must be nonzero?

21. *a.* How many seven-digit telephone numbers are possible if the first digit must be nonzero?

    b. How many international direct-dialing numbers are possible if each number consists of a three-digit area code (the first digit of which must be nonzero) and a number of the type described in (a)?

*22. A "lucky dollar" is one of the nine symbols that are printed on each reel of a slot machine with three reels. A player receives one of various payouts whenever one or more "lucky dollars" appear in the window of the machine. Find the number of winning combinations for which the machine gives a payoff. [*Hint:* (a) Compute the number of ways in which the nine symbols on the first, second, and third wheels can appear in the window slot and (b) compute the number of ways in which the eight symbols other than the "lucky dollar" can appear in the window slot. The difference $(a - b)$ is the number of ways in which the "lucky dollar" can appear in the window slot. Why?]

Solutions to Self-Check Exercises

6.3

1. A tourist has a choice of eight flights, five hotel accommodations, and eight tickets. By the generalized multiplication principle, there are $8 \cdot 5 \cdot 8$, or 320, travel packages.

2. There is a choice of two entrees, one dinner salad, one French roll, a choice of three vegetables, a choice of three wines, a choice of two nonalcoholic beverages, and a choice of six pastries. Therefore, by the generalized multiplication principle, there are $2 \cdot 1 \cdot 1 \cdot 3 \cdot 3 \cdot 2 \cdot 6$, or 216, combinations of dinner specials.

6.4

Permutations and Combinations

▶ Permutations

▶ Combinations

▶ Applications

▶ Permutations

In this section, we will apply the generalized multiplication principle to the solution of two types of counting problems. Both types involve the determination of the number of ways of arranging the elements of a set and both play an important role in the solution of problems in probability.

We begin by considering the permutations of a set. Specifically, given a set of (distinct) objects, a **permutation** of the set is a linear arrangement of these objects in a definite order.

EXAMPLE

31   Let $A = \{a, b, c\}$.

a. Find the number of permutations of A.

b. List all the permutations of A with the aid of a tree diagram.

SOLUTION

a. Each permutation of A consists of a sequence of the three letters a, b, c. Therefore we may think of such a sequence as being constructed by filling in each of the three blanks

$$\underline{} \quad \underline{} \quad \underline{}$$

with one of the three letters. Now, there are three ways in which we may fill the first blank—we may choose a, b, or c. Having selected a letter for the first blank, there are two letters left for the second blank. Finally, there is but one way left to fill the third blank. Schematically, we have

$$\underline{\quad 3 \quad} \quad \underline{\quad 2 \quad} \quad \underline{\quad 1 \quad}$$

Invoking the generalized multiplication principle, we conclude that there are $3 \cdot 2 \cdot 1$, or 6, permutations of the set A.

b. The tree diagram associated with this problem appears in Figure 6.15, and the six permutations of A are *abc, acb, bac, bca, cab,* and *cba.*

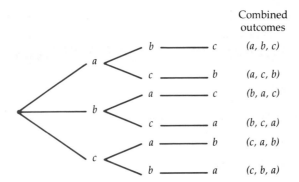

					Combined outcomes

Figure 6.15

EXAMPLE

32 Find the number of ways a baseball team consisting of nine people can arrange themselves in a line for a group picture.

SOLUTION We want to determine the number of permutations of the nine members of the baseball team. Each permutation in this situation consists of an arrangement of the nine team members in a line. The nine positions can be represented by nine blanks, thus

$$\underline{}\ \underline{}\ \underline{}\ \underline{}\ \underline{}\ \underline{}\ \underline{}\ \underline{}\ \underline{}$$
$$1 \quad 2 \quad 3 \quad 4 \quad 5 \quad 6 \quad 7 \quad 8 \quad 9$$

There are nine ways to choose from among the nine players to fill the first position. Next there are eight players left, which gives us eight ways to fill the second position. Proceeding in a similar manner, we find that there are seven ways to fill the third position, and so on. Schematically, we have

$$\underline{9}\ \ \underline{8}\ \ \underline{7}\ \ \underline{6}\ \ \underline{5}\ \ \underline{4}\ \ \underline{3}\ \ \underline{2}\ \ \underline{1}$$

Invoking the generalized multiplication principle, we conclude that there are $9 \cdot 8 \cdot 7 \cdot 6 \cdot 5 \cdot 4 \cdot 3 \cdot 2 \cdot 1$, or 362,880, ways the baseball team can be arranged for the picture. ◀

Pursuing the same line of argument used in solving the problems in the last two examples, it is easy to derive an expression for the number of ways of permuting a set A of n distinct objects taken n at a time. In fact, each permutation may be viewed as being obtained by filling each of n blanks with one and only one element from the set. There are n ways of filling the first blank, followed by $(n - 1)$ ways of filling the second blank, and so on, so by the generalized multiplication principle there are

$$n(n - 1)(n - 2) \cdots 3 \cdot 2 \cdot 1$$

ways of permuting the elements of the set A.

Before stating this result formally, let us introduce a notation that will enable us to write many of the expressions that follow in a compact form. We use the symbol $n!$ (read "n-factorial") to denote the product of the first n natural numbers.

---

n-factorial

For any natural number n

$$n! = n(n - 1)(n - 2) \ldots 3 \cdot 2 \cdot 1$$

and $\qquad 0! = 1.$

For example,

$$1! = 1$$
$$2! = 2 \cdot 1 = 2$$
$$3! = 3 \cdot 2 \cdot 1 = 6$$
$$4! = 4 \cdot 3 \cdot 2 \cdot 1 = 24$$
$$5! = 5 \cdot 4 \cdot 3 \cdot 2 \cdot 1 = 120$$
$$\vdots$$

and $\qquad 10! = 10 \cdot 9 \cdot 8 \cdot 7 \cdot 6 \cdot 5 \cdot 4 \cdot 3 \cdot 2 \cdot 1 = 3,628,800.$

Using this notation, we may express *the number of permutations of n distinct objects taken n at a time, P(n, n),* as

$$P(n, n) = n!$$

In many situations one is interested in determining the number of ways of permuting n distinct objects taken r at a time where $r \leq n$. In the special case where $r = n$, the result is obtained by computing $P(n, n) = n!$. To derive a formula for computing the number of ways of permuting a set consisting of n distinct objects taken r at a time, we observe that each such permutation may be viewed as being obtained by filling each of r blanks with precisely one element from the set. Now there are n ways of filling the first blank, followed by $(n - 1)$ ways of filling the second blank, and so on. Finally, there are $(n - r + 1)$ ways of filling the rth blank. We may represent this argument schematically as

Number of ways	n	$n - 1$	$n - 2$	$\cdots$	$n - r + 1$
Blank	1st	2nd	3rd		rth

Invoking the generalized multiplication principle, we conclude that *the number of ways of permuting n distinct objects taken r at a time, P(n, r), is given by*

$$P(n, r) = n(n - 1)(n - 2) \cdots (n - r + 1).$$

Since

$$n(n - 1)(n - 2) \cdots (n - r + 1) =$$

$$\frac{[n(n - 1)(n - 2) \cdots (n - r + 1)] [(n - r)(n - r - 1) \cdots 3 \cdot 2 \cdot 1]}{(n - r)(n - r - 1) \cdots 3 \cdot 2 \cdot 1}$$

$$= \frac{n!}{(n - r)!},$$

we have the following formula:

Permutations of *n* Distinct Objects

The number of **permutations** of *n* distinct objects taken *r* at a time is

$$P(n, r) = \frac{n!}{(n - r)!} \tag{6}$$

Note that when $n = r$, formula (6) reduces to

$$P(n, n) = \frac{n!}{0!} = \frac{n!}{1} = n! \quad \text{(Note that } 0! = 1\text{)}$$

as expected.

EXAMPLE

33 Let $A = \{a, b, c, d\}$

a. Use Formula (6) to compute the number of permutations of the set A taken two at a time.

b. Display the permutations of (a) with the aid of a tree diagram.

SOLUTION *a.* Here $n = 4$ and $r = 2$, so the required number of permutations is given by

$$P(4, 2) = \frac{4!}{(4 - 2)!} = \frac{4!}{2!} = \frac{4 \cdot 3 \cdot 2 \cdot 1}{2 \cdot 1} = 4 \cdot 3,$$

or 12.

b. The tree diagram associated with the problem is shown in Figure 6.16, and the permutations of A taken two at a time are

$$ab, ac, ad, ba, bc, bd, ca, cb, cd, da, db, \text{ and } dc$$

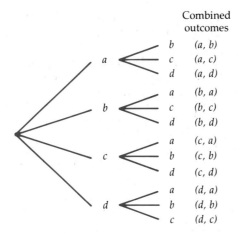

Combined
outcomes

b (a, b)
c (a, c)
d (a, d)

a (b, a)
c (b, c)
d (b, d)

a (c, a)
b (c, b)
d (c, d)

a (d, a)
b (d, b)
c (d, c)

Figure 6.16   ◀

EXAMPLE

34   Find the number of ways a chairman, a vice-chairman, a secretary, and a treasurer can be chosen from a committee of eight members.

SOLUTION   The problem is equivalent to finding the number of permutations of eight distinct objects taken four at a time. Therefore, the number of ways of choosing the four officials from the committee of eight members is given by

$$P(8, 4) = \frac{8!}{(8 - 4)!} = \frac{8!}{4!} = 8 \cdot 7 \cdot 6 \cdot 5 = 1680,$$

or 1680 ways.   ◀

EXAMPLE

35   In a State Lottery there are 15 finalists eligible for the Big Money Draw. In how many ways can the first, second, and third prizes be awarded if no ticket holder may win more than one prize?

SOLUTION   The number of ways the first three prizes can be awarded is given by

$$P(15, 3) = \frac{15!}{(15 - 3)!} = \frac{15!}{12!} = 15 \cdot 14 \cdot 13 = 2730,$$

or 2730 ways.   ◀

The permutations considered thus far have been those sets of *distinct* objects. In many situations we are interested in finding the number of permutations of a set of objects in which not all of the objects are distinct.

Permutations of n Objects, Not All Distinct

Given a set of n objects in which n_1 objects are alike and of one kind, n_2 objects are alike and of another kind, . . . , and, finally, n_r objects are alike and of yet another kind so that

$$n_1 + n_2 + \cdots + n_r = n$$

then the number of permutations of these n objects taken n at a time is given by

$$\frac{n!}{n_1!n_2! \ldots n_r!} \tag{7}$$

To establish Formula (7), let us denote the number of such permutations by x. Now, if we *think* of the n_1 objects as being distinct, then they may be permuted in $n_1!$ ways. Similarly, if we *think* of the n_2 objects as being distinct, then they may be permuted in $n_2!$ ways, and so on. Therefore, if we *think* of the n objects as being distinct, then, by the generalized multiplication principle, there are $x \cdot n_1! \cdot n_2! \cdot \ldots \cdot n_r!$ permutations of these objects. But, the number of permutations of a set of n distinct objects taken n at a time is just equal to $n!$. Therefore, we have

$$x(n_1! \cdot n_2! \cdot \ldots \cdot n_r!) = n!,$$

from which we deduce that

$$x = \frac{n!}{n_1!n_2! \cdots n_r!}.$$

EXAMPLE

36 Find the number of permutations that can be formed from all of the letters in the word *ATLANTA*.

SOLUTION There are seven objects (letters) involved, so $n = 7$. However, three of them are alike and of one kind (the three A's), two of them are alike and of another kind (the two T's), so that in this case, $n_1 = 3$, $n_2 = 2$, $n_3 = 1$, and $n_4 = 1$. Therefore, using Formula (7), the number of required permutations is given by

$$\frac{7!}{3!2!1!1!}$$

or 420. ◄

EXAMPLE

37   Weaver and Kline, a stock brokerage firm, has received nine inquiries regarding new accounts. In how many ways can these inquiries be directed to three of the firm's account executives if each account executive is to handle three inquiries?

SOLUTION   Here $n = 9$, $n_1 = n_2 = n_3 = 3$, so the number of ways of assigning the inquiries is given by

$$\frac{9!}{3!3!3!}$$

or 1680 ways.   ◀

▶ Combinations

Up to now, we have dealt with permutations of a set; that is, with arrangements of the objects of the set in which the *order* of the elements is taken into consideration. In many situations one is interested in determining the number of ways of selecting, say, r objects from a set of n objects without any regard to the order in which the objects are selected. Such a subset is called a **combination.** For example, if one is interested in knowing the number of five-card poker hands that can be dealt from a standard deck of 52 cards, then the order in which a poker hand is dealt is unimportant. In this situation, one is in fact interested in determining the number of combinations of five cards (objects) selected from a deck (set) of 52 cards (objects). The solution to this problem will be found later in Example 40.

To derive a formula for determining the number of combinations of n objects taken r at a time, written

$$C(n, r) \text{ or } \binom{n}{r},$$

we observe that each of the $C(n, r)$ combinations of r objects (Figure 6.17) can be permuted in $r!$ ways.

$$\underbrace{(r \text{ objects}) \quad (r \text{ objects}) \cdots (r \text{ objects})}_{\longleftarrow \quad C(n, r) \text{ of them} \quad \longrightarrow}$$

Figure 6.17

Thus, by the multiplication principle, the product $r!C(n, r)$ gives the number of permutations of n objects taken r at a time; that is,

$$r!C(n, r) = P(n, r),$$

from which we find

$$C(n, r) = \frac{P(n, r)}{r!},$$

or, using Formula (6),

$$C(n, r) = \frac{n!}{r!(n - r)!}.$$

Combinations of *n* Objects

The number of **combinations** of *n* distinct objects taken *r* at a time is given by

$$C(n, r) = \frac{n!}{r!(n - r)!} \quad \text{where } r \leq n \tag{8}$$

EXAMPLE

38 Compute (a) $C(4, 4)$ and (b) $C(4, 2)$ and interpret the results.

SOLUTION

a. $C(4, 4) = \dfrac{4!}{4!(4 - 4)!} = \dfrac{4!}{4!0!} = 1$ (Recall that $0! = 1$.)

This gives 1 as the number of combinations of four distinct objects taken four at a time.

b. $C(4, 2) = \dfrac{4!}{2!(4 - 2)!} = \dfrac{4!}{2!2!} = \dfrac{4 \cdot 3}{2} = 6$

This gives 6 as the number of combinations of four distinct objects taken two at a time. ◀

▶ Applications

EXAMPLE

39 A Senate investigation subcommittee of four members is to be selected from a Senate committee of ten members. Determine the number of ways this can be done.

SOLUTION

Since the order in which the members of the subcommittee are selected is unimportant, the number of ways of choosing the subcommittee is given by

$C(10, 4)$, the number of combinations of ten objects taken four at a time. But

$$C(10, 4) = \frac{10!}{4!(10 - 4)!} = \frac{10!}{4!6!} = \frac{10 \cdot 9 \cdot 8 \cdot 7}{4 \cdot 3 \cdot 2 \cdot 1} = 210,$$

so there are 210 ways of choosing such a subcommittee.   ◄

EXAMPLE

40   Find the number of ways a poker hand of 5 cards can be dealt from a standard deck of 52 cards.

SOLUTION   The order in which the 5 cards are dealt is not important, and the number of ways of dealing a poker hand of 5 cards from a standard deck of 52 cards is given by $C(52, 5)$, the number of combinations of 52 objects taken 5 at a time. Now,

$$C(52, 5) = \frac{52!}{5!(52 - 5)!} = \frac{52!}{5!47!}$$

$$= \frac{52 \cdot 51 \cdot 50 \cdot 49 \cdot 48}{5 \cdot 4 \cdot 3 \cdot 2 \cdot 1} = 2{,}598{,}960,$$

so there are 2,598,960 ways of dealing such a poker hand.   ◄

The next several examples show that solving a counting problem often involves the repeated application of Formula (6) and/or Formula (8), possibly in conjunction with the multiplication principle.

EXAMPLE

41   The members of a string quartet comprising two violinists, a violist, and a cellist are to be selected from a group of six violinists, three violists, and two cellists, respectively. In how many ways can the string quartet be formed?

SOLUTION   The violinist may be selected in $C(6, 2)$, or 15, ways; the violist may be selected in $C(3, 1)$ or 3 ways; and the cellist may be selected in $C(2, 1)$ or 2 ways. By the multiplication principle there are $15 \cdot 3 \cdot 2$, or 90, ways of forming the string quartet.   ◄

EXAMPLE

42   Refer to Example 41. In how many ways can the string quartet be formed if one of the violinists is to be designated as the first violinist and the other is to be designated as the second violinist?

SOLUTION The order in which the violinists are selected is important here. Consequently, the number of ways of selecting the violinists is given by $P(6, 2)$, or 30, ways. The number of ways of selecting the violist and the cellist are, of course, 3 and 2, respectively. Therefore, the number of ways in which the string quartet may be formed is given by $30 \cdot 3 \cdot 2$, or 180, ways. ◄

EXAMPLE

43 Refer to Example 30. Suppose that the investor has decided to purchase shares in the stocks of two aerospace companies, two companies involved in energy development, and two electronics companies. In how many ways may the investor select the group of six companies for the investment from the prepared list?

SOLUTION There are $C(5, 2)$ ways in which the investor may select the aerospace companies, $C(3, 2)$ ways in which she may select the companies involved in energy development, and $C(4, 2)$ ways in which she may select the electronics companies as investments. By the generalized multiplication principle there are

$$C(5, 2)C(3, 2)C(4, 2) = \frac{5!}{2!3!} \cdot \frac{3!}{2!1!} \cdot \frac{4!}{2!2!}$$
$$= \frac{5 \cdot 4}{2} \cdot 3 \cdot \frac{4 \cdot 3}{2}$$
$$= 180$$

ways of selecting the group of six companies for her investment. ◄

Self-Check Exercises

6.4

1. Evaluate *a.* 5!   *b.* $C(7, 4)$ *c.* $P(6, 2)$

2. A space shuttle crew consists of a shuttle commander, a pilot, 3 engineers, a scientist, and a civilian. The shuttle commander and pilot are to be chosen from 8 candidates, the 3 engineers from 12 candidates, the scientist from 5 candidates, and the civilian from 2 candidates. How many such space shuttle crews can be formed?

Solutions to Self-Check Exercises 6.4 can be found on page 322.

Exercises 6.4

In Exercises 1–13, evaluate the given expression:

1. $P(5, 2)$
2. $P(6, 6)$
3. $P(5, 3)$
4. $P(n, 1)$
5. $C(8, 8)$
6. $C(5, 0)$
7. $C(7, 4)$
8. $P(k, 2)$
9. $P(n, n - 2)$
10. $C(9, 3)$
11. $C(9, 6)$
12. $C(n, 2)$
13. $C(7, r)$

In Exercises 14–21, classify each problem according to whether it involves a permutation or a combination.

14. In how many ways can the letters of the word *glacier* be arranged?

15. In the eighth-grade dance class, there are 10 girls and 14 boys. In how many ways can these students be paired off to form dance couples consisting of one boy and one girl?

16. As part of a quality-control program, 3 record-o-phones are selected at random for testing from each 100 phones produced by the manufacturer. In how many ways can this test batch be chosen?

17. How many three-digit numbers can be formed using the numerals in the set {3, 2, 7, 9} if repetition is not allowed?

18. In how many ways can nine books be arranged on a shelf?

19. A member of a book club wishes to purchase two books from a selection of eight books recommended for a certain month. In how many ways can she choose them?

20. How many five-card poker hands can be dealt consisting of three queens and a pair?

21. Four couples are to be seated at an oval dining table. If each couple must be seated together, how many possible seating arrangements are there?

22. A group of five students studying for a bar exam have formed a study group. Each member of the group will be responsible for preparing a study outline for one of five courses. In how many different ways can the five courses be assigned to the members of the group?

23. In how many ways can a television programming director schedule six different commercials in the six time slots allocated to commercials during a one-hour program?

24. Seven people arrive at the ticket counter of a cinema at the same time. In how many ways can they line up to purchase their tickets?

25. In how many ways can a supermarket chain select 3 out of 12 possible sites for the construction of new supermarkets?

26. A student is given a reading list of ten books from which he must select two for an outside reading requirement. In how many ways can he make his selections?

27. In how many ways can a quality-control engineer select a sample of 3 transistors for testing from a batch of 100 transistors?

28. Weaver and Kline, a stock brokerage firm, has received 6 inquiries regarding new accounts. In how many ways can these inquiries be directed to its 12 account executives if each executive handles no more than 1 inquiry?

29. A company car that has a seating capacity of six is to be used by six employees who have formed a car pool. If only four of these employees can drive, how many possible seating arrangements are there for the group?

30. At a college library exhibition of faculty publications, three mathematics books, four social science books, and three biology books will be displayed on a shelf.

   a. In how many ways can the ten books be arranged on the shelf?

   b. In how many ways can the ten books be arranged on the shelf if books on the same subject matter are placed together?

31. In how many ways can four married couples attending a concert be seated in a row of eight seats

   a. if there are no restrictions?

   b. if each married couple is seated together?

   c. if the members of each sex are seated together?

32. The Futurists, a rock group, are planning a concert tour with performances to be given in five cities: San Francisco, Los Angeles, San Diego, Denver, and Las Vegas. In how many ways can they arrange their itinerary

   a. if there are no restrictions?

   b. if the three performances in California must be given consecutively?

33. Four items from 5 different departments of the Metro Department Store will be featured in a one-page newspaper advertisement as shown in the following diagram:

Advertisement

1	2	3	4
5	6	7	8
9	10	11	12
13	14	15	16
17	18	19	20

   a. In how many different ways can the 20 featured items be arranged on the page?

   b. If items from the same department must be in the same row, how many arrangements are possible?

34. The C & J Realty Company has received 12 inquiries from prospective home buyers. In how many ways can the inquiries be directed to 4 of the firm's real estate agents if each agent handles 3 inquiries?

35. In the women's tennis tournament at Wimbledon, two finalists, A and B, are competing for the title, which will be awarded to the first player to win two sets. In how many different ways can the match be completed?

36. The U.N. Security Council consists of 5 permanent members and 10 nonpermanent members. Decisions made by the council require 9 votes for passage. However, any permanent member may veto a measure and thus block its passage. In how many ways can a measure be passed

    a. if all 15 members of the Council vote (no abstentions)?

    b. if 2 particular permanent and 2 particular nonpermanent members of the Council abstain from voting?

37. In how many different ways can a panel of 12 jurors and 2 alternate jurors be chosen from a group of 30 prospective jurors?

38. Twelve graduate students have applied for three available teaching assistantships. In how many ways can the assistantships be awarded among these applicants

    a. if no preference is given to any student?

    b. if one particular student must be awarded an assistantship?

    c. if the group of applicants includes seven men and five women and it is stipulated that at least one woman must be awarded an assistantship?

39. A student taking an examination is required to answer 10 out of 15 questions.

    a. In how many ways can the 10 questions be selected?

    b. In how many ways can the 10 questions be selected if exactly 2 of the first 3 questions must be answered?

40. The U.B.S. Television Company is considering bids submitted by seven different firms for three different contracts. In how many ways can the contracts be awarded among these firms if no firm is to receive more than two contracts?

41. In how many ways can a subcommittee of four be chosen from a Senate committee of five Democrats and four Republicans

    a. if all members are eligible?

    b. if the subcommittee must consist of two Republicans and two Democrats?

42. A student planning her curriculum for the upcoming year must select one of five business courses, one of three mathematics courses, two of six elective courses, and either one of four history courses or one of three social science courses. How many different curricula are available for her consideration?

43. The J.C.L. Computer Company has five vacancies in its executive trainee program. In how many ways can the company select five trainees from a group of ten female and ten male applicants

    a. if the vacancies may be filled by any combination of men and women?

    b. if the vacancies must be filled by two men and three women?

44. The State Motor Vehicle Department requires learners to pass a written test on the motor vehicle laws of the state. The exam consists of ten true-or-false questions of which eight must be answered correctly to qualify for a permit. In how many different ways may a learner complete the exam and qualify for a permit?

45. The Goodman Tire Company has 32 tires of a particular size and grade in stock, 2 of which are defective. If a set of 4 tires is to be selected,

    a. how many different selections can be made?

    b. how many different selections can be made that do not include any defective tires?

46. The following is a schematic diagram of a city's street system between the points

A and *B*. The City Transit Authority is in the process of selecting a route from *A* to *B* along which to provide bus service. If the company's intention is to keep the route as short as possible, how many routes must be considered?

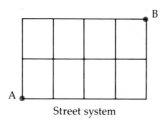

Street system

*47. In the World Series, one National League team and one American League team compete for the coveted title that is awarded to the first team to win four games. In how many different ways can the series of games be completed?

48. A quorum (minimum) of 6 voting members is required at all meetings of the Curtis Townhomes Owners Association. If there are a total of 12 voting members in the group, find the number of ways this quorum can be formed.

49. At the end of Section 3.3, we mentioned that a linear programming problem in 3 variables and 5 constraints may have up to 56 feasible corner points and that the determination of these feasible corner points calls for the solution of 56 3×3 systems of linear equations. Prove this assertion. [*Hint:* Don't forget the nonnegativity conditions $x \geq 0$, $y \geq 0$, and $z \geq 0$.

50. Refer to Exercise 49. Show that a linear programming problem in 5 variables and 10 constraints may have up to 3003 feasible corner points. This assertion was also made at the end of Section 3.3.

Solutions to Self-Check Exercises

6.4

1. *a.* $5! = 5 \cdot 4 \cdot 3 \cdot 2 \cdot 1 = 120$

   b. $C(7, 4) = \dfrac{7!}{4!3!} = \dfrac{7 \cdot 6 \cdot 5}{3 \cdot 2 \cdot 1} = 35$

   c. $P(6, 2) = \dfrac{6!}{4!} = 6 \cdot 5 = 30$

2. There are $P(8, 2)$ ways of picking the shuttle commander and pilot (the order *is* important here), $C(12, 3)$ ways of picking the engineers (the order is not important here), $C(5, 1)$ ways of picking the scientist, and $C(2, 1)$ ways of picking the civilian.

By the multiplication principle, there are

$$P(8, 2) \cdot C(12, 3) \cdot C(5, 1) \cdot C(2, 1)$$

$$= \frac{8!}{6!} \cdot \frac{12!}{9!3!} \cdot \frac{5!}{4!1!} \cdot \frac{2!}{1!1!}$$

$$= \frac{(8)(7)(12)(11)(10)(5)(2)}{(3)(2)}$$

$$= 123,200,$$

or 123,200 ways a crew could be selected.

Chapter 6 Review Exercises

1. List the elements of the following sets in roster notation.
   a. $\{x \mid 3x - 2 = 7; x, \text{ an integer}\}$
   b. $\{x \mid x \text{ is a letter of the word } TALLAHASSEE\}$
   c. The set whose elements are the even numbers between 3 and 11
   d. $\{x \mid (x - 3)(x + 4) = 0; x, \text{ a negative integer}\}$

2. Let $A = \{a, c, e, r\}$. Which of the following sets are equal to A?
   a. $\{r, e, c, a\}$
   b. $\{x \mid x \text{ is a letter of the word } career\}$
   c. $\{x \mid x \text{ is a letter of the word } racer\}$
   d. $\{x \mid x \text{ is a letter of the alphabet}\}$
   e. $\{x \mid x \text{ is a letter of the word } cares\}$

3. List all subsets of the set $A = \{e, f, g\}$. Indicate which of the subsets are proper.

4. Shade the portion of the figure shown below that represents each of the following sets.
   a. $A \cup (B \cap C)$ b. $(A \cap B \cap C)^c$ c. $A^c \cap B^c \cap C^c$

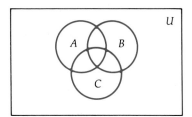

5. Let $U = \{a, b, c, d, e\}$, $A = \{a, b\}$, $B = \{b, c, d\}$, and $C = \{a, d, e\}$. Verify each of the following equations by direct computation.
   a. $A \cup (B \cup C) = (A \cup B) \cup C$ b. $A \cap (B \cap C) = (A \cap B) \cap C$
   c. $A \cap (B \cup C) = (A \cap B) \cup (A \cap C)$ d. $A \cup (B \cap C) = (A \cup B) \cap (A \cup C)$

6. Let A and B be subsets of a universal set U and suppose $n(U) = 350$, $n(A) = 120$, $n(B) = 80$, and $n(A \cap B) = 50$. Compute each of the following.

    a. $n(A \cup B)$ *b.* $n(A^c)$ *c.* $n(B^c)$

    d. $n(A^c \cap B)$ *e.* $n(A \cap B^c)$ *f.* $n(A^c \cap B^c)$

7. Evaluate:

    a. $C(20, 18)$ *b.* $C(n, r)$ *c.* $P(9, 7)$

    d. $P(n, n)$ *e.* $C(5, 3) \cdot P(4, 2)$

8. How many three-digit numbers can be formed from the numerals in the set {1, 2, 3, 4, 5}

    a. if repetition of digits is not allowed?

    b. if repetition of digits is allowed?

9. Find the number of distinguishable permutations that can be formed from the letters of each word.

    a. CINCINNATI   *b.* HONOLULU

10. A comparison of 5 major credit cards showed that

3 offered cash advances

3 offered extended payments for *all* goods and services purchased

2 required an annual fee of $35 or less

2 offered both cash advances and extended payments

1 offered extended payments and had an annual fee less than $35

no card had an annual fee less than $35 and offered both cash advances and extended payments

How many cards had an annual fee less than $35 and offered cash advances?

11. The Department of Foreign Languages of a liberal arts college conducted a survey of its recent graduates to determine the foreign language courses they had taken while undergraduates at the college. Of the 480 graduates

200 had at least one year of Spanish

178 had at least one year of French

140 had at least one year of German

33 had at least one year of Spanish and French

24 had at least one year of Spanish and German

18 had at least one year of French and German

3 had at least one year of all three languages

How many of the graduates had

    a. at least one year of at least one of the three languages?

    b. at least one year of exactly one of the three languages?

    c. less than a year of any of the three languages?

12. In an election being held by the Associated Students Organization there are six candidates for president, four for vice-president, five for secretary, and six for treasurer. How many different possible outcomes are there for this election?

13. From a standard 52-card deck, how many 5-card poker hands can be dealt consisting of

    a. 5 clubs?

    b. 3 kings and a pair?

14. In how many ways can seven students be assigned seats in a row containing seven desks

    a. if there are no restrictions?

    b. if two of the students must not be seated next to each other?

15. There are eight seniors and six juniors in the Math Club at Jefferson High School. In how many ways can a math team consisting of four seniors and two juniors be selected from the members of the Math Club?

16. A sample of 4 balls is to be selected at random from an urn containing 15 balls numbered 1 to 15. If 6 balls are green, 5 are white, and 4 are black

    a. how many different samples can be selected?

    b. how many samples can be selected that contain at least 1 white ball?

17. From a shipment of 60 transistors, 5 of which are defective, a sample of 4 transistors is selected at random.

    a. In how many different ways can the sample be selected?

    b. How many samples contain 3 defective transistors?

    c. How many samples do not contain any defective transistors?

Probability

Where did the defective picture tube come from? Picture tubes for the Pulsar 19-inch color television sets are manufactured in three locations and then shipped to the main plant of the Vista Vision Corporation for final assembly. Each location produces a certain number of the picture tubes with different degrees of reliability. In Example 32, page 393, we will determine the likelihood that a defective picture tube is manufactured in a particular location.

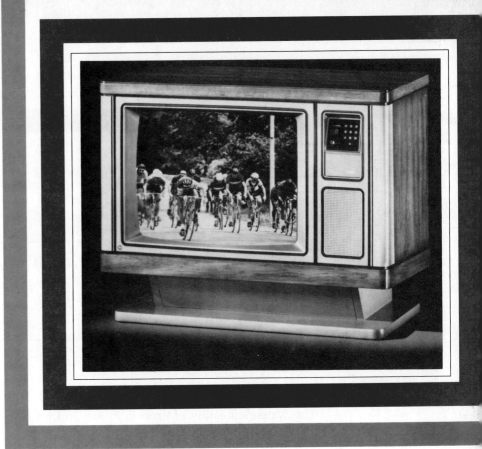

►CHAPTER

SEVEN

7.1

Experiments, Sample Spaces, and Events

▶ Terminology

▶ Applications

▶ Terminology

The systematic study of probability began in the seventeenth century; certain aristocrats of the time wanted to discover superior strategies to use in the gaming rooms of Europe. Some of the best mathematicians of the period were engaged in this pursuit. Since then, probability has evolved into an important branch of mathematics with widespread applications in virtually every sphere of human endeavor in which an element of uncertainty is present.

We will give the technical meaning of the term *probability* in Section 7.2, after we have introduced some of the basic terminology used in the study of the subject. The rest of this chapter will be devoted to the development of techniques for computing the probabilities of the occurrence of certain events.

There are a number of specialized terms used in the study of probability. We begin by defining the term *experiment*.

Experiment

An **experiment** is an activity with observable results.

The results of the experiment are called the **outcomes** of the experiment. Three examples of experiments are:

1. Tossing a coin and observing whether it falls "heads" or "tails."

2. Casting a die and observing which of the numbers 1, 2, 3, 4, 5, or 6 shows up.

3. Testing a spark plug from a batch of 1000 spark plugs and observing whether or not it is defective.

In our discussion of experiments, we will use the following terms:

Sample Point, Sample Space, and Event

Sample Point: an outcome of an experiment.

Sample Space: the set consisting of all possible sample points of an experiment.

Event: a subset of a sample space of an experiment.

The sample space of an experiment is a universal set whose elements are precisely the outcomes of the experiment, and the events of the experiment are the subsets of the universal set. A sample space associated with an experiment that has a finite number of possible outcomes (sample points) is called a **finite sample space.** This terminology is illustrated in the next several examples.

Since the events of an experiment are subsets of a universal set (the sample space of the experiment) we may use the set-theoretic results of Chapter 6 to aid us in the study of probability. We begin by explaining the role played by the empty set and a universal set when viewed as events associated with the experiment. The empty set, ϕ, is called the *impossible event*; it cannot occur since ϕ has no elements (outcomes). Next, the universal set S is referred to as the *certain event*; it must occur since S contains all the outcomes of the experiment.

EXAMPLE

1 Describe the sample space associated with the experiment of tossing a coin and observing whether it falls "heads" or "tails." What are the events of this experiment?

SOLUTION

The two outcomes are "heads" and "tails" and the required sample space is given by $S = \{H, T\}$ where H denotes the outcome "heads" and T denotes the outcome "tails." The events of the experiment, the subsets of S, follow:

$$\phi, \{H\}, \{T\}, S$$

Note that we have included the impossible event, ϕ, and the certain event, S. ◀

Since the events of an experiment are subsets of the sample space of the experiment, we may talk about the *union* and *intersection* of any two events; we can also consider the *complement* of an event with respect to the sample space.

The Union of Two Events

The **union** of the two events E and F is the event $E \cup F$.

Thus, the event $E \cup F$ comprises the set of outcomes of E and/or F.

The Intersection of Two Events

The **intersection** of the two events E and F is the event $E \cap F$.

Thus, the event $E \cap F$ comprises the set of outcomes of E and F.

The Complement of an Event

The **complement** of an event E is the event E^c.

Thus, the event E^c is the set comprising all the outcomes in the sample space S that are not in E.

Venn diagrams depicting the union, intersection, and complementation of events are shown in Figure 7.1.

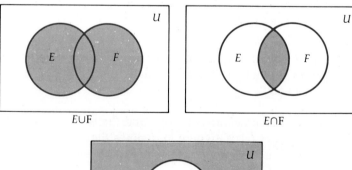

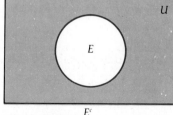

Figure 7.1

These concepts are illustrated in the following example.

EXAMPLE

2 Consider the experiment of casting a die and observing the number that falls uppermost. Let $S = \{1, 2, 3, 4, 5, 6\}$ denote the sample space of the experiment and $E = \{2, 4, 6\}$ and $F = \{1, 3\}$ be events of this experiment. Compute (a) $E \cup F$, (b) $E \cap F$, and (c) F^c. Interpret your results.

SOLUTION

a. $E \cup F = \{1, 2, 3, 4, 6\}$ and is the event that the outcome of the experiment is a 1, a 2, a 3, a 4, or a 6.

b. $E \cap F = \phi$ is the impossible event. This is clear since the number appearing uppermost when a die is cast cannot be both even and odd at the same time.

c. $F^c = \{2, 4, 5, 6\}$ is precisely the event that the event F cannot occur. ◀

Mutually Exclusive Events

E and F are **mutually exclusive** if $E \cap F = \phi$.

As before, we may use Venn diagrams to illustrate these events. In this case, the two mutually exclusive events are depicted as two nonintersecting circles (Figure 7.2).

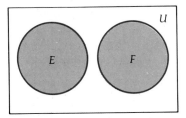

Figure 7.2

Mutually exclusive events

EXAMPLE

3 An experiment consists of tossing a coin three times and observing the resulting sequence of "heads" and "tails."

a. Describe the sample space S of the experiment.

b. Determine the event E that exactly two heads appear.

c. Determine the event F that at least one head appears.

SOLUTION

a. The sample points may be obtained with the aid of a tree diagram (see Figure 7.3). The required sample space S is given by

$$S = \{HHH, HHT, HTH, HTT, THH, THT, TTH, TTT\}$$

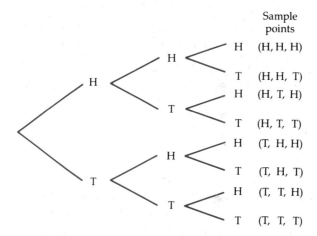

Sample points

Figure 7.3

b. By scanning the sample space S obtained in (a), it is easy to identify the outcomes in which exactly two heads appear, and this gives

$$E = \{HHT, HTH, THH\}.$$

c. Proceeding as in (b), we find

$$F = \{HHH, HHT, HTH, HTT, THH, THT, TTH\}. \qquad \blacktriangleleft$$

EXAMPLE

4 An experiment consists of casting a pair of dice and observing the number that falls uppermost on each die.

a. Describe the sample space S of the experiment.

b. Determine the events $E_2, E_3, E_4, \ldots, E_{12}$ that the sum of the numbers falling uppermost is 2, 3, 4, $\ldots$, 12, respectively.

SOLUTION

a. We may represent each outcome of the experiment by an ordered pair of numbers, the first representing the number that appears uppermost on the first die and the second representing the number that appears uppermost on the second die. In order to distinguish between the two dice, think of the first die as being red and the second as being green. Since there are six possible outcomes for each die, the multiplication principle implies that there are 6 · 6, or 36, elements in the sample space:

$$S = \{(1, 1), (1, 2), (1, 3), (1, 4), (1, 5), (1, 6),$$
$$(2, 1), (2, 2), (2, 3), (2, 4), (2, 5), (2, 6),$$
$$(3, 1), (3, 2), (3, 3), (3, 4), (3, 5), (3, 6),$$
$$(4, 1), (4, 2), (4, 3), (4, 4), (4, 5), (4, 6),$$
$$(5, 1), (5, 2), (5, 3), (5, 4), (5, 5), (5, 6),$$
$$(6, 1), (6, 2), (6, 3), (6, 4), (6, 5), (6, 6)\}$$

b. With the aid of the results of (a), we obtain the required results, which appear in Table 7.1.

Sum of Uppermost Numbers	Event
2	$E_2 = \{(1, 1)\}$
3	$E_3 = \{(1, 2), (2, 1)\}$
4	$E_4 = \{(1, 3), (2, 2), (3, 1)\}$
5	$E_5 = \{(1, 4), (2, 3), (3, 2), (4, 1)\}$
6	$E_6 = \{(1, 5), (2, 4), (3, 3), (4, 2), (5, 1)\}$
7	$E_7 = \{(1, 6), (2, 5), (3, 4), (4, 3), (5, 2), (6, 1)\}$
8	$E_8 = \{(2, 6), (3, 5), (4, 4), (5, 3), (6, 2)\}$
9	$E_9 = \{(3, 6), (4, 5), (5, 4), (6, 3)\}$
10	$E_{10} = \{(4, 6), (5, 5), (6, 4)\}$
11	$E_{11} = \{(5, 6), (6, 5)\}$
12	$E_{12} = \{(6, 6)\}$

Table 7.1   ◀

▶ Applications

EXAMPLE

5 The manager of a local cinema records the number of patrons attending a first-run movie at the 1:00, 3:00, 5:00, 7:30, and 10:00 P.M. screenings. The theater has a seating capacity of 500.

a. What is the sample space for this experiment?

b. Describe the event E that fewer than 50 people attended a screening.

c. Describe the event F that the theater is more than half full at a screening.

SOLUTION

a. The number of patrons at a particular screening (the outcome) could run from 0 to 500. Therefore, the required sample space S is given by

$$S = \{0, 1, 2, 3, \ldots, 500\}.$$

b. $E = \{0, 1, 2, 3, \ldots, 49\}$

c. $F = \{251, 252, 253, \ldots, 500\}$ ◀

The next example shows that sample spaces may be infinite.

EXAMPLE

6 The Ever-Brite Battery Company is developing a high-amperage, high-capacity battery as a source for powering electric cars. The battery is tested by installing it in a prototype electric car and running the car with a fully charged battery on a test track at a constant speed of 55 mph until the car runs out of power. The distance covered by the car is then observed.

a. What is the sample space for this experiment?

b. Describe the event E that the driving range under test conditions is less than 150 miles.

c. Describe the event F that the driving range is between 200 and 250 miles, inclusive.

SOLUTION

a. Since the distance d covered by the car in any run may be given by any nonnegative number, the sample space S is given by

$$S = \{d \mid d \geq 0\}.$$

b. The event E is given by

$$E = \{d \mid d < 150\}.$$

c. The event F is given by

$$F = \{d \mid 200 \leq d \leq 250\}.$$ ◀

Self-Check Exercises

7.1

1. A customer at Cavallaro's Fruit Stand picks a sample of 3 apples at random from a crate containing 80 apples of which 4 are rotten.

a. What is the sample space for this experiment?

b. Describe the event E that exactly one of the apples picked is rotten.

c. Describe the event F that the first apple picked is rotten.

> 2. Refer to Exercise 1.
>
>   a. Compute $E \cup F$.
>   b. Compute $E \cap F$.
>   c. Compute F^c.
>   d. Are the events E and F mutually exclusive?

Solutions to Self-Check Exercises 7.1 can be found on page 339.

▶

Exercises 7.1

1. Let $S = \{a, b, c\}$ be the sample space of an experiment with outcomes a, b, and c. Find all events of this experiment.

*2. Let S be a sample space of an experiment with n outcomes. Determine the number of events of this experiment.

In Exercises 3–7, let $S = \{a, b, c, d, e, f\}$ be the sample space of an experiment, and let $E = \{a, b\}$, $F = \{a, d, f\}$, and $G = \{b, c, e\}$ be events of this experiment.

3. Find the events $E \cup F$ and $E \cap F$.

4. Find the events $F \cup G$ and $F \cap G$.

5. Find the events F^c and $E \cap G^c$.

6. Are E and F mutually exclusive? F and G?

7. Are $E \cup F$ and $E^c \cap F^c$ mutually exclusive?

8. Let S be the sample space for an experiment. Show that if E is any event of an experiment, then E and E^c are mutually exclusive.

9. Let S be the sample space for an experiment, and let E and F be events of this experiment. Show that the events $E \cup F$ and $E^c \cap F^c$ are mutually exclusive. [*Hint:* Use De Morgan's law.]

10. Let $S = \{1, 2, 3, 4, 5, 6\}$, $E = \{2, 4, 6\}$, $F = \{1, 3, 5\}$, and $G = \{5, 6\}$.

    a. Find the event $E \cup F \cup G$.

    b. Find the event $E \cap F \cap G$.

    c. Are the events E and F mutually exclusive? the events F and G?

11. A bin contains ten transistors of which three are defective. An experiment consists of randomly selecting three of the transistors and noting whether they are defective (*d*) or nondefective (*n*). What is the sample space for this experiment?

12. An experiment consists of selecting a letter at random from the letters in the word *MASSACHUSETTS* and observing the outcomes.

    a. What is the sample space for this experiment?

    b. Describe the event "the letter selected is a vowel."

13. An experiment consists of spinning the hand of the numbered disc shown in the figure on the next page and observing the region in which the pointer stops. (If the needle stops on a line, the result is discounted and the needle is spun again.)

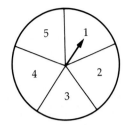

Figure for
Exercise 13

    a. What is the sample space for this experiment?

    b. Describe the event "the spinner points to the number 2."

    c. Describe the event "the spinner points to an odd number."

14. Human blood is classified by the presence or absence of three main antigens (A, B, and Rh). When a blood specimen is typed, the presence of the A and/or B antigen is indicated by listing the letter A and/or the letter B. If neither the A or B antigen is present, the letter O is used to represent this case. The presence or absence of the Rh antigen is indicated by the symbols $+$ or $-$, respectively. Thus, if a blood specimen is classified as AB$^+$, it contains the A and the B antigens as well as the Rh antigen. Similarly, O$^-$ blood contains none of the three antigens. Using this information, determine the sample space corresponding to the different blood groups.

15. An experiment consists of tossing a coin and casting a die and observing the outcomes.

    a. Describe the sample space for this experiment.

    b. Describe the event "a head is tossed and an even number is cast."

16. In a television game show, the winner is asked to select three prizes from five different prizes, A, B, C, D, and E.

    a. Describe a sample space of possible outcomes.

    b. How many points are there in the sample space corresponding to a selection that includes A?

    c. How many points are there in the sample space corresponding to a selection that includes A and B?

    d. How many points are there in the sample space corresponding to a selection that includes either A or B?

17. Let $S = \{1, 2, 3\}$ be a sample space associated with an experiment.

    a. List all events of this experiment.

    b. How many subsets of S contain the number 3?

    c. How many subsets of S contain either the number 2 or the number 3?

18. The manager of a local bank observes how long it takes a customer to complete his transactions at the automatic bank teller.

    a. Describe a sample space for this experiment.

    b. Describe the event that it takes a customer between 2 and 3 minutes to complete his transactions at the automatic bank teller.

In Exercises 19 and 20, let S be any sample space and E, F, and G be any three events associated with the experiment. Describe the given events using the symbols $\cup$, $\cap$, and ^{c}.

19. *a.* The event that E or F occurs.

b. The event that E and F occur.

c. The event that E but not F occurs.

20. *a.* The event that G does not occur.

b. The event that E, F, and G do not occur.

c. The event that E but not F and G occurs.

21. Mr. Robins purchased shares of a machine tool company and shares of an airline company. Let E be the event that the shares of the machine tool company increase in value over a six-month period, and let F be the event that the shares of the airline company increase in value over the next six months. Using the symbols $\cup$, $\cap$, and ^{c}, describe the following events.

a. The shares in the machine tool company do not increase in value.

b. The shares in both the machine tool company and the airline company do not increase in value.

c. The shares of at least one of the two companies increase in value.

d. The shares of only one of the two companies increase in value.

22. The customer service department of the Universal Instruments Company, manufacturer of the Galaxy home computer, conducted a survey among customers who had returned their purchase registration cards. Purchasers of its deluxe model home computer were asked to report the length of time (t) in days before service was required.

a. Describe a sample space corresponding to this survey.

b. Describe the event E that a home computer required service before a period of 90 days had elapsed.

c. Describe the event F that a home computer did not require service before a period of 1 year had elapsed.

23. A time study was conducted by the production manager of the Vista Vision Corporation to determine the length of time in minutes required by an assembly worker to complete a certain task during the assembly of its Pulsar color television sets.

a. Describe a sample space corresponding to this time study.

b. Describe the event E that an assembly worker took 2 minutes or less to complete the task.

c. Describe the event F that an assembly worker took more than 2 minutes to complete the task.

24. An opinion poll is conducted among the electorate in a state to determine the relationship between their income levels and their stands on a proposition aimed at reducing state income taxes. Voters are classified as belonging to either the low-, middle-, or upper-income group. They are asked whether they favor, oppose, or are undecided about the proposition. Let the letters L, M, and U represent the low-, middle-, and upper-income group, respectively, and let the letters f, o, and u represent the responses—favor, oppose, and undecided, respectively.

a. Describe the sample space corresponding to this poll.

b. Describe the event E_1 that the respondent favors the proposition.

c. Describe the event E_2 that the respondent opposes the proposition and does not belong to the low-income group.

d. Describe the event E_3 that the respondent does not favor the proposition and does not belong to the upper-income group.

25. As part of a quality-control procedure, an inspector at Bristol Farms randomly selects ten eggs from each consignment of eggs he receives and records the number of broken eggs.

   a. What is the sample space for this experiment?

   b. Describe the event *E* that at most three eggs are broken.

   c. Describe the event *F* that at least five eggs are broken.

26. In the opinion poll of Exercise 24, the voters were also asked to indicate their political affiliations—Democrat, Republican, or Independent. As before, let the letters *L*, *M*, and *U* represent the low-, middle-, and upper-income group, respectively, and let the letters *D*, *R*, and *I* represent Democrat, Republican, and Independent, respectively.

   a. Describe the sample space corresponding to this poll.

   b. Describe the event E_1 that the respondent is a Democrat.

   c. Describe the event E_2 that the respondent belongs to the upper-income group and is a Republican.

   d. Describe the event E_3 that the respondent belongs to the middle-income group and is not a Democrat.

27. A certain airport hotel operates a shuttle bus service between the hotel and the airport. The maximum capacity of a bus is 20 passengers. On alternate trips of the shuttle bus over a period of one week, the hotel manager kept a record of the number of passengers arriving at the hotel in each bus.

   a. What is the sample space for this experiment?

   b. Describe the event *E* that a shuttle bus carried fewer than ten passengers.

   c. Describe the event *F* that a shuttle bus arrived with a full capacity.

*28. Eight players, A, B, C, D, E, F, G, and H are competing in a series of elimination matches of a tennis tournament in which the winner of each preliminary match will advance to the semifinals and the winner of the semifinals will advance to the finals. An outline of the scheduled matches follows. Describe a sample space listing the possible participants in the finals.

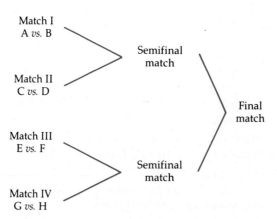

Figure for Exercise 28

Solutions to Self-Check Exercises

7.1

1. *a.* Let *g* denote a good apple and *r* a rotten apple. Thus, the required sample points may be obtained with the aid of a tree diagram (compare with Example 3). The required sample space is given by

$$S = \{ggg, ggr, grg, grr, rgg, rgr, rrg, rrr\}.$$

b. By scanning the sample space *S* obtained in (a), we identify the outcomes in which exactly one apple is rotten. We find

$$E = \{ggr, grg, rgg\}.$$

c. Proceeding as in (b), we find

$$F = \{rgg, rgr, rrg, rrr\}.$$

2. Using the results of Exercise 1, we find
   a. $E \cup F = \{ggr, grg, rgg, rgr, rrg, rrr\}$
   b. $E \cap F = \{rgg\}$
   c. F^c is the set of outcomes in *S* but not in *F*. Thus,

   $$F^c = \{ggg, ggr, grg, grr\}.$$

   d. Since $E \cap F \neq \phi$, we conclude that *E* and *F* are not mutually exclusive.

7.2

Definition of Probability

▶ Finding the Probability of an Event

▶ Applications

▶ Finding the Probability of an Event

Let us return to the coin-tossing experiment. The sample space of this experiment is given by $S = \{H, T\}$ where the sample points H and T correspond to the experiment's two possible outcomes, a *head* and a *tail*. If the coin is *unbiased*, then there is *one chance out of two* of obtaining a head (or a tail) and we say that the *probability* of tossing a head (tail) is 1/2, abbreviated

$$P(H) = \frac{1}{2}, \quad P(T) = \frac{1}{2}.$$

An alternative method of obtaining the values of $P(H)$ and $P(T)$ is based on continued experimentation and does not depend on the assumption that the two outcomes are equally likely. Table 7.2 summarizes the results of such an exercise.

Number of Tosses (n)	Number of Heads (m)	Relative Frequency of Heads (m/n)
10	4	0.4000
100	58	0.5800
1,000	492	0.4920
10,000	5,034	0.5034
20,000	10,024	0.5012
40,000	20,032	0.5008

Table 7.2

Observe that the relative frequencies (column 3) differ considerably when the number of trials is small, but as the number of trials becomes very large, the relative frequency approaches the number 0.5. This result suggests that we assign to $P(H)$ the value 1/2, as before.

The second method of arriving at the probability of obtaining a head on one coin-toss provides us with a popular interpretation of the probability of an event. More generally, consider an experiment that may be repeated over and over again under independent and similar conditions. Suppose that in n trials, an event E occurs m times. We call the ratio (m/n) the **relative frequency** of the event E after n repetitions. If this relative frequency approaches some value $P(E)$ as n becomes larger and larger, then $P(E)$ is called the **empirical probability** of E. According to this concept of probability, the probability $P(E)$ of an event occurring is a measure of the proportion of the time that the event E will occur in the long run. Observe that this method of computing the probability of a head occurring is effective even in the case when a biased coin is used in the experiment.

Since the **probability of an event** is a fraction a/b with $a \leq b$ or is obtained by taking the "limit" of fractions that lie between 0 and 1 inclusive, it is itself a number lying between 0 and 1. In general, the larger the probability of an event, the more likely the event will occur. Thus, an event with a probability of 0.8 is more likely to occur than an event with a probability of 0.6. An event with a probability of 1/2, or 0.5, has a "fifty-fifty" chance of occurring.

Now suppose we are given an experiment and wish to determine the probabilities associated with certain events of the experiment. This problem could be solved by computing $P(E)$ directly for each event E of interest. However, in practice, the number of events that we may be interested in is usually quite large, so this approach is not satisfactory.

The following approach is particularly suitable when the sample space of an experiment is finite.[†] Let S be a finite sample with n outcomes; that is,

$$S = \{s_1, s_2, s_3, \ldots , s_n\}.$$

[†] For the remainder of the chapter we will assume that all sample spaces are finite.

Then the events

$$\{s_1\}, \{s_2\}, \{s_3\}, \ldots, \{s_n\},$$

which consist of exactly one point, are called **elementary,** or **simple,** events of the experiment. They are elementary in the sense that any (nonempty) event of the experiment may be obtained by taking a finite union of suitable elementary events. The simple events of an experiment are also **mutually exclusive;** that is, given any two simple events of the experiment, only one can occur.

By assigning probabilities to each of the simple events, we obtain Table 7.3.

Table 7.3

Simple Event	Probability[†]
$\{s_1\}$	$P(s_1)$
$\{s_2\}$	$P(s_2)$
$\{s_3\}$	$P(s_3)$
$\vdots$	$\vdots$
$\{s_n\}$	$P(s_n)$

[†]For simplicity, we will use the notation $P(s_i)$ instead of the technically more correct $P(\{s_i\})$.

Table 7.3 is called a **probability distribution** for the experiment. The function P, which assigns a probability to each of the simple events, is called a **probability function.**

The numbers $P(s_1), P(s_2), \ldots, P(s_n)$ have the following properties:

1. $0 \le P(s_i) \le 1 \qquad (i = 1, 2, \ldots, n)$
2. $P(s_1) + P(s_2) + \cdots + P(s_n) = 1$
3. $P(\{s_i\} \cup \{s_j\}) = P(s_i) + P(s_j) \qquad i \ne j \qquad (i = 1, 2, \ldots, n; j = 1, 2, \ldots, n)$

The first property simply states that the probability of a simple event must be between 0 and 1 inclusive. The second property states that the sum of the probabilities of all simple events of the sample space is 1. This follows from the fact that the event S is certain to occur. The third property states that the probability of the union of two mutually exclusive events (note that two distinct simple events are mutually exclusive) is given by the sum of their respective probabilities.

As we saw earlier, there is no unique method for assigning probabilities to the simple events of an experiment or, equivalently, for determining the probability function P. In practice, the methods used to determine the probabilities of the simple events of an experiment may range from theoretical considerations of the problem on the one extreme to the reliance on "educated guesses" on the other.

Sample spaces in which the outcomes are equally likely are called **uniform sample spaces.** Assigning probabilities to the simple events in these spaces is relatively easy.

Probability of an Event in a Uniform Sample Space

If

$$S = \{s_1, s_2, \ldots, s_n\}$$

is the sample space for an experiment in which the outcomes are equally likely, then we assign the probabilities

$$P(s_1) = P(s_2) = \cdots = P(s_n) = \frac{1}{n}$$

to each of the simple events $s_1, s_2, \ldots, s_n$.

EXAMPLE

7 A fair die is cast and the number that falls uppermost is observed. Determine the probability distribution for the experiment.

SOLUTION The sample space for the experiment is $S = \{1, 2, 3, 4, 5, 6\}$ and the simple events are accordingly given by the sets $\{1\}$, $\{2\}$, $\{3\}$, $\{4\}$, $\{5\}$, and $\{6\}$. Since the die is assumed to be fair, the six outcomes are equally likely. We therefore assign a probability of 1/6 to each of the simple events and obtain the following probability distribution for the experiment:

Simple Event	Probability
{1}	1/6
{2}	1/6
{3}	1/6
{4}	1/6
{5}	1/6
{6}	1/6

◀

The next example shows how the *relative frequency* interpretation of probability lends itself to the computation of probabilities.

EXAMPLE

8 Refer to Example 6. The following data were obtained in tests involving 200 test runs. Each run was made with a fully charged battery.

Distance Covered in Miles (x)	Frequency of Occurrence
$0 \leq x \leq 50$	4
$50 < x \leq 100$	10
$100 < x \leq 150$	30
$150 < x \leq 200$	100
$200 < x \leq 250$	40
$250 < x$	16

a. Describe the sample space associated with the experiment.

b. Find the probability distribution for the experiment.

SOLUTION

a. Let s_1 denote the outcome that the distance covered by the car does not exceed 50 miles; let s_2 denote the outcome that the distance covered by the car is greater than 50 miles but does not exceed 100 miles, and so on. Finally, let s_6 denote the outcome that the distance covered by the car is greater than 250 miles. Then the required sample space S is given by

$$S = \{s_1, s_2, s_3, s_4, s_5, s_6\}.$$

b. To compute the probability distribution for the experiment, we turn to the relative frequency interpretation of probability. Accepting the inaccuracies inherent in a relatively small number of trials (200 runs), we will take the probability of s_1 occurring as

$$P(s_1) = \frac{\text{number of trials in which } s_1 \text{ occurs}}{\text{total number of trials}}$$

$$= \frac{4}{200}$$

$$= 0.02.$$

In a similar manner, we can assign probabilities to the other simple events, and obtain the following probability distribution for the experiment.

Simple Event	Probability
$\{s_1\}$	0.02
$\{s_2\}$	0.05
$\{s_3\}$	0.15
$\{s_4\}$	0.50
$\{s_5\}$	0.20
$\{s_6\}$	0.08

◀

We are now in a position to give a procedure for computing the probability $P(E)$ of an arbitrary event E of an experiment.

Finding the Probability of an Event *E*

1. Determine a sample space *S* associated with the experiment.

2. Assign probabilities to the simple events of *S*.

3. If *E* is the empty set, ϕ, then $P(E) = 0$. Otherwise, $P(E)$ is given by the sum of the probabilities of the simple events whose union is *E*. Thus, if $E = \{s_1, s_2, s_3, \ldots, s_n\}$ where $\{s_1\}, \{s_2\}, \{s_3\}, \ldots, \{s_n\}$ are simple events, then $P(E)$ is assigned the value

$$P(E) = P(s_1) + P(s_2) + P(s_3) + \cdots + P(s_n).$$

The principle stated in step (3) is called the **addition principle** and is a consequence of Property 3 of the probability function (page 341). This principle allows us to find the probabilities of all the other events once the probabilities of the simple events are known.

▶ Applications

EXAMPLE

9 A pair of fair dice is cast.

a. Calculate the probability that the two dice show the same number.

b. Calculate the probability that the sum of the numbers of the two dice is six.

SOLUTION

From the results of Example 4, we see that the sample space *S* of the experiment consists of 36 outcomes:

$$S = \{(1, 1), (1, 2), \ldots, (6, 5), (6, 6)\}.$$

(See Example 4 for a complete listing.) Since both dice are fair, each of the 36 outcomes is equally likely. Accordingly, we assign the probability of 1/36 to each simple event. We are now in a position to answer the questions posed.

a. The event that the two dice show the same number is given by

$$E = \{(1, 1), (2, 2), (3, 3), (4, 4), (5, 5), (6, 6)\}.$$

Therefore, by the addition principle, the probability that the two dice show the same number is given by

$$\begin{aligned}
P(E) &= P[(1, 1)] + P[(2, 2)] + \cdots + P[(6, 6)] \\
&= \frac{1}{36} + \frac{1}{36} + \cdots + \frac{1}{36} \quad \text{(Six terms)} \\
&= \frac{1}{6}.
\end{aligned}$$

b. The event that the sum of the numbers of the two dice is six is given by

$$E_6 = \{(1, 5), (2, 4), (3, 3), (4, 2), (5, 1)\} \quad \text{(See Table 7.1.)}$$

Therefore, the probability that the sum of the numbers on the two dice is six is given by

$$P(E_6) = P[(1, 5)] + P[(2, 4)] + P[(3, 3)] + P[(4, 2)] + P[(5, 1)]$$
$$= \frac{1}{36} + \frac{1}{36} + \cdots + \frac{1}{36} \quad \text{(Five terms)}$$
$$= \frac{5}{36}.$$

◀

EXAMPLE

10 Consider the experiment by the Ever-Brite Company in Example 8. What is the probability that the prototype car will travel more than 150 miles on a fully charged battery?

SOLUTION

Using the results of Example 8, we see that the event that the car will travel more than 150 miles on a fully charged battery is given by $E = \{s_4, s_5, s_6\}$. Therefore, the probability that the car will travel more than 150 miles on one charge is given by

$$P(E) = P(s_4) + P(s_5) + P(s_6),$$

or, using the probability distribution for the experiment obtained in Example 8,

$$P(E) = 0.50 + 0.20 + 0.08$$
$$= 0.78.$$

◀

Self-Check Exercises

7.2

1. A biased die was cast repeatedly and the results of the experiment are summarized in the following table.

Outcome	1	2	3	4	5	6
Frequency of Occurrence	142	173	158	175	162	190

Using the relative frequency interpretation of probability, find the probability distribution for this experiment.

2. In an experiment conducted to study the effectiveness of an eye-level third brake light in the prevention of rear-end collisions, 250 of the 500 highway patrol cars of a certain state were equipped with such lights. At the end of the one-year trial period, the records revealed that of those equipped with a third brake light, there were 14 incidents of rear-end collision. There were 22 such incidents involving the cars not equipped with the accessory. Based on these data, what is the probability that a highway patrol car equipped with a third brake light will be rear-ended within a one-year period? What is the probability that a car not so equipped will be rear-ended within a one-year period?

Solutions to Self-Check Exercises 7.2 can be found on page 350.

▶

Exercises 7.2

In Exercises 1–5, list the simple events associated with each of the given experiments.

1. A nickel and a dime are tossed and the result of heads or tails is recorded for each coin.

2. A card is selected at random from a standard 52-card deck and its suit—hearts (h), diamonds (d), spades (s), or clubs (c)—is recorded.

3. An opinion poll was conducted among a group of registered voters. Their political affiliation, Democratic (D), Republican (R), or Independent (I), and their sex, male (m) or female (f), were recorded.

4. As part of a quality-control procedure, eight circuit boards were checked and the number of defectives was recorded.

5. In a survey conducted to determine whether movie attendance was increasing (i), decreasing (d), or holding steady (s) among various sectors of the population, participants were classified as follows:

    Group 1: those aged 10–19

    Group 2: those aged 20–29

    Group 3: those aged 30–39

    Group 4: those aged 40–49

    Group 5: those 50 and over

The response of each participant and his or her age group were recorded.

In Exercises 6–9, find the number of simple events in each of the given experiments.

6. Two dice are cast and the number appearing uppermost on each die is recorded.

7. A meteorologist preparing a weather map classifies the expected average temperature in each of 50 states for the upcoming week as follows:

    a. More than 10° below average

    b. Normal to 10° below average

c. Higher than Normal to 10° above average

d. More than 10° above average

Using each state's abbreviation and the categories—(a), (b), (c), and (d)—the meteorologist records these data.

8. Blood tests are given as a part of the admission procedure at the Monterey Garden Community Hospital. The blood type of each patient (A, B, AB, or O) and the presence or absence of the Rh factor in each patient's blood (Rh$^+$ or Rh$^-$) are recorded.

9. Data concerning durable goods orders are obtained each month by an economist. A record is kept for a one-year period of any increase (i), decrease (d), or unchanged movement (u) in the number of durable goods orders for each month as compared to the number of such orders in the same month in the previous year.

10. The grade distribution for a certain class is shown in the following table. Find the probability distribution associated with these data.

Grade	A	B	C	D	F
Frequency of Occurrence	4	10	18	6	2

11. The percentage of the general population that has each blood type is shown in the following table.

Blood Type	A	B	AB	O
Percentage of Population	41%	12%	3%	44%

Determine the probability distribution associated with these data.

12. The number of cars entering a tunnel leading to an airport in a major city over a period of 200 peak hours was observed and the following data were obtained:

Number of Cars (x)	Frequency of Occurrence
$0 < x \le 200$	15
$200 < x \le 400$	20
$400 < x \le 600$	35
$600 < x \le 800$	70
$800 < x \le 1000$	45
$x > 1000$	15

a. Describe the sample space associated with this experiment.

b. Find the probability distribution for this experiment.

13. The Metro Telephone Company of Belmont compiled the following information during a service-utilization study pertaining to the number of customers using their

Dial-the-Time service from 7 A.M. to 9 A.M. on a certain weekday morning. Using these data, find the probability distribution associated with this experiment.

Number of Calls Received/Minute	Frequency of Occurrence
10	6
11	15
12	12
13	3
14	12
15	36
16	24
17	0
18	6
19	6

14. The results of a time study conducted by the production manager of the Ace Novelty Company are shown in the following table, where the number of space action-figures produced each quarter hour during an 8-hour workday have been tabulated. Find the probability distribution associated with this experiment.

Number of Figures Produced (in Dozens)	Frequency of Occurrence
30	4
31	0
32	6
33	8
34	6
35	4
36	4

15. The number of subscribers to five leading electronic mail services is shown in the following table.

Company	A	B	C	D	E
Subscribers	300,000	200,000	120,000	80,000	60,000

Find the probability distribution associated with these data.

16. A study conducted by the Corrections Department of a certain state revealed that 163,605 people out of a total adult population of 1,778,314 were under correctional supervision (on probation, parole, or in jail). What is the probability that a person selected at random from the adult population in that state is under correctional supervision?

17. If one card is drawn at random from a standard 52-card deck, what is the probability that the card drawn is

 a. a diamond?

 b. a black card?

 c. an ace?

18. One light bulb is selected at random from a lot of 120 light bulbs of which 5 percent are defective. What is the probability that the light bulb selected is defective?

19. A pair of fair dice is cast. What is the probability that

 a. the sum of the numbers shown uppermost is less than five?

 b. at least one 6 is cast?

20. If a ball is selected at random from an urn containing three red balls, two white balls, and five blue balls, what is the probability that it will be a white ball?

21. What is the probability of arriving at a traffic light when it is red if the red signal is flashed for 30 seconds, the yellow signal for 5 seconds, and the green signal for 45 seconds?

22. What is the probability that a roulette ball will come to rest on an even number other than 0 or 00? (Assume that there are 38 equally likely outcomes consisting of the numbers 1–36, 0, and 00.)

23. Refer to Exercise 12. What is the probability that more than 600 cars will enter the airport tunnel during a peak hour?

24. Refer to Exercise 10. What is the probability that a student selected at random from this class received a passing grade (D or better)?

25. One hundred thousand entries have been received in a sweepstakes sponsored by the Gemini Paper Products Company. If 1 grand prize, 5 first prizes, 25 second prizes, and 500 third prizes are to be awarded, what is the probability that

 a. a person who has submitted one entry will win the grand prize?

 b. a person who has submitted one entry will win a prize?

26. If a coin is tossed that is weighted in such a way that heads is four times as likely to occur as tails, what is the probability that heads will turn up?

In Exercises 27–30, determine whether the given experiment has a sample space with equally likely outcomes.

27. A loaded die is cast and the number appearing uppermost on the die is recorded.

28. Two dice are cast and the sum of the numbers appearing uppermost on each die is recorded.

29. A ball is selected at random from an urn containing seven black balls and eight red balls and the color of the ball is recorded.

30. A weighted coin is thown and the outcome of heads or tails is recorded.

31. Let $S = \{s_1, s_2, s_3, s_4, s_5\}$ be the sample space associated with an experiment having the following probability distribution.

Outcome	$\{s_1\}$	$\{s_2\}$	$\{s_3\}$	$\{s_4\}$	$\{s_5\}$
Probability	1/14	3/14	6/14	2/14	2/14

Find the probability of the event

    a. $A = \{s_1, s_2, s_4\}$ *b.* $B = \{s_1, s_5\}$ *c.* $C = S$

32. Let $S = \{s_1, s_2, s_3, s_4, s_5, s_6\}$ be the sample space associated with an experiment having the following probability distribution:

Outcome	$\{s_1\}$	$\{s_2\}$	$\{s_3\}$	$\{s_4\}$	$\{s_5\}$	$\{s_6\}$
Probability	1/12	1/4	1/12	1/6	1/3	1/12

Find the probability of the event

    a. $A = \{s_1, s_3\}$ *b.* $B = \{s_2, s_4, s_5, s_6\}$ *c.* S

33. An opinion poll was conducted among a group of registered voters in a certain state concerning a proposition aimed at limiting state and local taxes. Results of the poll indicated that 35 percent of the voters favored the proposition, 32 percent were against it, and the remaining group were undecided. If the results of the poll are assumed to be representative of the opinion of the electorate of the state, what is the probability that

    a. a registered voter selected at random from the electorate favors the proposition?

    b. a registered voter selected at random from the electorate is undecided about the proposition?

34. The number of planes in the fleets of five leading airlines that contain Airfones is shown in the following table.

Airline	Number of Planes with Airfones	Size of Fleet
A	50	295
B	40	325
C	31	167
D	29	50
E	25	248

    a. If a plane is selected at random from airline A, what is the probability that it contains an Airfone?

    b. If a plane is selected at random from the entire fleet of the five airlines, what is the probability that it contains an Airfone?

Solutions to Self-Check Exercises

7.2

1. $P(1) = \dfrac{\text{number of trials in which a 1 appears uppermost}}{\text{total number of trials}}$

    $= \dfrac{142}{1000} = 0.142$

Similarly, we compute $P(2), \ldots, P(6)$, obtaining the following probability distribution

Outcome	1	2	3	4	5	6
Probability	.142	.173	.158	.175	.162	.190

2. The probability that a highway patrol car equipped with a third brake light will be rear-ended within a one-year period is given by

$$\frac{\text{frequency of rear-end collisions involving cars equipped with a third brake light}}{\text{total number of such cars}}$$

$$= \frac{14}{250}$$

$$= 0.056.$$

The probability that a highway patrol car not equipped with a third brake light will be rear-ended within a one-year period is given by

$$\frac{\text{frequency of rear-end collisions involving cars not equipped with a third brake light}}{\text{total number of such cars}}$$

$$= \frac{22}{250}$$

$$= 0.088.$$

7.3

Rules of Probability

▶ Properties of the Probability Function and Their Applications

▶ Computations Involving the Rules of Probability

▶ Properties of the Probability Function and Their Applications

In this section we will examine some of the properties of the probability function and look at the role they play in solving certain problems. We begin by looking at the generalization of the three properties of the probability function, which were stated for simple events in the last section. Let S be a sample space of an experiment and suppose E and F are events of the experiment. Then we have

Property 1

$$P(E) \geq 0 \text{ for any } E.$$

Property 2

$$P(S) = 1.$$

Property 3

If E and F are mutually exclusive (that is, only one of them can occur, or equivalently, $E \cap F = \phi$), then

$$P(E \cup F) = P(E) + P(F).$$

Property 3 may be easily extended to the case involving any finite number of mutually exclusive events. Thus, if $E_1, E_2, \ldots, E_n$ are mutually exclusive events, then

$$P(E_1 \cup E_2 \cup \cdots \cup E_n) = P(E_1) + P(E_2) + \cdots + P(E_n).$$

EXAMPLE

11 The superintendent of a metropolitan school district has estimated the probabilities associated with the SAT verbal scores of students from that district. The results are shown in Table 7.4. If a student is selected at random, what is the probability that his or her SAT verbal score will be

a. more than 400?

b. less than or equal to 500?

c. greater than 400 but less than or equal to 600?

Score (x)	Probability
$x > 700$	0.01
$600 < x \leq 700$	0.07
$500 < x \leq 600$	0.19
$400 < x \leq 500$	0.23
$300 < x \leq 400$	0.31
$x \leq 300$	0.19

Table 7.4

SOLUTION

Let $A, B, C, D, E,$ and F denote, respectively, the event that the score is greater than 700, greater than 600 but less than or equal to 700, greater than 500 but less than or equal to 600, and so forth. Then, these events are mutually exclusive. Therefore *a.* the probability that the student's score will be more than 400 is given by

$$P(D \cup C \cup B \cup A) = P(D) + P(C) + P(B) + P(A)$$
$$= 0.23 + 0.19 + 0.07 + 0.01$$
$$= 0.5.$$

b. the probability that the student's score will be less than or equal to 500 is given by

$$P(D \cup E \cup F) = P(D) + P(E) + P(F)$$
$$= 0.23 + 0.31 + 0.19$$
$$= 0.73.$$

c. the probability that the student's score will be greater than 400 but less than or equal to 600 is given by

$$P(C \cup D) = P(C) + P(D)$$
$$= 0.19 + 0.23$$
$$= 0.42. \qquad \blacktriangleleft$$

Property 3 holds when and only when E and F are mutually exclusive. In the general case, we have

---

Property 4 (Addition Rule)

If E and F are any two events of an experiment, then

$$P(E \cup F) = P(E) + P(F) - P(E \cap F).$$

Observe that when E and F are mutually exclusive, that is, when $E \cap F = \phi$, then the equation of Property 4 reduces to that of Property 3.

EXAMPLE

12 A card is drawn from a well-shuffled deck of 52 playing cards. What is the probability that it is an ace or a spade?

SOLUTION

Let E denote the event that the card drawn is an ace, and let F denote the event that the card drawn is a spade. Then

$$P(E) = \frac{4}{52} \quad \text{and} \quad P(F) = \frac{13}{52}.$$

Furthermore, E and F are not mutually exclusive events. In fact $E \cap F$ is the event that the card drawn is an ace of spades. Consequently,

$$P(E \cap F) = \frac{1}{52}.$$

The event that a card drawn is an ace or a spade is $E \cup F$, with probability given by

$$\begin{aligned} P(E \cup F) &= P(E) + P(F) - P(E \cap F) \\ &= \frac{4}{52} + \frac{13}{52} - \frac{1}{52} \\ &= \frac{4}{13}. \end{aligned}$$

This result, of course, can be obtained by arguing that 16 of the 52 cards are either spades or aces of other suits. ◀

EXAMPLE

13 The quality control department of the Vista Vision Corporation, the manufacturer of the Pulsar 19-inch color television set, has determined from records obtained from the company's service centers that 3 percent of the sets sold experience video problems, 1 percent experience audio problems, and 0.1 percent experience both video as well as audio problems before the expiration of the 90-day warranty. Find the probability that a set purchased by a consumer will experience video or audio problems before the warranty expires.

SOLUTION Let E denote the event that a set purchased will experience video problems within 90 days, and let F denote the event that a set purchased will experience audio problems within 90 days. Then

$$P(E) = 0.03, \ P(F) = 0.01, \text{ and } P(E \cap F) = 0.001.$$

The event that a set purchased will experience video problems or audio problems before the warranty expires is $E \cup F$, and the probability of this event is given by

$$\begin{aligned} P(E \cup F) &= P(E) + P(F) - P(E \cap F) \\ &= 0.03 + 0.01 - 0.001 \\ &= 0.039. \end{aligned}$$

◀

Another property of the probability function, which is of considerable aid in computing probability, follows:

Property 5 (Rule of Complements)

If E is an event of an experiment, and E^c denotes the complement of E, then

$$P(E^c) = 1 - P(E).$$

Property 5 is an immediate consequence of Properties 2 and 3. Indeed, we have $E \cup E^c = S$ and $E \cap E^c = \phi$, so

$$1 = P(S) = P(E \cup E^c) = P(E) + P(E^c),$$

whence $P(E^c) = 1 - P(E).$

EXAMPLE

14 Refer to Example 13. What is the probability that a Pulsar 19-inch color television set bought by a consumer will not experience video or audio difficulties before the warranty expires?

SOLUTION Let E denote the event that a set bought by a consumer will experience video or audio difficulties before the warranty expires. Then the event that the set will not experience either problem before the warranty expires is given by E^c, with probability

$$P(E^c) = 1 - P(E)$$
$$= 1 - 0.039$$

◀

▶ Computations Involving the Rules of Probability

We close this section by looking at two further examples that illustrate the rules of probability.

EXAMPLE

15 Let E and F be two events that are mutually exclusive and suppose $P(E) = 0.1$ and $P(F) = 0.6$. Compute

a. $P(E \cap F)$
b. $P(E \cup F)$
c. $P(E^c)$
d. $P(E^c \cap F^c)$
e. $P(E^c \cup F^c)$

SOLUTION   *a.* Since the events E and F are mutually exclusive, that is, $E \cap F = \phi$, we have $P(E \cap F) = 0$.

b. $P(E \cup F) = P(E) + P(F)$ (Since E and F are mutually exclusive)

$\qquad\qquad = 0.1 + 0.6$

$\qquad\qquad = 0.7$

c. $P(E^c) = 1 - P(E)$ (Property 5)

$\qquad\quad = 1 - 0.1$

$\qquad\quad = 0.9$

d. Observe that, by De Morgan's Law, $E^c \cap F^c = (E \cup F)^c$, so

$$P(E^c \cap F^c) = P[(E \cup F)^c]$$
$$= 1 - P(E \cup F) \qquad \text{(Property 5)}$$
$$= 1 - 0.7 \qquad \text{[Using the result of (b)]}$$
$$= 0.3$$

e. Again, using De Morgan's Law, we find

$$P(E^c \cup F^c) = P[(E \cap F)^c]$$
$$= 1 - P(E \cap F)$$
$$= 1 - 0 \qquad \text{[Using the result of (a)]}$$
$$= 1 \qquad\qquad\qquad\qquad\qquad\qquad \blacktriangleleft$$

EXAMPLE

16 Let E and F be two events of an experiment with sample space S. Suppose $P(E) = 0.2$, $P(F) = 0.1$, and $P(E \cap F) = 0.05$. Compute

a. $P(E \cup F)$

b. $P(E^c \cap F^c)$

c. $P(E^c \cap F)$ [*Hint:* Draw a Venn diagram.]

SOLUTION   *a.* $P(E \cup F) = P(E) + P(F) - P(E \cap F)$ (Property 4)

$\qquad\qquad = 0.2 + 0.1 - 0.05$

$\qquad\qquad = 0.25$

b. Using De Morgan's Law, we have,

$$P(E^c \cap F^c) = P[(E \cup F)^c]$$
$$= 1 - P(E \cup F) \qquad \text{(Property 5)}$$
$$= 1 - 0.25 \qquad \text{[Using the result of (a)]}$$
$$= 0.75$$

c. From the Venn diagram describing the relationship between E, F, and S (Figure 7.4), we have (the shaded subset is the event $E^c \cap F$)

$$P(E^c \cap F) = 0.05.$$

This result may also be obtained by using the relationship

$$P(E^c \cap F) = P(F) - P(E \cap F)$$
$$= 0.1 - 0.05$$
$$= 0.05$$

as before.

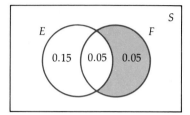

Figure 7.4

Self-Check Exercises

7.3

1. Let E and F be events of an experiment with sample space S. Suppose $P(E) = 0.4$, $P(F) = 0.5$, and $P(E \cap F) = 0.1$. Compute

   a. $P(E \cup F)$.

   b. $P(E \cap F^c)$.

2. Ms. Garcia wishes to sell or lease a condominium through a realty company. The realtor estimates that the probability of finding a buyer within a month of the date the property is listed for sale or lease is 0.3, the probability of finding a lessee is 0.8, and the probability of finding both a buyer and a lessee is 0.1. Determine the probability that the property will be sold or leased within one month from the date the property is listed for sale or lease.

Solutions to Self-Check Exercises 7.3 can be found on page 362.

Exercises 7.3

A pair of dice is cast and the number that appears uppermost on each die is observed. In Exercises 1–6, refer to this experiment and find the probability of the given event.

1. The sum of the numbers is an even number.

2. The sum of the numbers is either 7 or 11.

3. A pair of 1's is thrown.

4. A double is thrown.

5. One die shows a 6 and the other a number less than 3.

6. The sum of the numbers is at least 4.

An experiment consists of selecting a card at random from a 52-card deck. In Exercises 7–11, refer to this experiment and find the probability of the given event.

7. A king of diamonds is drawn.

8. A diamond or a king is drawn.

9. A face card is drawn.

10. A red face card is drawn.

11. An ace is not drawn.

12. Five hundred people have purchased lottery tickets. What is the probability that a person holding one ticket will win the first prize? What is the probability that he or she will not win the first prize?

In Exercises 13–21, explain why the given statement is incorrect.

13. The probability that a bus will not arrive on time at the Civic Center is 0.35, and the probability that it will either be on time or late is 0.60.

14. A 5-card poker hand is dealt from a 52-card deck. Let *A* denote the event that a flush is dealt and let *B* be the event that a straight is dealt. Then the events *A* and *B* are mutually exclusive.

15. A person participates in a weekly office pool in which he has one chance in ten of winning the purse. If he participates for five weeks in succession, the probability of winning at least one purse is 5/10.

16. The probability that a certain stock will increase in value over a period of one week is 0.6. Therefore, the probability that the stock will decrease in value is 0.4.

17. A red die and a green die are tossed. The probability that a 6 will appear uppermost on the red die is 1/6 and the probability that a 1 will appear uppermost on the green die is 1/6. Hence, the probability that the red die will show a 6 or the green die will show a 1 is 1/6 + 1/6.

18. Joanne, a high school senior, has applied for admission to four colleges, *A, B, C,* and *D.* She has estimated that the probability that she will be accepted for admission by *A, B, C,* and *D* is 0.5, 0.3, 0.1, and 0.08, respectively. Thus, the probability that she will be accepted for admission by at least one college is $P(A) + P(B) + P(C) + P(D) = 0.5 + 0.3 + 0.1 + 0.08 = 0.98$.

19. The sample space associated with an experiment is given by $S = \{a, b, c, d, e\}$. The events $E = \{a, b\}$ and $F = \{c, d\}$ are mutually exclusive. Hence, the events E^c and F^c are mutually exclusive.

20. Mr. Owens, an optician, estimates that the probability that a customer coming into his store will purchase one or more pairs of glasses but not contact lenses is 0.40, and the probability that he will purchase one or more pairs of contact lenses but not glasses is 0.25. Hence, Owens concludes that the probability that a customer coming into his store will purchase neither a pair of glasses nor a pair of contact lenses is 0.35.

21. There are eight grades in the Garfield Elementary School. If a student is selected at random from the school, then the probability that the student is in the first grade is 1/8.

22. An experiment consists of casting a die and observing the number that appears uppermost. Let E be the event that an even number is thrown and let F be the event that a number greater than 4 is thrown. Describe the following events in words.

    a. $E \cup F$ *b.* $E \cap F$ *c.* $E^c \cap F^c$ *d.* $E^c \cap F$

23. Refer to the following Venn diagram, in which the events A, B, and C are depicted along with their associated probabilities.

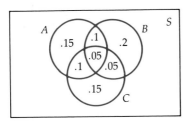

    a. Write the following events in set notation.

        i. Neither A nor B occurs.

        ii. A but not B occurs.

        iii. Both A and B but not C occur.

    b. Find the probability of each of the events described in (a).

24. Six balls numbered 1 through 6 are placed in an urn. Two of the balls are then selected at random. What is the probability that

    a. the balls numbered 2 and 5 are selected?

    b. neither of the balls numbered 2 and 5 are selected?

    c. at least one of the balls selected is numbered 5?

25. Let E and F be two events that are mutually exclusive and suppose $P(E) = 0.2$ and $P(F) = 0.5$. Compute

    a. $P(E \cap F)$ *b.* $P(E \cup F)$ *c.* $P(E^c)$

    d. $P(E^c \cap F^c)$ *e.* $P(E^c \cup F^c)$

26. Let $S = \{s_1, s_2, s_3, s_4\}$ be the sample space associated with an experiment having the probability distribution shown in the following table. If $A = \{s_1, s_2\}$ and $B = \{s_1, s_3\}$, find

    a. $P(A)$, $P(B)$ *b.* $P(A^c)$, $P(B^c)$ *c.* $P(A \cap B)$

    d. $P(A \cup B)$

Outcome	Probability
$\{s_1\}$	1/8
$\{s_2\}$	3/8
$\{s_3\}$	1/4
$\{s_4\}$	1/4

27. Let $S = \{s_1, s_2, s_3, s_4, s_5, s_6\}$ be the sample space associated with an experiment having the probability distribution shown in the table on the next page. If $A = \{s_1, s_2\}$ and $B = \{s_1, s_5, s_6\}$, find

a. $P(A)$, $P(B)$

b. $P(A^c)$, $P(B^c)$

c. $P(A \cap B)$

d. $P(A \cup B)$

e. $P(A^c \cap B^c)$

f. $P(A^c \cup B^c)$

Outcome	Probability
$\{s_1\}$	1/3
$\{s_2\}$	1/8
$\{s_3\}$	1/6
$\{s_4\}$	1/6
$\{s_5\}$	1/12
$\{s_6\}$	1/8

28. Let E and F be two events of an experiment with sample space S. Suppose $P(E) = 0.4$, $P(F) = 0.2$, and $P(E \cap F) = 0.1$. Compute

a. $P(E \cup F)$

b. $P(E^c \cap F^c)$

c. $P(E^c \cap F)$

29. Let E and F be two events of an experiment with sample space S. Suppose $P(E) = 0.6$, $P(F) = 0.4$, and $P(E \cap F) = 0.2$. Compute

a. $P(E \cup F)$

b. $P(E^c)$

c. $P(F^c)$

d. $P(E^c \cap F)$

e. $P(E \cap F^c)$

f. $P(E^c \cup F^c)$

30. Among 500 freshmen pursuing a business degree at a university, 320 are enrolled in an economics course, 225 are enrolled in a mathematics course, and 140 are enrolled in both an economics and a mathematics course. What is the probability that a freshman selected at random from this group is enrolled in

a. an economics and/or a mathematics course?

b. exactly one of these two courses?

c. neither an economics course nor a mathematics course?

31. A leading manufacturer of kitchen appliances advertised its products in two magazines: *Good Housekeeping* and the *Ladies Home Journal*. A survey of 500 customers revealed that 140 learned of its products from *Good Housekeeping*, 130 learned of its products from the *Ladies Home Journal*, and 80 learned of its products from both magazines. What is the probability that a person selected at random from this group saw the manufacturer's advertisement in

a. both magazines?

b. at least one of the two magazines?

c. exactly one magazine?

32. Students at a certain university were asked to state how many hours per week they spent studying in the library. Results of the survey revealed the following information:

Time Spent in Hours (x)	Percentage of Students
$0 \le x \le 1$	32.3
$1 < x \le 4$	40.7
$4 < x \le 10$	16.5
$x > 10$	10.5

Find the probability distribution associated with these data. What is the probability that a student selected at random at the university studied in the library

a. more than 4 hours per week?

b. no more than 10 hours per week?

33. A time study was conducted by the production manager of the Universal Instruments Company to determine how much time it took an assembly worker to complete a certain task during the assembly of its Galaxy home computers. Results of the study indicated that 20 percent of the workers were able to complete the task in less than 3 minutes, 60 percent of the workers were able to complete the task in 4 minutes or less, and 10 percent of the workers required more than 5 minutes to complete the task. If an assembly line worker is selected at random from this group, what is the probability that

a. he or she will be able to complete the task in 5 minutes or less?

b. he or she will not be able to complete the task within 4 minutes?

c. the time taken for the worker to complete the task will be between 3 and 4 minutes (inclusive)?

34. A poll was conducted among 250 residents of a certain city regarding tougher gun-control laws. The results of the poll are shown below.

	Own Only a Handgun	Own Only a Rifle	Own a Handgun and a Rifle	Own Neither	Total
Favor Tougher Laws	0	12	0	138	150
Oppose Tougher Laws	58	5	25	0	88
No Opinion	0	0	0	12	12
Total	58	17	25	150	250

If one of the participants in this poll is selected at random, what is the probability that he or she

a. favors tougher gun-control laws?

b. owns a handgun?

c. owns a handgun but not a rifle?

d. favors tougher gun-control laws and does not own a handgun?

35. Let S be a uniform sample space of an experiment, and suppose E and F are events of the experiment. Prove that

$$P(E \cup F) = P(E) + P(F) - P(E \cap F).$$

[Hint: Since E and F are subsets of S, we may use Formula 4, page 294: $n(E \cup F) = n(E) + n(F) - n(E \cap F)$. Divide both sides of this equation by $n(S)$, assuming the latter is different from zero, and interpret the quotients obtained.]

Solutions to Self-Check Exercises

7.3

1. *a.* Using Property 4, we find

$$P(E \cup F) = P(E) + P(F) - P(E \cap F)$$
$$= 0.4 + 0.5 - 0.1$$
$$= 0.8.$$

b. From the accompanying Venn diagram, in which the subset $E \cap F^c$ is shaded, we see that

$$P(E \cap F^c) = 0.3.$$

The result may also be obtained by using the relationship

$$P(E \cap F^c) = P(E) - P(E \cap F)$$
$$= 0.4 - 0.1$$
$$= 0.3.$$

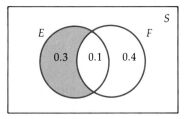

2. Let E denote the event that the property will be sold within one month of the date it is listed for sale or lease, and let F denote the event that the property will be leased within the same time period. Then

$$P(E) = 0.3,\ P(F) = 0.8,\ \text{and}\ P(E \cap F) = 0.1.$$

The event that the property will be sold or leased within one month of the date it is listed for sale or lease is given by

$$P(E \cup F) = P(E) + P(F) - P(E \cap F)$$
$$= 0.3 + 0.8 - 0.1$$
$$= 1,$$

that is, a certainty.

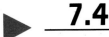

7.4

Use of Counting Techniques in Probability

▶ Further Applications of Counting Techniques

▶ The Birthday Problem

▶ Further Applications of Counting Techniques

As we have seen many times before, a problem in which the underlying sample space has a small number of elements may be solved by first determining all such sample points. For problems involving sample spaces with a large number of sample points, however, this approach is neither practical nor desirable.

In this section we will see how the counting techniques studied in Chapter 6 may be employed to help us solve problems in which the associated sample spaces contain large numbers of sample points. In particular, we will restrict our attention to the study of uniform sample spaces, that is, sample spaces in which the outcomes are equally likely. For such spaces we have the following result:

Computing the Probability of an Event in a Uniform Sample Space

Let S be a uniform sample space and let E be any event. Then,

$$P(E) = \frac{\text{number of favorable outcomes in } E}{\text{number of possible outcomes in } S} = \frac{n(E)}{n(S)} \qquad (1)$$

EXAMPLE

17 An unbiased coin is tossed six times.

 a. What is the probability that the coin will land heads all six times?

 b. What is the probability that the coin will land heads on the first and the last toss?

SOLUTION

 a. Each outcome of the experiment may be represented as a sequence of heads and tails. Using the generalized multiplication principle, we see that the number of outcomes of this experiment is given by 2^6, or 64 (compare with Example 28, Chapter 6). Let E denote the event that the coin lands heads all six times.

Clearly, there is only one way this can occur, so the probability of this event is given by

$$P(E) = \frac{n(E)}{n(S)} = \frac{1}{64} \quad (S, \text{ sample space of the experiment})$$

b. Let F denote the event that the coin lands heads on the first and the last toss. Then $n(F) = 1 \cdot 2 \cdot 2 \cdot 2 \cdot 2 \cdot 1 = 2^4$, so the probability that this event occurs is

$$P(F) = \frac{2^4}{2^6}$$

$$= \frac{1}{2^2}$$

$$= \frac{1}{4}.$$

◀

EXAMPLE

18 Two cards are selected at random from a well-shuffled pack of 52 playing cards.

   a. What is the probability that they are both aces?

   b. What is the probability that neither of them is an ace?

SOLUTION

a. The experiment consists of selecting 2 cards from a pack of 52 playing cards. Since the order in which the cards is selected is immaterial, the sample points are combinations of 52 cards taken 2 at a time. Now, there are $C(52, 2)$ ways of selecting 52 cards taken 2 at a time, so the number of elements in the sample space S is given by $C(52, 2)$. Next, we observe that there are $C(4, 2)$ ways of selecting 2 aces from the 4 in the deck. Therefore, if E denotes the event that the cards selected are both aces, then

$$P(E) = \frac{n(E)}{n(S)}$$

$$= \frac{C(4, 2)}{C(52, 2)} = \frac{\dfrac{4!}{2!2!}}{\dfrac{52!}{2!50!}}$$

$$= \frac{1}{221}.$$

b. Let F denote the event that neither of the two cards selected is an ace. Since there are $C(48, 2)$ ways of selecting two cards neither of which is an ace, we find that

$$P(F) = \frac{n(F)}{n(S)} = \frac{C(48, 2)}{C(52, 2)} = \frac{\dfrac{48!}{2!46!}}{\dfrac{52!}{2!50!}} = \frac{48 \cdot 47}{2} \cdot \frac{2}{52 \cdot 51}$$

$$= \frac{188}{221}.$$ ◀

EXAMPLE

19 A bin in the hi-fi department of Building 20, a bargain outlet, contains 100 blank cassette tapes of which 10 are known to be defective. If a customer selects 6 of these cassette tapes, determine the probability that

a. 2 of them are defective.

b. at least 1 of them is defective.

SOLUTION

a. There are $C(100, 6)$ ways of selecting a set of 6 cassette tapes from the 100, and this gives $n(S)$, the number of outcomes in the sample space associated with the experiment. Next, we observe that there are $C(10, 2)$ ways of selecting a set of 2 defective cassette tapes from the 10 defective cassette tapes and $C(90, 4)$ ways of selecting a set of 4 nondefective cassette tapes from the 90 nondefective cassette tapes (see Figure 7.5).

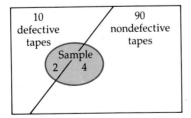

Figure 7.5

Thus, by the multiplication principle, there are $C(10, 2) \cdot C(90, 4)$ ways of selecting 2 defective and 4 nondefective cassette tapes. Therefore, the probability of selecting 6 cassette tapes of which 2 are defective is given by

$$\frac{C(10, 2) \cdot C(90, 4)}{C(100, 6)} = \frac{\dfrac{10!}{2!8!} \dfrac{90!}{4!86!}}{\dfrac{100!}{6!94!}}$$

$$= \frac{10 \cdot 9}{2} \cdot \frac{90 \cdot 89 \cdot 88 \cdot 87}{4 \cdot 3 \cdot 2 \cdot 1} \cdot \frac{6 \cdot 5 \cdot 4 \cdot 3 \cdot 2 \cdot 1}{100 \cdot 99 \cdot 98 \cdot 97 \cdot 96 \cdot 95}$$

$$\approx 0.097.$$

b. Let E denote the event that none of the cassette tapes selected is defective. Then E^c gives the event that at least 1 of the cassette tapes is defective. But, by Property 5 (see Section 7.3),

$$P(E^c) = 1 - P(E).$$

To compute $P(E)$, we observe that there are $C(90, 6)$ ways of selecting a set of 6 cassette tapes that are nondefective. Therefore,

$$P(E) = \frac{C(90, 6)}{C(100, 6)}$$

and

$$P(E^c) = 1 - \frac{C(90, 6)}{C(100, 6)}$$

$$= 1 - \frac{\dfrac{90!}{6!84!}}{\dfrac{100!}{6!94!}}$$

$$= 1 - \frac{90 \cdot 89 \cdot 88 \cdot 87 \cdot 86 \cdot 85}{6 \cdot 5 \cdot 4 \cdot 3 \cdot 2 \cdot 1} \cdot \frac{6 \cdot 5 \cdot 4 \cdot 3 \cdot 2 \cdot 1}{100 \cdot 99 \cdot 98 \cdot 97 \cdot 96 \cdot 95}$$

$$\approx 0.48. \quad \blacktriangleleft$$

▶ The Birthday Problem

EXAMPLE

20 A group of 5 people is selected at random. What is the probability that at least 2 of them have the same birthday?

SOLUTION

For simplicity we will assume that none of the 5 people was born on February 29 of a leap year. Since the 5 people were selected at random, we may also assume that each of them is equally likely to have any of the 365 days of a year as his or her birthday. If we let A, B, C, D, and E represent the 5 people, then it is easy to see that an outcome of the experiment may be represented by (a, b, c, d, e), where the numbers $a, b, c, d,$ and e give the birthdays of A, B, C, D, and E, respectively. We first observe that since there are 365 possibilities for each of the dates a, b, c, d, and e, the multiplication principle implies that there are

$$365 \cdot 365 \cdot 365 \cdot 365 \cdot 365, \quad \text{or} \quad 365^5$$

outcomes of the experiment; that is

$$n(S) = 365^5,$$

where S denotes the sample space of the experiment. Next, let E denote the event that 2 or more of the 5 people have the same birthday. It is now necessary to compute $P(E)$. However, a direct computation of $P(E)$ is relatively difficult. It is much easier to compute $P(E^c)$ where E^c is, of course, the event that no 2 of the 5 people have the same birthday and then use the relation

$$P(E) = 1 - P(E^c).$$

To this end, observe that there are 365 ways (corresponding to the 365 dates) on which A's birthday can occur followed by 364 ways on which B's birthday could occur if B were not to have the same birthday as A, and so on, so by the generalized multiplication principle,

$$n(E^c) = 365 \cdot 364 \cdot 363 \cdot 362 \cdot 361.$$

Thus,
$$P(E^c) = \frac{n(E^c)}{n(S)}$$
$$= \frac{365 \cdot 364 \cdot 363 \cdot 362 \cdot 361}{365^5}$$

and
$$P(E) = 1 - P(E^c)$$
$$= 1 - \frac{365 \cdot 364 \cdot 363 \cdot 362 \cdot 361}{365^5}$$
$$\approx 0.027. \qquad \blacktriangleleft$$

The result obtained in Example 20 may be easily extended to the general case involving r people. In fact, if E denotes the event that at least 2 of the r people have the same birthday, an argument similar to that used in Example 20 leads to the result

$$P(E) = 1 - \frac{365 \cdot 364 \cdot 363 \cdots (365 - r + 1)}{365^r}.$$

By letting r take on the values 5, 10, 15, 20, . . . , 50, in turn, we obtain the probabilities that at least two of 5, 10, 15, 20, . . . , 50 people, respectively, have the same birthday. These results are summarized in Table 7.5. The results show

Table 7.5

r	5	10	15	20	22	23	25	30	40	50
$P(E)$	0.027	0.117	0.253	0.411	0.476	0.507	0.569	0.706	0.891	0.970

that the chances are greater than 50 percent that in a group of 23 randomly selected people, at least 2 of them will have the same birthday. In a group of 50 people, it is an excellent bet that at least 2 people in the group will have the same birthday.

Self-Check Exercises

7.4

1. Four balls are selected at random without replacement from an urn containing 10 white balls and 8 red balls. What is the probability that all the chosen balls are white?

2. A box contains 20 microchips of which 4 are substandard. If 2 of the chips are taken from the box, what is the probability that they are both substandard?

Solutions to Self-Check Exercises 7.4 can be found on page 372.

▶

Exercises 7.4

1. An unbiased coin is tossed 5 times. What is the probability that the coin will land heads

 a. all 5 times?

 b. exactly once?

 c. at least once?

2. Two cards are selected at random from a well-shuffled pack of 52 playing cards. What is the probability that

 a. a pair is drawn?

 b. a pair is not drawn?

3. Five balls are selected at random without replacement from an urn containing 6 white balls and 9 blue balls. What is the probability that

 a. all of the balls are blue?

 b. exactly 2 are white?

 c. 2 or 3 are white?

4. An exam consists of 10 true-or-false questions. If a student guesses at every answer, what is the probability that he or she will answer exactly 6 questions correctly?

5. What is the probability that in a family of 3 children

 a. the first 2 are girls? (Assume that the probability of a boy being born is the same as the probability of a girl being born.)

    b. 2 are boys and 1 is a girl?

    c. at least 1 is a girl?

6. Jacobs & Johnson, Inc., an accounting firm, employs 14 accountants of whom 8 are C.P.A.s. If a delegation of 3 accountants is randomly selected from the firm to attend a conference, what is the probability that 3 C.P.A.'s will be selected?

7. Two light bulbs are selected at random from a lot of 24 of which 4 are defective. What is the probability that

    a. both of the light bulbs are defective?

    b. at least 1 of the light bulbs is defective?

8. A customer at Cavallaro's Fruit Stand picks a sample of 3 oranges at random from a crate containing 60 oranges of which 4 are rotten. What is the probability that the sample contains 1 or more rotten oranges?

9. A shelf in the Metro Department Store contains 80 typewriter cartridge ribbons for a popular portable typewriter. Six of the cartridges are defective. If a customer selects 2 of these cartridges at random from the shelf, determine the probability that

    a. both are defective.

    b. at least 1 is defective.

10. Electronic baseball games manufactured by Tempco Electronics are shipped in lots of 24. Before shipping, a quality-control inspector randomly selects a sample of 8 from each lot for testing. If the sample contains any defective games, the entire lot is rejected. What is the probability that a lot containing 2 defective games will still be shipped?

11. The City Transit Authority of a certain city plans to hire 12 new bus drivers. From a group of 100 qualified applicants of which 60 are men and 40 are women, 12 names are to be selected by lot. If Mary and John Lewis are among the 100 qualified applicants,

    a. what is the probability that Mary's name will be selected? that both Mary's and John's names will be selected?

    b. and it is stipulated that an equal number of men and women are to be selected (6 men from the group of 60 men and 6 women from the group of 40 women), what is the probability that Mary's name will be selected? that Mary's and John's names will be selected?

12. The City Housing Authority has received 50 applications from qualified applicants for 8 low-income apartments. Three of the apartments are on the north side of town and 5 are on the south side. If the apartments are to be assigned by means of a lottery, what is the probability that

    a. a qualified applicant will be selected for one of these apartments?

    b. two specific qualified applicants will be selected for apartments on the same side of town?

13. A student studying for a vocabulary test knows the meaning of 12 words from a list of 20 words. If the test contains 10 words from the study list, what is the probability that at least 8 of the words on the test are words that the student knows?

14. Five different written driving tests are administered by the City Motor Vehicle Department. One of these 5 tests is selected at random for each applicant for a driver's license. If a group consisting of 2 women and 3 men apply for a license, what is the probability that

    a. 2 of the 5 will take the same test?

b. the 2 women will take the same test?

c. all 5 people take a different test?

15. A druggist wishes to select three brands of aspirin to sell in his store. He has five major brands to choose from: brands A, B, C, D, and E. If he selects the three brands at random, what is the probability that he will select

    a. brand B?

    b. brands B and C?

    c. at least one of the two brands, B and C?

16. In the game of Blackjack a 2-card hand consisting of an ace and a face card or a ten is called a blackjack.

    a. If a player is dealt 2 cards from a standard deck of 52 well-shuffled cards, what is the probability that the player will receive a blackjack?

    b. If a player is dealt 2 cards from 2 well-shuffled standard decks, what is the probability that the player will receive a blackjack?

17. Refer to Problem 22, Exercises 6.3, where the "Lucky Dollar" slot machine was described. What is the probability that the 3 "Lucky Dollar" symbols will appear in the window of the slot machine?

18. In 1959 a world record was set for the longest run on an ungaffed (fair) roulette wheel at the El San Juan Hotel in Puerto Rico. The number 10 appeared 6 times in a row. What is the probability of the occurrence of this event? (Assume that there are 38 equally likely outcomes consisting of the numbers 1–36, 0, and 00.)

*19. In "The Numbers Game," a state lottery, 4 numbers are drawn with replacement from an urn containing the digits 0–9 inclusive. Prizes are given to ticket holders with the selected numbers in exact order for (a) all 4 digits, (b) the first or last 3 digits, (c) any 2 digits, and (d) any 1 digit. What is the probability of the occurrence of event (a)? event (b)? event (c)? event (d)? (A ticket holder may claim only 1 prize for each ticket.)

*20. Refer to Exercise 19. In "The Numbers Game," prizes are also given to ticket holders with the selected numbers in any order for (a) all 4 digits, (b) the first 3 digits, and (c) the last 3 digits. What is the probability of the occurrence of event (a)? event (b)? event (c)? (A ticket holder may claim only 1 prize for each ticket.)

A list of poker hands ranked in order from the highest to the lowest is shown in the following table along with a description and example of each hand. Use the table to answer Exercises 21–26.

Hand	Description	Example
Straight flush	5 cards in sequence in the same suit	A ♥ 2 ♥ 3 ♥ 4 ♥ 5 ♥
Four of a kind	4 cards of the same rank and any other card	K ♥ K ♦ K ♠ K ♣ 2 ♥
Full house	3 of a kind and a pair	3 ♥ 3 ♦ 3 ♠ 7 ♥ 7 ♦
Flush	5 cards of the same suit that are not all in sequence	5 ♥ 6 ♥ 9 ♥ J ♥ K ♥

Hand	Description	Example
Straight	5 cards in sequence but not all of the same suit	10 ♥ J ♦ Q ♣ K ♠ A ♥
Three of a kind	3 cards of the same rank and 2 unmatched cards	K ♥ K ♦ K ♠ 2 ♥ 4 ♦
Two pair	2 cards of the same rank and 2 cards of any other rank with an unmatched card	K ♥ K ♦ 2 ♥ 2 ♠ 4 ♣
One pair	2 cards of the same rank and 3 unmatched cards	K ♥ K ♦ 5 ♥ 2 ♠ 4 ♥

If a 5-card poker hand is dealt from a well-shuffled deck of 52 cards, what is the probability of being dealt

21. a straight flush? (Note that an ace may be played as either a high or a low card in a straight sequence; that is, A, 2, 3, 4, 5 or 10, J, Q, K, A. Hence, there are 10 possible sequences for a straight in one suit.)

22. a straight (but not a straight flush)?

23. a flush (but not a straight flush)?

24. four of a kind?

25. a full house?

26. two pairs?

27. There are 12 signs of the Zodiac: Aries, Taurus, Gemini, Cancer, Leo, Virgo, Libra, Scorpio, Sagittarius, Capricorn, Aquarius, and Pisces. Each sign corresponds to a different calendar period of approximately one month. Assuming that a person is just as likely to be born under one sign as another, what is the probability that in a group of 5 people

   a. at least 2 of the people have the same sign?

   b. at least 2 of the people were born under the sign of Aries?

28. What is the probability that at least 2 justices of the U.S. Supreme Court have the same birthday?

29. Fifty people are selected at random. What is the probability that none of the people in this group have the same birthday?

*30. During an episode of the *Tonight Show*, Johnny Carson related "The Birthday Problem" to the audience—noting that, in a group of 50 or more people, probabilists have calculated that the probability of at least 2 people having the same birthday is very high. To illustrate this point, he proceeded to conduct his own experiment. A person selected at random from the audience was asked to state his birthday. Mr. Carson then asked if anyone in the audience had the same birthday. The response was negative. He repeated the experiment. Once again, the response was negative. These results, observed Mr. Carson, were contrary to expectations. In a later episode of the show Mr. Carson explained why this experiment had been improperly conducted. Explain why his experiment failed to illustrate the point he was trying to make.

Solutions to Self-Check Exercises

7.4

1. The probability that all 4 balls selected are white is given by

$$\frac{\text{the number of ways of selecting 4 white balls from the 10 in the urn}}{\text{the number of ways of selecting any 4 balls from the 18 balls in the urn}}$$

$$= \frac{C(10, 4)}{C(18, 4)}$$

$$= \frac{\dfrac{10!}{4!6!}}{\dfrac{18!}{4!14!}}$$

$$= \frac{10 \cdot 9 \cdot 8 \cdot 7}{4 \cdot 3 \cdot 2} \cdot \frac{4 \cdot 3 \cdot 2}{18 \cdot 17 \cdot 16 \cdot 15}$$

$$= 0.069.$$

2. The probability that both chips are substandard is given by

$$\frac{\text{the number of ways of choosing any 2 of the 4 substandard chips}}{\text{the number of ways of choosing any 2 of the 20 chips}}$$

$$= \frac{C(4, 2)}{C(20, 2)}$$

$$= \frac{\dfrac{4!}{2!2!}}{\dfrac{20!}{2!18!}}$$

$$= \frac{4 \cdot 3}{2} \cdot \frac{2}{20 \cdot 19}$$

$$= 0.032.$$

▶ ## 7.5

Conditional Probability and Independent Events

▶ Conditional Probability

▶ More on Tree Diagrams

▶ Independent Events

▶ Conditional Probability

The probability of an event is often affected by the occurrence of other events and/or by the knowledge of information relevant to the event. Basically, the injection of conditions into a problem modifies the underlying sample space of the original problem. This in turn leads to a change in the probability of the event.

EXAMPLE

21 Two cards are drawn without replacement from a well-shuffled deck of 52 playing cards.

a. What is the probability that the first card drawn is an ace?

b. What is the probability that the second card drawn is an ace given that the first card drawn was not an ace?

c. What is the probability that the second card drawn is an ace given that the first card drawn was an ace?

SOLUTION

a. The sample space here consists of 52 equally likely outcomes, 4 of which are aces. Therefore, the probability that the first card drawn is an ace is 4/52.

b. Having drawn the first card, there are 51 cards left in the deck. In other words, for the second phase of the experiment, we are working in a *reduced* sample space. If the first card drawn was not an ace, then this modified sample space of 51 points contains 4 "favorable" outcomes (the 4 aces), so the probability that the second card drawn is an ace is given by 4/51.

c. If the first card drawn was an ace, then there are 3 aces left in the deck of 51 playing cards, so the probability that the second card drawn is an ace is given by 3/51.   ◀

Observe that in Example 21 the occurrence of the first event reduces the size of the original sample space. The information concerning the first card drawn also leads us to the consideration of modified sample spaces: in (b) the deck contained 4 aces, and in (c), the deck contained 3 aces.

The probability found in (b) or (c) of Example 21 is known as a **conditional probability,** since it is the probability of an event occurring given that another event has already occurred. For example, in (b), we computed the probability of the event that the second card drawn is an ace, given the event that the first card drawn was not an ace. In general, given two events A and B of an experiment, one may, under certain circumstances, compute the probability of the event B given that the event A has already occurred. This probability, denoted by $P(B \mid A)$, is called the *conditional probability of B given A.*

A formula for computing the conditional probability of B given A may be discovered with the aid of a Venn diagram. Consider an experiment with a uniform sample space S and suppose A and B are two events of the experiment (see Figure 7.6).

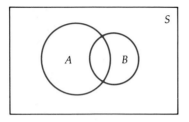

Figure 7.6   Original sample space   Reduced sample space

The condition that the event A has occurred tells us that the possible outcomes of the experiment in the second phase are restricted to those outcomes (elements) in the set A. In other words, we may work with the reduced sample space A instead of the original sample space S in the experiment. Next, we observe that, with respect to the reduced sample space A, the outcomes favorable to the event B are precisely those elements in the set $A \cap B$. Consequently, the conditional probability of B given A is

$$P(B \mid A) = \frac{\text{number of elements in } A \cap B}{\text{number of elements in } A}$$

$$= \frac{n(A \cap B)}{n(A)} \quad (n(A) \neq 0)$$

Dividing the numerator and the denominator by $n(S)$, the number of elements in S, we have

$$P(B \mid A) = \frac{\dfrac{n(A \cap B)}{n(S)}}{\dfrac{n(A)}{n(S)}},$$

which is equivalent to the following formula:

Conditional Probability of an Event

If A and B are events in an experiment and $P(A) \neq 0$, then the **conditional probability** that the event B will occur given that the event A has already occurred is

$$P(B \mid A) = \frac{P(A \cap B)}{P(A)} \tag{2}$$

EXAMPLE

22 A pair of fair dice is cast. What is the probability that the sum of the numbers falling uppermost is 7 if it is known that one of the numbers is a 5?

SOLUTION Let A denote the event that the sum of the numbers falling uppermost is 7 and let B denote the event that one of the numbers is a 5. From the results of Example 4, we find that

$$A = \{(6, 1), (5, 2), (4, 3), (3, 4), (2, 5), (1, 6)\}$$
and
$$B = \{(5, 1), (5, 2), (5, 3), (5, 4), (5, 5), (5, 6), (1, 5), (2, 5),$$
$$(3, 5), (4, 5), (6, 5)\}$$
so that $A \cap B = \{(5, 2), (2, 5)\}$.

Since the dice are fair, each outcome of the experiment is equally likely; therefore

$$P(A \cap B) = \frac{2}{36} \quad \text{and} \quad P(B) = \frac{11}{36} \quad \text{(Recall that } n(S) = 36.\text{)}$$

Thus, the probability that the sum of the numbers falling uppermost is 7 given that one of the numbers is a 5 is, by virtue of Formula (2),

$$P(A \mid B) = \frac{\dfrac{2}{36}}{\dfrac{11}{36}} = \frac{2}{11}.$$

◀

EXAMPLE

23 In a recently conducted test by the U.S. Army, it was found that of 1000 new recruits, 600 men and 400 women, 50 of the men and 4 of the women were red-green colorblind. Given that a recruit selected at random from this group is red-green colorblind, what is the probability that the recruit is a male?

SOLUTION Let C denote the event that a randomly selected subject is red-green colorblind and let M denote the event that the subject is a male recruit. Since 54 out of the 1000 subjects are colorblind, we may take

$$P(C) = \frac{54}{1000} = 0.054.$$

Therefore, by Formula (2), the probability that a subject is male given that the subject is red-green colorblind is

$$P(M \mid C) = \frac{P(M \cap C)}{P(C)}$$

$$= \frac{0.05}{0.054} = \frac{25}{27}.$$ ◀

In certain problems, the probability of an event B occurring given that A has occurred, written $P(B \mid A)$, is known and we wish to find the probability of A *and* B occurring. The solution to such a problem is facilitated by the use of the following formula:

The Product Rule

$$P(A \cap B) = P(A) \cdot P(B \mid A) \qquad (3)$$

This formula is obtained from Equation (2) by multiplying both sides of the equation by $P(A)$. We illustrate the use of the product rule in the next several examples.

EXAMPLE

24 There are 300 seniors in Jefferson High School of which 140 are males. It is known that 80 percent of the males and 60 percent of the females have their driver's license. If a student is selected at random from this senior class,

a. what is the probability that the student is a male and has a driver's license?

b. what is the probability that the student is a female who does not have a driver's license?

SOLUTION   *a.* Let M denote the event that the student is a male and let D denote the event that the student has a driver's license. Then

$$P(M) = \frac{140}{300} \quad \text{and} \quad P(D \mid M) = 0.8.$$

Now, the event that the student selected at random is a male and has a driver's license is $M \cap D$, and, by the product rule, the probability of this event occurring is given by

$$P(M \cap D) = P(M) \cdot P(D \mid M)$$
$$= \left(\frac{140}{300}\right)(0.8) = \frac{28}{75}.$$

b. Let F denote the event that the student is a female and let D be as before. Then D^c is the event that the student does not have a driver's license. We have

$$P(F) = \frac{160}{300} \quad \text{and} \quad P(D^c \,|\, F) = 1 - 0.6 = 0.4.$$

Note that we have used the complement rule in the computation of $P(D^c \,|\, F)$. Now, the event that the student selected at random is a female and does not have a driver's license is $F \cap D^c$, and, by the product rule, the probability of this event occurring is given by

$$P(F \cap D^c) = P(F) \cdot P(D^c \,|\, F)$$
$$= \left(\frac{160}{300}\right)(0.4) = \frac{16}{75}. \qquad \blacktriangleleft$$

EXAMPLE

25 Two cards are drawn without replacement from a well-shuffled deck of 52 playing cards. What is the probability that the first card drawn is an ace and the second card drawn is a face card?

SOLUTION

Let A denote the event that the first card drawn is an ace and let F denote the event that the second card drawn is a face card. Then $P(A) = 4/52$. After drawing the first card, there are 51 cards left in the deck of which 12 are face cards. Therefore, the probability of drawing a face card given that the first card drawn was an ace is given by

$$P(F \,|\, A) = \frac{12}{51}.$$

By the product rule, the probability that the first card drawn is an ace and the second card drawn is a face card is given by

$$P(A \cap F) = P(A) \cdot P(F \,|\, A)$$
$$= \left(\frac{4}{52}\right)\left(\frac{12}{51}\right) = \frac{4}{221}. \qquad \blacktriangleleft$$

The product rule may be generalized to the case involving any finite number of events. For example, in the case involving the three events E, F, and G, it may be shown that

$$P(E \cap F \cap G) = P(E) \cdot P(F \,|\, E) \cdot P(G \,|\, E \cap F) \tag{4}$$

▶ More on Tree Diagrams

Formula (4) and its generalizations may be used to aid us in solving problems that involve *finite stochastic processes.* More specifically, a **finite stochastic process** is an experiment consisting of a finite number of stages in which the outcomes and associated probabilities of each stage depend on the outcomes and associated probabilities of the preceding stages. Experiments involving finite stochastic processes may be described by tree diagrams, which may also help in solving problems involving such processes. Consider, for example, the experiment consisting of drawing 2 cards without replacement from a well-shuffled deck of 52 playing cards. What is the probability that the second card drawn is a face card?

We may think of this experiment as a stochastic process with two stages. The events associated with the first stage are F, that the card drawn is a face card, and F^c, that the card drawn is not a face card. Since there are 12 face cards, we have

$$P(F) = \frac{12}{52} \quad \text{and} \quad P(F^c) = 1 - \frac{12}{52} = \frac{40}{52}.$$

The outcomes of this trial together with the associated probabilities may be represented along two "branches" of a tree diagram as shown in Figure 7.7.

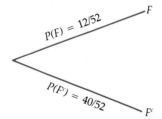

Figure 7.7

In the second trial, we again have two events: G, that the card drawn is a face card, and G^c, that the card drawn is not a face card. But the outcome of the second trial depends on the outcome of the first trial. For example, if the first card drawn was a face card, then the event G that the second card drawn is a face card has probability given by the *conditional probability* $P(G \mid F)$. Since the occurrence of a face card in the first draw leaves 11 face cards in a deck of 51 cards for the second draw, we see that

$$P(G \mid F) = \frac{11}{51}.$$

Similarly, the occurrence of a face card in the first draw leaves 40 that are other than face cards in a deck of 51 cards for the second draw. Therefore, the probability of drawing other than a face card in the second draw given that the first card drawn is a face card is

$$P(G^c \mid F) = \frac{40}{51}.$$

The results just obtained allow us to extend the tree diagram of Figure 7.7 by displaying another two branches of the tree growing from its upper branch (see Figure 7.8).

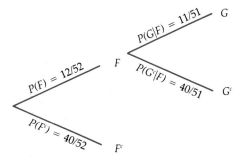

Figure 7.8

To complete the tree diagram we compute $P(G \mid F^c)$ and $P(G^c \mid F^c)$, the conditional probabilities that the second card drawn is a face card and other than a face card, respectively, given that the first card drawn is not a face card. We find that

$$P(G \mid F^c) = \frac{12}{51} \quad \text{and} \quad P(G^c \mid F^c) = \frac{39}{51}.$$

This leads to the completion of the tree diagram shown in Figure 7.9, where the branches of the tree that lead to the two outcomes of interest have been highlighted.

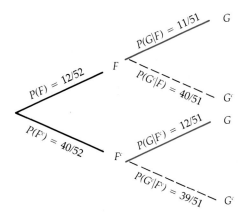

Figure 7.9

Having constructed the tree diagram associated with the problem, we are now in a position to answer the question posed earlier, namely, "What is the probability of the second card being a face card?" Observe that Figure 7.9 shows the two ways in which a face card may result in the second draw, namely the two Gs on the extreme right of the diagram.

Now, by the product rule, the probability that the second card drawn is a face card given that the first card drawn is a face card (this is represented by the upper branch) is

$$P(G \cap F) = P(F) \cdot P(G \mid F).$$

Similarly, the probability that the second card drawn is a face card given that the first card drawn is other than a face card (this corresponds to the other branch) is

$$P(G \cap F^c) = P(F^c) \cdot P(G \mid F^c).$$

Observe that each of these probabilities is obtained by taking the *product of the probabilities appearing on the respective branch.* Since $G \cap F$ and $G \cap F^c$ are mutually exclusive events (why?), the probability that the second card drawn is a face card is given by

$$P(G \cap F) + P(G \cap F^c) = P(F) \cdot P(G \mid F) + P(F^c) \cdot P(G \mid F^c),$$

or, upon replacing the probabilities on the right of the expression by their numerical values,

$$P(G \cap F) + P(G \cap F^c) = \left(\frac{12}{52}\right)\left(\frac{11}{51}\right) + \left(\frac{40}{52}\right)\left(\frac{12}{51}\right)$$

$$= \frac{3}{13}.$$

EXAMPLE

26 The picture tubes for the Pulsar 19-inch color television sets are manufactured in three locations and then shipped to the main plant of the Vista Vision Corporation for final assembly. Plants A, B, and C supply 50 percent, 30 percent, and 20 percent, respectively, of the picture tubes used by the company. The quality control department of the company has determined that 1 percent of the picture tubes produced by plant A are defective, whereas 2 percent of the picture tubes produced by plants B and C are defective. What is the probability that a randomly selected Pulsar 19-inch color television set will have a defective picture tube?

SOLUTION

Let A, B, and C denote the events that the set chosen has a picture tube manufactured in plant A, plant B, and plant C, respectively. Also, let D denote the event that a set has a defective picture tube. Using the given information, we draw the tree diagram shown in Figure 7.10. (The events that result in a set with a defective picture tube being selected are circled.) Taking the product of the probabilities along each branch leading to such an event and adding them yields the probability that a set chosen at random has a defective picture tube. Thus, the required probability is given by

$$(0.5)(0.01) + (0.3)(0.02) + (0.2)(0.02)$$
$$= 0.005 + 0.006 + 0.004$$
$$= 0.015.$$

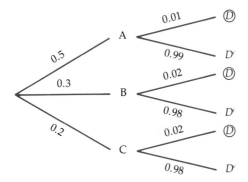

Figure 7.10

EXAMPLE

27 A box contains 8 nine-volt transistor batteries of which 2 are known to be defective. The batteries are selected one at a time without replacement and tested until a nondefective one is found. What is the probability that the number of batteries tested is (a) 1, (b) 2, (c) 3?

SOLUTION We may view this experiment as a multistage process with up to three stages. In the first stage, a battery is selected with a probability of 6/8 of it being nondefective and a probability of 2/8 of it being defective. If the battery selected is good, the experiment is terminated. Otherwise, a second battery is selected with probabilities of 6/7 and 1/7, respectively, of being nondefective and defective. If the second battery selected is good, the experiment is terminated. Otherwise, a third battery is selected with probabilities of 1 and 0, respectively, of being nondefective and defective. The tree diagram associated with this experiment is shown in Figure 7.11, where N denotes the event that the battery selected is nondefective, and D denotes the event that the battery selected is defective.

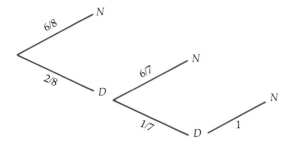

Figure 7.11

With the aid of the tree diagram we see that (a) the probability that only 1 battery is selected is 6/8, (b) the probability that 2 batteries are selected is (2/8)(6/7) or 3/14, and (c) the probability that 3 batteries are selected is (2/8)(1/7)(1) = 1/28. ◀

▶ Independent Events

Returning to Example 25 we see that the event of a face card being drawn was affected by the first outcome—the event of an ace being drawn. In fact, the drawing of the ace reduced the original sample space from 52 elements to one with 51 elements and, in this manner, affected the probability of a face card being drawn next. In certain situations, the occurrence or nonoccurrence of an event is not affected by the occurrence or nonoccurrence of another event. A simple example of such a situation is provided by the experiment of tossing a coin twice and observing the outcomes. It is intuitively clear that the outcome of the second toss is in no way affected by the outcome of the first toss; in this case we say that the two events are independent. In general, two events A and B are **independent** if the outcome of one does not affect the outcome of the other. Symbolically, two events A and B are independent if

$$P(A \mid B) = P(A) \quad \text{or} \quad P(B \mid A) = P(B).$$

In fact, the first equation reads "the probability of A given B is equal to the probability of A," thereby implying the independence of A from B. The second equation has a similar implication.

Using the product rule, it is easy to find a simple test for the independence of two events. Indeed, if A and B are independent, then

$$P(B \mid A) = P(B),$$

so by the product rule, we have

$$P(A \cap B) = P(A) \cdot P(B \mid A) = P(A) \cdot P(B).$$

Conversely, if the above equation holds, then it may be seen that $P(B \mid A) = P(B)$; that is, A and B are independent. Accordingly, we have the following *test for the independence of two events.*

Test for the Independence of Two Events

Two events A and B are **independent** if and only if

$$P(A \cap B) = P(A) \cdot P(B) \tag{5}$$

EXAMPLE

28 Consider the experiment consisting of tossing a fair coin twice and observing the outcomes. Show that the event of "heads" in the first toss and "tails" in the second toss are independent events.

SOLUTION   Let A denote the event that the outcome of the first toss is a *head* and let B denote the event that the outcome of the second toss is a *tail*. The sample space of the experiment is

$$S = \{(H, H), (H, T), (T, H), (T, T)\}$$

and

$$A = \{(H, H), (H, T)\}, \qquad B = \{(H, T), (T, T)\}$$

so that

$$A \cap B = \{(H, T)\}.$$

Next, we compute

$$P(A \cap B) = \frac{1}{4}, \qquad P(A) = \frac{1}{2}, \quad \text{and} \quad P(B) = \frac{1}{2}$$

and observe that Equation (5) is satisfied in this case so that A and B are independent events, as we set out to show.   ◀

EXAMPLE

29   A survey conducted by an independent agency for the National Lung Society found that of 2000 women, 680 were heavy smokers and 50 had emphysema. Of those who had emphysema, 42 were also heavy smokers. Using the data in this survey, determine whether the events "being a heavy smoker" and "having emphysema" are independent events.

SOLUTION   Let A denote the event that a woman is a heavy smoker and let B denote the event that a woman has emphysema. Then the probability that a woman is a heavy smoker and has emphysema is given by

$$P(A \cap B) = \frac{42}{2000} = 0.021.$$

Next,

$$P(A) = \frac{680}{2000} = 0.34 \quad \text{and} \quad P(B) = \frac{50}{2000} = 0.025$$

so that,

$$P(A) \cdot P(B) = (0.34)(0.025) = 0.0085.$$

Since $P(A \cap B) \neq P(A) \cdot P(B)$, we conclude that A and B are not independent events.   ◀

The solution of many practical problems involves more than two independent events. In such cases it is true that

$$P(E_1 \cap E_2 \cap \cdots \cap E_n) = P(E_1) \cdot P(E_2) \cdot \cdots \cdot P(E_n) \qquad (6)$$

where $E_1, E_2, \ldots, E_n$ are known to be independent events. Formula (6) states that the probability of the simultaneous occurrence of n independent events is equal to the product of the probabilities of the n events.

It is important to note that the mere requirement that the n events $E_1, E_2, \ldots, E_n$ satisfy Formula (6) is not sufficient to guarantee that the n events are indeed independent. However, a criterion does exist for determining the independence of n events and may be found in more advanced texts on probability.

EXAMPLE

30   It is known that the three events A, B, and C are independent and $P(A) = 0.2$, $P(B) = 0.4$, and $P(C) = 0.5$.

   a. Compute $P(A \cap B)$.

   b. Compute $P(A \cap B \cap C)$.

SOLUTION   Using Formula (6), we find

a. $P(A \cap B) = P(A) \cdot P(B)$
$= (0.2)(0.4) = 0.08.$

b. $P(A \cap B \cap C) = P(A) \cdot P(B) \cdot P(C)$
$= (0.2)(0.4)(0.5) = 0.04$ ◀

EXAMPLE

31   The Acrosonic model F loudspeaker system has four loudspeaker components: a woofer, a midrange, a tweeter, and an electrical crossover. The quality control manager of Acrosonic has determined that on the average 1 percent of the woofers, 0.8 percent of the midranges, and 0.5 percent of the tweeters are defective, while 1.5 percent of the electrical crossovers are defective. Determine the probability that an Acrosonic model F loudspeaker system selected at random coming off the assembly line and before final inspection is not defective. Assume that the defects in the manufacturing of the components are unrelated.

SOLUTION   Let A, B, C, and D denote, respectively, the events that the woofer, the midrange, the tweeter, and the electrical crossover are defective. Then,

$$P(A) = 0.01, \quad P(B) = 0.008, \quad P(C) = 0.005, \quad \text{and} \quad P(D) = 0.015$$

and the probabilities of the corresponding complementary events are

$$P(A^c) = 0.99, \quad P(B^c) = 0.992, \quad P(C^c) = 0.995, \quad \text{and} \quad P(D^c) = 0.985$$

The event that a loudspeaker system selected at random is not defective is given by $A^c \cap B^c \cap C^c \cap D^c$, and since the events A, B, C, and D (and therefore also A^c, B^c, C^c, and D^c) are assumed to be independent, we find that the required probability is given by

$$P(A^c \cap B^c \cap C^c \cap D^c) = P(A^c) \cdot P(B^c) \cdot P(C^c) \cdot P(D^c)$$
$$= (0.99)(0.992)(0.995)(0.985)$$
$$\approx 0.96.$$ ◀

Self-Check Exercises

7.5

1. Let A and B be events in a sample space S such that $P(A) = 0.4$, $P(B) = 0.8$, and $P(A \cap B) = 0.3$. Find

    a. $P(A \mid B)$ *b.* $P(B \mid A)$

2. Three friends—Alice, Betty, and Cathy—are unmarried and are 30, 35, and 40 years old, respectively. While they were having tea one afternoon, Alice recalled reading an article by Yale sociologist Neil Bennett in which he concluded that women who are still unmarried at age 30 have only a 20 percent chance of marrying, those who are still unmarried at 35 have a 5.4 percent chance, and those who are still unmarried at 40 have a 1.3 percent chance. Betty wondered what the probability was that all three of them would eventually "tie the knot." Cathy, a statistician, pulled out her pocket calculator and answered Betty's question. What was her answer?

Solutions to Self-Check Exercises 7.5 can be found on page 390.

▶

Exercises 7.5

1. Let A and B be events in a sample space S such that $P(A) = 0.6$, $P(B) = 0.5$, and $P(A \cap B) = 0.2$. Find

    a. $P(A \mid B)$ *b.* $P(B \mid A)$

2. Let A and B be two events in a sample space S such that $P(A) = 0.6$ and $P(B \mid A) = 0.5$. Find $P(A \cap B)$.

3. Let A and B be the events described in Exercise 1. Find

    a. $P(A \mid B^c)$ *b.* $P(B \mid A^c)$ [*Hint:* $(A \cap B^c) \cup (A \cap B) = A$]

In Exercises 4–6, determine whether the given events A and B are independent.

4. $P(A) = 0.5$, $P(B) = 0.7$, $P(A \cup B) = 0.85$

5. $P(A^c) = 0.3$, $P(B^c) = 0.4$, $P(A \cap B) = 0.42$

6. $P(A) = 0.6$, $P(B) = 0.8$, $P(A \cap B) = 0.2$

7. If A and B are independent events and $P(A) = 0.4$ and $P(B) = 0.6$, find

    a. $P(A \cap B)$ *b.* $P(A \cup B)$

8. A pair of fair dice is cast. What is the probability that the sum of the numbers falling uppermost is less than nine, if it is known that one of the numbers is a 6?

9. Two cards are drawn without replacement from a well-shuffled deck of 52 playing cards.

    a. What is the probability that the first card drawn is a heart?

    b. What is the probability that the second card drawn is a heart given that the first card drawn was not a heart?

   c. What is the probability that the second card drawn is a heart given that the first card drawn was a heart?

10. Three cards are drawn without replacement from a well-shuffled deck of 52 playing cards. What is the probability that the third card drawn is a diamond?

11. A coin is tossed three times. What is the probability that the coin will land

   a. heads at least twice?

   b. heads on the second toss given that heads were thrown on the first toss?

   c. heads on the third toss given that tails were thrown on the first toss?

12. Five black balls and four white balls are placed in an urn. Two balls are then drawn in succession. What is the probability that the second ball drawn is a white ball

   a. if the second ball is drawn without replacing the first?

   b. if the first ball is replaced before the second is drawn?

13. The following tree diagram represents an experiment consisting of two trials.

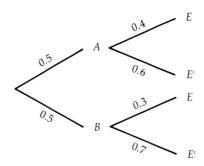

Use the diagram to find

   a. $P(A)$ *b.* $P(E \mid A)$

   c. $P(A \cap E)$ *d.* $P(E)$

   e. Does $P(A \cap E) = P(A) \cdot P(E)$?   *f.* Are A and E independent events?

14. The following tree diagram represents an experiment consisting of two trials. Use the diagram to find

   a. $P(A)$ *b.* $P(E \mid A)$

   c. $P(A \cap E)$ *d.* $P(E)$

   e. Does $P(A \cap E) = P(A) \cdot P(E)$?   *f.* Are A and E independent events?

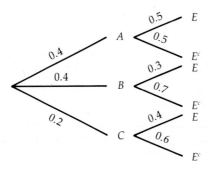

15. An experiment consists of two trials. The outcomes of the first trial are A and B with probabilities of occurring of 0.4 and 0.6. There are also two outcomes C and D in the second trial with probabilities of 0.3 and 0.7. Draw a tree diagram representing this experiment. Use this diagram to find

    a. $P(A)$ *b.* $P(C \mid A)$ *c.* $P(A \cap C)$ *d.* $P(C)$

    e. Does $P(A \cap C) = P(A) \cdot P(C)$?

    f. Are A and C independent events?

16. An experiment consists of two trials. The outcomes of the first trial are A, B, and C with probabilities of occurring of 0.2, 0.5, and 0.3, respectively. The outcomes of the second trial are E and F with probabilities of occurring of 0.6 and 0.4. Draw a tree diagram representing this experiment. Use this diagram to find

    a. $P(B)$ *b.* $P(F \mid B)$ *c.* $P(B \cap F)$ *d.* $P(F)$

    e. Does $P(B \cap F) = P(B) \cdot P(F)$?

    f. Are B and F independent events?

17. Two cards are drawn without replacement from a well-shuffled deck of 52 cards. Let A be the event that the first card drawn is a heart and let B be the event that the second card drawn is a red card. Show that the events A and B are dependent events.

18. A tax specialist has estimated the probability that a tax return selected at random will be audited is 0.02. Furthermore, he estimates the probability that an audited return will result in additional assessment being levied on the taxpayer is 0.60. What is the probability that a tax return selected at random will result in additional assessments being levied on the taxpayer?

19. In a three-child family, what is the probability that all three children are girls given that one of the children is a girl? (Assume that the probability of a boy being born is the same as the probability of a girl being born.)

20. The probability that a battery will last 10 hours or more is 0.80 and the probability that it will last 15 hours or more is 0.15. Given that a battery has lasted 10 hours, find the probability that it will last 15 hours or more.

21. At a certain medical school, 1/7 of the students are from a minority group. Of those students who belong to a minority group, 1/3 are black.

    a. What is the probability that a student selected at random from this medical school is black?

    b. What is the probability that a student selected at random from this medical school is black if it is known that the student is a member of a minority group?

22. In a survey of 1000 eligible voters selected at random, it was found that 80 had a college degree. Additionally, it was found that 80 percent of those who had a college degree voted in the last presidential election, whereas 55 percent of the people who did not have a college degree voted in the last presidential election. Assuming that the poll is representative of all eligible voters, find the probability that an eligible voter selected at random

    a. had a college degree and voted in the last presidential election.

    b. did not have a college degree and did not vote in the last presidential election.

    c. voted in the last presidential election.

    d. did not vote in the last presidential election.

23. The data on the next page were obtained from the financial aid office of a certain university.

	Receiving Financial Aid	Not Receiving Financial Aid	Total
Undergraduates	4,222	3,898	8,120
Graduates	1,879	731	2,610
Total	6,101	4,629	10,730

What is the probability that a student selected at random at this university

    a. is not receiving financial aid i. it is known that the student is in graduate school?

    b. is in graduate school if it is known that the student is receiving financial aid?

24. The personnel department of the Franklin National Life Insurance Company compiled the following data regarding the income and education of its employees.

	Income $30,000 or Below	Income Above $30,000
Noncollege Graduate	2040	840
College Graduate	400	720

Let A be the event that an employee has a college degree and B the event that an employee's income is more than $30,000.

    a. Find each of the following probabilities:

        1 $P(A)$ **2** $P(B)$ **3** $P(A \cap B)$ **4** $P(B \mid A)$ **5** $P(B \mid A^c)$

    b. Are the events A and B independent events?

25. Suppose the probability that your mail will be delivered before 2:00 P.M. on a delivery day is 0.90. What is the probability that your mail will be delivered before 2:00 P.M.

    a. for two consecutive delivery days?

    b. for two out of three consecutive delivery days?

26. Suppose that a box contains two defective Christmas tree lights that have been inadvertently mixed with eight nondefective lights. If the lights are selected one at a time without replacement and tested until both defective lights are found, what is the probability that both defective lights will be found after three trials?

27. A nationwide survey conducted by the National Cancer Society revealed the following information: Of 10,000 people surveyed, 3,200 were "heavy coffee drinkers" and 160 had cancer of the pancreas. Of those who had cancer of the pancreas, 132 were heavy coffee drinkers. Using the data in this survey, determine whether the events "being a heavy coffee drinker" and "having cancer of the pancreas" are independent events.

28. The proprietor of Cunningham's Hardware Store has decided to install floodlights on the premises as a measure against vandalism and theft. If the probability is 0.01 that a certain brand of floodlight will burn out within a year, find the minimum number of floodlights that must be installed to insure the probability that at least one

of them will remain functional within the year is at least 0.99999. (Assume that the floodlights operate independently.)

29. It is estimated that 0.80 percent of a large consignment of eggs in a certain supermarket are broken.

a. What is the probability that a customer who randomly selects a dozen of these eggs receives at least one broken egg?

b. What is the probability that a customer who selects these eggs at random will have to check three cartons before finding a carton without any broken eggs? (Each carton contains a dozen eggs.)

30. Copykwik, Inc. has four photocopy machines, A, B, C, and D. The probability that a given machine will break down on a particular day is as follows:

$$P(A) = \frac{1}{50}, \quad P(B) = \frac{1}{60}, \quad P(C) = \frac{1}{75}, \quad P(D) = \frac{1}{40}$$

Assuming independence, what is the probability on a particular day that

a. all four machines will break down?

b. none of the machines will break down?

c. exactly one machine will break down?

31. An automobile manufacturer obtains the microprocessors used to regulate fuel consumption in its automobiles from three microelectronic firms, A, B, and C. The quality control department of the company has determined that 1 percent of the microprocessors produced by firm A are defective, 2 percent of those produced by firm B are defective, and 1.5 percent of those produced by firm C are defective. Firms A, B, and C supply 45 percent, 25 percent, and 30 percent, respectively, of the microprocessors used by the company. What is the probability that a randomly selected automobile manufactured by the company will have a defective microprocessor?

32. Before being allowed to enter a maximum-security area at a military installation, a person must pass three identification tests: a voice-pattern test, a fingerprint test, and a handwriting test. If the reliability of the first test is 97 percent, the reliability of the second test is 98.5 percent, and that of the third is 98.5 percent, what is the probability that this security system will allow an improperly identified person to enter the maximum security area?

33. Figures obtained from a city's police department seem to indicate that, of all motor vehicles reported as stolen, 64 percent were stolen by professionals, whereas 36 percent were stolen by amateurs (primarily for joy rides). Of those vehicles presumed stolen by professionals, 24 percent were recovered within 48 hours, 16 percent were recovered after 48 hours, and 60 percent were never recovered. Of those vehicles presumed stolen by amateurs, 38 percent were recovered within 48 hours, 58 percent were recovered after 48 hours, and 4 percent were never recovered.

a. Draw a tree diagram representing these data.

b. What is the probability that a vehicle stolen by a professional in this city will be recovered within 48 hours?

c. What is the probability that a vehicle stolen in this city will never be recovered?

34. The chief loan officer of the La Crosse Home Mortgage Company summarized the housing loans extended by the company in 1986 according to type and term of the loan. Her list shows that 70 percent of the loans were fixed-rate mortgages (F), 25

percent were adjustable-rate mortgages (A), and 5 percent belong to some other category (O) [mostly second trust-deed loans and loans extended under the graduated payment plan]. Of the fixed-rate mortgages, 80 percent were 30-year loans and 20 percent were 15-year loans; of the adjustable-rate mortgages, 40 percent were 30-year loans and 60 percent were 15-year loans; finally, of the other loans extended, 30 percent were 20-year loans, 60 percent were 10-year loans, and 10 percent were for a term of 5 years or less.

    a. Draw a tree diagram representing this experiment.

    b. What is the probability that a home loan extended by La Crosse has an adjustable rate and is for a term of 15 years?

    c. What is the probability that a home loan extended by La Crosse is for a term of 15 years?

35. The admissions office of a private university released the following admission data for the preceding academic year: From a pool of 3900 male applicants, 40 percent were accepted by the university and of these, 40 percent subsequently enrolled. Additionally, from a pool of 3600 female applicants, 45 percent were accepted by the university and of these, 40 percent subsequently enrolled. Find the probability that

    a. a male applicant will be accepted by and subsequently will enroll in the university.

    b. a student who applies for admission will be accepted by the university.

    c. a student who applies for admission will be accepted by the university and subsequently will enroll.

Solutions to Self-Check Exercises

7.5

1. a. $P(A \mid B) = \dfrac{P(A \cap B)}{P(B)}$

             $= \dfrac{0.3}{0.8} = \dfrac{3}{8}$

b. $P(B \mid A) = \dfrac{P(A \cap B)}{P(A)}$

             $= \dfrac{0.3}{0.4} = \dfrac{3}{4}$

2. Let A, B, and C denote the events that three women who are still unmarried at ages 30, 35, and 40 *will* marry eventually. Then it seems reasonable to assume that these events are independent, with $P(A) = 0.20$, $P(B) = 0.054$, and $P(C) = 0.013$. Based on these figures, the probability that all three women will be married eventually is given by

$$P(A)P(B)P(C) = (0.2)(0.054)(0.013)$$
$$= 0.00014.$$

7.6

Bayes' Theorem

▶ *A Posteriori* Probabilities

▶ Applications

▶ *A Posteriori* Probabilities

In the previous sections we were concerned with calculating probabilities that give the likelihood that an event *will* occur. Such probabilities are called ***a priori* probabilities** since they are calculated *prior* to observing the results of the experiments. In this section, we will consider probabilities that are calculated *after* the outcomes of experiments have been observed. Such probabilities are called *a posteriori* **probabilities.**

Suppose three machines, A, B, and C, produce similar engine components. Machine A produces 45 percent of the total components, machine B produces 30 percent, and machine C, 25 percent. For the usual production schedule, 6 percent of the components produced by machine A do not meet established specifications; for machine B and machine C, the corresponding figures are 4 percent and 3 percent. One component is selected at random from the total output and is found to be defective. What is the probability that the component selected was produced by machine A?

The answer to this question is found by determining the *a posteriori* probability for the event that the component selected was produced by machine A. To this end, let *A*, *B*, and *C* denote the event that a component is produced by machine A, machine B, and machine C, respectively. We may represent this experiment with a Venn diagram (Figure 7.12).

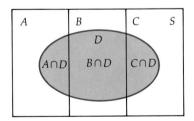

Figure 7.12

The three mutually exclusive events *A*, *B*, and *C* form a **partition** of the sample space *S*. That is, aside from being mutually exclusive, their union is precisely *S*. The event *D* that a component is defective is the shaded area. Again referring to Figure 7.12 we see that

1. The event *D* may be expressed as

$$D = (A \cap D) \cup (B \cap D) \cup (C \cap D).$$

2. The event that a component is defective and is produced by machine A is given by $A \cap D$.

Thus, the *a posteriori* probability that a defective component selected was produced by machine A is given by

$$P(A \mid D) = \frac{n(A \cap D)}{n(D)}$$

or, upon dividing both the numerator and the denominator by $n(S)$ and observing that the events $A \cap D$, $B \cap D$, and $C \cap D$ are mutually exclusive, we obtain

$$P(A \mid D) = \frac{P(A \cap D)}{P(D)}$$

$$= \frac{P(A \cap D)}{P(A \cap D) + P(B \cap D) + P(C \cap D)} \tag{7}$$

Next, using the product rule, we may express

$$P(A \cap D) = P(A) \cdot P(D \mid A)$$
$$P(B \cap D) = P(B) \cdot P(D \mid B)$$
and
$$P(C \cap D) = P(C) \cdot P(D \mid C)$$

so that Formula (7) may be expressed in the form

$$P(A \mid D) = \frac{P(A) \cdot P(D \mid A)}{P(A) \cdot P(D \mid A) + P(B) \cdot P(D \mid B) + P(C) \cdot P(D \mid C)} \tag{8}$$

which is a special case of a result known as **Bayes' Theorem.** Observe that the expression on the right of Formula (8) involves the probabilities $P(A)$, $P(B)$, $P(C)$ and the conditional probabilities $P(D \mid A)$, $P(D \mid B)$, and $P(D \mid C)$, all of which may be calculated in the usual fashion. In fact, by displaying these quantities on a tree diagram, we obtain Figure 7.13.

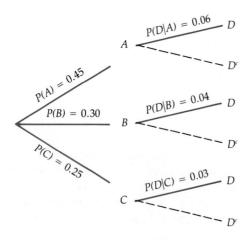

Figure 7.13

We may compute the required probability by substituting the relevant quantities into Formula (8), or we may make use of the following device: *the numerator is given by the product of the probabilities along the limb through A, whereas the denominator is just the sum of the products of the probabilities along each limb terminating at D.* In either case, we obtain

$$P(A \mid D) = \frac{(0.45)(0.06)}{(0.45)(0.06) + (0.3)(0.04) + (0.25)(0.03)}$$
$$= 0.58.$$

Before looking at any further examples, let us state the general form of Bayes' Theorem.

Bayes' Theorem

Let $A_1, A_2, \ldots, A_n$ be a partition of a sample space S (that is, $A_1, A_2, \ldots, A_n$ are mutually exclusive events and $S = A_1 \cup A_2 \cup \cdots \cup A_n$), and let E be an event of the experiment such that $P(E) \neq 0$. Then the *a posteriori* probability $P(A_i \mid E)(1 \leq i \leq n)$ is given by

$$P(A_i \mid E) = \frac{P(A_i) \cdot P(E \mid A_i)}{P(A_1) \cdot P(E \mid A_1) + P(A_2) \cdot P(E \mid A_2) + \cdots + P(A_n) \cdot P(E \mid A_n)} \qquad (9)$$

▶ Applications

EXAMPLE

32 The picture tubes for the Pulsar 19-inch color television sets are manufactured in three locations and then shipped to the main plant of the Vista Vision Corporation for final assembly. Plants A, B, and C supply 50 percent, 30 percent, and 20 percent, respectively, of the picture tubes used by Vista Vision. The quality control department of the company has determined that 1 percent of the picture tubes produced by plant A are defective, whereas 2 percent of the picture tubes produced by plants B and C are defective. If a Pulsar 19-inch color television set is selected at random and the picture tube is found to be defective, what is the probability that the picture tube was manufactured in plant C? Compare with Example 26, page 380.

SOLUTION

Let A, B, and C denote the event that the set chosen has a picture tube manufactured in plant A, plant B, and plant C, respectively. Also, let D denote the event that a set has a defective picture tube. Using the given information we may draw the tree diagram shown in Figure 7.14.

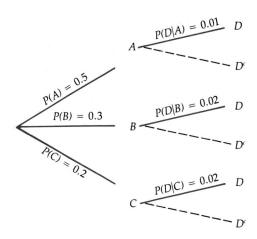

Figure 7.14

Next, using Formula (9), we find that the required *a posteriori* probability is given by

$$P(C \mid D) = \frac{P(C) \cdot P(D \mid C)}{P(A) \cdot P(D \mid A) + P(B) \cdot P(D \mid B) + P(C) \cdot P(D \mid C)}$$

$$= \frac{(0.2)(0.02)}{(0.5)(0.01) + (0.3)(0.02) + (0.2)(0.02)}$$

$$\approx 0.27. \qquad \blacktriangleleft$$

EXAMPLE

33 A study was conducted in a large metropolitan area to determine the annual incomes of married couples in which the husbands were the sole providers and of those in which the husbands and wives were both employed. Table 7.6 gives the results of this study.

Annual Family Income ($)	% of Married Couples	% of Income Group with Both Spouses Working
75,000 and over	4	65
50,000–74,999	10	73
35,000–49,999	21	68
25,000–34,999	24	63
15,000–24,999	30	43
Under 15,000	11	28

Table 7.6

a. What is the probability that a couple selected at random from this area has two incomes?

b. If a randomly chosen couple has two incomes, what is the probability that the annual income of this couple is over $75,000?

c. If a randomly chosen couple has two incomes, what is the probability that the annual income of this couple is greater than $24,999?

SOLUTION Let A denote the event that the annual income of the couple is $75,000 and over; let B denote the event that the annual income is between $50,000 and $74,999; let C denote the event that the annual income is between $35,000 and $49,999, and so on; finally, let F denote the event that the annual income is less than $15,000.

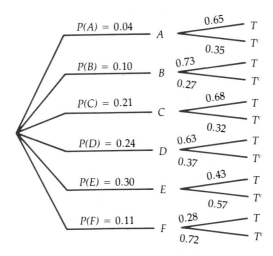

Figure 7.15

a. The probability that a couple selected at random from this group has two incomes is given by

$$P(T) = P(A) \cdot P(T \mid A) + P(B) \cdot P(T \mid B) + P(C) \cdot P(T \mid C) + P(D) \cdot P(T \mid D)$$
$$+ P(E) \cdot P(T \mid E) + P(F) \cdot P(T \mid F)$$
$$= (0.04)(0.65) + (0.10)(0.73) + (0.21)(0.68) + (0.24)(0.63)$$
$$+ (0.30)(0.43) + (0.11)(0.28)$$
$$= 0.5528.$$

b. Using the parts of (a) and Bayes' Theorem, we find that the probability that a randomly chosen couple has an annual income over $75,000 given that both spouses are working is

$$P(A \mid T) = \frac{P(A) \cdot P(T \mid A)}{P(T)} = \frac{(0.04)(0.65)}{0.5528} = 0.047.$$

c. The probability that a randomly chosen couple has an annual income greater than \$24,999 given that both spouses are working is

$$P(A \mid T) + P(B \mid T) + P(C \mid T) + P(D \mid T)$$

$$= \frac{P(A) \cdot P(T \mid A) + P(B) \cdot P(T \mid B) + P(C) \cdot P(T \mid C) + P(D) \cdot P(T \mid D)}{P(T)}$$

$$= \frac{(0.04)(0.65) + (0.1)(0.73) + (0.21)(0.68) + (0.24)(0.63)}{0.5528}$$

$$= 0.711. \qquad \blacktriangleleft$$

Self-Check Exercises

7.6

1. The following tree diagram represents a two-stage experiment. Use the diagram to find $P(B \mid D)$.

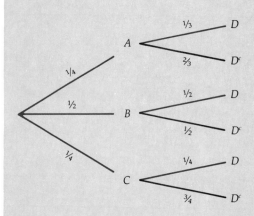

2. In a recent presidential election, it was estimated that the probability that the Republican candidate would be elected was 3/5 and therefore the probability that the Democratic candidate would be elected was 2/5 (the two independent candidates were given little chance of being elected). It was also estimated that if the Republican candidate were elected, then the probability that research for a new manned bomber would continue was 4/5. But if the Democratic candidate were successful, then the probability that the research would continue was 3/10. Research was terminated shortly after the successful presidential candidate took office. What is the probability that the Republican candidate won that election?

Solutions to Self-Check Exercises 7.6 can be found on page 403.

Exercises 7.6

In Exercises 1–3, refer to the Venn diagram that follows. An experiment in which the three mutually exclusive events A, B, and C form a partition of the uniform sample space S is depicted in the diagram.

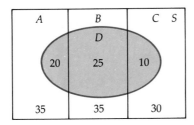

Figure for
Exercises 1–3

1. Draw a tree diagram using the information given in the Venn diagram illustrating the probabilities of the events *A*, *B*, *C*, and *D*.

2. Find: *a.* $P(D)$ *b.* $P(A \mid D)$

3. Find: *a.* $P(D^c)$ *b.* $P(B \mid D^c)$

In Exercises 4–6, refer to the Venn diagram that follows. An experiment in which the three mutually exclusive events A, B, and C form a partition of the uniform sample space S is depicted in the diagram.

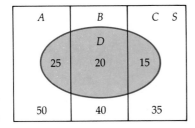

Figure for
Exercises 4–6

4. Find: *a.* $P(D)$ *b.* $P(A \mid D)$

5. Find: *a.* $P(D)$ *b.* $P(B \mid D)$

6. Find: *a.* $P(D^c)$ *b.* $P(B \mid D^c)$

7. The tree diagram on the next page represents a two-stage experiment. Use the diagram to find:

     a. $P(A) \cdot P(D \mid A)$ *b.* $P(B) \cdot P(D \mid B)$ *c.* $P(A \mid D)$

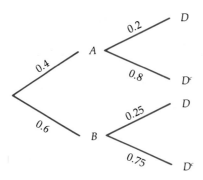

8. The tree diagram that follows represents a two-stage experiment. Use the diagram to find

a. $P(A) \cdot P(D \mid A)$ b. $P(B) \cdot P(D \mid B)$
c. $P(C) \cdot P(D \mid C)$ d. $P(A \mid D)$

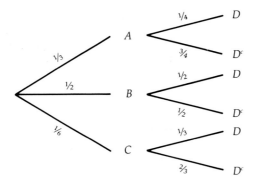

9. The tree diagram that follows represents a two-stage experiment. Use this diagram to find

a. $P(A \cap D)$ b. $P(B \cap D)$ c. $P(C \cap D)$ d. $P(D)$
e. Verify:

$$P(A \mid D) = \frac{P(A \cap D)}{P(D)} = \frac{P(A) \cdot P(D \mid A)}{P(A) \cdot P(D \mid A) + P(B) \cdot P(D \mid B) + P(C) \cdot P(D \mid C)}$$

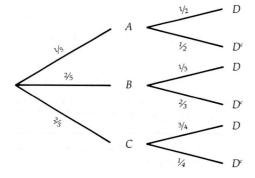

In Exercises 10–12, refer to the following experiment. Two cards are drawn in succession without replacement from a standard deck of 52 cards.

10. What is the probability that the first card is a heart given that the second card is a heart?

11. What is the probability that the first card is a jack given that the second card is an ace?

12. What is the probability that the first card is a face card given that the second card is an ace?

In Exercises 13–15, refer to the following experiment: Urn A contains four white and six black balls. Urn B contains three white and five black balls. A ball is drawn from urn A and then transferred to urn B. A ball is then drawn from urn B.

13. Represent the probabilities associated with this two-stage experiment in the form of a tree diagram.

14. What is the probability that the transferred ball was white given that the second ball drawn was white?

15. What is the probability that the transferred ball was black given that the second ball was white?

16. Jansen Electronics has four machines that produce an identical component for use in its videocassette players. The proportion of the components produced by each machine and the probability of that component being defective are shown in the following table. What is the probability that

  a. a component selected at random is defective?

  b. a defective component was produced by machine I?

  c. a defective component was produced by machine II?

Machine	Proportion of Components Produced	Probability of Defective Component
I	0.15	0.04
II	0.30	0.02
III	0.35	0.02
IV	0.20	0.03

17. An experiment consists of randomly selecting one of three coins, tossing it, and observing the outcome—heads or tails. The first coin is a two-headed coin, the second is a biased coin such that $P(H) = 0.75$, and the third is a fair coin.

  a. What is the probability that the coin that is tossed will show heads?

  b. If the coin selected shows heads, what is the probability that this coin is the fair coin?

18. A medical test has been designed to detect the presence of a certain disease. Among those who have the disease, the probability that the disease will be detected by the test is 0.95. However, the probability that the test will erroneously indicate the presence of the disease in those who do not actually have it is 0.04. It is estimated that 4 percent of the population who take this test have the disease.

  a. If the test administered to an individual is positive, what is the probability that the person actually has the disease?

b. If an individual takes the test twice and both times the test is positive, what is the probability that the person actually has the disease?

19. Refer to Exercise 18. Suppose that 20 percent of the people who were referred to a clinic for the test did in fact have the disease. If the test administered to an individual from this group is positive, what is the probability that the person actually has the disease?

20. A desk lamp produced by the Luminar Company was found to be defective. The company has three factories where the lamps are manufactured. The percentage of the total number of desk lamps produced by each factory and the probability that a lamp manufactured by that factory is defective are shown in the following table. What is the probability that the defective lamp was manufactured in factory III?

Factory	Percentage of Total Production	Probability of Defective Component
I	0.35	0.015
II	0.35	0.01
III	0.30	0.02

21. An insurance company has compiled the following data relating the age of drivers and the accident rate (the probability of being involved in an accident during a one-year period) for drivers within that group.

Age Group	Percentage of Insured Drivers	Accident Rate
Under 25	0.16	0.055
25–44	0.40	0.025
45–64	0.30	0.02
65 and over	0.14	0.04

a. What is the probability that an insured driver will be involved in an accident?

b. What is the probability that an insured driver who is involved in an accident is under 25?

22. Data compiled by the Highway Patrol Department regarding the use of seat belts by drivers in a certain area after the passage of a compulsory seat belt law is shown in the following table:

Drivers	Percentage of Drivers in Group	Percentage of Group Stopped for Moving Violation
Group I (using seat belts)	0.64	0.002
Group II (not using seat belts)	0.36	0.005

If a driver in that area is stopped for a moving violation, what is the probability that

a. he or she will have a seat belt on?

b. he or she will not have a seat belt on?

23. A poll was conducted among 500 registered voters in a certain area regarding their position on a national lottery to raise revenue for the government. The results of the poll are shown in the following table.

Sex	Percentage of Voters Polled	Percentage Favoring Lottery	Percentage Not Favoring Lottery	Percentage Expressing No Opinion
Male	0.51	0.62	0.32	0.06
Female	0.49	0.68	0.28	0.04

What is the probability that a registered voter

   a. who favored a national lottery was a woman?

   b. who expressed no opinion regarding the lottery was a woman?

24. The sales department of the Thompson Drug Company released the following data concerning the sales of a certain pain reliever manufactured by the company.

Pain Reliever	Percentage of Drug Sold	Percentage of Group Sold in Extra-Strength Dosage
Group I (capsule form)	0.57	0.38
Group II (tablet form)	0.43	0.31

If a customer purchased the extra-strength dosage of this drug, what is the probability that it was in capsule form?

25. In a past presidential election, it was estimated that the probability that the Republican candidate would be elected was 3/5 and therefore the probability that the Democratic candidate would be elected was 2/5 (the two independent candidates were given little chance of being elected). It was also estimated that if the Republican candidate were elected, the probabilities that a conservative, moderate, or liberal judge would be appointed to the Supreme Court (one retirement was expected during the presidential term) were 1/2, 1/3, and 1/6, respectively. If the Democratic candidate were elected, the probabilities that a conservative, moderate, or liberal judge would be appointed to the Supreme Court would be 1/8, 3/8, and 1/2, respectively. A conservative judge *was* appointed to the Supreme Court during the presidential term. What is the probability that the Democratic candidate was elected?

26. Applicants for temporary office work at the Carter Temporary Help Agency who have successfully completed a typing test are then placed in suitable positions by Ms. Dwyer and Ms. Newberg. Employers who hire temporary help through the agency return a card indicating satisfaction or dissatisfaction with the work performance of those hired. From past experience it is known that 80 percent of the employees placed by Ms. Dwyer are rated as satisfactory, whereas 70 percent of those placed by Ms. Newberg are rated as satisfactory. Ms. Newberg places 55 percent of the temporary office help at the agency and Ms. Dwyer the remaining 45 percent. If a Carter office worker is rated unsatisfactory, what is the probability that he or she was placed by Ms. Newberg?

27. The Office of Admissions and Records of a large western university released the following information concerning the contemplated majors of its freshman class.

Major	Percentage of Freshmen Choosing This Major	Percentage of Females	Percentage of Males
Business	0.24	0.38	0.62
Humanities	0.08	0.60	0.40
Education	0.08	0.66	0.34
Social science	0.07	0.58	0.42
Natural sciences	0.09	0.52	0.48
Other	0.44	0.48	0.52

a. What is the probability that a student selected at random from the freshman class is a female?

b. What is the probability that a business student selected at random from the freshman class is a male?

c. What is the probability that a female student selected at random from the freshman class is majoring in business?

28. A study was conducted among a certain group of union members whose health insurance policies required second opinions prior to surgery. Of those members whose doctors advised them to have surgery, 20 percent were informed by a second doctor that no surgery was needed. Of these, 70 percent took the second doctor's opinion and did not go through with the surgery. Of the members who were advised to have surgery by both doctors, 95 percent went through with the surgery. What is the probability that a union member who had surgery was advised to do so by a second doctor?

29. A study conducted by the Metro Housing Agency in a midwestern city revealed the following information concerning the age distribution of renters within the city.

Age	Percentage of Adult Population	Percentage of Group Who Are Renters
21–44	0.51	0.58
45–64	0.31	0.45
65 and over	0.18	0.60

a. What is the probability that an adult selected at random from this population is a renter?

b. If a renter is selected at random, what is the probability that he or she is in the 21–44 age bracket?

c. If a renter is selected at random, what is the probability that he or she is 45 years of age or older?

Solutions to Self-Check Exercises

7.6

1. By Bayes' Theorem, we have, using the probability given in the tree diagram,

$$P(B \mid D) = \frac{P(B)P(D \mid B)}{P(A)P(D \mid A) + P(B)P(D \mid B) + P(C)P(D \mid C)}$$

$$= \frac{\left(\frac{1}{2}\right)\left(\frac{1}{2}\right)}{\left(\frac{1}{4}\right)\left(\frac{1}{3}\right) + \left(\frac{1}{2}\right)\left(\frac{1}{2}\right) + \left(\frac{1}{4}\right)\left(\frac{1}{4}\right)}$$

$$= \frac{12}{19}.$$

2. Let R and D, respectively, denote the event that the Republican and the Democratic candidate won the presidential election. Then $P(R) = 3/5$ and $P(D) = 2/5$. Also, let C denote the event that research for the new manned bomber would continue. These data may be exhibited in the following tree diagram.

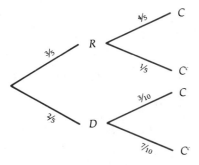

Using Bayes' Theorem, we find that the probability that the Republican candidate had won the election is given by

$$P(R \mid C^c) = \frac{P(R)P(C^c \mid R)}{P(R)P(C^c \mid R) + P(D)P(C^c \mid D)}$$

$$= \frac{\left(\frac{3}{5}\right)\left(\frac{1}{5}\right)}{\left(\frac{3}{5}\right)\left(\frac{1}{5}\right) + \left(\frac{2}{5}\right)\left(\frac{7}{10}\right)}$$

$$= \frac{3}{10}.$$

▶

Chapter 7 Review Exercises

1. Let E and F be two events that are mutually exclusive and suppose $P(E) = 0.4$ and $P(F) = 0.2$. Compute

    a. $P(E \cap F)$ *b.* $P(E \cup F)$ *c.* $P(E^c)$

    d. $P(E^c \cap F^c)$ *e.* $P(E^c \cup F^c)$

2. Let E and F be two events of an experiment with sample space S. Suppose $P(E) = 0.3$, $P(F) = 0.2$, and $P(E \cap F) = 0.15$. Compute

    a. $P(E \cup F)$ *b.* $P(E^c \cap F^c)$ *c.* $P(E^c \cap F)$

3. Suppose a die is loaded and it was determined that the probability distribution associated with the experiment of casting the die and observing which number falls uppermost is given by

Simple Event	Probability
{1}	0.20
{2}	0.12
{3}	0.16
{4}	0.18
{5}	0.15
{6}	0.19

    a. What is the probability of the number being even?

    b. What is the probability of the number being either a 1 or a 6?

    c. What is the probability of the number being less than 4?

4. An urn contains six red, five black, and four green balls. If two balls are selected at random without replacement from the urn, what is the probability that a red ball and a black ball will be selected?

5. The quality control department of the Starr Communications Company, the manufacturer of video-game cartridges, has determined from records that 1.5 percent of the cartridges sold have video defects, 0.8 percent have audio defects, and 0.4 percent have both audio and video defects. What is the probability that a cartridge purchased by a customer

    a. will have a video or audio defect?

    b. will not have a video or audio defect?

6. An experiment consists of tossing a fair coin three times and observing the outcomes. Let A be the event that at least one head is thrown and B the event that at most two tails are thrown.

    a. Find $P(A)$.

    b. Find $P(B)$.

    c. Are A and B independent events?

7. In a group of 20 ballpoint pens on a shelf in the stationery department of the Metro Department Store, 2 are known to be defective. If a customer selects 3 of these pens, what is the probability that

   a. at least 1 is defective?

   b. no more than 1 is defective?

  8. Five people are selected at random. What is the probability that none of the people in this group were born on the same day of the week?

  9. A pair of fair dice is cast. What is the probability that the sum of the numbers falling uppermost is 8 if it is known that the two numbers are different?

  10. Two cards are drawn at random from a well-shuffled deck of 52 playing cards. What is the probability that the second card is an ace

   a. if the first card is replaced before the second card is drawn?

   b. if the first card is not replaced before the second card is drawn?

  11. Of 320 male and 280 female employees at the home office of the Gibraltar Insurance Company, 160 of the men and 190 of the women are on flex-time (flexible working hours). Given that an employee selected at random from this group is on flex-time, what is the probability that the employee is a male?

  12. The tree diagram that follows represents an experiment consisting of two trials. Use the diagram to find

   a. $P(A \cap E)$ *b.* $P(B \cap E)$ *c.* $P(C \cap E)$

   d. $P(E)$ *e.* $P(A \mid E)$

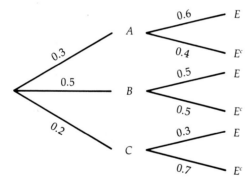

  13. In a manufacturing plant, three machines, A, B and C, produce 40 percent, 35 percent, and 25 percent, respectively, of the total production. The quality control department of the company has determined that 1 percent of the items produced by machine A, 1.5 percent of the items produced by machine B, and 2 percent of the items produced by machine C are defective. If an item is selected at random and found to be defective, what is the probability that it was produced by machine B?

  14. Applicants who wish to be admitted to a certain professional school in a large university are required to take a screening test that was devised by an educational testing service. From past results, the testing service has estimated that 70 percent of all applicants are eligible for admission and that 92 percent of those who are eligible for admission pass the exam, whereas 12 percent of those who are ineligible for admission pass the exam. Using these results, find the probability that

   a. an applicant for admission who passed the exam was actually ineligible.

   b. an applicant for admission passed the exam.

Probability Distributions and Statistics

Where do we go from here? Some of the top 10 percent of this senior class will further their education at one of the campuses of the State University system. In Example 34, page 473, we will determine the minimum grade point average that a senior needs to be eligible for admission to one of the State Universities.

▶**CHAPTER**

EIGHT

8.1

Distributions of Random Variables

▶ Random Variables

▶ Probability Distributions of Random Variables

▶ Histograms

Statistics is that branch of mathematics concerned with the collection, analysis, and interpretation of data. In the next three sections we will take a look at descriptive statistics. In these sections our interest is in the description and presentation of data in the form of tables, graphs, and so on. In the rest of the chapter we will take a glimpse at inductive statistics, and we will see how mathematical tools such as those developed in the last chapter may be used in conjunction with these data to help us draw certain conclusions and to make forecasts.

▶ Random Variables

In many situations, it is desirable to assign numerical values to the outcomes of an experiment. For example, if an experiment consists of casting a die and observing the face that lands uppermost, then it is natural to assign the numbers 1, 2, 3, 4, 5, and 6, respectively, to the outcomes *one, two, three, four, five,* and *six* of the experiment. If we let X denote the outcome of the experiment, then X assumes one of the numbers. Because the values assumed by X depend on the outcomes of a chance experiment, the outcome X is referred to as a *random variable.*

Random Variable

A **random variable** is a rule that assigns a number to each outcome of a chance experiment.

More precisely, a random variable is a function with domain given by the set of outcomes of a chance experiment and range contained in the set of real numbers.

EXAMPLE

1 A coin is tossed three times. Let the random variable X denote the number of heads that occur in the three tosses.

a. List the outcomes of the experiment; that is, find the domain of the function X.

b. Find the value assigned to each outcome of the experiment by the random variable X.

c. Find the event comprising the outcomes to which a value of 2 has been assigned by X. This event is written $(X = 2)$ and is the event comprising the outcomes in which two heads occur.

SOLUTION *a.* From the result of Example 3, Chapter 7 (page 331) we see that the set of outcomes of the experiment is given by the sample space

$$S = \{HHH, HHT, HTH, THH, HTT, THT, TTH, TTT\}$$

of the experiment.

b. The outcomes of the experiment are displayed in the first column of Table 8.1. The corresponding value assigned to each such outcome by the random variable X (the number of heads) appears in the second column.

Outcome	Value of X
HHH	3
HHT	2
HTH	2
THH	2
HTT	1
THT	1
TTH	1
TTT	0

Table 8.1

c. With the aid of Table 8.1, we see that the event $(X = 2)$ is given by the set

$$\{HHT, HTH, THH\} \qquad \blacktriangleleft$$

EXAMPLE

2 A coin is tossed repeatedly until a head occurs. Let the random variable Y denote the number of tosses of the coin in the experiment. What are the values of Y?

SOLUTION   The outcomes of the experiment comprise the infinite set

$$S = \{H, TH, TTH, TTTH, TTTTH, \ldots\}.$$

These outcomes of the experiment are displayed in the first column of Table 8.2. The corresponding values assumed by the random variable Y (the number of tosses) appear in the second column.

Outcome	Value of Y
H	1
TH	2
TTH	3
TTTH	4
TTTTH	5
⋮	⋮

Table 8.2

◀

EXAMPLE

3   A disposable flashlight is turned on until its battery runs out. Let the random variable Z denote the length (in hours) of the life of the battery. What values may Z assume?

SOLUTION   The values assumed by Z may be any nonnegative real numbers; that is, the possible values of Z comprise the interval $0 \le Z < \infty$.   ◀

The advantage of working with random variables rather than working directly with the outcomes of an experiment is that random variables are functions that may be added, subtracted, and multiplied. Because of this, results developed in the field of algebra and other areas of mathematics may be used freely to help us solve problems in the fields of probability and statistics.

A random variable is classified into three categories depending on the set of values it assumes. A random variable is called **finite discrete** if it assumes only finitely many values. For example, the random variable X of Example 1 is finite discrete since it may assume only values from the finite set $\{0, 1, 2, 3\}$ of numbers. Next, a random variable is said to be **infinite discrete** if it takes on infinitely many values, which may be arranged in a sequence. For example, the random variable Y of Example 2 is infinite discrete since it assumes values from the set $\{1, 2, 3, 4, 5, \ldots\}$, which has been arranged in the form of an infinite sequence. Finally, a **random variable** is called **continuous** if the values it may assume comprise an interval of real numbers. For example, the random variable Z of Example 3 is continuous since the values it may assume comprise the interval of nonnegative real numbers. For the remainder of this section *all random variables will be assumed to be finite discrete.*

▶ Probability Distributions of Random Variables

In Section 7.2 we learned how to construct the probability distribution for an experiment. There, the probability distribution took the form of a table that gave the probabilities associated with the outcomes of an experiment. Since the random variable associated with an experiment is related to the outcomes of the experiment, it is clear that we should be able to construct a probability distribution associated with the *random variable* rather than one associated with the outcomes of the experiment. Such a distribution is called the *probability distribution of a random variable* and may be given in the form of a formula or displayed in a table, which gives the distinct (numerical) values of the random variable X and the probabilities associated with these values. Thus, if $x_1, x_2, \ldots, x_n$ are the values assumed by the random variable X with associated probabilities $P(X = x_1)$, $P(X = x_2), \ldots, P(X = x_n)$, respectively, then the required probability distribution of the random variable X is given by

x	$P(X = x)$
x_1	p_1
x_2	p_2
x_3	p_3
$\vdots$	$\vdots$
x_n	p_n

The next several examples illustrate the construction of certain probability distributions.

EXAMPLE

4 Find the probability distribution of the random variable associated with the experiment of Example 1.

SOLUTION

From the results of Example 1, we see that the values assumed by the random variable X are 0, 1, 2, and 3, corresponding to the events of 0, 1, 2, and 3 heads occurring, respectively. Referring to Table 8.1 once again, we see that the outcome associated with the event $(X = 0)$ is given by the set {TTT}. Consequently, the probability associated with the random variable X when it assumes the value 0 is given by

$$P(X = 0) = \frac{1}{8} \quad \text{(Note that } n(S) = 8\text{)}$$

Next, observe that the event $(X = 1)$ is given by the set {HTT, THT, TTH}, so

$$P(X = 1) = \frac{3}{8}.$$

In a similar manner we may compute $P(X = 2)$ and $P(X = 3)$, which gives the following probability distribution (Table 8.3).

x	$P(X = x)$
0	1/8
1	3/8
2	3/8
3	1/8

Table 8.3

◀

EXAMPLE

5 Let X denote the random variable that gives the sum of the faces that fall upper-most when two fair dice are cast. Find the probability distribution of X.

SOLUTION The values assumed by the random variable X are 2, 3, 4, . . . , 12, corresponding to the events E_2, E_3, E_4, . . . , E_{12} (see Example 4, Section 7.1). Next, the proba-bilities associated with the random variable X when X assumes the values 2, 3, 4, . . . , 12 are precisely the probabilities $P(E_2)$, $P(E_3)$, . . . , $P(E_{12})$, respectively, and may be computed in much the same way as the solution to Example 9, Section 7.2. Thus,

$$P(X = 2) = P(E_2) = \frac{1}{36}, \qquad P(X = 3) = P(E_3) = \frac{2}{36}, \quad \text{and so on.}$$

The required probability distribution of X is given by Table 8.4.

x	2	3	4	5	6	7	8	9	10	11	12
$P(X = x)$	1/36	2/36	3/36	4/36	5/36	6/36	5/36	4/36	3/36	2/36	1/36

Table 8.4

◀

EXAMPLE

6 The following data give the number of cars observed waiting in line at the beginning of every 2 minutes between 3 and 5 P.M. on a certain Friday at the drive-in teller of the Westwood Savings Bank and the corresponding frequency of occurrence. Find the probability distribution of the random variable X, where X denotes the number of cars observed waiting in line.

Number of Cars	Frequency of Occurrence
0	2
1	9
2	16
3	12
4	8
5	6
6	4
7	2
8	1

SOLUTION   Dividing each number on the right of the given table by 60 (the sum of these numbers) gives the respective probabilities (here, we use the relative frequency interpretation of probability) associated with the random variable X when X assumes the values 0, 1, 2, . . . , 8. For example,

$$P(X = 0) = \frac{2}{60} \approx 0.03,$$

$$P(X = 1) = \frac{9}{60} = 0.15, \quad \text{and so on.}$$

The resulting probability distribution is shown in Table 8.5.

x	$P(X = x)$
0	0.03
1	0.15
2	0.27
3	0.20
4	0.13
5	0.10
6	0.07
7	0.03
8	0.02

Table 8.5   ◀

▶ Histograms

A probability distribution of a random variable may be exhibited graphically by means of a **histogram**. To construct a histogram of a particular probability distribution, first locate the values of the random variable on the number line. Then, above each such number, erect a rectangle with width 1 and height equal to the probability associated with that value of the random variable. For example, the histogram of the probability distribution appearing in Table 8.3 is shown in Figure 8.1. The histograms of the probability distributions of Examples 5 and 6 are constructed in a similar manner and are displayed in Figures 8.2 and 8.3, respectively.

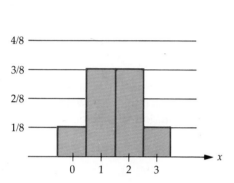

Figure 8.1

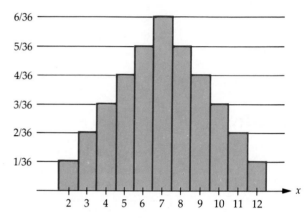

Figure 8.2

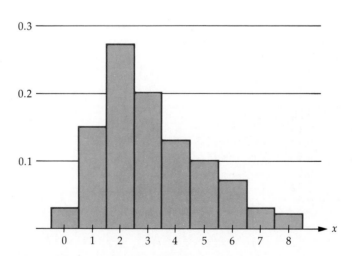

Figure 8.3

Observe that in each of the histograms of probability distributions of random variables, the area of a rectangle associated with a value of a random variable X gives precisely the probability associated with that value of X. This follows because each such rectangle, by construction, has width 1 and height corresponding to the probability associated with the value of the random variable. Another consequence arising from the method of construction of a histogram is that *the probability associated with more than one value of the random variable X is given by the sum of the areas of the rectangles associated with those values of X*. For example, in the coin-tossing experiment of Example 1, the event of obtaining at least two heads, which corresponds to the event $(X = 2)$ or $(X = 3)$, is given by

$$P(X = 2) + P(X = 3)$$

and may be obtained from the histogram depicted in Figure 8.1 by adding the areas associated with the values 2 and 3, respectively, of the random variable X. We obtain

$$P(X = 2) + P(X = 3) = (1)\left(\frac{3}{8}\right) + (1)\left(\frac{1}{8}\right)$$

$$= \frac{1}{2}.$$

This result provides us with a method of computing the probabilities of events associated with an experiment directly from the knowledge of a histogram of the probability distribution of the random variable associated with the experiment.

EXAMPLE

7   Suppose that the probability distribution of a random variable X is represented by the histogram shown in Figure 8.4. Identify that part of the histogram whose area gives the probability $P(10 \le X \le 20)$. Do not evaluate the result.

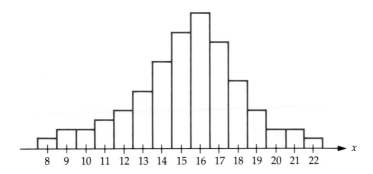

Figure 8.4   8   9   10   11   12   13   14   15   16   17   18   19   20   21   22

SOLUTION

The event $(10 \le X \le 20)$ is the event comprising outcomes related to the values 10, 11, 12, . . . , 20 of the random variable X. The probability of this event $P(10 \le X \le 20)$ is therefore given by the shaded area of the histogram in Figure 8.5.

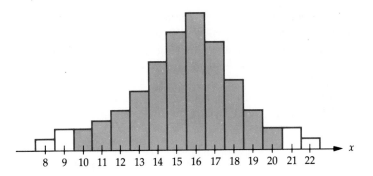

Figure 8.5

Self-Check Exercises

8.1

1. Three balls are selected at random without replacement from an urn containing four black balls and five white balls. Let the random variable X denote the number of black balls drawn.

a. List the outcomes of the experiment.

b. Find the value assigned to each outcome of the experiment by the random variable X.

c. Find the event comprising the outcomes to which a value of 2 has been assigned by X.

2. The following data, extracted from the records of the Dover Public Library, give the number of books borrowed by the library's members over a one-month period.

Number of Books	0	1	2	3	4	5	6	7	8
Frequency of Occurrence	780	300	412	205	98	54	57	30	6

a. Find the probability distribution of the random variable X, where X denotes the number of books checked out by the library's members.

b. Draw the histogram representing this probability distribution.

Solutions to Self-Check Exercises 8.1 can be found on page 419.

Exercises 8.1

1. A die is cast repeatedly until a 6 falls uppermost. Let the random variable X denote the number of times the die is cast. What are the values that X may assume?

2. Let X denote the random variable that gives the sum of the faces that fall uppermost when two fair dice are cast. Find $P(X = 7)$.

3. Three balls are selected at random without replacement from an urn containing four green balls and six red balls. Let the random variable X denote the number of green balls drawn.

   a. List the outcomes of the experiment.

   b. Find the value assigned to each outcome of the experiment by the random variable X.

   c. Find the event comprising the outcomes to which a value of 3 has been assigned by X.

4. Two dice are cast. Let the random variable X denote the number that falls uppermost on the first die and let Y denote the number that falls uppermost on the second die.

   a. Find the probability distributions of X and Y.

   b. Find the probability distribution of $X + Y$.

In Exercises 5–10, give the range of values that the random variable X may assume and classify the random variable as finite discrete, infinite discrete, or continuous.

5. $X = $ the number of times a die is thrown until a 2 appears.

6. $X = $ the number of defective watches in a sample of eight watches.

7. $X = $ the distance a commuter travels to work.

8. $X = $ the number of hours a child watches television on a given day.

9. $X = $ the number of times an accountant takes the C.P.A. examination before passing.

10. $X = $ the number of boys in a four-child family.

11. The probability distribution of the random variable X is shown in the following table:

x	-10	-5	0	5	10	15	20
$P(X = x)$	0.20	0.15	0.05	0.1	0.25	0.1	0.15

Find:

   a. $P(X = -10)$ b. $P(X \geq 5)$ c. $P(-5 \leq X \leq 5)$
   d. $P(X \leq 20)$

12. The probability distribution of the random variable X is shown in the following table.

x	-5	-3	-2	0	2	3
$P(X = x)$	0.17	0.13	0.33	0.16	0.11	0.10

Find:

   a. $P(X \leq 0)$ b. $P(X \leq -3)$ c. $P(-2 \leq X \leq 2)$

13. Suppose that a probability distribution of a random variable X is represented by the following histogram. Shade that part of the histogram whose area gives the probability $P(17 \le X \le 20)$.

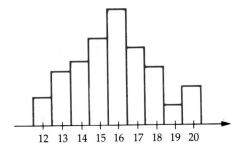

14. An examination consisting of 10 true-or-false questions was taken by a class of 100 students. The probability distribution of the random variable X, where X denotes the number of questions answered correctly by a student, is represented by the following histogram. The rectangle with base centered on the number 8 is missing. What should the height of this rectangle be?

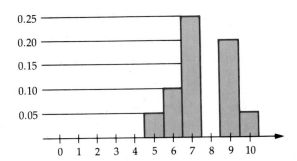

15. The following data were obtained in a study conducted by the manager of the Sav-More Supermarket. In this study the number of customers waiting in line at the express checkout at the beginning of each 3-minute interval between 9 A.M. and 12 noon on Saturday was observed.

Number of Customers	0	1	2	3	4	5	6	7	8	9	10
Frequency of Occurrence	1	4	2	7	14	8	10	6	3	4	1

a. Find the probability distribution of the random variable X where X denotes the number of customers observed waiting in line.

b. Draw the histogram representing this probability distribution.

16. A survey was conducted by the Public Housing Authority in a certain community

among 1000 families to determine the distribution of families by size. The results follow.

Family Size	2	3	4	5	6	7	8
Frequency of Occurrence	350	200	245	125	66	10	4

a. Find the probability distribution of the random variable X where X denotes the number of persons in a family.

b. Draw the histogram corresponding to the probability distribution found in (a).

17. After the private screening of a new television pilot, members of the audience were asked to rate the new show on a scale of 1 to 10 (10 being the highest rating). From a group of 140 people, the following responses were obtained:

Rating	1	2	3	4	5	6	7	8	9	10
Frequency of Occurrence	1	4	3	11	23	21	28	29	16	4

Let the random variable X denote the rating that was given to the show by a member of the audience. Find the probability distribution associated with these data.

18. The rates paid by 30 financial institutions on a certain day for money-market deposit accounts is shown in the following table.

Rate (%)	6	6.25	6.55	6.56	6.58	6.60	6.65	6.85
Number of Institutions	1	7	7	1	1	8	3	2

Let the random variable X denote the interest paid on money-market deposit accounts and find the probability distribution associated with these data.

Solutions to Self-Check Exercises

8.1

1. a. Using the accompanying tree diagram, we see that the outcomes of the experiment are

$S = \{(B, B, B), (B, B, W), (B, W, B), (B, W, W), (W, B, B), (W, B, W), (W, W, B), (W, W, W)\}$

Outcomes

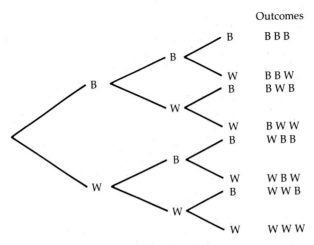

Outcomes
B B B
B B W
B W B
B W W
W B B
W B W
W W B
W W W

b. Using the results of (a), we obtain the values assigned to the outcomes of the experiment as follows:

Outcome	BBB	BBW	BWB	BWW	WBB	WBW	WWB	WWW
Value	3	2	2	1	2	1	1	0

c. The required event is {BBW, BWB, WBB}.

2. *a.* We divide each number in the bottom row of the given table by 1942 (the sum of these numbers) to obtain the probabilities associated with the random variable X when X takes on the values 0, 1, 2, 3, 4, 5, 6, 7, and 8. For example,

$$P(X = 0) = \frac{780}{1942} \approx 0.402$$

$$P(X = 1) = \frac{300}{1942} \approx 0.155$$

The required probability distribution is as follows:

x	0	1	2	3	4	5	6	7	8
$P(X = x)$	0.402	0.155	0.212	0.106	0.051	0.028	0.029	0.016	0.003

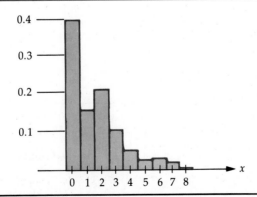

8.2

Expected Value

▶ Mean

▶ Expected Value

▶ Applications

▶ Fair Odds

▶ Mean

The average value of a set of numbers is a familiar notion to most people. For example, to compute the average of the four numbers

$$12, 16, 23, 37$$

we simply add these numbers and divide the resulting sum by 4, giving the required average as

$$\frac{12 + 16 + 23 + 37}{4} = \frac{88}{4},$$

or 22. In general, we have the following definition:

Average or Mean

The **average,** or **mean,** of the n numbers

$$x_1, x_2, \ldots, x_n$$

is $\bar{x}$ (read "x bar") where

$$\bar{x} = \frac{x_1 + x_2 + \cdots + x_n}{n}$$

EXAMPLE

8 Refer to Example 6. Find the average number of cars waiting in line at the drive-in teller of the bank observed at the beginning of each 2-minute interval during the period in question.

SOLUTION

The number of cars together with its corresponding frequency of occurrence are reproduced in Table 8.6.

Number of Cars	0	1	2	3	4	5	6	7	8
Frequency of Occurrence	2	9	16	12	8	6	4	2	1

Table 8.6

Observe from the table that the number 0 (of cars) occurs twice, the number 1 occurs 9 times, and so on. There are altogether

$$2 + 9 + 16 + 12 + 8 + 6 + 4 + 2 + 1,$$

or 60, numbers to be averaged. Therefore, the required average is given by

$$\frac{0 \cdot 2 + 1 \cdot 9 + 2 \cdot 16 + 3 \cdot 12 + 4 \cdot 8 + 5 \cdot 6 + 6 \cdot 4 + 7 \cdot 2 + 8 \cdot 1}{60} \approx 3.1, \quad (1)$$

or approximately 3.1 cars. ◄

► Expected Value

Let us reconsider the expression (1) that gives the average of the frequency distribution shown in Table 8.6. Dividing each term by the denominator, the expression may be cast in the form

$$0 \cdot \left(\frac{2}{60}\right) + 1 \cdot \left(\frac{9}{60}\right) + 2 \cdot \left(\frac{16}{60}\right) + 3 \cdot \left(\frac{12}{60}\right) + 4 \cdot \left(\frac{8}{60}\right) + 5 \cdot \left(\frac{6}{60}\right)$$

$$+ 6 \cdot \left(\frac{4}{60}\right) + 7 \cdot \left(\frac{2}{60}\right) + 8 \cdot \left(\frac{1}{60}\right).$$

Observe that each term in the sum is a product of two factors; the first factor is the value assumed by the random variable X where X denotes the number of cars observed waiting in line, and the second factor is just the probability associated with that value of the random variable (see Example 6). This observation suggests the following general method for calculating the *expected value* (that is, the average, or mean) of a random variable X that assumes a finite number of values from the knowledge of its probability distribution:

Expected Value of a Random Variable X

Let X denote a random variable that assumes the values $x_1, x_2, \ldots, x_n$ with associated probabilities $p_1, p_2, \ldots, p_n$, respectively. Then the **expected value** of X, $E(X)$, is given by

$$E(X) = x_1 p_1 + x_2 p_2 + \cdots + x_n p_n \quad (2)$$

EXAMPLE

9 Using Table 8.5, the probability distribution associated with the experiment, solve Example 8 again.

SOLUTION Let X denote the number of cars observed waiting in line. Then the average number of cars observed waiting in line is given by the expected value of X, that is, by

$$E(X) = 0 \cdot (0.03) + 1 \cdot (0.15) + 2 \cdot (0.27) + 3 \cdot (0.20) + 4 \cdot (0.13)$$
$$+ 5 \cdot (0.10) + 6 \cdot (0.07) + 7 \cdot (0.03) + 8 \cdot (0.02)$$
$$= 3.1,$$

or 3.1 cars, which agrees with the result obtained earlier. ◄

The expected value of a random variable X is a measure of central tendency of the probability distribution associated with X. In repeated trials of an experiment with random variable X, the observed values of X get closer and closer to the expected value of X as the number of trials gets larger and larger. Geometrically, the expected value of a random variable X has the following simple interpretation: If a laminate is made of the histogram of a probability distribution associated with a random variable X, then the expected value of X corresponds to the point on the base of the laminate at which the latter will balance perfectly when the point is directly over a fulcrum (see Figure 8.6).

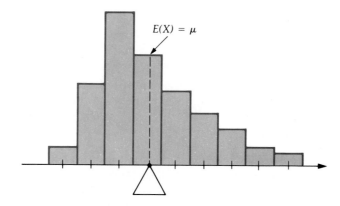

Figure 8.6

EXAMPLE

10 Let X denote the random variable that gives the sum of the faces that fall uppermost when two fair dice are cast. Find the expected value, $E(X)$, of X.

SOLUTION The probability distribution of X, which is reproduced in the following table, was found in Example 5.

x	$P(X = x)$
2	1/36
3	2/36
4	3/36
5	4/36
6	5/36
7	6/36
8	5/36
9	4/36
10	3/36
11	2/36
12	1/36

Using this result we find

$$E(X) = 2\left(\frac{1}{36}\right) + 3\left(\frac{2}{36}\right) + 4\left(\frac{3}{36}\right) + 5\left(\frac{4}{36}\right) + 6\left(\frac{5}{36}\right) + 7\left(\frac{6}{36}\right) + 8\left(\frac{5}{36}\right)$$
$$+ 9\left(\frac{4}{36}\right) + 10\left(\frac{3}{36}\right) + 11\left(\frac{2}{36}\right) + 12\left(\frac{1}{36}\right)$$
$$= 7.$$

Note that, because of the symmetry of the histogram of the probability distribution with respect to the vertical line $x = 7$, the result could have been obtained by merely inspecting Figure 8.2. ◀

▶ Applications

The next example shows how the concept of expected value is used to help us make the best decision in an investment situation.

EXAMPLE

11 A group of private investors intends to purchase one of two motels that are currently being offered for sale in a certain city. The terms of sale of the two motels are similar, although the Regina Inn has 52 rooms and is in a slightly better location than the Merlin Motor Lodge, which has 60 rooms. Records obtained for each of the motels reveal that the occupancy rates, with corresponding probabilities, during the May–September tourist season are:

Regina Inn

Occupancy Rate	0.80	0.85	0.90	0.95	1.00
Probability	0.19	0.22	0.31	0.23	0.05

Occupancy Rate	0.75	0.80	0.85	0.90	0.95	1.00
Probability	0.35	0.21	0.18	0.15	0.09	0.02

Merlin
Motor Lodge

The average profit per day for each occupied room at the Regina Inn is $10, whereas the average profit per day for each occupied room at the Merlin Motor Lodge is $9.

a. Find the average number of rooms occupied per day at each motel.

b. If the investors' objective is to purchase the motel that generates the higher daily profit, which motel should they purchase? (Compare the expected daily profit of the two motels.)

SOLUTION

a. Let X denote the occupancy rate at the Regina Inn. Then the average daily occupancy rate at the Regina Inn is given by the expected value of X, that is, by

$$E(X) = (.80)(.19) + (.85)(.22) + (.90)(.31) + (.95)(.23) + (1.00)(.05)$$
$$= .8865.$$

The average number of rooms occupied per day at the Regina is

$$(.8865)(52) \approx 46.1,$$

or 46.1 rooms. Similarly, letting Y denote the occupancy rate at the Merlin Motor Lodge, we have

$$E(Y) = (.75)(.35) + (.80)(.21) + (.85)(.18) + (.90)(.15)$$
$$+ (.95)(.09) + (1.00)(.02)$$
$$= .8240.$$

The average number of rooms occupied per day at the Merlin is

$$(.8240)(60) \approx 49.4,$$

or 49.4 rooms.

b. The expected daily profit at the Regina is given by

$$46.1(10) = 461,$$

or $461. The expected daily profit at the Merlin is given by

$$(49.4)(9) = 444.6,$$

or $444.60. From these results we conclude that the investors should purchase the Regina Inn, which is expected to yield a higher daily profit. ◀

EXAMPLE

12 The Island Club is holding a fund-raising raffle. Ten thousand tickets have been sold for $2 each. There will be a first prize of $3000, three second prizes of $1000 each, five third prizes of $500 each, and twenty consolation prizes of $100 each. Letting X denote the net winnings (that is, winnings less the cost of the ticket) associated with the tickets, find $E(X)$. Interpret your results.

SOLUTION The values assumed by X are $(0 - 2)$, $(100 - 2)$, $(500 - 2)$, $(1000 - 2)$, and $(3000 - 2)$, that is, -2, 98, 498, 998, and 2998, which correspond, respectively, to the value of a losing ticket, a consolation prize, a third prize, and so on. The probability distribution of X may be calculated in the usual manner and appears in the following table. Using the table, we find

$$E(X) = (-2)(0.9971) + 98(0.002) + 498(0.0005) + 998(0.0003) + 2998(0.0001)$$
$$= -0.95.$$

This expected value gives the long-run average loss (negative gain) of a holder of one ticket; that is, if one participated in such a raffle by purchasing one ticket each time, in the long run, one may expect to lose, on the average, $.95 per raffle.

x	$P(X = x)$
-2	0.9971
98	0.0020
498	0.0005
998	0.0003
2998	0.0001

◄

EXAMPLE

13 In the game of roulette as played in Las Vegas casinos, the wheel is divided into 38 compartments numbered 1 through 36, 0, and 00. One-half of the numbers 1 through 36 are red, the other half black, and 0 and 00 are green. Of the many types of bets that may be placed, one type involves betting on the outcome of the color of the winning number. For example, one may place a certain sum of money on *red*. If the winning number is red, one wins an amount equal to the bet placed and loses the bet otherwise. Find the expected value of the winnings on a $1 bet placed on *red*.

SOLUTION Let X be a random variable whose values are 1 and -1, which correspond to a win and a loss. The probabilities associated with the values 1 and -1 are 18/38 and 20/38, respectively. Therefore, the expected value is given by

$$E(X) = 1\left(\frac{18}{38}\right) + (-1)\left(\frac{20}{38}\right)$$

$$= -\frac{2}{38}$$

$$\approx -0.053.$$

Thus, if one places a $1 bet on *red* over and over again, one may expect to lose, on the average, approximately $.05 per bet in the long run. ◀

The last two examples illustrate games that are not "fair." Of course, most participants in such games are aware of this fact and participate in them for other reasons. In a fair game, neither party has an advantage, a condition that translates into the condition that $E(X) = 0$, where X takes on the values of the winnings of a player.

EXAMPLE

14 Mike and Bill play a card game with a standard deck of 52 cards. Mike selects a card from a well-shuffled deck and receives A dollars from Bill if the card selected is a diamond; otherwise, Mike pays Bill a dollar. Determine the value of A if the game is to be fair.

SOLUTION

Let X denote a random variable whose values are associated with Mike's winnings. Then X takes on the value A with probability $P(X = A) = 1/4$ (since there are 13 diamonds in the deck of cards) if Mike wins and takes on the value -1 with probability $P(X = -1) = 3/4$ if Mike loses. Since the game is to be a fair one, the expected value $E(X)$ of Mike's winnings is to be equal to zero, that is,

$$E(X) = A\left(\frac{1}{4}\right) + (-1)\left(\frac{3}{4}\right) = 0.$$

Solving the above equation for A gives $A = 3$. Thus, the card game will be fair if Bill makes a $3 payoff for a winning bet of $1 placed by Mike. ◀

▶ Fair Odds

In the last example, we saw that the card game could be made fair if Bill agreed to pay Mike $3 for a $1 winning bet placed by Mike. The ratio 3 : 1 is called the *fair odds* for the game. In general, odds of $a : b$ mean that a player who bets b and wins will receive a payoff of a (in addition to the b that was bet). The fair odds for a game may be computed from the knowledge of the probability of winning the bet. To this end, suppose the probability of winning the bet is p. Let X be a random variable associated with the winnings of a player who places a bet of $\$b$ in the game. Then the values taken on by X are a and $-b$, corresponding to a win or a loss with respective probabilities $P(X = a) = p$ and $P(X = -b) = 1 - p$. Setting $E(X) = 0$ for a fair game, we find

$$E(X) = ap + (-b)(1 - p) = 0$$

or
$$\frac{a}{b} = \frac{1 - p}{p} \tag{3}$$

Observe that Formula (3) states that the odds $a : b$ for a fair game is just the ratio of the probability of losing to the probability of winning.

EXAMPLE

15 Find the fair odds for a bet on *red* in American roulette.

SOLUTION The probability of winning a bet here—the probability that the ball lands on a red compartment—is given by $p = 18/38$. Therefore, the fair odds are

$$\frac{a}{b} = \frac{1 - 18/38}{18/38}$$

$$= \frac{20}{18}$$

$$= \frac{10}{9},$$

that is, $10 : 9$. Thus, a player should receive $\$10$ on a winning $\$9$ bet or equivalently $\$1.11$ on a winning $\$1$ bet. ◄

Solving Equation (3) for p in terms of a and b gives

$$p = \frac{b}{a + b} \tag{4}$$

Equation (4) enables us to compute the probability of winning a bet given the fair odds for the bet.

EXAMPLE

16 The odds for a fair game are known to be $5 : 4$. Find the probability of winning a bet in the game.

SOLUTION The fair odds for the game are $a : b = 5 : 4$ or, equivalently, $a = 5$ when $b = 4$.

Using these values of a and b and Formula (4), we find

$$p = \frac{4}{5 + 4},$$

or approximately 0.44.   ◄

In concluding this section, we should mention that in addition to the mean there are two other measures of central tendency of a set of numerical data, the median and the mode of a set of numbers. The **median** is the middle value in a set of data arranged in increasing or decreasing order (when there are an odd number of entries). If there are an even number of entries, the median is the mean of the two middle numbers. The **mode** is the value that occurs most frequently in a set of data. (See Problems 27 and 28, Exercises 8.2.)

Self-Check Exercises
8.2

1. Find the expected value of a random variable X having the following probability distribution.

x	-4	-3	-1	0	1	2
$P(X = x)$	0.10	0.20	0.25	0.10	0.25	0.10

2. The developer of the Shoreline Condominiums has provided the following estimate of the probability that 20, 25, 30, 35, 40, 45, or 50 of the townhouses will be sold within the first month they are offered for sale.

Number of Units	20	25	30	35	40	45	50
Probability	0.05	0.10	0.30	0.25	0.15	0.10	0.05

How many townhouses can the developer expect to sell within the first month they are put on the market?

Solutions to Self-Check Exercises 8.2 can be found on page 434.

▶

Exercises 8.2

1. During the first year at a university that uses a four-point grading system, a freshman took ten 3-credit courses and received two A's, three B's, four C's, and one D.

    a. Compute this student's grade-point average.

    b. Let the random variable X denote the number of points corresponding to a given letter grade. Find the probability distribution of the random variable X and compute $E(X)$, the expected value of X.

2. Records kept by the chief dietitian at the university cafeteria over a 30-week period show the following weekly consumption of milk (in gallons).

Milk (in Gallons)	200	205	210	215	220	225	230	235	240
Number of Weeks	3	4	6	5	4	3	2	2	1

    a. Find the average number of gallons of milk consumed per week in the cafeteria.

    b. Let the random variable X denote the number of gallons of milk consumed in a week at the cafeteria. Find the probability distribution of the random variable X and compute $E(X)$, the expected value of X.

3. Find the expected value of a random variable X having the following probability distribution.

x	-5	-1	0	1	5	8
$P(X = x)$	0.12	0.16	0.28	0.22	0.12	0.1

4. Find the expected value of a random variable X having the following probability distribution.

x	0	1	2	3	4	5
$P(X = x)$	1/8	1/4	3/16	1/4	1/16	1/8

5. The daily earnings X of an employee who works on a commission basis are given by the following probability distribution. Find the employee's expected earnings.

x (in \$)	0	25	50	75	100	125	150
$P(X = x)$	0.07	0.12	0.17	0.14	0.28	0.18	0.04

6. In a four-child family, what is the expected number of boys? (Assume that the probability of a boy being born is the same as the probability of a girl being born.)

7. Based on past experience, the manager of the Video-Rama Store has compiled the following table, which gives the probabilities that a customer who enters the Video-Rama Store will buy 0, 1, 2, 3, or 4 videocassettes. How many videocassettes can a customer entering this store be expected to buy?

Number of Videocassettes	0	1	2	3	4
Probability	0.42	0.36	0.14	0.05	0.03

8. If a sample of 3 batteries is selected from a lot of ten of which two are defective, what is the expected number of defective batteries?

9. The number of accidents that occur at a certain intersection known as "Five Corners" on a Friday afternoon between the hours of 3:00 and 6:00 P.M. along with the corresponding probabilities are shown in the following table. Find the expected number of accidents during the period in question.

Number of Accidents	0	1	2	3	4
Probability	0.935	0.03	0.02	0.01	0.005

10. The owner of a newsstand in a college community estimates the weekly demand for a certain magazine is as follows:

Quantity Demanded	10	11	12	13	14	15
Probability	0.05	0.15	0.25	0.30	0.20	0.05

Find the number of issues of the magazine that the owner of the newsstand can expect to sell per week.

11. A bank has two automatic tellers at its main office and two at each of its three branches. The number of these machines that break down on a given day along with the corresponding probabilities are shown in the following table:

Number of Machines That Break Down	0	1	2	3	4	5	6	7	8
Probability	0.43	0.19	0.12	0.09	0.04	0.03	0.03	0.02	0.05

Find the expected number of machines that will break down on a given day.

12. In a lottery, 5000 tickets are sold for $1 each. One first prize of $2000, 1 second prize of $500, 3 third prizes of $100, and 10 consolation prizes of $25 are to be awarded. What are the expected net earnings of a person who buys one ticket?

13. A man wishes to purchase a five-year term life insurance policy that will pay the beneficiary $20,000 in the event that the man's death occurs during the next five years. Using life insurance tables, he determines that the probability that he will live another

five years is 0.96. What is the minimum amount that he can expect to pay for his premium?

14. A woman purchased a $10,000, one-year term life insurance policy for $130. Assuming that the probability that she will live for another year is 0.992, find the company's expected gain.

15. As a fringe benefit, Mr. Taylor receives a $25,000 life insurance policy from his employer. The probability that Mr. Taylor will live for another year is .9935. If he purchases the same coverage for himself, find the minimum amount that he could expect to pay for the policy.

16. The management of Multi-Vision, Inc., a cable T.V. company, intends to submit a bid for the cable television rights in one of two cities, A or B. If the company obtains the rights to city A, the probability of which is 0.2, the estimated profit over the next ten years is $10,000,000; if the company obtains the rights to city B, the probability of which is 0.3, the estimated profit over the next ten years is $7,000,000. The cost of submitting a bid for rights in city A is $250,000 and that of city B is $200,000. By comparing the expected profits for each venture, determine whether the company should bid for the rights in city A or city B.

17. Mr. Hunt intends to purchase one of two car dealerships that are currently for sale in a certain city. Records obtained from each of the two dealers reveal that their weekly volume of sales, with corresponding probabilities, are as follows:

Dahl Motors

Number of Cars Sold Per Week	5	6	7	8	9	10	11	12
Probability	0.05	0.09	0.14	0.24	0.18	0.14	0.11	0.05

Farthington Auto Sales

Number of Cars Sold Per Week	5	6	7	8	9	10
Probability	0.08	0.21	0.31	0.24	0.10	0.06

The average profit per car at Dahl Motors is $362, and the average profit per car at Farthington Auto Sales is $436.

  a. Find the average number of cars sold per week at each dealership.

  b. If Mr. Hunt's objective is to purchase the dealership that generates the higher weekly profit, which dealership should he purchase? (Compare the expected weekly profit for each dealership.)

18. Ms. Leonard, a real estate broker, is relocating in a large metropolitan area where she has received job offers from realty company A and realty company B. The number of houses she expects to sell in a year at each firm and the associated probabilities are shown in the following tables:

Company A

Number of Houses Sold	12	13	14	15	16	17	18	19	20	21	22	23	24
Probability	0.02	0.03	0.05	0.07	0.07	0.16	0.17	0.13	0.11	0.09	0.06	0.03	0.01

Number of Houses Sold	6	7	8	9	10	11	12	13	14	15	16	17	18
Probability	0.01	0.04	0.07	0.06	0.11	0.12	0.19	0.17	0.13	0.04	0.03	0.02	0.01

Company B

The average price of a house in the locale of company A is $104,000, whereas the average price of a house in the locale of company B is $177,000. Based on these expectations and the fact that Ms. Leonard will receive a commission of 3 percent on sales at both companies, which job offer should she accept?

19. A player casts a fair die and receives a dollar amount equal to the number appearing uppermost on the face of the die. If this is a fair game, determine the amount that the player should be willing to pay to participate in the game.

20. If the probability that it will rain tomorrow is 0.3,

    a. what are the odds that it will rain tomorrow?

    b. what are the odds that it will not rain tomorrow?

21. Find the probabilities corresponding to the given odds in a fair game:

    a. 9 to 5   *b.* 1 to 3   *c.* 4 to 7

22. If a sports forecaster states that the odds of a certain boxer winning a match are 4 to 3, what is the probability that the boxer will win the match?

23. In American roulette as described in Example 13, a player may bet on a split (two adjacent numbers). In this case, if the player bets $1 and either number comes up, the player wins $17 and gets his $1 back. If neither number comes up, he loses his $1 bet. Find the expected value of the winnings on a $1 bet placed on a split.

24. If a player placed a $1 bet on red and a $1 bet on black in a single play in American roulette, what would the expected value of his winnings be?

25. In European roulette the wheel is divided into 37 compartments numbered 1 through 36 and 0. (In American roulette there are 38 compartments numbered 1 through 36, 0, and 00.) Find the expected value of the winnings on a $1 bet placed on red in European roulette.

26. *a.* Show that for any number c,

$$E(cX) = cE(X).$$

    b. Use this result to find the expected loss if a gambler bets $300 on red in a single play in American roulette.

27. In an examination given to a class of 20 students, the following test scores were obtained:

    40, 45, 50, 50, 55, 60, 60, 75, 75, 80, 80, 85, 85, 85, 85, 90, 90, 95, 95, 100

    a. Find the mean, or average, score, the mode, and the median score.

    b. Which of these three measures of central tendency do you think is the least representative of the set of scores?

28. The frequency distribution of the hourly wage rates among blue-collar workers in a certain factory is given in the table on page 434. Find the mean, or average, wage rate, the mode, and the median wage rate of these workers.

Wage Rate (in $)	5.70	5.80	5.90	6.00	6.10	6.20
Frequency	60	90	75	120	60	45

Solutions to Self-Check Exercises

8.2

1. $E(X) = (-4)(0.10) + (-3)(0.20) + (-1)(0.25) + (0)(0.10) + (1)(0.25) + (2)(0.10)$
   $= -0.8$

2. Let X denote the number of townhouses that will be sold within one month of being put on the market. Then, the number of townhouses the developer expects to sell within one month is given by the expected value of X, that is, by

   $E(X) = 20(0.05) + 25(0.10) + 30(0.30) + 35(0.25) + 40(0.15) + 45(0.10) + 50(0.05)$
   $= 34.25,$

   or 34 townhouses.

8.3

Variance and Standard Deviation

▶ Variance

▶ Standard Deviation

▶ Chebychev's Inequality

▶ Variance

The mean, or expected value, of a random variable, introduced in the last section, enables us to express an important property of the probability distribution associated with the random variable in terms of a single number. But the knowledge of the location, or central tendency, of a probability distribution alone is usually not enough to give a reasonably accurate picture of the probability distribution. Consider, for example, the two probability distributions whose histograms appear in Figure 8.7. Both distributions have the same expected value, or mean, $\mu = 4$ (the Greek letter μ is read "mu"). Note that the probability distribution with histogram shown in Figure 8.7(a) is closely concentrated about

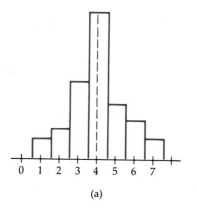

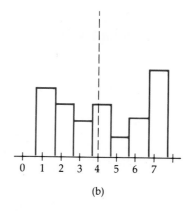

Figure 8.7

(a) (b)

its mean μ, whereas the one with histogram shown in Figure 8.7(b) is widely dispersed or spread about its mean.

As another example, suppose Mr. Alan Horowitz, host of the popular television show "The Consumer Advocate," decides to demonstrate the accuracy of the weights of two popular brands of potato chips. Ten packages of potato chips of each brand are selected at random and weighed carefully. The results are as follows:

Weight in Ounces

Brand A	16.1	16	15.8	16	15.9	16.1	15.9	16	16	16.2
Brand B	16.3	15.7	15.8	16.2	15.9	16.1	15.7	16.2	16	16.1

It may be readily verified that the mean weights for each of the two brands is 16 ounces (see Example 19). However, a cursory examination of the data shows that the weights of the brand B packages exhibit much greater dispersion about the mean than those of brand A.

One measure of the degree of dispersion, or spread, of a probability distribution about its mean is given by the variance of the random variable associated with the probability distribution. A probability distribution with a small spread about its mean will have a small variance, whereas one with a larger spread will have a larger variance. Thus, the variance of the random variable associated with the probability distribution whose histogram appears in Figure 8.7(a) is smaller than the variance of the random variable associated with the probability distribution whose histogram is shown in Figure 8.7(b) [see Example 17]. Also, as we will see in Example 19, the variance of the random variable associated with the weights of the brand A potato chips is smaller than that of the random variable associated with the weights of the brand B potato chips. (This observation was made earlier.)

We now define the variance of a random variable.

Variance of a Random Variable X

Suppose a random variable has the following probability distribution:

x	x_1	x_2	x_3	$\ldots$	x_n
$P(X = x)$	p_1	p_2	p_3	$\ldots$	p_n

and expected value

$$E(X) = \mu.$$

Then the **variance** of the random variable X is

$$\text{Var } (X) = p_1 (x_1 - \mu)^2 + p_2(x_2 - \mu)^2 + \cdots + p_n(x_n - \mu)^2 \qquad (5)$$

An explanation of Formula (5) is in order. First of all, let us note that the numbers

$$x_1 - \mu, x_2 - \mu, \ldots, x_n - \mu \qquad (6)$$

measure the **deviations** of $x_1, x_2, \ldots, x_n$ from μ, respectively. Thus the numbers

$$(x_1 - \mu)^2, (x_2 - \mu)^2, \ldots, (x_n - \mu)^2 \qquad (7)$$

measure the squares of the deviations of $x_1, x_2, \ldots, x_n$ from μ, respectively. Next, by multiplying each of the numbers in (7) by the respective probability associated with each value of the random variable X, the numbers are weighted accordingly so that their sum is a measure of the variance of X about its mean. An attempt to define the variance of a random variable about its mean in a similar manner using the deviations (6) rather than their squares is not fruitful, since some of the deviations may be positive whereas others may be negative and (because of cancellations) the sum does not give a satisfactory measure of the variance of the random variable.

EXAMPLE

17 Find the variance of the random variable X and of the random variable Y whose probability distributions follow. These are the probability distributions associated with the histograms shown in Figure 8.7(a) and Figure 8.7(b).

x	$P(X = x)$	y	$P(Y = y)$
1	0.05	1	0.2
2	0.075	2	0.15
3	0.2	3	0.1
4	0.375	4	0.15
5	0.15	5	0.05
6	0.1	6	0.1
7	0.05	7	0.25

SOLUTION   The mean of the random variable X is given by

$$\mu_x = (1)(0.05) + (2)(0.075) + (3)(0.2) + (4)(0.375) + (5)(0.15)$$
$$+ (6)(0.1) + (7)(0.05)$$
$$= 4.$$

Therefore, using Formula (5) and the data from the probability distribution of X, we find that the variance of X is given by

$$\text{Var}\,(X) = (0.05)(1 - 4)^2 + (0.075)(2 - 4)^2 + (0.2)(3 - 4)^2 + (0.375)(4 - 4)^2$$
$$+ (0.15)(5 - 4)^2 + (0.1)(6 - 4)^2 + (0.05)(7 - 4)^2$$
$$= 1.95.$$

Next, we find that the mean of the random variable Y is given by

$$\mu_Y = (1)(0.2) + (2)(0.15) + (3)(0.1) + (4)(0.15) + (5)(0.05)$$
$$+ (6)(0.1) + (7)(0.25)$$
$$= 4,$$

and so the variance of Y is given by

$$\text{Var}\,(Y) = (0.2)(1 - 4)^2 + (0.15)(2 - 4)^2 + (0.1)(3 - 4)^2 + (0.15)(4 - 4)^2$$
$$+ (0.05)(5 - 4)^2 + (0.1)(6 - 4)^2 + (0.25)(7 - 4)^2$$
$$= 5.2.$$

Note that Var (X) is smaller than Var (Y), which confirms the earlier observations about the spread, or dispersion, of the probability distribution of X and Y, respectively. ◄

► Standard Deviation

Because Formula (5), which gives the variance of the random variable X, involves the square of the deviations, the unit of measurement of Var (X) is the square of the unit of measurement of the values of X. For example, if the values assumed by the random variable X are measured in units of a gram, then Var (X) will be measured in units involving the *square* of a gram. To remedy this situation, one normally works with the square root of Var (X) rather than Var (X) itself. The former is called the standard deviation of X.

Standard Deviation of a Random Variable X

The **standard deviation** of a random variable X, σ (pronounced "sigma"), is defined by

$$\sigma = \sqrt{\text{Var}(X)}$$
$$= \sqrt{p_1(x_1 - \mu)^2 + p_2(x_2 - \mu)^2 + \cdots + p_n(x_n - \mu)^2} \qquad (8)$$

where $x_1, x_2, \ldots, x_n$ denote the values assumed by the random variable X and $p_1 = P(X = x_1)$, $p_2 = P(X = x_2)$, $\ldots$, $p_n = P(X = x_n)$.

EXAMPLE

18 Find the standard deviations of the random variables X and Y of Example 17.

SOLUTION

From the results of Example 17, we have Var $(X) = 1.95$ and Var $(Y) = 5.2$. Extracting their respective square roots, we have

$$\sigma_X = \sqrt{1.95}$$
$$\approx 1.40$$

and

$$\sigma_Y = \sqrt{5.2}$$
$$\approx 2.28.$$ ◀

EXAMPLE

19 Let X and Y denote the random variables whose values are the weights of the brand A and brand B potato chips, respectively (see page 435). Compute the means and standard deviations of X and Y and interpret your results.

SOLUTION

The probability distributions of X and Y may be computed from the given data as follows:

Brand A

x	Relative Frequency of Occurrence	$P(X = x)$
15.8	1	0.1
15.9	2	0.2
16.0	4	0.4
16.1	2	0.2
16.2	1	0.1

Brand B

y	Relative Frequency of Occurrence	$P(Y = y)$
15.7	2	0.2
15.8	1	0.1
15.9	1	0.1
16.0	1	0.1
16.1	2	0.2
16.2	2	0.2
16.3	1	0.1

The means of X and Y are given by

$$\mu_X = (0.1)(15.8) + (0.2)(15.9) + (0.4)(16.0) + (0.2)(16.1)$$
$$+ (0.1)(16.2)$$
$$= 16$$
$$\mu_Y = (0.2)(15.7) + (0.1)(15.8) + (0.1)(15.9) + (0.1)(16.0)$$
$$+ (0.2)(16.1) + (0.2)(16.2) + (0.1)(16.3)$$
$$= 16.$$

Therefore,

$$\text{Var}\,(X) = (0.1)(15.8 - 16)^2 + (0.2)(15.9 - 16)^2 + (0.4)(16 - 16)^2$$
$$+ (0.2)(16.1 - 16)^2 + (0.1)(16.2 - 16)^2$$
$$= 0.012$$

and

$$\text{Var}\,(Y) = (0.2)(15.7 - 16)^2 + (0.1)(15.8 - 16)^2 + (0.1)(15.9 - 16)^2$$
$$+ (0.1)(16 - 16)^2 + (0.2)(16.1 - 16)^2 + (0.2)(16.2 - 16)^2$$
$$+ (0.1)(16.3 - 16)^2$$
$$= 0.042,$$

so that the required standard deviations are

$$\sigma_X = \sqrt{\text{Var}\,(X)}$$
$$= \sqrt{0.012}$$
$$\approx 0.11$$

and

$$\sigma_Y = \sqrt{\text{Var}\,(Y)}$$
$$= \sqrt{0.042}$$
$$\approx 0.21.$$

The mean of X and that of Y are both equal to 16. Therefore, the average weight of a package of potato chips of either brand is 16 ounces. However, the standard deviation of Y is greater than that of X. This tells us that the weights of the packages of brand B potato chips are more widely dispersed about the common mean of 16 than are those of brand A. ◀

The standard deviation of a random variable X may be used in statistical estimations. For example, the following result, derived by the Russian mathematician P. L. Chebychev (1821–1894), gives the proportion of the values of X lying within k standard deviations of the expected value of X.

▶ Chebychev's Inequality

Chebychev's Inequality

Let X be a random variable with expected value μ and standard deviation σ. Then the probability that a randomly chosen outcome of the experiment lies between $\mu - k\sigma$ and $\mu + k\sigma$ is at least $1 - (1/k^2)$, that is,

$$P(\mu - k\sigma \le X \le \mu + k\sigma) \ge 1 - \frac{1}{k^2} \qquad (9)$$

EXAMPLE

20 A probability distribution has a mean of 10 and a standard deviation of 1.5. Use the Chebychev inequality to estimate the probability that an outcome of the experiment lies between 7 and 13.

SOLUTION Here $\mu = 10$ and $\sigma = 1.5$. Next, to determine the value of k, note that $\mu - k\sigma = 7$ and $\mu + k\sigma = 13$. Substituting the appropriate values for μ and σ, we find $k = 2$. Using the Chebychev inequality (9), we see that the probability that an outcome of the experiment lies between 7 and 13 is given by

$$P(7 \le X \le 13) \ge 1 - \left(\frac{1}{2^2}\right)$$

$$= \frac{3}{4},$$

that is, at least 75 percent. ◀

EXAMPLE

21 The Great Northwest Lumber Company employs 400 workers in its mills. It has been estimated that X, the random variable measuring the number of mill workers who have industrial accidents during a one-year period, is normally distributed with a mean of 40 and a standard deviation of 6. Using the Chebychev inequality, estimate the probability that the number of workers who will have an industrial accident over a one-year period is between 30 and 50, inclusive.

SOLUTION Here $\mu = 40$ and $\sigma = 6$. We wish to estimate $P(30 \le X \le 50)$. In order to use the Chebychev inequality (9), we first determine the value of k from the equation

$$\mu - k\sigma = 30 \quad \text{or} \quad \mu + k\sigma = 50.$$

Since $\mu = 40$ and $\sigma = 6$ in this case, we see that k satisfies

$$40 - 6k = 30 \quad \text{and} \quad 40 + 6k = 50,$$

from which we deduce that $k = 5/3$. Thus, the probability that the number of mill workers who will have an industrial accident during a one-year period is between 30 and 50 is given by

$$P(30 \leq X \leq 50) \geq 1 - \frac{1}{(5/3)^2}$$

$$= \frac{16}{25},$$

that is, at least 64 percent. ◀

Self-Check Exercises

8.3

1. Compute the mean, variance, and standard deviation of the random variable X with probability distribution as follows.

x	-4	-3	-1	0	2	5
$P(X = x)$	0.1	0.1	0.2	0.3	0.1	0.2

2. James recorded the following travel times (the length of time in minutes it took him to drive to work) on ten consecutive days:

$$55 \quad 50 \quad 52 \quad 48 \quad 50 \quad 52 \quad 46 \quad 48 \quad 50 \quad 51$$

Calculate the mean and standard deviation of these times.

Solutions to Self-Check Exercises 8.3 can be found on page 445.

▶

Exercises 8.3

In Exercises 1–6, the probability distribution of a random variable X is given. Compute the mean, variance, and standard deviation of X.

1.

x	1	2	3	4
$P(X = x)$	0.4	0.3	0.2	0.1

2.

x	−4	−2	0	2	4
$P(X = x)$	0.1	0.2	0.3	0.1	0.3

3.

x	−2	−1	0	1	2
$P(X = x)$	1/16	4/16	6/16	4/16	1/16

4.

x	10	11	12	13	14	15
$P(X = x)$	1/8	2/8	1/8	2/8	1/8	1/8

5.

x	430	480	520	565	580
$P(X = x)$	0.1	0.2	0.4	0.2	0.1

6.

x	−198	−195	−193	−188	−185
$P(X = x)$	0.15	0.30	0.10	0.25	0.20

7. An experiment consists of casting an eight-sided die (numbered 1 through 8) and observing the number that appears uppermost. Find the mean and variance of this experiment.

8. The following histograms represent the probability distributions of the random variables X and Y. Determine by inspection which one of the probability distributions has the larger variance.

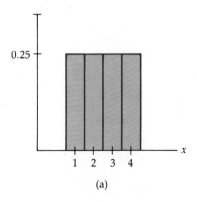

(a)

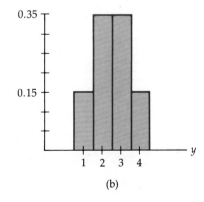

(b)

9. The following histograms represent the probability distributions of the random variables X and Y. Determine by inspection which one of the probability distributions has the larger variance.

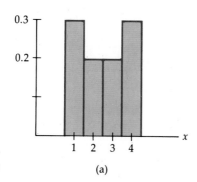

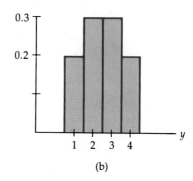

Figure for
Exercise 9

(a)

(b)

10. The minimum age requirement for a regular driver's license differs from state to state. The frequency distribution for this age requirement in the fifty states is given in the following table.

Minimum Age	15	16	17	18	19	21
Frequency of Occurrence	1	15	4	28	1	1

a. Describe a random variable X that is associated with these data.

b. Find the probability distribution for the random variable X.

c. Compute the mean, variance, and standard deviation of X.

11. The birth rate in the United States for the years 1975–1984 is given in the following table. (The birth rate is the number of live births per 1000 population.)

Year	1975	1976	1977	1978	1979	1980	1981	1982	1983	1984
Birth Rate (Number/1000)	14.8	14.8	15.4	15.3	15.9	15.9	15.9	15.5	15.5	15.7

a. Describe a random variable X that is associated with these data.

b. Find the probability distribution for the random variable X.

c. Compute the mean, variance, and standard deviation of X.

12. Mr. Hunt is considering two business ventures. The anticipated returns (in thousands of dollars) of each venture are described by the following probability distributions.

Earnings	Probability
−20	0.3
40	0.4
50	0.3

Venture A

Earnings	Probability
−15	0.2
30	0.5
40	0.3

Venture B

a. Compute the mean and variance for each venture.

b. Which investment would provide Mr. Hunt with the higher expected return (the greater mean)?

c. In which investment would the element of risk be less (that is, which probability distribution has the smaller variance)?

13. The distribution of the number of chocolate chips in a cookie is shown in the following table. Find the mean and the variance of the number of chocolate chips in a cookie.

Number of Chocolate Chips (x)	0	1	2	3	4	5	6	7	8
$P(X = x)$	.01	.03	.05	.11	.13	.24	.22	.16	.05

14. Formula (5) can also be expressed in the form

$$\text{Var}(X) = (p_1 x_1^2 + p_2 x_2^2 + \cdots + p_n x_n^2) - \mu^2.$$

Find the variance of the distribution of Exercise 1 using this formula.

15. Find the variance of the distribution of Exercise 13 using the formula

$$\text{Var}(X) = (p_1 x_1^2 + p_2 x_2^2 + \cdots + p_n x_n^2) - \mu^2.$$

16. A survey was conducted by the market research department of the National Real Estate Company among 500 prospective buyers in a large metropolitan area to determine the maximum price that a prospective buyer would be willing to pay for a house. From the data collected, the distribution that follows was obtained. Compute the mean, variance, and standard deviation of the maximum price that these buyers were willing to pay for a house.

Maximum Price Considered, x (in thousands of dollars)	$P(X = x)$
50	10/500
60	20/500
70	75/500
80	85/500
90	70/500
100	90/500
120	90/500
150	55/500
200	5/500

17. A probability distribution has a mean of 20 and a standard deviation of 3. Use the Chebychev inequality to estimate the probability that an outcome of the experiment lies between

    a. 15 and 25   *b.* 10 and 30

18. A probability distribution has a mean of 42 and a standard deviation of 2. Use the Chebychev inequality to estimate the probability that an outcome of the experiment lies between

    a. 38 and 46   *b.* 32 and 52

19. A probability distribution has a mean of 50 and a standard deviation of 1.4. Use the Chebychev inequality to find the value of c, which guarantees that the probability is at least 96 percent that an outcome of the experiment lies between $50 - c$ and $50 + c$.

20. Suppose X is a random variable with mean μ and standard deviation σ. If a large number of trials are observed, at least what percentage of these values is expected to lie between $\mu - 2\sigma$ and $\mu + 2\sigma$?

21. A Christmas tree light has an expected life of 200 hours and a standard deviation of 2 hours.

    a. Estimate the probability that one of these Christmas tree lights will last between 190 and 210 hours.

    b. Suppose 150,000 of these Christmas tree lights are used by a large city as part of its Christmas decorations. Estimate the number of lights that will require replacement between 180 and 220 hours of use.

22. The expected lifetime of the deluxe model hair dryer produced by the Roland Electric Company has a mean life of 24 months and a standard deviation of 3 months. Find the probability that one of these hair dryers will last between 20 and 28 months.

23. The mean annual starting salary of a new graduate in a certain profession is $25,000 with a standard deviation of $500. What is the probability that the starting salary of a new graduate in this profession will be between $23,000 and $27,000?

24. Sugar packaged by a certain machine has a mean weight of 5 pounds and a standard deviation of 0.02 pounds. For what value of c can the manufacturer of the machinery claim that the sugar packaged by this machine has a weight between $5 - c$ and $5 + c$ pounds with probability at least 96 percent?

Solutions to Self-Check Exercises

8.3

1. The mean of the random variable X is

$$\mu = (-4)(0.1) + (-3)(0.1) + (-1)(0.2) + (0)(0.3) + (2)(0.1) + (5)(0.2)$$
$$= 0.3.$$

The variance of X is

$$\text{Var}\,(X) = (0.1)(-4 - 0.3)^2 + (0.1)(-3 - 0.3)^2 + (0.2)(-1 - 0.3)^2 + (0.3)(0 - 0.3)^2$$
$$+ (0.1)(2 - 0.3)^2 + (0.2)(5 - 0.3)^2$$
$$= 8.01.$$

The standard deviation of X is

$$\sigma = \sqrt{\text{Var}\,(X)} = \sqrt{8.01}$$
$$\approx 2.83.$$

2. We first compute the probability distribution of X from the given data as follows:

x	Relative Frequency of Occurrence	$P(X = x)$
46	1	0.1
48	2	0.2
50	3	0.3
51	1	0.1
52	2	0.2
55	1	0.1

The mean of X is

$$\mu = (0.1)(46) + (0.2)(48) + (0.3)(50) + (0.1)(51) + (0.2)(52) + (0.1)(55)$$
$$= 50.2.$$

The variance of X is

$$\text{Var}\,(X) = (0.1)(46 - 50.2)^2 + (0.2)(48 - 50.2)^2 + (0.3)(50 - 50.2)^2$$
$$+ (0.1)(51 - 50.2)^2 + (0.2)(52 - 50.2)^2 + (0.1)(55 - 50.2)^2$$
$$= 5.76,$$

from which we deduce the standard deviation

$$\sigma = \sqrt{5.76}$$
$$= 2.4.$$

8.4

The Binomial Distribution

▶ Bernoulli Trials

▶ Probabilities in Bernoulli Trials

▶ Applications

▶ Bernoulli Trials

An important class of experiments are those that have (or may be viewed as having) two outcomes. For example, in a coin-tossing experiment, the two outcomes are *heads* and *tails*. In the card game played by Mike and Bill (see Example 14) one may view the selection of a diamond as a *win* (for Mike) and the selection of a card of another suit as a *loss* for Mike. For a third example, consider the experiment in which a person is inoculated with a flu vaccine. Here, the vaccine may be classified as being "effective" or "ineffective" with respect to that particular person.

In general, experiments with two outcomes, as in these examples, are called *Bernoulli trials*, or *binomial trials*. It is standard practice to label one of the outcomes of a binomial trial a *success* and the other a *failure*. For example, in a coin-tossing experiment, the outcome *a head* may be called a success, in which case the outcome *a tail* will be called a failure. Note that by using the terms *success* and *failure* in this way, we depart from their usual connotations.

A sequence of Bernoulli (binomial trials) is called a *binomial experiment*. More precisely, we have the following definition:

Binomial Experiment

A **binomial experiment** is characterized by the following properties:

1. The number of trials in the experiment is fixed.
2. There are two outcomes to the experiment: "success" and "failure."
3. The probability of success in each trial is the same.
4. The trials are independent of each other.

In a binomial experiment, it is customary to denote the probability of a success by the letter p and the probability of a failure by the letter q. Because the event of a success and the event of a failure are complementary events, we have the relationship

$$p + q = 1$$

or equivalently,

$$q = 1 - p.$$

The properties of a binomial experiment are illustrated in the following example.

EXAMPLE

22 A fair die is cast four times. Compute the probability of obtaining exactly one 6 in the four throws.

SOLUTION

There are four trials in this experiment. Each trial consists of casting the die once and observing the face that lands uppermost. We may view each trial as an experiment with two outcomes: a success (S) if the face that lands uppermost is a 6 and a failure (F) if it is any of the other five numbers. Letting p and q denote the probability of success and failure, respectively, of a single trial of the experiment, we find that

$$p = \frac{1}{6} \quad \text{and} \quad q = 1 - \frac{1}{6} = \frac{5}{6}.$$

Furthermore, the trials of this experiment may be assumed to be independent. Thus, we have a binomial experiment.

With the aid of the multiplication principle, we see that the experiment has 2^4, or 16, outcomes. We also obtain these outcomes by constructing the tree diagrams associated with the experiment (see Table 8.7, where they are listed according to the number of successes).

0 Success	*1 Success*	*2 Successes*	*3 Successes*	*4 Successes*
FFFF	SFFF	SSFF	SSSF	SSSS
	FSFF	SFSF	SSFS	
	FFSF	SFFS	SFSS	
	FFFS	FSSF	FSSS	
		FSFS		
		FFSS		

Table 8.7

From Table 8.7, we see that the event of obtaining exactly one success in four trials is given by

$$E = \{SFFF, FSFF, FFSF, FFFS\}$$

with probability given by

$$P(E) = P(SFFF) + P(FSFF) + P(FFSF) + P(FFFS) \tag{10}$$

Since the trials (throws) are independent, the terms on the right-hand side of (10) may be computed as follows:

$$P(SFFF) = P(S)P(F)P(F)P(F) = p \cdot q \cdot q \cdot q = pq^3$$
$$P(FSFF) = P(F)P(S)P(F)P(F) = q \cdot p \cdot q \cdot q = pq^3$$
$$P(FFSF) = P(F)P(F)P(S)P(F) = q \cdot q \cdot p \cdot q = pq^3$$
$$P(FFFS) = P(F)P(F)P(F)P(S) = q \cdot q \cdot q \cdot p = pq^3$$

Therefore, upon substituting these values in Equation (10), we obtain

$$P(E) = pq^3 + pq^3 + pq^3 + pq^3$$
$$= 4pq^3$$
$$= 4\left(\frac{1}{6}\right)\left(\frac{5}{6}\right)^3$$
$$= 0.386.$$

◀

▶ Probabilities in Bernoulli Trials

Let us re-examine the computation performed in the last example. There it was found that the probability of obtaining exactly one success in a binomial experiment with four independent trials with probability of success in a single trial p was given by

$$P(E) = 4pq^3 \quad \text{where } q = 1 - p \tag{11}$$

Observe that the coefficient 4 of pq^3 appearing in Equation (11) is precisely the number of outcomes of the experiment with exactly one success and three failures, the outcomes being

SFFF, FSFF, FFSF, and FFFS.

Another way of obtaining this coefficient is to think of the outcomes as arrangements of the letter S and F. Then, the number of ways of selecting one position for S from four possibilities is given by

$$C(4, 1) = \frac{4!}{1!(4-1)!}$$
$$= 4.$$

Next, observe that, because the trials are independent, each of the four outcomes of the experiment has the same probability, given by

$$pq^3$$

where the exponents 1 and 3 of p and q, respectively, correspond to exactly one success and three failures in the trials that make up each outcome.

As a result of the above discussion, we may write Equation (11) as

$$P(E) = C(4, 1)pq^3 \tag{12}$$

We are also in a position to generalize this result. Suppose that in a binomial experiment the probability of success in any trial is p. What is the probability of obtaining exactly x successes in n independent trials? We start by counting the number of outcomes of the experiment, each of which has exactly x successes. Now, one such outcome involves x successive successes followed by $(n - x)$ failures; that is,

$$\underbrace{SS \cdots S}_{x} \underbrace{FF \cdots F}_{n-x} \tag{13}$$

The other outcomes, each of which has exactly x successes, are obtained by rearranging the S's (x of them) and F's ($n - x$ of them). But there are $C(n, x)$ ways of arranging these letters. Next, arguing as in Example 22, we see that each such outcome has probability given by

$$p^x q^{n-x}.$$

For example, for the outcome (13), we find

$$P(\underbrace{SS \cdots S}_{x} \underbrace{FF \cdots F}_{(n-x)}) = \underbrace{P(S)P(S) \cdots P(S)}_{x} \underbrace{P(F)P(F) \cdots P(F)}_{(n-x)}$$

$$= \underbrace{pp \cdots p}_{x} \underbrace{qq \cdots q}_{n-x}$$

$$= p^x q^{n-x}.$$

Let us state this important result formally:

Computation of Probabilities in Bernoulli Trials

In a binomial experiment in which the probability of success in any trial is p, the probability of exactly x successes in n independent trials is given by

$$C(n, x)p^x q^{n-x}.$$

If we let X be the random variable that gives the number of successes in a binomial experiment (such a random variable is called a **binomial random variable**), then the probability of exactly x successes in n independent trials may be written

$$P(X = x) = C(n, x)p^x q^{n-x} \quad (x = 0, 1, 2, \ldots n) \tag{14}$$

The probability distribution of the binomial random variable is called a **binomial distribution.**

EXAMPLE

23 A fair die is cast five times. If a 1 or a 6 lands uppermost in a trial, then the throw is considered a success. Otherwise, the throw is considered a failure.

a. Find the probability of obtaining exactly 0, 1, 2, 3, 4, and 5 successes, respectively, in this experiment.

b. Using the results obtained in the solution to (a), construct the binomial distribution for this experiment and draw the histogram associated with it.

SOLUTION

a. This is a binomial experiment with X, the binomial random variable, taking on each of the values 0, 1, 2, 3, 4, and 5 corresponding to exactly 0, 1, 2, 3, 4, and 5 successes, respectively, in five trials. Since the die is fair, the probability of a 1 or a 6 landing uppermost in any trial is given by $p = 2/6 = 1/3$, from which it also follows that $q = 1 - p = 2/3$. Finally, $n = 5$, since there are five trials (throws of the die) in this experiment. Using Formula (14), we find that the required probabilities are

$$P(X = 0) = C(5, 0)\left(\frac{1}{3}\right)^0 \left(\frac{2}{3}\right)^5 = \frac{5!}{0!5!} \cdot 1 \cdot \frac{32}{243} \approx 0.132$$

$$P(X = 1) = C(5, 1)\left(\frac{1}{3}\right)^1 \left(\frac{2}{3}\right)^4 = \frac{5!}{1!4!} \cdot \frac{16}{243} \approx 0.329$$

$$P(X = 2) = C(5, 2)\left(\frac{1}{3}\right)^2 \left(\frac{2}{3}\right)^3 = \frac{5!}{2!3!} \cdot \frac{8}{243} \approx 0.329$$

$$P(X = 3) = C(5, 3)\left(\frac{1}{3}\right)^3 \left(\frac{2}{3}\right)^2 = \frac{5!}{3!2!} \cdot \frac{4}{243} \approx 0.165$$

$$P(X = 4) = C(5, 4)\left(\frac{1}{3}\right)^4 \left(\frac{2}{3}\right)^1 = \frac{5!}{4!1!} \cdot \frac{2}{243} \approx 0.041$$

$$P(X = 5) = C(5, 5)\left(\frac{1}{3}\right)^5 \left(\frac{2}{3}\right)^0 = \frac{5!}{5!0!} \cdot \frac{1}{243} \approx 0.004$$

b. Using these results, we find that the required binomial distribution associated with this experiment is given in Table 8.8. Using Table 8.8, we may construct the histogram associated with the probability distribution (Figure 8.8).

x	$P(X = x)$
0	0.132
1	0.329
2	0.329
3	0.165
4	0.041
5	0.004

Table 8.8

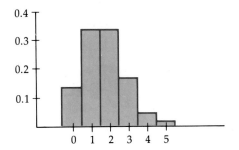

Figure 8.8

EXAMPLE

24 A fair die is cast five times. If a 1 or a 6 lands uppermost in a trial, then the throw is considered a success. Use the results from Example 23 to answer the following questions:

a. What is the probability of obtaining 0 or 1 success in the experiment?

b. What is the probability of obtaining at least 1 success in the experiment?

SOLUTION Interpreting the probability associated with the random variable X, when X assumes the value $X = a$, as the area of the rectangle centered about $X = a$ (see Figure 8.8), or otherwise, we find that

a. the probability of obtaining 0 or 1 success in the experiment is given by

$$P(X = 0) + P(X = 1) = 0.132 + 0.329$$
$$= 0.461$$

and

b. the probability of obtaining at least 1 success in the experiment is given by

$$P(X = 1) + P(X = 2) + P(X = 3) + P(X = 4) + P(X = 5)$$
$$= 0.329 + 0.329 + 0.165 + 0.041 + 0.004$$
$$= 0.868.$$ ◀

The following formulas (which we state without proof) will be useful in the solution of problems involving binomial experiments.

Mean, Variance, and Standard Deviation of a Random Variable X

If X is a binomial random variable associated with a binomial experiment consisting of n trials with probability of success p and probability of failure q, then the **mean** (expected value), **variance**, and **standard deviation** of X are:

$$\mu = E(X) = np \tag{15}$$
$$\text{Var}(X) = npq \tag{16}$$
$$\sigma_X = \sqrt{npq} \tag{17}$$

EXAMPLE

25 For the experiment in Examples 23 and 24, compute the mean, the variance, and the standard deviation of X, (a) using Formulas (15), (16), and (17), respectively, and (b) by direct computation using their respective definitions (see Sections 8.2 and 8.3).

SOLUTION a. We see Formulas (15), (16), and (17), with $p = 1/3$, $q = 2/3$, and $n = 5$, obtaining

$$\mu = E(X) = (5)\left(\frac{1}{3}\right) = \frac{5}{3}$$
$$\approx 1.67$$

$$\text{Var}(X) = (5)\left(\frac{1}{3}\right)\left(\frac{2}{3}\right) = \frac{10}{9}$$
$$\approx 1.11$$

and
$$\sigma_X = \sqrt{\text{Var}(X)}$$
$$= \sqrt{1.11}$$
$$\approx 1.05.$$

We leave it to you to interpret the results.

b. Using Equation (2) and the values of the probability distribution shown in Table 8.8, we find

$$\mu = E(X) = (0)(0.132) + (1)(0.329) + (2)(0.329)$$
$$+ (3)(0.165) + (4)(0.041) + (5)(0.004)$$
$$\approx 1.67,$$

which agrees with the result obtained in (a). Next, using Equation (5) and the knowledge that $\mu = 1.67$, we find

$$\text{Var}(X) = (0.132)(-1.67)^2 + (0.329)(-0.67)^2 + (0.329)(0.33)^2$$
$$+ (0.165)(1.33)^2 + (0.041)(2.33)^2 + (0.004)(3.33)^2$$
$$\approx 1.11$$

and
$$\sigma_X \approx \sqrt{1.11}$$
$$\approx 1.05,$$

which again agrees with the results obtained earlier. ◄

▶ Applications

We close this section by looking at several examples involving binomial experiments. In what follows, a pocket calculator may be used, or you may consult Table 2, Appendix C.

EXAMPLE

26 A division of the Solaron Corporation manufactures photovoltaic cells to use in the company's solar energy converters. It is estimated that 5 percent of the cells manufactured are defective. If a random sample of 20 is selected from a large lot of cells manufactured by the company, what is the probability that it will contain at most 2 defective cells?

SOLUTION This is a binomial experiment with $p = 0.05$, $q = 0.95$, and $n = 20$. Letting X denote the number of defective cells, we find that the probability of finding at most 2 defective cells in the sample of 20 is given by

$$P(X = 0) + P(X = 1) + P(X = 2)$$
$$= C(20, 0)(0.05)^0(0.95)^{20} + C(20, 1)(0.05)^1(0.95)^{19} + C(20, 2)(0.05)^2(0.95)^{18}$$
$$\approx 0.3585 + 0.3774 + 0.1887$$
$$= 0.9246.$$

Thus, for lots of photovoltaic cells manufactured by the company, approximately 93 percent of the samples will have at most 2 defective cells; equivalently, approximately 7 percent of the samples will contain more than 2 defective cells. ◀

EXAMPLE

27 The probability that a heart transplant performed at the Medical Center is successful (that is, the patient survives a year or more after undergoing such an operation) is 0.7. Of six patients who have recently undergone such an operation, what is the probability that a year from now

a. none of the heart recipients will be alive?

b. exactly three will be alive?

c. at least three will be alive?

d. all will be alive?

SOLUTION

Here $n = 6$, $p = 0.7$, and $q = 0.3$. Let X denote the number of successful operations. Then,

a. the probability that no heart recipients will be alive after one year is given by

$$P(X = 0) = C(6, 0)(0.7)^0(0.3)^6$$
$$= \frac{6!}{0!6!} \cdot 1 \cdot (0.3)^6$$
$$\approx 0.001.$$

b. the probability that exactly three will be alive after one year is given by

$$P(X = 3) = C(6, 3)(0.7)^3(0.3)^3$$
$$= \frac{6!}{3!3!} (0.7)^3(0.3)^3$$
$$\approx 0.19.$$

c. the probability that at least three will be alive after one year is given by

$$P(X = 3) + P(X = 4) + P(X = 5) + P(X = 6)$$
$$= C(6, 3)(0.7)^3(0.3)^3 + C(6, 4)(0.7)^4(0.3)^2$$
$$\quad + C(6, 5)(0.7)^5(0.3)^1 + C(6, 6)(0.7)^6(0.3)^0$$
$$= \frac{6!}{3!3!} (0.7)^3(0.3)^3 + \frac{6!}{4!2!} (0.7)^4(0.3)^2 + \frac{6!}{5!1!} (0.7)^5(0.3)^1$$
$$\quad + \frac{6!}{6!0!} (0.7)^6 \cdot 1$$
$$\approx 0.93.$$

d. the probability that all will be alive after one year is given by

$$P(X = 6) = C(6, 6)(0.7)^6(0.3)^0$$

$$= \frac{6!}{6!0!} (0.7)^6$$

$$\approx 0.12.$$

◀

EXAMPLE

28 The P.A.R. Bearings Company manufactures ball bearings, which are packaged in lots of a hundred each. The quality control department of the company has determined that 2 percent of the ball bearings manufactured do not meet the specifications imposed by a buyer. Find the average number of ball bearings per package that fails to meet with the specifications imposed by the buyer.

SOLUTION

The experiment under consideration is binomial. The average number of ball bearings per package that fails to meet with the specifications is therefore given by the expected value of the associated binomial random variable. Using Formula (15), we find that the required number is given by

$$\mu = E(X) = np$$
$$= (100)(0.02)$$
$$= 2,$$

or 2 substandard ball bearings in a package of 100.

◀

Self-Check Exercises

8.4

1. A binomial experiment consists of four independent trials. The probability of success in each trial is 0.2.

a. Find the probability of obtaining exactly 0, 1, 2, 3, and 4 successes, respectively, in this experiment.

b. Construct the binomial distribution and draw the histogram associated with this experiment.

c. Compute the mean and the standard deviation of the random variable associated with this experiment.

2. A recent survey shows that 60 percent of the households in a large metropolitan area have microwave ovens. If ten households are selected at random, what is the probability that five or fewer of these households have microwave ovens?

Solutions to Self-Check Exercises 8.4 can be found on page 458.

Exercises 8.4

In Exercises 1–5, determine whether the given experiment is a binomial experiment. Justify your answer.

1. Casting a fair die three times and observing the number of times a 6 is thrown

2. Casting a fair die and observing the number of times the die is thrown until a 6 appears uppermost

3. Casting a fair die three times and observing the number that appears uppermost

4. A card is selected from a deck of 52 cards and its color is observed. A second card is then drawn (without replacement) and its color is observed

5. Recording the number of accidents that occur at a given intersection on four clear days and one rainy day

In Exercises 6–10, use the formula $C(n, x)p^x q^{n-x}$ to determine the probability of the given event.

6. The probability of exactly three successes in six trials of a binomial experiment in which $p = 1/2$

7. The probability of at least three successes in six trials of a binomial experiment in which $p = 1/2$

8. The probability of no successful outcomes in five trials of a binomial experiment in which $p = 1/3$

9. The probability of no failures in five trials of a binomial experiment in which $p = 1/3$

10. The probability of at least one failure in five trials of a binomial experiment in which $p = 1/3$

11. A binomial experiment consists of five independent trials. The probability of success in each trial is 0.4.

    a. Find the probability of obtaining exactly 0, 1, 2, 3, 4, and 5 successes, respectively, in this experiment.

    b. Construct the binomial distribution and draw the histogram associated with this experiment.

    c. Compute the mean and the standard deviation of the random variable associated with this experiment.

12. Let the random variable X denote the number of girls in a five-child family. If the probability of a female birth is 0.5,

    a. find the probability of 0, 1, 2, 3, 4, and 5 girls in a five-child family.

    b. construct the binomial distribution and draw the histogram associated with this experiment.

    c. compute the mean and the standard deviation of the random variable X.

13. The probability that a fuse produced by a certain manufacturing process will be defective is 1/50. Is it correct to infer from this statement that there is at most 1 defective fuse in each lot of 50 produced by this process? Justify your answer.

14. Let X be the number of successes in five independent trials of a binomial experiment in which the probability of success is $p = 2/5$. Find

    a. $P(X = 4)$ *b.* $P(2 \le X \le 4)$

15. A fair die is cast four times. Calculate the probability of obtaining exactly two 6's.

16. If the probability that a certain tennis player will serve an ace is 1/4, what is the probability that he will serve exactly two aces out of five services?

17. Mayco, a mail order department store, has six telephone lines available for customers who wish to place their orders. If the probability that during business hours any one of the six telephone lines is engaged is 1/4, find the probability that when a customer calls to place an order all six lines will be in use.

18. From experience, the manager of Kramer's Book Mart knows that 40 percent of the people who are browsing in the store will make a purchase. What is the probability that among ten people who are browsing in the store, at least three will make a purchase?

19. An advertisement for brand A chicken noodle soup claims that 60 percent of all consumers prefer brand A over brand B, the chief competitor's product. To test this claim, Mr. Alan Horowitz, the host of "The Consumer Advocate," selected ten people at random from the audience. After tasting both soups, each person was asked to state his or her preference. Find the probability that

a. the company's claim was supported by the experiment; that is, six or more people stated a preference for brand A.

b. the company's claim was not supported by the experiment; that is, less than six people stated a preference for brand A.

20. In a certain congressional district, it is known that 40 percent of the registered voters classify themselves as conservatives. If ten registered voters are selected at random from this district, what is the probability that four of them will be conservatives?

21. It is estimated that one-third of the general population has blood type A^+. If a sample of nine people are selected at random, what is the probability that

a. exactly three of them have blood type A^+?

b. at most three of them have blood type A^+?

22. A biology quiz consists of eight multiple-choice questions. Five must be answered correctly to receive a passing grade. If each question has five possible answers of which only one is correct, what is the probability that a student who guesses at random on each question will pass the examination?

23. A psychology quiz consists of ten true-or-false questions. If a student knows the correct answer to six of the questions but determines the answer to the remaining questions by flipping a coin, what is the probability that she will obtain a score of at least 90 percent?

24. The probability that a video-disc player produced by the V.C.A. Television Company is defective is estimated to be 0.02. If a sample of ten sets is selected at random, what is the probability that the sample contains

a. no defectives?

b. at most two defectives?

25. As part of its quality-control program, the video cartridges produced by the Starr Communications Company are subjected to a final inspection before shipment. A sample of six cartridges is selected at random from each lot of cartridges produced and the lot is rejected if the sample contains one or more defective cartridges. If 1.5 percent of the cartridges produced by the company are defective, find the probability that a shipment will be accepted.

26. An automobile manufacturing company uses ten industrial robots as welders on its assembly line. On a given working day, the probability that a robot will be inoperative is .05. What is the probability that on a given working day

   a. exactly two robots are inoperative?

   b. more than two robots are inoperative?

27. The probability that an airplane engine will fail in a transcontinental flight is 0.001. Assuming that engine failures are independent of each other, what is the probability that, on a certain transcontinental flight, a four-engine plane will experience

   a. exactly one engine failure?

   b. exactly two engine failures?

   c. more than two engine failures? [*Note:* In this event, the airplane will crash!]

28. The manager of Toy World has decided to accept a shipment of electronic games if none of a random sample of 20 is found to be defective.

   a. What is the probability that he will accept the shipment if 10 percent of the electronic games are defective?

   b. What is the probability that he will accept the shipment if 5 percent of the electronic games are defective?

29. Refer to Exercise 28. If the manager's criterion for accepting shipment is that there be no more than 1 defective electronic game in a random sample of 20, what is the probability that he will accept the shipment if 10 percent of the electronic games are defective?

30. Refer to Exercise 28. If the manager of the store changes his sample size to ten and decides to accept shipment if none of the games is defective, what is the probability that he will accept the shipment if 10 percent of the games are defective?

31. How many times must a person toss a coin if the chances of obtaining at least one head are 99 percent or better?

32. A new drug has been found to be effective in treating 75 percent of the people afflicted by a certain disease. If the drug is administered to 500 people who have this disease, what are the mean and the standard deviation of the number of people for whom the drug can be expected to be effective?

33. At a certain university the probability that an entering freshman will graduate within 4 years is 0.6. From an incoming class of 2000 freshmen, find

   a. the expected number of students who will graduate within 4 years.

   b. the standard deviation of the number of students who will graduate within 4 years.

Solutions to Self-Check Exercises

8.4

1. *a.* We use Formula (14) with $n = 4$, $p = 0.2$, and $q = 1 - 0.2 = 0.8$, obtaining

$$P(X = 0) = C(4, 0)(0.2)^0(0.8)^4 = \frac{4!}{0!4!} \cdot 1 \cdot (0.8)^4 \approx 0.410$$

$$P(X = 1) = C(4, 1)(0.2)^1(0.8)^3 = \frac{4!}{1!3!}(0.2)(0.8)^3 \approx 0.410$$

$$P(X = 2) = C(4, 2)(0.2)^2(0.8)^2 = \frac{4!}{2!2!}(0.2)^2(0.8)^2 \approx 0.154$$

$$P(X = 3) = C(4, 3)(0.2)^3(0.8) = \frac{4!}{3!1!}(0.2)^3(0.8) \approx 0.026$$

$$P(X = 4) = C(4, 4)(0.2)^4(0.8)^0 = \frac{4!}{4!0!}(0.2)^4 \cdot 1 \approx 0.002$$

b. The required binomial distribution and histogram are as follows:

x	0	1	2	3	4
$P(X = x)$	0.410	0.410	0.154	0.026	0.002

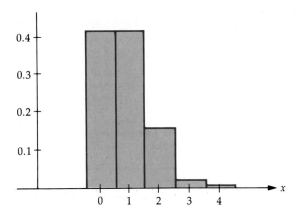

c. The mean is

$$\mu = E(X) = np = (4)(0.2)$$
$$= 0.8,$$

and the standard deviation is

$$\sigma = \sqrt{npq} = \sqrt{(4)(0.2)(0.8)}$$
$$= 0.8.$$

2. This is a binomial experiment with $n = 10$, $p = 0.6$, and $q = 0.4$. Let X denote the number of households that have microwave ovens. Then, the probability that 5 or fewer households have microwave ovens is given by

$$P(X = 0) + P(X = 1) + P(X = 2) + P(X = 3) + P(X = 4) + P(X = 5)$$
$$= C(10, 0)(0.6)^0(0.4)^{10} + C(10, 1)(0.6)^1(0.4)^9 + C(10, 2)(0.6)^2(0.4)^8$$
$$+ C(10, 3)(0.6)^3(0.4)^7 + C(10, 4)(0.6)^4(0.4)^6 + C(10, 5)(0.6)^5(0.4)^5$$
$$\approx 0 + 0.002 + 0.011 + 0.042 + 0.111 + 0.201$$
$$\approx 0.37.$$

8.5

The Normal Distribution

▶ Probability Density Functions

▶ Normal Distributions

▶ Computations of Probabilities Associated with Normal Distributions

▶ Probability Density Functions

The probability distributions discussed in the preceding sections were all associated with finite random variables; that is, random variables that take on finitely many values. Such probability distributions are referred to as **finite probability distributions.** In this section, we will consider probability distributions associated with a continuous random variable; that is, a random variable that may take on any value lying in an interval of real numbers. Such probability distributions are called *continuous probability distributions.*

Unlike a finite probability distribution, which may be exhibited in the form of a table, a continuous probability distribution is defined by a function f whose domain coincides with the interval of values taken on by the random variable associated with the experiment. Such a function f is called the **probability density function** associated with the probability distribution and has the following properties:

1. $f(x)$ is nonnegative for all values of x.

2. The area of the region between the graph of f and the x-axis is equal to 1. (See Figure 8.9.)

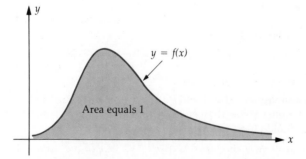

Figure 8.9

Now suppose we are given a continuous probability distribution defined by some probability density function f (see Figure 8.10). Then the probability that the random variable X assumes a value in an interval $a < x < b$ is given by the area of the region between the graph of f and the x-axis from $x = a$ to $x = b$. We denote the value of this probability by $P(a < X < b)$. Because the area under one point of the graph of f is equal to zero, we see immediately that, in fact, $P(a < X < b) = P(a < X \le b) = P(a \le X < b) = P(a \le X \le b)$. Observe that Property 2 of the probability density function states that the probability that a continuous random variable takes on a value lying in its range is 1, a certainty, which is expected. Note the analogy between the areas under the probability density curves and the histograms associated with finite probability distributions (see Section 8.1).

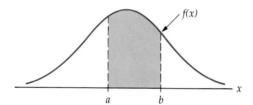

Figure 8.10

▶ Normal Distributions

The mean μ and the standard deviation σ of a continuous probability distribution have roughly the same meanings as the mean and standard deviation of a finite probability distribution. Thus, the mean of a continuous probability distribution is a measure of central tendency of the probability distribution, and the standard deviation of the probability distribution measures its spread about its mean. Both these numbers will play an important role in the following discussion.

For the remainder of this section we will discuss a special class of continuous probability distributions known as **normal distributions.** The normal distribution is without doubt the most important of all the probability distributions. Many phenomena such as the heights of people in a given population, the weights of newborn infants, the IQ of college students, the actual weights of 16-ounce packages of cereals, and so on, have probability distributions that are normal. The normal distribution also provides us with an accurate approximation to the distributions of many random variables associated with random-sampling problems. In fact, in the next section we will see how a normal distribution may be used to approximate a binomial distribution under certain conditions.

The graph of a normal distribution, which is bell shaped, is called a **normal curve** (see Figure 8.11).

The normal curve (and therefore the corresponding normal distribution) is completely determined by its mean μ and standard deviation σ. In fact, the

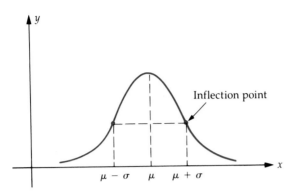

Figure 8.11

normal curve has the following characteristics, described in terms of these two parameters:[†]

1. The curve has a peak at $x = \mu$.

2. The curve is symmetrical with respect to the vertical line $x = \mu$.

3. The curve is *concave downward* (it bends downward) between $x = \mu - \sigma$ and $x = \mu + \sigma$ and *concave upward* (it bends upward) for $x < \mu - \sigma$ and for $x > \mu + \sigma$.

4. The curve has two *inflection points* (points where its concavity changes—see Property 3) located a distance of σ units on either side of its mean.

5. The curve is *asymptotic* to the x-axis; that is, as $|x|$ grows without bound the curve gets closer and closer to the x-axis but always stays above it.

Figure 8.12 shows two normal distributions with different means μ_1 and μ_2 but the same standard deviation. Next, Figure 8.13 shows two normal distributions with the same mean but different standard deviations σ_1 and σ_2. (Which number is smaller?)

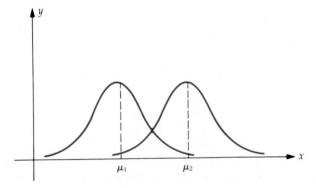

Figure 8.12

[†]The probability density function associated with this normal curve is given by

$$y = \frac{1}{\sigma\sqrt{2\pi}} e^{-(1/2)[(x-\mu)/\sigma]^2},$$

but its direct use will not be required in our discussion of the normal distribution.

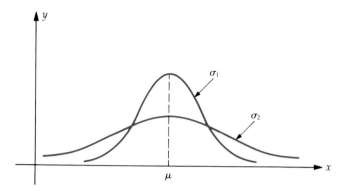

Figure 8.13

In general, the mean μ of a normal distribution determines where the center of the curve is located, whereas the standard deviation σ of a normal distribution determines the sharpness (or flatness) of the curve.

As this discussion reveals, there are infinitely many normal curves corresponding to different choices of the parameters μ and σ, which characterize such curves. Fortunately, any normal curve may be easily transformed into any other normal curve (as we will see later on), so in the study of normal curves (and their corresponding normal distributions) it suffices to single out one such particular curve for special attention. Such a normal curve is chosen to have mean $\mu = 0$ and standard deviation $\sigma = 1$ and is called the **standard normal curve.** The corresponding distribution is called the **standard normal distribution.** The random variable itself is called the *standard normal variable* and is commonly denoted by Z.

▶ Computations of Probabilities Associated with Normal Distributions

Areas under the standard normal curve have been extensively computed and tabulated. Table 3, Appendix C gives the areas of the regions under the standard normal curve to the left of the number z; these areas correspond, of course, to probabilities of the form $P(Z < z)$ or $P(Z \leq z)$. The next several examples illustrate the use of this table in computations involving the probabilities associated with the standard normal variable.

EXAMPLE

29 Let Z be the standard normal variable. By first making a sketch of the appropriate region under the standard normal curve, find the values of

a. $P(Z < 1.24)$

b. $P(Z > 0.5)$

c. $P(0.24 < Z < 1.48)$

d. $P(-1.65 < Z < 2.02)$

SOLUTION

a. The region under the standard normal curve associated with the probability $P(Z < 1.24)$ is shown in Figure 8.14. To find the area of the required region using Table 3, Appendix C, we first locate the number 1.2 in the column and the number 0.04 in the row, both headed by z, and read off the number 0.8925 appearing in the body of the table. Thus,

$$P(Z < 1.24) = 0.8925.$$

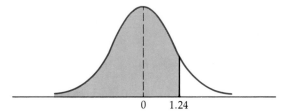

Figure 8.14

b. The region under the standard normal curve associated with the probability $P(Z > 0.5)$ is shown in Figure 8.15(a). Observe, however, that the required area is, by virtue of the symmetry of the standard normal curve, equal to the shaded area shown in Figure 8.15(b).

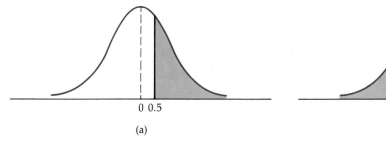

(a)   (b)

Figure 8.15

Thus,

$$P(Z > 0.5) = P(Z < -0.5)$$
$$= 0.3085.$$

c. The probability $P(0.24 < Z < 1.48)$ is equal to the shaded area shown in Figure 8.16. But this area is obtained by subtracting the area under the curve to the left of $z = 0.24$ from the area under the curve to the left of $z = 1.48$; that is,

$$P(0.24 < Z < 1.48) = P(Z < 1.48) - P(Z < 0.24)$$
$$= 0.9306 - 0.5948$$
$$= 0.3358.$$

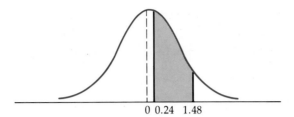

Figure 8.16

0   0.24   1.48

d. The probability $P(-1.65 < Z < 2.02)$ is given by the shaded area shown in Figure 8.17. We have

$$P(-1.65 < Z < 2.02) = P(Z < 2.02) - P(Z < -1.65)$$
$$= 0.9783 - 0.0495$$
$$= 0.9288.$$

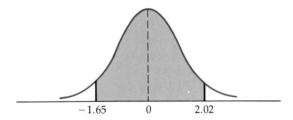

Figure 8.17

−1.65   0   2.02   ◄

EXAMPLE

30 Let Z be the standard normal variable. Find the value of z if z satisfies

a. $P(Z < z) = 0.9474$
b. $P(Z > z) = 0.9115$
c. $P(-z < Z < z) = 0.7888$

SOLUTION   *a.* Refer to Figure 8.18. We want the value of Z such that the area of the region under the standard normal curve and to the left of $Z = z$ is 0.9474. Locating the number 0.9474 in Table 3, Appendix C, and reading back, we find that $z = 1.62$.

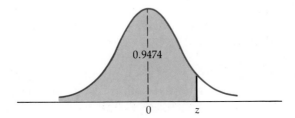

Figure 8.18

0.9474

0   z

b. Since $P(Z > z)$, or equivalently, the area of the region to the right of z is greater than 0.5, z must be negative (see Figure 8.19).

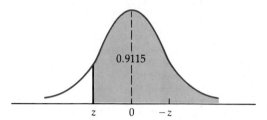

0.9115

z 0 $-z$

Figure 8.19

Therefore $-z$ is positive. Furthermore, the area of the region to the right of z is the same as the area of the region to the left of $-z$. Therefore

$$P(Z > z) = P(Z < -z)$$
$$= 0.9115.$$

Looking up the table, we find $-z = 1.35$, so $z = -1.35$.

c. The region associated with $P(-z < Z < z)$ is shown in Figure 8.20. Observe that the area of the region to the left of z is equal to the area to the left of $-z$ plus the area of the region between $-z$ and z. Thus,

$$P(Z < z) = P(Z < -z) + P(-z < Z < z)$$
$$= [1 - P(Z < z)] + P(-z < Z < z) \qquad \text{Why?}$$

Therefore, $2P(Z < z) = 1 + P(-z < Z < z)$

and $P(Z < z) = \dfrac{1}{2}[1 + P(-z < Z < z)]$

$$= \frac{1}{2}(1 + 0.7888)$$

$$= 0.8944.$$

Consulting the table, we find $z = 1.25$.

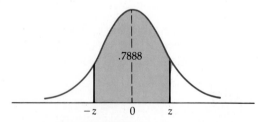

.7888

$-z$ 0 z

Figure 8.20

We now turn our attention to the computation of probabilities associated with normal distributions whose means and standard deviations are not necessarily equal to 0 and 1, respectively. As mentioned earlier, any normal curve may be transformed into the standard normal curve by means of a suitable transformation. In particular, it may be shown that if X is a normal random variable with mean μ and standard deviation σ, then it can be transformed into the standard normal random variable Z by means of the transformation

$$Z = \frac{X - \mu}{\sigma} \tag{18}$$

Under the transformation (18) the numbers a and b are mapped onto the numbers $(a - \mu)/\sigma$ and $(b - \mu)/\sigma$, respectively (see Figure 8.21). Moreover, the transformation (18) has the important property of being area preserving. The area of the region under the normal curve (with random variable X) between $x = a$ and $x = b$ is *equal* to the area of the region under the standard normal curve between $z = (a - \mu)/\sigma$ and $z = (b - \mu)/\sigma$. In terms of probabilities associated with these distributions, we have

$$P(a < X < b) = P\left(\frac{a - \mu}{\sigma} < Z < \frac{b - \mu}{\sigma}\right).$$

Thus, with the help of the transformation (18), computations of probabilities associated with any normal distribution may be reduced to the computations of areas of regions under the standard normal curve.

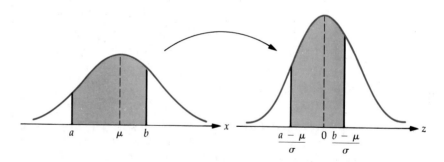

Figure 8.21

EXAMPLE

31 Suppose X is a normal random variable with $\mu = 100$ and $\sigma = 20$. Find the values of

a. $P(X < 120)$ *b.* $P(X > 70)$ *c.* $P(75 < X < 110)$

SOLUTION Using the transformation (18) and the table of values of Z, we have

a.
$$P(X < 120) = P\left(Z < \frac{120 - 100}{20}\right)$$
$$= P(Z < 1)$$
$$= 0.8413.$$

b.
$$P(X > 70) = P\left(Z > \frac{70 - 100}{20}\right)$$
$$= P(Z > -1.5)$$
$$= P(Z < 1.5)$$
$$= 0.9332.$$

c.
$$P(75 < X < 110) = P\left(\frac{75 - 100}{20} < Z < \frac{110 - 100}{20}\right)$$
$$= P(-1.25 < Z < 0.5)$$
$$= P(Z < 0.5) - P(Z < -1.25) \quad \text{(See Figure 8.22)}$$
$$= 0.6915 - 0.1056$$
$$= 0.5859.$$

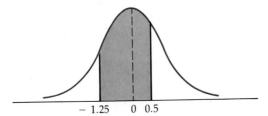

Figure 8.22
$-1.25 \quad 0 \quad 0.5$

Self-Check Exercises

8.5

1. Let Z be a standard normal variable.
   a. Find the value of $P(-1.2 < Z < 2.1)$ by first making a sketch of the appropriate region under the standard normal curve.
   b. Find the value of z if z satisfies $P(-z < Z < z) = 0.8764$.
2. Let X be a normal random variable with $\mu = 80$ and $\sigma = 10$. Find the values of
   a. $P(X < 100)$ b. $P(X > 60)$ c. $P(70 < X < 90)$

Solutions to Self-Check Exercises 8.5 can be found on page 470.

▶ Exercises 8.5

In Exercises 1–6, find the value of the probability of the standard normal variable Z corresponding to the shaded area under the standard normal curve.

1. $P(Z < 1.45)$

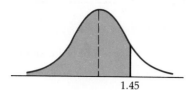

1.45

2. $P(Z > 1.11)$

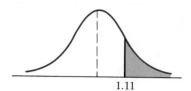

1.11

3. $P(Z < -1.75)$

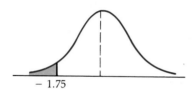

-1.75

4. $P(0.3 < Z < 1.83)$

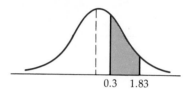

0.3   1.83

5. $P(-1.32 < Z < 1.74)$

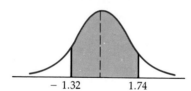

-1.32 1.74

6. $P(-2.35 < Z < -0.51)$

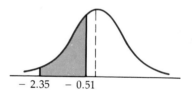

-2.35 -0.51

In Exercises 7–14, (a) make a sketch of the area under the standard normal curve corresponding to the given probability and (b) find the value of the probability of the standard normal variable Z corresponding to this area.

7. $P(Z < 1.37)$

8. $P(Z > 2.24)$

9. $P(Z < -0.65)$

10. $P(0.45 < Z < 1.75)$

11. $P(Z > -1.25)$

12. $P(-1.48 < Z < 1.54)$

13. $P(0.68 < Z < 2.02)$

14. $P(-1.41 < Z < -0.24)$

15. Let Z be the standard normal variable. Find the values of z if z satisfies

   a. $P(Z < z) = 0.8907$

   b. $P(Z < z) = 0.2090$

16. Let Z be the standard normal variable. Find the values of z if z satisfies

   a. $P(Z > z) = 0.9678$

   b. $P(-z < Z < z) = 0.8354$

17. Let Z be the standard normal variable. Find the values of z if z satisfies

   a. $P(Z > -z) = 0.9713$

   b. $P(Z < -z) = 0.9713$

18. Suppose X is a normal random variable with $\mu = 380$ and $\sigma = 20$. Find the value of

   a. $P(X < 405)$ **b.** $P(400 < X < 430)$ **c.** $P(X > 400)$

19. Suppose X is a normal random variable with $\mu = 50$ and $\sigma = 5$. Find the value of

   a. $P(X < 60)$ **b.** $P(X > 43)$ **c.** $P(46 < X < 58)$

20. Suppose X is a normal random variable with $\mu = 500$ and $\sigma = 75$. Find the value of

   a. $P(X < 750)$ **b.** $P(X > 350)$ **c.** $P(400 < X < 600)$

Solutions to Self-Check Exercises

8.5

1. a. The probability $P(-1.2 < Z < 2.1)$ is given by the shaded area in the following figure.

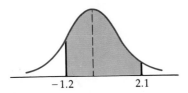

-1.2 2.1

We have

$$P(-1.2 < Z < 2.1) = P(Z < 2.1) - P(Z < -1.2)$$
$$= 0.9821 - 0.1151$$
$$= 0.867.$$

b. The region associated with $P(-z < Z < z)$ is shown in the following figure.

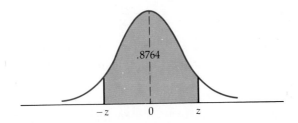

.8764

$-z$ 0 z

Observe that we have the following relationship:

$$P(Z < z) = \tfrac{1}{2}[1 + P(-z < Z < z)]$$

[See Example 30(c)]. With $P(-z < Z < z) = 0.8764$, we find

$$P(Z < z) = \tfrac{1}{2}(1 + 0.8764)$$
$$= 0.9382.$$

Consulting the table, we find $z = 1.54$.

2. Using the transformation (18) and the table of values of Z, we have

a. $P(X < 100) = P\left(Z < \dfrac{100 - 80}{10}\right)$
$$= P(Z < 2)$$
$$= 0.9772.$$

b. $P(X > 60) = P\left(Z > \dfrac{60 - 80}{10}\right)$

$\qquad\qquad = P(Z > -2)$

$\qquad\qquad = P(Z < 2)$

$\qquad\qquad = 0.9772.$

c. $P(70 < X < 90) = P\left(\dfrac{70 - 80}{10} < Z < \dfrac{90 - 80}{10}\right)$

$\qquad\qquad\qquad = P(-1 < Z < 1)$

$\qquad\qquad\qquad = P(Z < 1) - P(Z < -1)$

$\qquad\qquad\qquad = 0.8413 - 0.1587$

$\qquad\qquad\qquad = 0.6826.$

8.6

Applications of the Normal Distribution

▶ Applications Involving Normal Random Variables

▶ Approximating Binomial Distributions

▶ Applications Involving Binomial Random Variables

▶ Applications Involving Normal Random Variables

In this section we will look at some applications involving the normal distribution.

EXAMPLE

32 The medical records of infants delivered at the Kaiser Memorial Hospital show that the infants' birth weights in pounds are normally distributed with a mean of 7.4 and a standard deviation of 1.2. Find the probability that an infant selected at random from among those delivered at the hospital weighed more than 9.2 pounds at birth.

SOLUTION

Let X be the normal random variable denoting the birth weights of infants delivered at the hospital. Then the probability that an infant selected at random has a birth weight of more than 9.2 pounds is given by $P(X > 9.2)$. To compute

$P(X > 9.2)$ we use the transformation (18), keeping in mind that, in this case, $\mu = 7.4$ and $\sigma = 1.2$. We find

$$P(X > 9.2) = P\left(Z > \frac{9.2 - 7.4}{1.2}\right)$$
$$= P(Z > 1.5)$$
$$= P(Z < -1.5)$$
$$= 0.0668.$$

Thus, the probability that an infant delivered at the hospital weighs more than 9.2 pounds is 0.0668. ◀

EXAMPLE

33 The Idaho Natural Produce Corporation ships potatoes to its distributors in bags with a mean weight of 50 pounds and standard deviation of 0.5 pounds. If a bag of potatoes is selected at random from a shipment, what is the probability that it weighs

a. more than 51 pounds?

b. less than 48 pounds?

c. between 49 and 51 pounds?

SOLUTION

Let X denote the weight of potatoes packed by the company. Then the mean and standard deviation of X are $\mu = 50$ and $\sigma = 0.5$, respectively.

a. The probability that a bag selected at random weighs more than 51 pounds is given by

$$P(X > 51) = P\left(Z > \frac{51 - 50}{0.5}\right)$$
$$= P(Z > 2)$$
$$= P(Z < -2)$$
$$= 0.0228.$$

b. The probability that a bag selected at random weighs less than 48 pounds is given by

$$P(X < 48) = P\left(Z < \frac{48 - 50}{0.5}\right)$$
$$= P(Z < -4)$$
$$= 0.$$

c. The probability that a bag selected at random weighs between 49 and 51 pounds is given by

$$P(49 < X < 51) = P\left(\frac{49 - 50}{0.5} < Z < \frac{51 - 50}{0.5}\right)$$
$$= P(-2 < Z < 2)$$
$$= P(Z < 2) - P(Z < -2)$$
$$= 0.9772 - 0.0228$$
$$= 0.9544.$$
◄

EXAMPLE

34 The grade-point average of the senior class of Jefferson High School is normally distributed with a mean of 2.7 and a standard deviation of 0.4 points. If a senior in the top 10 percent of his or her class is eligible for admission to any of the nine campuses of the State University system, what is the minimum grade-point average that a senior should have to ensure eligibility for admission to the State University system?

SOLUTION Let X denote the grade-point average of a randomly selected senior at Jefferson High School and let x denote the minimum grade-point average to ensure his or her eligibility for admission to the university. Since only the top 10 percent are eligible for admission, x must satisfy the equation

$$P(X \geq x) = 0.1.$$

Using transformation (18) with $\mu = 2.7$ and $\sigma = 0.4$, we find

$$P(X \geq x) = P\left(Z \geq \frac{x - 2.7}{0.4}\right) = 0.1.$$

But, this is equivalent to the equation

$$P\left(Z \leq \frac{x - 2.7}{0.4}\right) = 0.9 \qquad \text{Why?}$$

Consulting Table 3, Appendix C, we find

$$\frac{x - 2.7}{0.4} = 1.28.$$

(Compare results with Example 30.) Upon solving for x, we obtain

$$x = (1.28)(0.4) + 2.7$$
$$\approx 3.2.$$

Thus, to ensure eligibility for admission to one of the nine campuses of the State University system, a senior at Jefferson High School should have a minimum 3.2 grade-point average. ◄

▶ Approximating Binomial Distributions

As mentioned in the last section, one important application of the normal distribution is that, under certain conditions, it provides us with an accurate approximation of other continuous probability distributions. We will now show how a binomial distribution may be approximated by a suitable normal distribution. This technique leads to a convenient and simple solution to certain problems involving binomial probabilities.

Recall that a binomial distribution is a probability distribution of the form

$$P(X = x) = C(n, x)p^x q^{n-x} \quad (x = 0, 1, 2, \ldots, n) \tag{19}$$

(See Section 8.4.) For small values of n, the arithmetic computations of the binomial probabilities may be done with relative ease. However, if n is large, then the work involved becomes prodigious, even when tables of $P(X = x)$ are available. For example, if $n = 50$, $p = 0.3$, then the probability of ten or more successes is given by

$$P(X \geq 10) = P(X = 10) + P(X = 11) + \cdots + P(X = 50)$$
$$= \frac{50!}{10!40!}(0.3)^{10}(0.7)^{40} + \frac{50!}{11!39!}(0.3)^{11}(0.7)^{39} + \cdots$$
$$+ \frac{50!}{50!0!}(0.3)^{50}(0.7)^0.$$

To see how the normal distribution helps us in such situations, let us consider the coin-tossing experiment. Suppose a fair coin is tossed 20 times and we wish to compute the probability of obtaining 10 or more heads. The solution to this problem may be obtained, of course, by computing

$$P(X \geq 10) = P(X = 10) + P(X = 11) + \cdots + P(X = 20).$$

The inconvenience of this approach for solving the problems at hand has already been pointed out. As an alternative solution, let us begin by interpreting the solution in terms of finding the area of suitable rectangles of the histogram for the distribution associated with the problem. Using Formula (19), we compute the probability of obtaining exactly x heads in 20 tosses of the coin. The results lead to the binomial distribution displayed in Table 8.9.

x	$P(X = x)$	x	$P(X = x)$
0	0.0000	8	0.1201
1	0.0000	9	0.1602
2	0.0002	10	0.1762
3	0.0011	11	0.1602
4	0.0046	12	0.1201
5	0.0148		
6	0.0370	⋮	⋮
7	0.0739	20	0.0000

Table 8.9

Using the data from Table 8.9, we next construct the histogram for the distribution (see Figure 8.23). The probability of obtaining 10 or more heads in 20 tosses of the coin is equal to the sum of the areas of the shaded rectangles of the histogram of the binomial distribution shown in Figure 8.24.

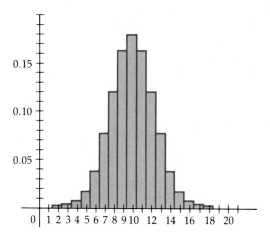

Figure 8.23

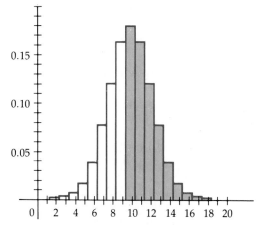

Figure 8.24

Next, let us observe that the shape of the histogram suggests that the binomial distribution under consideration may be approximated by a suitable normal distribution. Since the mean and standard deviation of the binomial distribution are given by

$$\mu = np$$
$$= (20)(0.5)$$
$$= 10$$

and

$$\sigma = \sqrt{npq}$$
$$= \sqrt{(20)(0.5)(0.5)}$$
$$= 2.24,$$

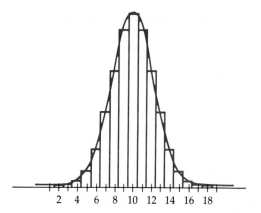

Figure 8.25

respectively (see Section 8.4), the natural choice of a normal curve for this purpose is one with a mean of 10 and standard deviation of 2.24. Figure 8.25 shows such a normal curve superimposed on the histogram of the binomial distribution.

The good fit suggests that the sum of the areas of the rectangles representing $P(X \geq 10)$, the probability of obtaining 10 or more heads in 20 tosses of the coin, may be approximated by the area of an appropriate region under the normal curve. To determine this region let us note that the base of the portion of the histogram representing the required probability extends from $x = 9.5$ on, since the base of the leftmost rectangle is centered at $x = 10$ and the base of each rectangle has length 1. (See Figure 8.26.) Therefore the required region under the normal curve should also have $x \geq 9.5$. Letting Y denote the continuous normal variable, we have

$$
\begin{aligned}
P(X \geq 10) &\approx P(Y \geq 9.5) \\
&= P(Y > 9.5) \\
&= P\left(Z > \frac{9.5 - 10}{2.24}\right) \\
&= P(Z > -0.22) \\
&= P(Z < 0.22) \\
&= 0.5871.
\end{aligned}
$$

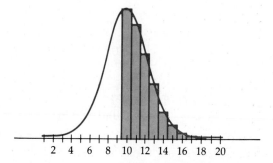

Figure 8.26

The exact value of $P(X \geq 10)$ may be found by computing

$$P(X = 10) + P(X = 11) + \cdots + P(X = 20)$$

in the usual fashion and is equal to 0.5881. Thus, the normal distribution with suitably chosen mean and standard deviation does provide us with a good approximation of the binomial distribution.

In the general case, the following result, which is a special case of the *central limit theorem*, guarantees the accuracy of the approximation of a binomial distribution by a normal distribution under certain conditions.

Theorem

Suppose we are given a binomial distribution associated with a (binomial) experiment involving n trials each with a probability of success p and probability of failure q. Then if n is large and p is not close to 0 or 1, the binomial distribution may be approximated by a normal distribution with

$$\mu = np \quad \text{and} \quad \sigma = \sqrt{npq}$$

In fact, it can be shown that if both np and nq are greater than 5, then the error resulting from this approximation is negligible.

▶ Applications Involving Binomial Random Variables

EXAMPLE

35   An automobile manufacturer receives the microprocessors used to regulate fuel consumption in its automobiles in shipments of 1000 each from a certain supplier. It has been estimated that, on the average, 1 percent of the microprocessors manufactured by the supplier are defective. Determine the probability that more than 20 of the microprocessors in a single shipment are defective.

SOLUTION   Let X denote the number of defective microprocessors in a single shipment. Then X has a binomial distribution with $n = 1000$, $p = 0.01$, and $q = 0.99$, so

$$\mu = (1000)(0.01) = 10$$

and

$$\sigma = \sqrt{(1000)(0.01)(0.99)}$$
$$\approx 3.15.$$

Approximating the binomial distribution by a normal distribution with a mean of 10 and a standard deviation of 3.15, we find that the probability that more than 20 microprocessors in a shipment are defective is given by

$$P(X > 20) \approx P(Y > 20.5) \quad \text{(Where } Y \text{ denotes the normal random variable)}$$
$$= P\left(Z > \frac{20.5 - 10}{3.15}\right)$$
$$= P(Z > 3.33)$$
$$= P(Z < -3.33)$$
$$= 0.0004.$$

In other words, approximately 0.04 percent of the shipments containing 1000 microprocessors each will contain 20 or more defective units. ◀

EXAMPLE

36 The probability that a heart transplant performed at the Medical Center is successful (that is, the patient survives a year or more after undergoing the surgery) is 0.7. Of 100 patients who had undergone such an operation, what is the probability that

a. less than 75 would survive a year or more after the operation?

b. between 80 and 90, inclusive, would survive a year or more after the operation?

SOLUTION

Let X denote the number of patients who survive a year or more after undergoing a heart transplant at the Medical Center. Then X is a binomial random variable. Also, $n = 100$, $p = 0.7$, and $q = 0.3$, so

$$\mu = (100)(0.7) = 70$$
and
$$\sigma = \sqrt{(100)(0.7)(0.3)}$$
$$\approx 4.58.$$

Approximating the binomial distribution by a normal distribution with a mean of 70 and a standard deviation of 4.58, we find upon letting Y denote the associated normal random variable:

a. The probability that less than 75 patients would survive a year or more is given by

$$P(X < 75) \approx P(Y < 74.5) \quad \text{(Why?)}$$
$$= P\left(Z < \frac{74.5 - 70}{4.58}\right)$$
$$= P(Z < 0.98)$$
$$= 0.8365.$$

b. The probability that between 80 and 90, inclusive, of the patients would survive a year or more is given by

$$P(80 \leq X \leq 90) \approx P(79.5 < Y < 90.5)$$
$$= P\left(\frac{79.5 - 70}{4.58} < Z < \frac{90.5 - 70}{4.58}\right)$$
$$= P(2.07 < Z < 4.48)$$
$$= P(Z < 4.48) - P(Z < 2.07)$$
$$= 1 - 0.9808 \quad [Note: P(Z < 4.48) \approx 1]$$
$$= 0.0192.$$

Self-Check Exercises

8.6

1. The serum cholesterol levels in mg/dl in a current Mediterranean population are found to be normally distributed with a mean of 160 and a standard deviation of 50. Scientists at the National Heart, Lung, and Blood Institute consider this pattern ideal for a minimal risk of heart attacks. Find the percentage of the population having blood cholesterol levels between 160 and 180 mg/dl.

2. It has been estimated that 4 percent of the luggage manufactured by The Luggage Company fail to meet the standards established by the company and are sold as "seconds" to discount and outlet stores. What is the probability that in a production of 500 bags, more than 30 will be classified as "seconds"?

Solutions to Self-Check Exercises 8.6 can be found on page 482.

Exercises 8.6

1. The medical records of infants delivered at the Kaiser Memorial Hospital show that the infants' lengths at birth (in inches) are normally distributed with a mean of 20 and a standard deviation of 2.6. Find the probability that an infant selected at random from among those delivered at the hospital measured

   a. more than 22 inches.

   b. less than 18 inches.

   c. between 19 and 21 inches.

2. According to the data released by the Chamber of Commerce of a certain city, the weekly wages of factory workers are normally distributed with a mean of $300 and a standard deviation of $50. Find the probability that a worker selected at random from the city makes a weekly wage of

   a. less than $200.

   b. more than $360.

   c. between $250 and $350.

3. The TKK Products Corporation manufactures electric light bulbs in the 50, 60, 75, and 100 watt range. Laboratory tests show that the lives of these light bulbs are normally distributed with a mean of 750 hours and a standard deviation of 75 hours. What is the probability that a TKK light bulb selected at random will burn

   a. for more than 900 hours?

   b. for less than 600 hours?

   c. between 750 and 900 hours?

   d. between 600 and 800 hours?

4. On the average, a student takes 100 words per minute midway through an advanced shorthand course at the American Institute of Stenography. Assuming that the dictation speeds of the students are normally distributed and that the standard deviation is 20 words per minute, find the probability that a student randomly selected from the course could take dictation at a speed of

   a. more than 120 words per minute.

   b. between 80 and 120 words per minute.

   c. less than 80 words per minute.

5. The IQ's of students at the Wilson Elementary School were measured recently and found to be normally distributed with a mean of 100 and a standard deviation of 15. What is the probability that a student selected at random will have an IQ of

   a. 140 or higher?

   b. 120 or higher?

   c. between 100 and 120?

   d. 90 or less?

6. The tread lives of the Super Titan radial tires under normal driving conditions are normally distributed with a mean of 40,000 miles and a standard deviation of 2,000 miles. What is the probability that a tire selected at random will have a tread life of more than 35,000 miles? If four new tires are installed in a car and they experience even wear, determine the probability that all four tires still have useful tread lives after 35,000 miles of driving.

7. According to data released by the Chamber of Commerce of a certain city, the weekly wages (in dollars) of female factory workers are normally distributed with a mean of 275 and a standard deviation of 50. Find the probability that a female factory worker selected at random from the city makes a weekly wage of $250 to $350.

8. To be eligible for further consideration, applicants for certain Civil Service positions must first pass a written qualifying examination, on which a score of 70 or more must be obtained. In a recent examination, it was found that the scores were normally distributed with a mean of 60 points and a standard deviation of 10 points. Determine the percentage of applicants who passed the written qualifying examination.

9. The general manager of the Service Department of the MCA Television Company has estimated that the time that elapses between the dates of purchase and the dates on which the 19-inch sets manufactured by the company first require service is normally distributed with a mean of 22 months and a standard deviation of 4 months. If the company gives a one-year warranty on parts and labor for these sets, determine the percentage of sets manufactured and sold by the company that may require service before the warranty period runs out.

10. The scores on an economics examination are normally distributed with a mean of

72 and a standard deviation of 16. If the instructor assigns a grade of A to 10 percent of the class, what is the lowest score a student may have and still obtain an A?

11. The scores on a sociology examination are normally distributed with a mean of 70 and a standard deviation of 10. If the instructor assigns A's to 15 percent, B's to 25 percent, C's to 40 percent, D's to 15 percent, and F's to 5 percent of the class, find the cutoff points for these grades.

In Exercises 12–22, use the appropriate normal distributions to approximate the resulting binomial distributions.

12. A fair coin is tossed 20 times. Find the probability of obtaining

   a. less than 8 heads.

   b. more than 6 heads.

   c. between 6 and 10 heads inclusive.

13. A coin is weighted so that the probability of obtaining a head in a single toss is 0.4. If the coin is tossed 25 times, determine the probability of obtaining

   a. less than 10 heads.

   b. between 10 and 12 heads inclusive.

   c. more than 15 heads.

14. A basketball player has a 75 percent chance of making a free throw. What is the probability of his making 100 or more free throws in 120 trials?

15. A marksman's chance of hitting a target with each of his shots is 60 percent. If he fires 30 shots, what is the probability of his hitting the target

   a. at least 20 times?

   b. less than 10 times?

   c. between 15 and 20 times inclusive?

16. The P.A.R. Bearings Company is the principal supplier of ball bearings for the Sperry Gyroscope Company. It has been determined that 6 percent of the ball bearings shipped are rejected because they fail to meet tolerance requirements. What is the probability that a shipment of 200 ball bearings contains more than 10 rejects?

17. The manager of C & R Clothiers, a major manufacturer of men's shirts, has determined that 3 percent of C & R's shirts do not meet with company standards and are sold as "seconds" to discount and outlet stores. What is the probability that in a day's production of 200 dozen shirts, less than 10 dozen will be classified as "seconds"?

18. The Colorado Mining and Mineral Company has 800 employees engaged in its mining operations. It has been estimated that the probability of a worker meeting with an accident during a one-year period is 0.1. What is the probability that more than 70 workers will meet with an accident during the one-year period?

19. An experiment was conducted to test the effectiveness of a new drug in treating a certain disease. The drug was administered to 50 mice that had been previously exposed to the disease. It was found that 35 mice subsequently recovered from the disease. It was determined that the natural recovery rate from the disease is 0.5.

   a. Determine the probability that 35 or more of the mice not treated with the drug would recover from the disease.

   b. Using the results obtained in (a), comment on the effectiveness of the drug in the treatment of the disease.

20. The manager of the Madison Finance Company has estimated that, because of a recession year, 5 percent of its 400 loan accounts will be delinquent. If the manager's estimate is correct, what is the probability that 25 or more of the accounts will be delinquent?

*21. Because of late cancellations, Neptune Lines, an operator of cruise ships, has a policy of accepting more reservations than there are accommodations available. From experience, 8 percent of the bookings for the 90-day around-the-world cruise on the S.S. *Drion*, which has accommodations for 2000 passengers, are subsequently canceled. If the management of Neptune Lines has decided, for public relations reasons, that a person who has made a reservation should have a probability of 0.99 of obtaining accommodation on the ship, determine the largest number of reservations that should be taken for a cruise on the S.S. *Drion*.

*22. The Preview Showcase, a research firm, screens pilots of new TV shows before a randomly selected audience and then solicits their opinions of the shows. Based on past experience, 20 percent of those who get complimentary tickets are "no-shows." The theater has a seating capacity of 500. Management has decided, for public relations reasons, that a person who has been solicited for a screening should have a probability of 0.99 of being seated. How many tickets should the company send out to prospective viewers for each screening?

Solutions to Self-Check Exercises

8.6

1. Let X be the normal random variable denoting the serum cholesterol levels in mg/dl in the current Mediterranean population under consideration. Thus, the percentage of the population having blood cholesterol levels between 160 and 180 mg/dl is given by $P(160 < X < 180)$. To compute $P(160 < X < 180)$, we use (18) with $\mu = 160$ and $\sigma = 50$. We find

$$
\begin{aligned}
P(160 < X < 180) &= P\left(\frac{160 - 160}{50} < Z < \frac{180 - 160}{50}\right) \\
&= P(0 < Z < 0.4) \\
&= P(Z < 0.4) - P(Z < 0) \\
&= 0.6554 - 0.5000 \\
&= 0.1554,
\end{aligned}
$$

so approximately 15.5 percent of the population has blood cholesterol levels between 160 and 180 mg/dl.

2. Let X denote the number of substandard bags in the production. Then X has a binomial distribution with $n = 500$, $p = 0.04$, and $q = 0.96$, so

$$
\mu = (500)(0.04) = 20
$$

and

$$
\begin{aligned}
\sigma &= \sqrt{(500)(0.04)(0.96)} \\
&= 4.38.
\end{aligned}
$$

Approximating the binomial distribution by a normal distribution with a mean of 20 and standard deviation of 4.38, we find that the probability that more than 30 bags in the production of 500 will be substandard is given by

$$P(X > 30) \approx P(Y > 30.5) \qquad \text{(Where } Y \text{ denotes the normal random variable)}$$
$$= P\left(Z > \frac{30.5 - 20}{4.38}\right)$$
$$= P(Z > 2.40)$$
$$= P(Z < -2.40)$$
$$= 0.0082,$$

or approximately 0.8 percent.

▶

Chapter 8 Review Exercises

1. A man purchased a $25,000, one-year term life insurance policy for $375. Assuming that the probability that he will live for another year is 0.989, find the company's expected gain.

2. The probability distribution of a random variable X is shown in the following table.

x	$P(X = x)$
0	0.1
1	0.1
2	0.2
3	0.3
4	0.2
5	0.1

   a. Compute $P(1 \le X \le 4)$.

   b. Compute the mean and standard deviation of X.

3. A binomial experiment consists of four trials in which the probability of success in any one trial is 2/5.

   a. Construct the probability distribution for the experiment.

   b. Compute the mean and standard deviation of the probability distribution.

4. Let Z be the standard normal variable. Make a rough sketch of the appropriate region under the standard normal curve and find

   a. $P(Z < 0.5)$ *b.* $P(Z < -0.75)$ *c.* $P(-0.75 < Z < 0.5)$

5. Let Z be the standard normal variable. Find z if z satisfies

   a. $P(Z < z) = 0.9922$ *b.* $P(Z < z) = 0.1469$

   c. $P(Z > z) = 0.9788$ *d.* $P(-z < Z < z) = 0.8444$

6. Let X be a normal random variable with $\mu = 10$ and $\sigma = 2$. Find the values of

    a. $P(X < 11)$ *b.* $P(X > 8)$ *c.* $P(7 < X < 9)$

7. If the probability that a bowler will bowl a strike is 0.7, what is the probability that he will get exactly two strikes in four attempts? at least two strikes in four attempts?

8. The heights of 4000 women who participated in a recent survey were found to be normally distributed with a mean of 64.5 inches and a standard deviation of 2.5 inches. What percentage of these women have heights of 67 inches or greater?

9. Refer to Exercise 8. Use the Chebychev inequality to estimate the probability that the height of a woman who participated in the survey will fall within two standard deviations of the mean; that is, that her height will be between 59.5 and 69.5 inches.

10. The proprietor of a hardware store will accept a shipment of ceramic wall tiles if no more than 2 of a random sample of 20 are found to be defective. What is the probability that he will accept shipment if 10 percent of the tiles in a certain shipment are defective?

11. The Dayton Iron Works Company manufactures steel rods to a specification of 1 inch diameter. These rods are accepted by the buyer if they fall within the tolerance limits of 0.995 and 1.005. Assuming that the diameter of the rods are normally distributed about a mean of 1 inch and a standard deviation of 0.002 inches, estimate the percentage of rods that will be rejected by the buyer.

12. A coin is biased so that the probability of it landing heads is 0.6. If the coin is tossed 100 times, what is the probability that heads will appear more than 50 times in the 100 tosses?

13. A division of the Soloron Corporation manufactures photovoltaic cells for use in the company's solar energy converters. It is estimated that 5 percent of the cells manufactured are defective. In a batch of 200 cells manufactured by the company, what is the probability that it will contain at most 20 defective units?

Markov Chains

and the

Theory of Games

Will the flowers be red? In a certain species a plant will produce red, pink, or white flowers, depending on its genetic makeup. In Example 14, page 516, we will show that if the offspring of two plants are crossed successively with plants of a certain genetic makeup only, then in the long run all the flowers produced by the plants will be red.

▶CHAPTER

NINE

9.1

Markov Chains

▶ Transitional Probabilities

▶ Distribution Vectors

▶ Transitional Probabilities

In this chapter we will look at two important applications of mathematics that are based primarily on matrix theory and the theory of probability. Both of these applications, Markov chains and the theory of games, are relatively recent developments in the field of mathematics and have wide applications in many practical areas.

A finite stochastic process, you may recall, is an experiment consisting of a finite number of stages in which the outcomes and associated probabilities at each stage depend on the outcomes and associated probabilities of the *preceding stages*. In this chapter we will be concerned with a special class of stochastic processes, namely, those in which the probabilities associated with the outcomes at any stage of the experiment depend only on the outcomes of the *preceding stage*. Such a process is called a **Markov process,** or a **Markov chain,** named after the Russian mathematician A. A. Markov (1856–1922).

The outcome at any stage of the experiment in a Markov process is called the **state** of the experiment. In particular, the outcome at the current stage of the experiment is called the **current state** of the process. Following is a typical problem involving a Markov chain. Starting from some state of the process (the current state), determine the probability that the process will be at a particular state some time in the future.

EXAMPLE

1   An analyst at Weaver and Kline, a stock brokerage firm, observes that the closing price of the preferred stock of an airline company over a short span of time depends only on its previous closing price. At the end of each trading day he makes a note of the stock's performance for that day, recording the closing price

as "higher," "unchanged," or "lower" according to whether the stock closes higher, unchanged, or lower than the previous day's closing price. This sequence of observations may be viewed as a Markov chain. ◀

The transition from one state to another in a Markov chain may be studied with the aid of a tree diagram, as in the next example.

EXAMPLE

2 Refer to Example 1. If on a certain day the closing price of the stock is higher than that of the previous day, then the probability that it closes higher, unchanged, or lower on the next trading day is 0.2, 0.3, and 0.5, respectively. Next, if the closing price of the stock is unchanged from the previous day, then the probability that it closes higher, unchanged, or lower on the next trading day is 0.5, 0.2, and 0.3, respectively. Finally, if the closing price of the stock is lower than that of the previous day, then the probability that it closes higher, unchanged, or lower on the next trading day is 0.4, 0.4, and 0.2, respectively. Describe the transition between states and the probabilities associated with these transitions with the aid of a tree diagram.

SOLUTION

The Markov chain being described has three states: higher, unchanged, or lower. If the current state is higher, then the transition to the other states from this state may be displayed by constructing a tree diagram in which the associated probabilities are shown on the appropriate limbs [see Figure 9.1(a)]. Tree diagrams describing the transition from each of the other two possible current states to the other states may be constructed in a similar manner [Figures 9.1(b) and (c)].

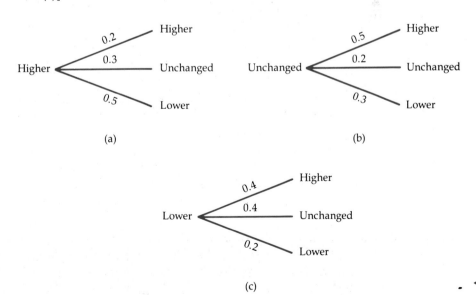

Figure 9.1

◀

The probabilities encountered in this example are called **transition probabilities** because they are associated with the transition from one state to the next in the Markov process. These transition probabilities may be conveniently represented in the form of a matrix. Suppose for simplicity that we have a Markov chain with three possible outcomes at each stage of the experiment. Let us refer to these outcomes as state 1, state 2, and state 3. Then the transition probabilities associated with the transition from state 1 to each of the states 1, 2, and 3 in the next phase of the experiment are precisely the respective conditional probabilities that the outcome is state 1, state 2, and state 3 *given* that the outcome state 1 has occurred. In short, the desired transition problems are $P(\text{state 1} \mid \text{state 1})$, $P(\text{state 2} \mid \text{state 1})$, and $P(\text{state 3} \mid \text{state 1})$, respectively. Let us write

$$a_{11} = P(\text{state 1} \mid \text{state 1})$$
$$a_{12} = P(\text{state 2} \mid \text{state 1})$$

and
$$a_{13} = P(\text{state 3} \mid \text{state 1})$$

Note that the first subscript in this notation refers to the current state and the second subscript refers to the state in the next stage of the experiment. Using a tree diagram, we have the following representation:

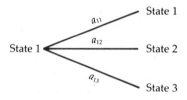

Similarly, the transition probabilities associated with the transition from state 2 and state 3 to each of the states 1, 2, and 3 are

$$
\begin{array}{ll}
a_{21} = P(\text{state 1} \mid \text{state 2}) & a_{31} = P(\text{state 1} \mid \text{state 3}) \\
a_{22} = P(\text{state 2} \mid \text{state 2}) \quad \text{and} \quad & a_{32} = P(\text{state 2} \mid \text{state 3}) \\
a_{23} = P(\text{state 3} \mid \text{state 2}) & a_{33} = P(\text{state 3} \mid \text{state 3})
\end{array}
$$

These observations lead to the following matrix representation of the transition probabilities:

		Next state		
		State 1	State 2	State 3
	State 1	a_{11}	a_{12}	a_{13}
Current state	State 2	a_{21}	a_{22}	a_{23}
	State 3	a_{31}	a_{32}	a_{33}

EXAMPLE

3 Use a matrix to represent the transition probabilities obtained in Example 2.

SOLUTION There are three states at each stage of the Markov chain under consideration. Letting state 1, state 2, and state 3 denote the states "higher," "unchanged," and "lower," respectively, we find that

$$a_{11} = 0.2, a_{12} = 0.3, a_{13} = 0.5, \text{ and so on,}$$

so the required matrix representation is given by

$$T = \begin{bmatrix} .2 & .3 & .5 \\ .5 & .2 & .3 \\ .4 & .4 & .2 \end{bmatrix}. \qquad \blacktriangleleft$$

The matrix obtained in the preceding example is a transition matrix. In the general case, we have the following definition:

Transition Matrix

A **transition matrix** associated with a Markov chain with n states is an $n \times n$ matrix T with entries a_{ij} $(1 \le i \le n; 1 \le j \le n)$

Next state

	State 1	State 2	$\cdots$	State j	$\cdots$	State n
State 1	a_{11}	a_{12}	$\cdots$	a_{1j}	$\cdots$	a_{1n}
State 2	a_{21}	a_{22}	$\cdots$	a_{2j}	$\cdots$	a_{2n}
$\vdots$	$\vdots$	$\vdots$		$\vdots$		$\vdots$
State i	a_{i1}	a_{i2}	$\cdots$	a_{ij}	$\cdots$	a_{in}
$\vdots$	$\vdots$	$\vdots$		$\vdots$		$\vdots$
State n	a_{n1}	a_{n2}	$\cdots$	a_{nj}	$\cdots$	a_{nn}

$T = $ Current state

having the following properties:

1. $a_{ij} \ge 0$ for all i and j.

2. The sum of the entries in each row of T is 1.

Since $a_{ij} = P(\text{state } j \mid \text{state } i)$ is the probability of the occurrence of an event, it must be nonnegative, and this is precisely what Property 1 implies. Property 2 follows from the fact that the transition from any one of the current states must terminate in one of the n states in the next stage of the experiment. Any square matrix satisfying Properties 1 and 2 is referred to as a **stochastic matrix**.

An advantage in representing the transition probabilities in the form of a matrix is that we may use the results from matrix theory to help us solve problems involving Markov processes, a fact that will be demonstrated repeatedly in the next several sections.

Next, for simplicity, let us consider the following Markov process where each stage of the experiment has precisely two possible states.

EXAMPLE

4 Because of the continued successful implementation of an urban renewal program, it is expected that each year 3 percent of the population currently residing in the city will move to the suburbs, and 6 percent of the population currently residing in the suburbs will move into the city. At present, 65 percent of the total population of the metropolitan area live in the city itself, while the remaining 35 percent live in the suburbs. Assuming that the total population of the metropolitan area remains constant, what will the distribution of the population be like one year from now?

SOLUTION

This problem may be solved with the aid of a tree diagram and the techniques of Chapter 7. The required tree diagram describing this process is shown in Figure 9.2. Using the method of Section 7.5, we find the probability that a person selected at random will be a city dweller one year from now is given by

$$(0.65)(0.97) + (0.35)(0.06),$$

or 0.6515. In a similar manner we find that the probability that a person selected at random will reside in the suburbs one year from now is given by

$$(0.65)(0.03) + (0.35)(0.94),$$

or 0.3485. Thus, the population of the area one year from now may be expected to be distributed as follows: 65.15 percent living in the city and 34.85 percent residing in the suburbs.

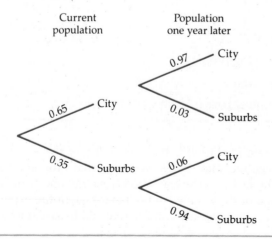

Current
population

Population
one year later

0.97 — City

0.65 — City

0.03 — Suburbs

0.35 — Suburbs

0.06 — City

0.94 — Suburbs

Figure 9.2

Let us re-examine the solution to this problem. First of all, as noted earlier, the process under consideration may be viewed as a Markov chain with two possible states at each stage of the experiment: "living in the city" (state 1) and "living in the suburbs" (state 2). The transition matrix associated with this Markov chain is

$$T = \begin{matrix} \\ \text{State 1} \\ \text{State 2} \end{matrix} \begin{matrix} \text{State 1} & \text{State 2} \\ \begin{bmatrix} .97 & .03 \\ .06 & .94 \end{bmatrix} \end{matrix}$$

Next, observe that the initial (current) probability distribution of the population may be summarized in the form of a row vector of dimension 2 (that is, a 1×2 matrix). Thus,

$$\begin{matrix} \text{State 1} & \text{State 2} \\ X_0 = [.65 & .35] \end{matrix}$$

The population distribution one year later may be written, using the results of Example 4, as

$$\begin{matrix} \text{State 1} & \text{State 2} \\ X_1 = [.6515 & .3485] \end{matrix}$$

You may now verify that

$$X_0 T = [.65 \quad .35] \begin{bmatrix} .97 & .03 \\ .06 & .94 \end{bmatrix}$$

$$= [.6515 \quad .3485] = X_1,$$

so this problem may be solved using matrix multiplication.

EXAMPLE

5 Refer to Example 4. What is the population distribution of the city after two years? after three years?

SOLUTION

Let X_2 be the row vector representing the probability population distribution of the city after two years. We may view X_1, the vector representing the probability population distribution of the metropolitan area after one year, as representing the "initial" probability distribution in this part of our calculation. Thus,

$$X_2 = X_1 T$$

$$= [.6515 \quad .3485] \begin{bmatrix} .97 & .03 \\ .06 & .94 \end{bmatrix}$$

$$= [.6529 \quad .3471].$$

The vector representing the probability distribution of the metropolitan area after three years is given by

$$X_3 = X_2 T = [.6529 \quad .3471] \begin{bmatrix} .97 & .03 \\ .06 & .94 \end{bmatrix}$$

$$= [.6541 \quad .3459].$$

That is, after three years, the population will be distributed as follows: 65.41 percent will live in the city and 34.59 percent will live in the suburbs. ◀

▶ Distribution Vectors

Observe that, in the foregoing computations, we have $X_1 = X_0 T$, $X_2 = X_1 T = X_0 T^2$ and $X_3 = X_2 T = X_0 T^3$. These results are easily generalized. To see this, suppose we have a Markov process in which there are n possible states at each stage of the experiment. Suppose further that the probability of the system being in state 1, state 2, ..., state n, initially, is given by $p_1, p_2, ..., p_n$, respectively. This distribution may be represented as an n-dimensional vector

$$X_0 = [p_1, p_2, ..., p_n],$$

called a **distribution vector**. If T represents the $n \times n$ transition matrix associated with the Markov process, then the probability distribution of the system after m observations is given by

$$X_m = X_0 T^m \tag{1}$$

EXAMPLE

6 In order to keep track of the location of its cabs, the Zephyr Cab Company has divided a town into three zones: zone I, zone II, and zone III. Zephyr's management has determined from company records that of the passengers picked up in zone I, 60 percent are discharged in the same zone, 30 percent are discharged in zone II, and 10 percent are discharged in zone III. Of those picked up in zone II, 40 percent are discharged in zone I, 30 percent are discharged in zone II, and 30 percent are discharged in zone III. Of those picked up in zone III, 30 percent are discharged in zone I, 30 percent are discharged in zone II, and 40 percent are discharged in zone III. Suppose that at the beginning of the day 80 percent of the cabs are in zone I, 15 percent are in zone II, and 5 percent are in zone III and that a taxi without a passenger will cruise within the zone it is currently in until a pickup is made.

a. Find the transition matrix for the Markov chain that describes the successive locations of a cab.

b. What is the distribution of the cabs after all of them have made one pickup and discharge?

c. What is the distribution of the cabs after all of them have made two pickups and discharges?

SOLUTION   Let zone I, zone II, and zone III correspond to state 1, state 2, and state 3 of the Markov chain.

a. The required transition matrix is given by

$$T = \begin{bmatrix} .6 & .3 & .1 \\ .4 & .3 & .3 \\ .3 & .3 & .4 \end{bmatrix}.$$

b. The initial distribution vector associated with the problem is

$$X_0 = [.8 \quad .15 \quad .05].$$

If X_1 denotes the distribution vector after one observation, that is, after all the cabs have made one pickup and discharge, then

$$X_1 = X_0 T$$

$$= [.8 \quad .15 \quad .05] \begin{bmatrix} .6 & .3 & .1 \\ .4 & .3 & .3 \\ .3 & .3 & .4 \end{bmatrix}$$

$$= [.555 \quad .3 \quad .145];$$

that is, 55.5 percent of the cabs are in zone I, 30 percent are in zone II, and 14.5 percent are in zone III.

c. Let X_2 denote the distribution vector after all the cabs have made two pickups and discharges. Then

$$X_2 = X_1 T$$

$$= [.555 \quad .3 \quad .145] \begin{bmatrix} .6 & .3 & .1 \\ .4 & .3 & .3 \\ .3 & .3 & .4 \end{bmatrix}$$

$$= [.4965 \quad .3 \quad .2035];$$

that is, 49.65 percent of the cabs are in zone I, 30 percent are in zone II, and 20.35 percent are in zone III. You should verify that the same result may be obtained by computing $X_0 T^2$.   ◄

Self-Check Exercises

9.1

1. Three supermarkets serve a certain section of a city. During the upcoming year, supermarket A is expected to retain 80 percent of its customers, lose 5 percent of

its customers to supermarket B, and lose 15 percent to supermarket C. Supermarket B is expected to retain 90 percent of its customers and lose 5 percent of its customers to each supermarket A and supermarket C. Supermarket C is expected to retain 75 percent of its customers, lose 10 percent to supermarket A, and lose 15 percent to supermarket B. Construct the transition matrix for the Markov chain that describes the change in the market share of the three supermarkets.

2. Refer to Exercise 1. Currently the market shares of supermarket A, supermarket B, and supermarket C are 0.4, 0.3, and 0.3, respectively.

  a. Find the initial distribution vector for this Markov chain.

  b. What share of the market will be held by each supermarket after one year? Assuming that the trend continues, what will the market share be after two years?

Solutions to Self-Check Exercises 9.1 can be found on page 500.

▶

Exercises 9.1

In Exercises 1–9, determine which of the given markets are stochastic.

1. $\begin{bmatrix} .4 & .6 \\ .7 & .3 \end{bmatrix}$

2. $\begin{bmatrix} .8 & .3 \\ .2 & .7 \end{bmatrix}$

3. $\begin{bmatrix} \frac{1}{4} & \frac{3}{4} \\ \frac{1}{8} & \frac{7}{8} \end{bmatrix}$

4. $\begin{bmatrix} \frac{1}{3} & \frac{1}{2} & \frac{1}{4} \\ 0 & 1 & 0 \\ \frac{1}{2} & 0 & \frac{1}{2} \end{bmatrix}$

5. $\begin{bmatrix} .3 & .4 & .3 \\ .2 & .7 & .1 \\ .4 & .3 & .2 \end{bmatrix}$

6. $\begin{bmatrix} \frac{1}{3} & \frac{1}{3} & \frac{1}{3} \\ \frac{1}{4} & 0 & \frac{3}{4} \\ \frac{1}{2} & -\frac{1}{2} & \frac{1}{2} \end{bmatrix}$

7. $\begin{bmatrix} .1 & .7 & .2 \\ .4 & .2 & .4 \\ .3 & .1 & .6 \end{bmatrix}$

8. $\begin{bmatrix} 1 & 0 & 0 \\ 0 & 0 & 1 \\ 0 & 1 & 0 \end{bmatrix}$

9. $\begin{bmatrix} .2 & .3 & .5 \\ .3 & .1 & .6 \end{bmatrix}$

10. The transition matrix for a Markov process is given by

$$\begin{array}{cc} & \text{State} \\ & \begin{array}{cc} 1 & 2 \end{array} \\ T = \text{State} \begin{array}{c} 1 \\ 2 \end{array} & \begin{bmatrix} .6 & .4 \\ .2 & .8 \end{bmatrix} \end{array}$$

and the initial state distribution vector is given by

$$\begin{array}{c} \text{State} \\ \begin{array}{cc} 1 & 2 \end{array} \\ X_0 = \begin{bmatrix} .5 & .5 \end{bmatrix} \end{array}$$

Find X_0T and interpret your result with the aid of a tree diagram.

11. The transition matrix for a Markov process is given by

State

1 2

$$T = \text{State } \begin{array}{c} 1 \\ 2 \end{array} \begin{bmatrix} \tfrac{1}{2} & \tfrac{1}{2} \\ \tfrac{3}{4} & \tfrac{1}{4} \end{bmatrix}$$

and the initial state distribution vector is given by

State

1 2

$$X_0 = [\tfrac{1}{3} \quad \tfrac{2}{3}]$$

Find X_0T and interpret your result with the aid of a tree diagram.

In Exercises 12–15, find X_2 (the probability distribution of the system after two obser-vations) for the given distribution vector X_0 and the given transition matrix T.

12. $X_0 = [.6 \quad .4] \qquad T = \begin{bmatrix} .4 & .6 \\ .8 & .2 \end{bmatrix}$

13. $X_0 = [1/2 \quad 1/2 \quad 0] \qquad T = \begin{bmatrix} \tfrac{1}{2} & 0 & \tfrac{1}{2} \\ \tfrac{1}{3} & \tfrac{1}{3} & \tfrac{1}{3} \\ \tfrac{1}{2} & \tfrac{1}{4} & \tfrac{1}{4} \end{bmatrix}$

14. $X_0 = [1/4 \quad 1/2 \quad 1/4] \qquad T = \begin{bmatrix} \tfrac{1}{4} & \tfrac{1}{4} & \tfrac{1}{2} \\ \tfrac{1}{4} & \tfrac{1}{2} & \tfrac{1}{4} \\ \tfrac{1}{2} & \tfrac{1}{2} & 0 \end{bmatrix}$

15. $X_0 = [.25 \quad .40 \quad .35] \qquad T = \begin{bmatrix} .1 & .8 & .1 \\ .1 & .7 & .2 \\ .3 & .2 & .5 \end{bmatrix}$

16. A psychologist conducts an experiment in which a mouse is placed in a T-maze where it has a choice at the T-junction of turning left and receiving a reward (cheese) or turning right and receiving a mild electric shock. At the end of each trial a record

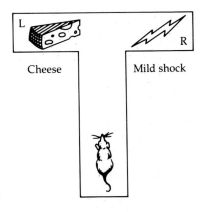

Cheese   Mild shock

Figure for
Exercise 16

is kept of the mouse's response. It is observed that the mouse is as likely to turn left (state 1) as right (state 2) during the first trial. In subsequent trials, however, the observation is made that if the mouse had turned left in the previous trial, then on the next trial the probability that it will turn left is 0.8, whereas the probability that it will turn right is 0.2. If the mouse had turned right in the previous trial then the probability that it will turn right on the next trial is 0.1, whereas the probability that it will turn left is 0.9.

a. Using a tree diagram, describe the transitions between states and the probabilities associated with these transitions.

b. Represent the transition probabilities obtained in (a) in terms of a matrix.

c. What is the initial state probability vector?

d. Use the results of (b) and (c) to find the probability that a mouse will turn left on the second trial.

17. Twenty percent of the commuters within a large metropolitan area currently use the public transportation system, whereas the remaining 80 percent commute via automobile. The city has recently revitalized and expanded its public transportation system. It is expected that six months from now 30 percent of those who are now commuting to work via automobile will switch to public transportation, and 70 percent will continue to commute via automobile. At the same time, it is expected that 20 percent of those now using public transportation will commute via automobile, and 80 percent will continue to use public transportation.

a. Construct the transition matrix for the Markov chain that describes the change in the mode of transportation used by these commuters.

b. Find the initial distribution vector for this Markov chain.

c. What percentage of the commuters are expected to use public transportation six months from now?

18. The Morris Polling Group conducted a poll six months before an election in a state in which a Democrat and a Republican were running for governor and found that 60 percent of the voters intended to vote for the Democrat and 40 percent intended to vote for the Republican. In a poll conducted three months later it was found that 70 percent of those who had earlier stated a preference for the Democratic candidate still maintained that preference, whereas 30 percent of these voters now preferred the Republican candidate. Of those who had earlier stated a preference for the Republican, 80 percent still maintained that preference, whereas 20 percent now preferred the Democratic candidate.

a. If the election were held at this time, who would win?

b. Assuming that this trend continues, which candidate is expected to win the election?

19. Refer to Example 4. If the initial probability distribution is

$$X_0 = \begin{bmatrix} \text{City} & \text{Suburb} \\ .80 & .20 \end{bmatrix}$$

what will the population distribution of the city be after one year? after two years?

20. Refer to Example 6. If the initial distribution vector for the location of the taxis is

$$
\begin{array}{cccc}
& \text{Zone I} & \text{Zone II} & \text{Zone III} \\
X_0 = & [.6 & .2 & .2]
\end{array}
$$

what will the distribution be after all of them have made one pickup and discharge?

21. At a certain university, three bookstores—the University Bookstore, the Campus Bookstore, and the Book Mart—currently serve the university community. From a survey conducted at the beginning of the fall quarter it was found that the University Bookstore and the Campus Bookstore each had 40 percent of the market, whereas the Book Mart had 20 percent of the market. Each quarter, the University Bookstore retains 80 percent of its customers but loses 10 percent to the Campus Bookstore and 10 percent to the Book Mart. The Campus Bookstore retains 75 percent of its customers but loses 10 percent to the University Bookstore and 15 percent to the Book Mart. The Book Mart retains 90 percent of its customers but loses 5 percent to the University Bookstore and 5 percent to the Campus Bookstore. What percentage of the market will each store have at the beginning of the second quarter? the third quarter?

22. A study conducted by the Urban Energy Commission in a large metropolitan area indicates the probabilities that homeowners within the area will use certain heating fuels or solar energy during the next ten years as the major source of heat for their homes. The transition matrix representing the transition probabilities from one state to another is shown below.

$$
\begin{array}{l}
\quad\quad\quad\quad\quad\quad \text{Elec.} \quad \text{Gas} \quad \text{Oil} \quad \text{Solar} \\
\begin{array}{l}
\text{Electricity} \\
\text{Natural gas} \\
\text{Fuel oil} \\
\text{Solar energy}
\end{array}
\left[
\begin{array}{cccc}
.70 & .15 & .05 & .10 \\
0 & .90 & .02 & .08 \\
0 & .20 & .75 & .05 \\
0 & .05 & 0 & .95
\end{array}
\right]
\end{array}
$$

Among homeowners within the area, 20 percent currently use electricity, 35 percent use natural gas, 40 percent use oil, and 5 percent use solar energy as the major source of heat for their homes. What is the expected distribution of the homeowners that will be using each type of heating fuel or solar energy within the next decade?

23. Records compiled by the Admissions Office at a state university indicating the percentage of students that change their major each year are shown in the following transition matrix. Of the freshmen now at the university, 30 percent have chosen their major field in business, 30 percent in the humanities, 20 percent in education, and 20 percent in the natural sciences and other fields. Assuming that this trend continues, find the percentage of these students that will be majoring in each of the given areas in their senior year. [*Hint:* Find $X_0 T^3$.]

$$
\begin{array}{l}
\quad\quad\quad\quad\quad\quad\quad\quad\quad\quad\quad\quad\quad\quad\quad\quad\quad\quad\quad\quad\quad \text{Nat. sci.} \\
\quad\quad\quad\quad\quad\quad\quad\quad\quad\quad\quad\quad\quad \text{Bus.} \quad \text{Hum.} \quad \text{Educ.} \quad \text{and others} \\
\begin{array}{l}
\text{Business} \\
\text{Humanities} \\
\text{Education} \\
\text{Natural sciences and others}
\end{array}
\left[
\begin{array}{cccc}
.80 & .10 & .05 & .05 \\
.10 & .70 & .10 & .10 \\
.20 & .10 & .60 & .10 \\
.10 & .05 & .05 & .80
\end{array}
\right]
\end{array}
$$

Solutions to Self-Check Exercises

9.1

1. The required transition matrix is

$$T = \begin{bmatrix} .80 & .05 & .15 \\ .05 & .90 & .05 \\ .10 & .15 & .75 \end{bmatrix}.$$

2. *a.* The initial distribution vector is

$$X_0 = [.4 \quad .3 \quad .3].$$

b. The vector representing the market share of supermarket A after one year is

$$X_1 = X_0 T$$

$$= [.4 \quad .3 \quad .3] \begin{bmatrix} .80 & .05 & .15 \\ .05 & .90 & .05 \\ .10 & .15 & .75 \end{bmatrix}$$

$$= [.365 \quad .335 \quad 0.3]$$

That is, after one year supermarket A will command a 36.5 percent market share, supermarket B will have a 33.5 percent share, and supermarket C will have a 30 percent market share.

The vector representing the market share of the supermarkets after two years is

$$X_2 = X_1 T = [.365 \quad .335 \quad .3] \begin{bmatrix} .80 & .05 & .15 \\ .05 & .90 & .05 \\ .10 & .15 & .75 \end{bmatrix}$$

$$= [.3388 \quad .3648 \quad .2965]$$

That is, two years later the market shares of supermarkets A, B, and C will be 33.88 percent, 36.48 percent, and 29.65 percent, respectively.

9.2

Regular Markov Chains

▶ Steady-State Distribution Vectors

▶ Regular Markov Chains

▶ Steady-State Distribution Vectors

In the last section we derived a formula for computing the likelihood that a physical system will be in any one of the possible states associated with each stage of a Markov process describing the system. In this section, we will use this formula to help us investigate the long-term trends of certain Markov processes.

EXAMPLE

7 A survey conducted by the National Commission on the Educational Status of Women reveals that 70 percent of the daughters of women who have completed two or more years of college have also completed two or more years of college, whereas 20 percent of the daughters of women who have had less than two years of college have completed two or more years of college. If this trend continues, determine, in the long run, the percentage of women in the population who would have completed at least two years of college given that currently only 20 percent of the women have completed at least two years of college.

SOLUTION This problem may be viewed as a Markov process with two possible states: "completed two or more years of college" (state 1) and "completed less than two years of college" (state 2). The transition matrix associated with this Markov chain is given by

$$T = \begin{bmatrix} .7 & .3 \\ .2 & .8 \end{bmatrix}.$$

The initial distribution vector is given by

$$X_0 = [.2 \quad .8].$$

In order to study the long-term trend pertaining to this particular aspect of the educational status of women, let us compute $X_1, X_2, \ldots$, the distribution vectors associated with the Markov process under consideration. These vectors give the percentage of women with two or more years of college and that of women with less than two years of college after one generation, after two generations, ... With the aid of Formula (1), we find that (to four decimal places)

$$X_1 = X_0 T = [.2 \quad .8] \begin{bmatrix} .7 & .3 \\ .2 & .8 \end{bmatrix}$$
$$= [.3 \quad .7]$$
$$X_2 = X_1 T = [.3 \quad .7] \begin{bmatrix} .7 & .3 \\ .2 & .8 \end{bmatrix}$$
$$= [.35 \quad .65]$$
$$X_3 = X_2 T = [.35 \quad .65] \begin{bmatrix} .7 & .3 \\ .2 & .8 \end{bmatrix}$$
$$= [.375 \quad .625].$$

Proceeding further, we obtain the following sequence of vectors:

$$X_4 = [.3875 \quad .6125]$$
$$X_5 = [.3938 \quad .6062]$$

$$X_6 = [.3969 \quad .6031]$$
$$X_7 = [.3985 \quad .6015]$$
$$X_8 = [.3993 \quad .6008]$$
$$X_9 = [.3997 \quad .6004]$$
$$X_{10} = [.3999 \quad .6002]$$

From the results of this computation, we see that as m increases, the probability distribution vector X_m approaches the probability distribution vector

$$[.4 \quad .6] \quad \text{or} \quad [\tfrac{2}{5} \quad \tfrac{3}{5}]$$

Such a vector is called the **limiting** or **steady-state distribution vector** for the system. The interpretation of the results is as follows: Initially, 20 percent of the women in the population have completed two or more years of college, whereas 80 percent have completed less than two years of college. After one generation, the former has increased to 30 percent of the population and the latter has dropped to 70 percent of the population. The trend continues, and eventually, 40 percent of all women in future generations would have completed two or more years of college, whereas 60 percent would have completed less than two years of college. ◀

In order to explain the foregoing result, let us evaluate Formula (1) more closely. Now, the initial distribution vector X_0 is a constant; that is, it remains fixed throughout our computation of $X_1, X_2, \ldots$. It appears reasonable, therefore, to conjecture that this phenomenon is a result of the behavior of the powers, T^m, of the transition matrix T. Pursuing this line of investigation, we compute

$$T^2 = \begin{bmatrix} .7 & .3 \\ .2 & .8 \end{bmatrix} \begin{bmatrix} .7 & .3 \\ .2 & .8 \end{bmatrix} = \begin{bmatrix} .55 & .45 \\ .3 & .7 \end{bmatrix}$$

$$T^3 = \begin{bmatrix} .7 & .3 \\ .2 & .8 \end{bmatrix} \begin{bmatrix} .55 & .45 \\ .3 & .7 \end{bmatrix} = \begin{bmatrix} .475 & .525 \\ .35 & .65 \end{bmatrix}$$

Proceeding further, we obtain the following sequence of matrices:

$$T^4 = \begin{bmatrix} .4375 & .5625 \\ .375 & .625 \end{bmatrix} \quad T^5 = \begin{bmatrix} .4188 & .5813 \\ .3875 & .6125 \end{bmatrix} \quad T^6 = \begin{bmatrix} .4094 & .5907 \\ .3938 & .6063 \end{bmatrix}$$

$$T^7 = \begin{bmatrix} .4047 & .5954 \\ .3969 & .6032 \end{bmatrix} \quad T^8 = \begin{bmatrix} .4024 & .5977 \\ .3985 & .6016 \end{bmatrix} \quad T^9 = \begin{bmatrix} .4012 & .5988 \\ .3993 & .6008 \end{bmatrix}$$

$$T^{10} = \begin{bmatrix} .4006 & .5994 \\ .3997 & .6004 \end{bmatrix} \quad T^{11} = \begin{bmatrix} .4003 & .5997 \\ .3999 & .6002 \end{bmatrix}$$

These results show that the powers T^m of the transition matrix T tend toward a fixed matrix as m gets larger and larger. In this case, the "limiting matrix" is the matrix

$$L = \begin{bmatrix} .40 & .60 \\ .40 & .60 \end{bmatrix} \quad \text{or} \quad \begin{bmatrix} \tfrac{2}{5} & \tfrac{3}{5} \\ \tfrac{2}{5} & \tfrac{3}{5} \end{bmatrix}$$

Such a matrix is called the **steady-state matrix** for the system. Thus, as suspected, the long-term behavior of a Markov process such as the one in this example depends on the behavior of the limiting matrix of the powers of the transition matrix—the steady-state matrix for the system. In view of this, the long-term (steady-state) distribution vector for this problem may be found by computing the product

$$X_0 L = [.2 \quad .8] \begin{bmatrix} .40 & .60 \\ .40 & .60 \end{bmatrix} = [.40 \quad .60],$$

which agrees with the result obtained earlier.

Next, since the transition matrix T in this situation seems to have a stabilizing effect over the long term, we are led to wonder whether the steady state would be reached regardless of the initial state of the system. To answer this question, suppose that the initial distribution vector were

$$X_0 = [p \quad 1 - p].$$

Then, as before, the steady-state distribution vector is given by

$$X_0 L = [p \quad 1 - p] \begin{bmatrix} .40 & .60 \\ .40 & .60 \end{bmatrix} = [.40 \quad .60].$$

Thus, the steady state is reached regardless of the initial state of the system!

▶ Regular Markov Chains

The transition matrix T of Example 7 has several important properties, which we have emphasized in the foregoing discussion. First of all, the sequence T, T^2, T^3, . . . approaches a steady-state matrix in which the rows of the limiting matrix are all equal. Such a transition matrix is said to be a **regular stochastic matrix.** In general, it can be shown that *a stochastic matrix* T *is regular if some power of* T *has entries that are all positive.* Second, as in the case of Example 7, a Markov chain with a regular transition matrix (called a *regular Markov chain*) has a steady-state distribution vector whose elements coincide with those of a row (since they are all the same) of the steady-state matrix; thus, this steady-state distribution vector is always reached regardless of the initial distribution vector.

We will return to computations involving regular Markov chains, but for the moment let us see how one may determine whether a given matrix is indeed regular.

EXAMPLE

8 Determine which of the following matrices are regular:

$$a. \begin{bmatrix} .7 & .3 \\ .2 & .8 \end{bmatrix} \qquad b. \begin{bmatrix} .4 & .6 \\ 1 & 0 \end{bmatrix} \qquad c. \begin{bmatrix} 0 & 1 \\ 1 & 0 \end{bmatrix}$$

SOLUTION

a. Since all the entries of the matrix are positive, the given matrix is regular. Note that this is the transition matrix of Example 7.

b. In this case, one of the entries of the given matrix is equal to zero. Let us compute

$$\begin{bmatrix} .4 & .6 \\ 1 & 0 \end{bmatrix}^2 = \begin{bmatrix} .4 & .6 \\ 1 & 0 \end{bmatrix}\begin{bmatrix} .4 & .6 \\ 1 & 0 \end{bmatrix} = \begin{bmatrix} .76 & .24 \\ .4 & .6 \end{bmatrix}$$

Since the second power of the matrix has entries that are all positive, we conclude that the given matrix is in fact regular.

c. Denote the given matrix by *A*. Then

$$A = \begin{bmatrix} 0 & 1 \\ 1 & 0 \end{bmatrix}, \qquad A^2 = \begin{bmatrix} 0 & 1 \\ 1 & 0 \end{bmatrix}\begin{bmatrix} 0 & 1 \\ 1 & 0 \end{bmatrix} = \begin{bmatrix} 1 & 0 \\ 0 & 1 \end{bmatrix}$$

$$A^3 = \begin{bmatrix} 0 & 1 \\ 1 & 0 \end{bmatrix}\begin{bmatrix} 1 & 0 \\ 0 & 1 \end{bmatrix} = \begin{bmatrix} 0 & 1 \\ 1 & 0 \end{bmatrix}$$

Observe that $A^3 = A$. It follows, therefore, that $A^4 = A^2$, $A^5 = A$, and so on. In other words, any power of *A* must coincide with either *A* or A^2. Since not all the entries of *A* and A^2 are positive, the same is true of any power of *A*. We conclude, accordingly, that the given matrix is not regular. ◀

We now return to the study of regular Markov chains. In Example 7, we found the steady-state distribution vector associated with a regular Markov chain by studying the limiting behavior of a sequence of distribution vectors. Alternatively, as pointed out in the subsequent discussion, the steady-state distribution vector may also be obtained by first determining the steady-state matrix associated with the regular Markov chain, which is obtained by studying the limiting behavior of a sequence of transition matrices associated with the process.

Fortunately, there is a relatively simple procedure for finding the steady-state distribution vector associated with a regular Markov process. It does not involve the rather tedious computations required to obtain the sequences in Example 7. The procedure follows:

Finding the Steady-State Distribution Vector

Let *T* be a regular stochastic matrix. Then the steady-state distribution vector *X* may be found by solving the vector equation

$$XT = X$$

together with the condition that the sum of the elements of the vector *X* be equal to 1.

A justification of the foregoing procedure is given in Problem 26, Exercises 9.2.

EXAMPLE

9 Find the steady-state matrix for the regular Markov chain whose transition matrix is

$$T = \begin{bmatrix} .7 & .3 \\ .2 & .8 \end{bmatrix} \quad \text{(See Example 7.)}$$

SOLUTION Let $X = [x \quad y]$ be the steady-state distribution vector associated with the Markov process, where the numbers x and y are to be determined. The condition $XT = X$ translates into the matrix equation

$$[x \quad y] \begin{bmatrix} .7 & .3 \\ .2 & .8 \end{bmatrix} = [x \quad y]$$

or equivalently, the following system of linear equations

$$0.7x + 0.2y = x$$
$$0.3x + 0.8y = y.$$

But each of the equations that make up this system of equations is equivalent to the single equation

$$0.3x - 0.2y = 0.$$

Next, the condition that the sum of the elements of X add up to 1 gives

$$x + y = 1.$$

Thus, the fulfillment of the two conditions simultaneously implies that x and y be the solutions of the system

$$0.3x - 0.2y = 0$$
$$x + y = 1.$$

Solving the first equation for x we obtain

$$x = \frac{2}{3}y,$$

which, upon substitution into the second, yields

$$\frac{2}{3}y + y = 1$$

or

$$y = \frac{3}{5}.$$

Thus, $x = 2/5$ and the required steady-state matrix is given by $X = [2/5 \quad 3/5]$, which agrees with the result obtained earlier. ◀

EXAMPLE

10 In Example 6, we showed that the transition matrix that described the movement of taxis from zone to zone was given by the regular stochastic matrix

$$T = \begin{bmatrix} .6 & .3 & .1 \\ .4 & .3 & .3 \\ .3 & .3 & .4 \end{bmatrix}.$$

Use this information to determine the long-term distribution of the taxis in the three zones.

SOLUTION

Let $X = [x \ \ y \ \ z]$ be the steady-state distribution vector associated with the Markov process under consideration, where x, y, and z are to be determined. The condition $XT = X$ translates into the matrix equation

$$[x \ \ y \ \ z] \begin{bmatrix} .6 & .3 & .1 \\ .4 & .3 & .3 \\ .3 & .3 & .4 \end{bmatrix} = [x \ \ y \ \ z]$$

or, equivalently, the following system of linear equations:

$$0.6x + 0.4y + 0.3z = x$$
$$0.3x + 0.3y + 0.3z = y$$
$$0.1x + 0.3y + 0.4z = z.$$

This system simplifies into

$$4x - 4y - 3z = 0$$
$$3x - 7y + 3z = 0$$
$$x + 3y - 6z = 0.$$

Since $x + y + z = 1$ as well, we are required to solve the system

$$x + y + z = 1$$
$$4x - 4y - 3z = 0$$
$$3x - 7y + 3z = 0$$
$$x + 3y - 6z = 0.$$

Using the Gauss–Jordan elimination procedure of Chapter 2, we find that

$$x = \frac{33}{70}, \quad y = \frac{3}{10}, \quad \text{and} \quad z = \frac{8}{35}$$

or $x \approx 0.47$, $y = 0.30$, and $z \approx 0.23$. Thus, in the long run, 47 percent of the taxis will be in zone I, 30 percent in zone II, and 23 percent in zone III. ◀

Self-Check Exercises
9.2

1. Find the steady-state matrix for the regular Markov chain whose transition matrix is

$$T = \begin{bmatrix} .5 & .5 \\ .8 & .2 \end{bmatrix}.$$

2. Three supermarkets serve a certain section of a city. During the year, super-market A is expected to retain 80 percent of its customers, lose 5 percent of its customers to supermarket B, and lose 15 percent to supermarket C. Supermarket B is expected to retain 90 percent of its customers and lose 5 percent to each of supermarket A and supermarket C. Supermarket C is expected to retain 75 percent of its customers, lose 10 percent to supermarket A, and lose 15 percent to super-market B. In the long run, what will the market share of each supermarket be?

Solutions to Self-Check Exercises 9.2 can be found on page 510.

▶

Exercises 9.2

In Exercises 1–8, determine which of the given matrices are regular.

1. $\begin{bmatrix} \frac{2}{5} & \frac{3}{5} \\ \frac{3}{4} & \frac{1}{4} \end{bmatrix}$

2. $\begin{bmatrix} 0 & 1 \\ .3 & .7 \end{bmatrix}$

3. $\begin{bmatrix} 1 & 0 \\ .8 & .2 \end{bmatrix}$

4. $\begin{bmatrix} \frac{1}{3} & \frac{2}{3} \\ 0 & 1 \end{bmatrix}$

5. $\begin{bmatrix} \frac{1}{2} & \frac{1}{2} & 0 \\ \frac{3}{4} & 0 & \frac{1}{4} \\ 0 & \frac{1}{2} & \frac{1}{2} \end{bmatrix}$

6. $\begin{bmatrix} 1 & 0 & 0 \\ .3 & .4 & .3 \\ .1 & .8 & .1 \end{bmatrix}$

7. $\begin{bmatrix} .7 & .3 & 0 \\ .2 & .8 & 0 \\ .3 & .3 & .4 \end{bmatrix}$

8. $\begin{bmatrix} 0 & 1 & 0 \\ 0 & 0 & 1 \\ \frac{1}{4} & 0 & \frac{3}{4} \end{bmatrix}$

In Exercises 9–16, find the steady-state vector for the given transition matrix.

9. $\begin{bmatrix} \frac{1}{3} & \frac{2}{3} \\ \frac{1}{4} & \frac{3}{4} \end{bmatrix}$

10. $\begin{bmatrix} .8 & .2 \\ .6 & .4 \end{bmatrix}$

11. $\begin{bmatrix} .5 & .5 \\ .2 & .8 \end{bmatrix}$

12. $\begin{bmatrix} .9 & .1 \\ 1 & 0 \end{bmatrix}$

13. $\begin{bmatrix} 0 & 1 & 0 \\ \frac{1}{8} & \frac{5}{8} & \frac{1}{4} \\ 1 & 0 & 0 \end{bmatrix}$

14. $\begin{bmatrix} .6 & .4 & 0 \\ .3 & .4 & .3 \\ 0 & .6 & .4 \end{bmatrix}$

15. $\begin{bmatrix} .2 & 0 & .8 \\ 0 & .6 & .4 \\ .3 & .4 & .3 \end{bmatrix}$ **16.** $\begin{bmatrix} .1 & .1 & .8 \\ .2 & .2 & .6 \\ .3 & .3 & .4 \end{bmatrix}$

17. A psychologist conducts an experiment in which a mouse is placed in a T-maze where it has a choice at the T-junction of turning left and receiving a reward (cheese) or turning right and receiving a mild shock. At the end of each trial a record is kept of the mouse's response. It is observed that the mouse is as likely to turn left (state 1) as right (state 2) during the first trial. In subsequent trials, however, the observation is made that if the mouse had turned left in the previous trial, then the probability that it will turn left in the next trial is 0.8, whereas the probability that it will turn right is 0.2. If the mouse had turned right in the previous trial, then the probability that it will turn right in the next trial is 0.1, whereas the probability that it will turn left is 0.9. In the long run, what percent of the time will the mouse turn left at the T-junction?

18. Twenty percent of the commuters within a large metropolitan area currently use the public transportation system, whereas the remaining 80 percent commute via automobile. The city has recently revitalized and expanded its public transportation system. It is expected that six months from now 30 percent of those who are now commuting to work via automobile will switch to public transportation, and 70 percent will continue to commute via automobile. At the same time, it is expected that 20 percent of those now using public transportation will commute via automobile, and 80 percent will continue to use public transportation. In the long run, what percentage of the commuters will be using public transportation?

19. From data compiled over a ten-year period by Manpower, Inc., in a statewide study of married couples in which at least one spouse was working, the following transition matrix was constructed. It gives the transitional probabilities for one and two wage earners among married couples.

		Next State	
		1 Wage Earner	2 Wage Earners
Current State	1 Wage Earner	.72	.28
	2 Wage Earners	.12	.88

At the present time, 48 percent of the married couples (in which at least one spouse is working) have one wage earner and 52 percent have two wage earners. Assuming that this trend continues, what will the distribution of one- and two-wage-earner families be among married couples in this area (a) ten years from now? (b) over the long run?

20. From data compiled over a five-year period by the *Women's Daily* in a study of the number of women in the professions, the following transition matrix was constructed. It gives the transitional probabilities for the number of men and women in the professions.

		Next State	
		Men	Women
Current State	Men	.95	.05
	Women	.04	.96

As of the beginning of 1981, 52.9 percent of professional jobs were held by men.

    a. Assuming that this trend continues through 1986, show that women will become the "new majority" in 1986.

    b. If this trend continues, what will the percentage of women holding professional jobs be in the long run?

21. From data collected by the Association of Realtors of a certain city, the following transition matrix was obtained. The matrix describes the buying pattern of home buyers who buy single-family homes (S) or condominiums (C).

		Next State	
		S	C
Current State	S	.85	.15
	C	.35	.65

Currently, 80 percent of the homeowners live in single-family homes, whereas 20 percent live in condominiums. If this trend continues, determine the percentage of homeowners in this city who will live in single-family homes and condominiums (a) two years from now and (b) in the long run.

22. A study conducted by the Urban Energy Commission in a large metropolitan area indicates the probabilities that homeowners within the area will use certain heating fuels or solar energy during the next ten years as the major source of heat for their homes. The following transition matrix represents the transition probabilities from one state to another.

	Elec.	Gas	Oil	Solar
Electricity	.70	.15	.05	.10
Natural gas	0	.90	.02	.08
Fuel oil	0	.20	.75	.05
Solar energy	0	.05	0	.95

Among the homeowners within the area, 20 percent currently use electricity, 35 percent use natural gas, 40 percent use oil, and 5 percent use solar energy as their major source of heat for their homes. In the long run, what percentage of homeowners within the area will be using solar energy as their major source of heating fuel?

23. A television poll was conducted among regular viewers of the national news in a certain region where the three national networks share the same time slot for the evening news. Results of the poll indicate that 30 percent of the viewers watched the ABC evening news, 40 percent watched the CBS evening news, and 30 percent watched the NBC evening news. Furthermore, it was found that of those viewers who watched the ABC evening news during one week, 80 percent would again watch the ABC evening news during the next week, 10 percent would watch the CBS news, and 10 percent would watch the NBC news. Of those viewers who watched the CBS evening news during one week, 85 percent would again watch the CBS evening news during the next week, 10 percent would watch the ABC news, and 5 percent would watch the NBC news. Of those viewers who watched the NBC evening news during one week, 85 percent would again watch the NBC news during the next week, 10 percent would watch ABC, and 5 percent would watch CBS.

a. What share of the audience consisting of regular viewers of the national news will each network command after two weeks?

b. In the long run, what share of the audience will each network command?

24. Refer to Exercise 23. If the initial distribution vector was

$$
\begin{array}{ccc}
\text{ABC} & \text{CBS} & \text{NBC} \\
X_0 = [.40 & .40 & .20]
\end{array}
$$

what would be the long-run distribution of viewers?

25. In a certain species of roses, a plant with genotype (genetic makeup) AA has red flowers, a plant with genotype Aa has pink flowers, and a plant with genotype aa has white flowers, where A is the dominant gene and a is the recessive gene for color. If a plant with one genotype is crossed with another plant, then the color of the offspring's flowers is determined by the genotype of the parent plants. If a plant of each genotype is crossed with a pink-flowered plant, then the transition matrix used to determine the color of the offspring's flowers is given by

Offspring

$$
\begin{array}{c}
 \\
\text{Parent} \\
 \\
\end{array}
\begin{array}{c}
\text{Red (AA)} \\
\text{Pink (Aa) or (aA)} \\
\text{White (aa)}
\end{array}
\begin{array}{ccc}
\text{Red} & \text{Pink} & \text{White} \\
\begin{bmatrix} \frac{1}{2} & \frac{1}{2} & 0 \\ \frac{1}{4} & \frac{1}{2} & \frac{1}{4} \\ 0 & \frac{1}{2} & \frac{1}{2} \end{bmatrix}
\end{array}
$$

If the offspring of each generation are crossed only with pink-flowered plants, in the long run what percentage of the plants will have red flowers? pink flowers? white flowers?

*26. Let T be a regular stochastic matrix. Show that the steady-state distribution vector X may be found by solving the vector equation $XT = X$ together with the condition that the sum of the elements of X be equal to 1. [*Hint:* Take the initial distribution vector to be X, the steady-state distribution vector. Then when n is large, $X \approx XT^n$. Why? Multiply both sides of the last equation by T (on the right) and consider the resulting equation when n is large.]

Solutions to Self-Check Exercises

9.2

1. Let $X = [x \ \ y]$ be the steady-state distribution vector associated with the Markov process, where the numbers x and y are to be determined. The condition $XT = X$ translates into the matrix equation

$$
[x \ \ y] \begin{bmatrix} .5 & .5 \\ .8 & .2 \end{bmatrix} = [x \ \ y],
$$

which is equivalent to the system of linear equations

$$
0.5x + 0.8y = x
$$
$$
0.5x + 0.2y = y.
$$

Each equation in the system is equivalent to the equation

$$0.5x - 0.8y = 0.$$

Next, the condition that the sum of the elements of X add up to 1 gives

$$x + y = 1.$$

Thus, the fulfillment of the two conditions simultaneously implies that x and y be the solutions of the system

$$0.5x - 0.8y = 0$$
$$x + y = 1.$$

Solving the first equation for x, we obtain

$$x = \frac{8}{5}y,$$

which upon substitution into the second yields

$$\frac{8}{5}y + y = 1$$

or

$$y = \frac{5}{13}.$$

Therefore, $x = 8/13$, and the required steady-state matrix is [8/13 5/13].

2. The transition matrix for the Markov process under consideration is

$$T = \begin{bmatrix} .80 & .05 & .15 \\ .05 & .90 & .05 \\ .10 & .15 & .75 \end{bmatrix}.$$

Now, let $X = [x \ y \ z]$ be the steady-state distribution vector associated with the Markov process under consideration, where x, y, and z are to be determined. The condition $XT = X$ is

$$[x \ y \ z] \begin{bmatrix} .80 & .05 & .15 \\ .05 & .90 & .05 \\ .10 & .15 & .75 \end{bmatrix} = [x \ y \ z]$$

or, equivalently, the following system of linear equations:

$$0.8\ x + 0.05y + 0.10z = x$$
$$0.05x + 0.9\ y + 0.15z = y$$
$$0.15x + 0.05y + 0.75z = z.$$

This system simplifies into

$$4x - y - 2z = 0$$
$$x - 2y + 3z = 0$$
$$3x + y - 5z = 0.$$

Since $x + y + z = 1$ as well, we are required to solve the system

$$
\begin{aligned}
4x - y - 2z &= 0 \\
x - 2y + 3z &= 0 \\
3x + y - 5z &= 0 \\
x + y + z &= 1.
\end{aligned}
$$

Using the Gauss-Jordan elimination procedure, we find

$$
x = \frac{1}{4}, \quad y = \frac{1}{2}, \quad \text{and} \quad z = \frac{1}{4}
$$

Therefore, in the long run, supermarkets A and C will each have one-quarter of the business, and supermarket B will have half of the business.

9.3

Absorbing Markov Chains

▶ Absorbing Markov Chains

▶ Applications

▶ Absorbing Markov Chains

In this section we will investigate the long-term trends of a certain class of Markov chains that involve transition matrices that are not regular. In particular, we will study Markov chains in which the transition matrices, known as absorbing stochastic matrices, have the special properties to be described presently.

Consider the stochastic matrix

$$
\begin{bmatrix}
1 & 0 & 0 & 0 \\
0 & 1 & 0 & 0 \\
.2 & .3 & .5 & 0 \\
0 & 1 & 0 & 0
\end{bmatrix}
$$

associated with a Markov process. Interpreting it in the usual fashion we see that after one observation, the probability is 1 (a certainty) that an object previously in state 1 will remain in state 1. Similarly we see that an object previously in state 2 must remain in state 2. Next, we find that the probability that an object previously in state 3 has a probability of 0.2 of going to state 1, a probability of 0.3 of going to state 2, a probability of 0.5 of remaining in state 3, and no chance of going to state 4. Finally, an object previously in state 4 must, after one observation, end up in state 2.

This stochastic matrix exhibits certain special characteristics. First of all, as observed earlier, an object in state 1 or state 2 must stay in state 1 or state 2, respectively. Such states are called absorbing states. In general, an **absorbing state** is one from which it is impossible for an object to leave. To identify the absorbing states of a stochastic matrix we simply examine each row of the matrix. If Row i has a 1 in the a_{ii} position (that is, on the main diagonal of the matrix) and zeros elsewhere in that row, then, and only then, is state i an absorbing state.

Second, observe that states 3 and 4, although not absorbing states, have the property that an object in each of these states has a possibility of going to an absorbing state. For example, an object currently in state 3 has a probability of 0.2 of ending up in state 1, an absorbing state, and an object in state 4 must end up in state 2, also an absorbing state, after one transition.

Absorbing Stochastic Matrix

An absorbing stochastic matrix has the following properties:

1. There is at least one absorbing state.

2. It is possible to go from any nonabsorbing state to any absorbing state in one or more stages.

A Markov chain is said to be an *absorbing Markov chain* if the transition matrix associated with the process is an absorbing stochastic matrix.

EXAMPLE

11 Determine whether the following matrices are absorbing stochastic matrices.

$$a. \begin{bmatrix} .7 & 0 & .3 & 0 \\ 0 & 1 & 0 & 0 \\ .1 & .5 & .2 & .2 \\ 0 & 0 & 0 & 1 \end{bmatrix} \qquad b. \begin{bmatrix} 1 & 0 & 0 & 0 \\ 0 & 1 & 0 & 0 \\ 0 & 0 & .5 & .5 \\ 0 & 0 & .4 & .6 \end{bmatrix}$$

SOLUTION

a. States 2 and 4 are both absorbing states. Furthermore, even though state 1 is not an absorbing state there is a possibility (with probability 0.3) for an object to go from this state to state 3. State 3 itself is nonabsorbing, but an object in that state has a probability of 0.5 of going to the absorbing state 2 and a probability of 0.2 of going to the absorbing state 4. Thus, the given matrix is an absorbing stochastic matrix.

b. States 1 and 2 are absorbing states. However, it is impossible for an object to go from the nonabsorbing states 3 and 4 to either or both of the absorbing states. Thus, the given matrix is not an absorbing stochastic matrix. ◀

Given an absorbing stochastic matrix, it is always possible, by suitably reordering the states if necessary, to rewrite it so that the absorbing states appear first. Then the resulting matrix can be partitioned into four submatrices

$$
\begin{array}{l}
\text{Absorbing} \quad \{ \\
\\
\text{Nonabsorbing} \quad \{
\end{array}
\left[
\begin{array}{c|c}
I & O \\
\hline
R & S
\end{array}
\right]
$$

where I is an identity matrix whose order is determined by the number of absorbing states and O is a zero matrix. The submatrices R and S correspond to the nonabsorbing states. As an example, the absorbing stochastic matrix of Example 11(a)

$$
\begin{array}{c}
\\
1 \\
2 \\
3 \\
4
\end{array}
\begin{array}{cccc}
1 & 2 & 3 & 4 \\
\left[\begin{array}{cccc}
.7 & 0 & .3 & 0 \\
0 & 1 & 0 & 0 \\
.1 & .5 & .2 & .2 \\
0 & 0 & 0 & 1
\end{array}\right]
\end{array}
\quad \text{may be written as} \quad
\begin{array}{c}
\\
4 \\
2 \\
\\
1 \\
3
\end{array}
\begin{array}{cccc}
4 & 2 & 1 & 3 \\
\left[\begin{array}{cc|cc}
1 & 0 & 0 & 0 \\
0 & 1 & 0 & 0 \\
\hline
0 & 0 & .7 & .3 \\
.2 & .5 & .1 & .2
\end{array}\right]
\end{array}
$$

upon reordering the states as indicated.

▶ Applications

EXAMPLE

12 *Gambler's ruin.* John has decided to risk \$2 in the following game of chance. He places a \$1 bet on each repeated play of the game in which the probability of his winning \$1 is 0.4, and he continues to play until he has accumulated a total of \$3 or he has lost all of his money. Write the transition matrix for the related absorbing Markov chain.

SOLUTION

There are four states in this Markov chain, which correspond to John accumulating a total of \$0, \$1, \$2, and \$3. Since the first and last states listed are absorbing states, we will list these states first, resulting in the transition matrix

$$
\begin{array}{c}
\\
\$0 \\
\$3 \\
\\
\$1 \\
\$2
\end{array}
\begin{array}{cccc}
\$0 & \$3 & \$1 & \$2 \\
\left[\begin{array}{cc|cc}
1 & 0 & 0 & 0 \\
0 & 1 & 0 & 0 \\
\hline
.6 & 0 & 0 & .4 \\
0 & .4 & .6 & 0
\end{array}\right]
\end{array}
$$

which is constructed as follows: Since the state "\$0" is an absorbing state, we see that $a_{11} = 1$, $a_{12} = a_{13} = a_{14} = 0$. Similarly, the state "$3" is an absorbing state, so $a_{22} = 1$ and $a_{21} = a_{23} = a_{24} = 0$. To construct the row corresponding to the nonabsorbing state "$1," we note that there is a probability of 0.6 (John loses) in going from an accumulated amount of $1 to $0, so $a_{31} = 0.6$; $a_{32} = a_{33} = 0$ because it is not feasible to go from an accumulated amount of $1 to either an accumulated amount of $3 or $1 in one transition (play). Finally, there is a probability of 0.4 (John wins) in going from an accumulated amount of $1 to an accumulated amount of $2, so $a_{34} = 0.4$. The last row of the transition matrix is constructed by reasoning in a similar manner. ◀

The following question arises in connection with the last example. If John continues to play the game as he had decided, what is the probability that he would depart from the game victorious, that is, leave with an accumulated amount of $3?

In order to answer this question we have to look at the long-term trend of the relevant Markov chain. Taking a cue from our work in the last section, we may compute the powers of the transition matrix associated with the Markov chain. Just as in the case of regular stochastic matrices, it turns out that the powers of an absorbing stochastic matrix approach a steady-state matrix. However, instead of demonstrating this we will make use of the following result, which we state without proof, for computing the steady-state matrix:

Suppose an absorbing stochastic matrix A has been partitioned into submatrices

$$A = \left[\begin{array}{c|c} I & O \\ \hline R & S \end{array} \right]$$

Then the *steady-state matrix* of A is given by

$$\left[\begin{array}{c|c} I & O \\ \hline (I - S)^{-1} R & O \end{array} \right]$$

where the order of the identity matrix appearing in the expression $(I - S)^{-1}$ is chosen to have the same order as S.

EXAMPLE

13 *Gambler's ruin continued.* Refer to Example 12. If John continues to play the game until he has accumulated a sum of $3 or until he has lost all his money, what is the probability that he will accumulate $3?

SOLUTION   The transition matrix associated with the Markov process is (see Example 12)

$$A = \begin{bmatrix} 1 & 0 & 0 & 0 \\ 0 & 1 & 0 & 0 \\ \hline .6 & 0 & 0 & .4 \\ 0 & .4 & .6 & 0 \end{bmatrix}$$

We need to find the steady-state matrix of A. In this case,

$$S = \begin{bmatrix} 0 & .4 \\ .6 & 0 \end{bmatrix},$$

so

$$I - S = \begin{bmatrix} 1 & 0 \\ 0 & 1 \end{bmatrix} - \begin{bmatrix} 0 & .4 \\ .6 & 0 \end{bmatrix} = \begin{bmatrix} 1 & -.4 \\ -.6 & 1 \end{bmatrix}.$$

Using the technique of Section 2.5, we find that

$$(I - S)^{-1} = \begin{bmatrix} 1.32 & .53 \\ .79 & 1.32 \end{bmatrix}$$

and so

$$(I - S)^{-1} R = \begin{bmatrix} 1.32 & .53 \\ .79 & 1.32 \end{bmatrix} \begin{bmatrix} .6 & 0 \\ 0 & .4 \end{bmatrix} = \begin{bmatrix} .79 & .21 \\ .47 & .53 \end{bmatrix}.$$

Therefore, the required steady-state matrix of A is given by

$$\begin{bmatrix} I & \vdots & O \\ \hline (I - S)^{-1} R & \vdots & O \end{bmatrix} = \begin{array}{c} \\ \$0 \\ \$3 \\ \$1 \\ \$2 \end{array} \begin{bmatrix} 1 & 0 & 0 & 0 \\ 0 & 1 & 0 & 0 \\ \hline .79 & .21 & 0 & 0 \\ .47 & .53 & 0 & 0 \end{bmatrix}$$

From this result we see that starting with $2, the probability is 0.53 that John will leave the game with an accumulated amount of $3, that is, that he wins $1. ◀

Our last example shows an application of Markov chains in the field of genetics.

EXAMPLE

14   In a certain species of roses, a plant of genotype (genetic makeup) AA has red flowers, a plant of genotype Aa has pink flowers, and a plant of genotype aa has white flowers, where A is the dominant gene and a is the recessive gene for color. If a plant of one genotype is crossed with another plant, then the color of the offspring's flowers is determined by the genotype of the parent plants. If the offspring are crossed successively with plants of genotype AA only, show that in the long run all the flowers produced by the plants will be red.

SOLUTION   First of all, let us construct the transition matrix associated with the resulting Markov chain. In crossing a plant of genotype AA with another of the same genotype AA, the offspring will inherit one dominant gene from each parent and thus will have genotype AA. Therefore, the probabilities of the offspring being genotype AA, Aa, and aa are 1, 0, and 0, respectively.

Next, in crossing a plant of genotype AA with one of genotype Aa, the probability of the offspring having genotype AA (inheriting an A gene from the first parent and an A from the second) is 1/2; the probability of the offspring being of genotype Aa (inheriting an A gene from the first parent and an a gene from the second parent) is 1/2; finally, the probability of the offspring being of genotype aa is 0, since this is clearly impossible.

A similar argument shows that in crossing a plant of genotype AA with one of genotype aa, the probabilities of the offspring having genotype AA, Aa, and aa are 0, 1, and 0, respectively.

The required transition matrix is thus given by

$$T = \begin{array}{c} \\ AA \\ Aa \\ aa \end{array} \begin{array}{c} \begin{array}{ccc} AA & Aa & aa \end{array} \\ \begin{bmatrix} 1 & 0 & 0 \\ \frac{1}{2} & \frac{1}{2} & 0 \\ 0 & 1 & 0 \end{bmatrix} \end{array}$$

Observe that the state AA is an absorbing state. Furthermore, it is possible to go from each of the other two nonabsorbing states to the absorbing state AA. Thus, the Markov chain is an absorbing Markov chain. To determine the long-term effects of this experiment, let us compute the steady-state matrix of T. Partitioning T in the usual manner, we find

$$T = \begin{bmatrix} 1 & 0 & 0 \\ \hline \frac{1}{2} & \frac{1}{2} & 0 \\ 0 & 1 & 0 \end{bmatrix}$$

so that

$$R = \begin{bmatrix} \frac{1}{2} \\ 0 \end{bmatrix} \quad \text{and} \quad S = \begin{bmatrix} \frac{1}{2} & 0 \\ 1 & 0 \end{bmatrix}.$$

Next, we compute

$$I - S = \begin{bmatrix} 1 & 0 \\ 0 & 1 \end{bmatrix} - \begin{bmatrix} \frac{1}{2} & 0 \\ 1 & 0 \end{bmatrix} = \begin{bmatrix} \frac{1}{2} & 0 \\ -1 & 1 \end{bmatrix}$$

and, using the technique of Section 2.5,

$$(I - S)^{-1} = \begin{bmatrix} 2 & 0 \\ 2 & 1 \end{bmatrix},$$

so

$$(I - S)^{-1} R = \begin{bmatrix} 2 & 0 \\ 2 & 1 \end{bmatrix} \begin{bmatrix} \frac{1}{2} \\ 0 \end{bmatrix} = \begin{bmatrix} 1 \\ 1 \end{bmatrix}.$$

Therefore, the steady-state matrix of T is given by

$$
\begin{bmatrix}
I & | & O \\
\hline
(I - S)^{-1} R & | & O
\end{bmatrix}
=
\begin{array}{c}
\\
AA \\
Aa \\
aa
\end{array}
\begin{array}{ccc}
AA & Aa & aa
\end{array}
\begin{bmatrix}
1 & | & 0 & 0 \\
\hline
1 & | & 0 & 0 \\
1 & | & 0 & 0
\end{bmatrix}
$$

Interpreting the steady-state matrix of T, we see that the long-term result of crossing the offspring with plants of genotype AA only leads to the absorbing state AA. In other words, such a procedure will result in the production of plants that will bear only red flowers, as we set out to demonstrate. ◄

Self-Check Exercises

9.3

1. Let

$$
T = \begin{bmatrix}
.2 & .3 & .5 \\
0 & 1 & 0 \\
0 & .6 & .4
\end{bmatrix}.
$$

a. Show that T is an absorbing stochastic matrix.

b. Rewrite T so that the absorbing states appear first, partition the resulting matrix, and identify the submatrices R and S.

c. Compute the steady-state matrix of T.

2. Recently there has been a trend toward increased use of computer-aided transcription (CAT) and electronic recording (ER) as alternatives to manual transcription (MT) of court proceedings by court stenographers in a certain state. Suppose that the following stochastic matrix is the transition matrix associated with the Markov process.

$$
\begin{array}{c}
\\
CAT \\
T = ER \\
MT
\end{array}
\begin{array}{ccc}
CAT & ER & MT
\end{array}
\begin{bmatrix}
1 & 0 & 0 \\
.3 & .6 & .1 \\
.2 & .3 & .5
\end{bmatrix}
$$

Determine the probability that a court now using electronic recording or transcribing its proceedings manually will eventually change to computer-aided transcription.

Solutions to Self-Check Exercises 9.3 can be found on page 521.

▶
Exercises 9.3

In Exercises 1–8, determine whether the given matrix is an absorbing stochastic matrix.

1. $\begin{bmatrix} \frac{2}{5} & \frac{3}{5} \\ 0 & 1 \end{bmatrix}$

2. $\begin{bmatrix} 1 & 0 \\ 0 & 1 \end{bmatrix}$

3. $\begin{bmatrix} 1 & 0 & 0 \\ .5 & 0 & .5 \\ 0 & 1 & 0 \end{bmatrix}$

4. $\begin{bmatrix} 1 & 0 & 0 \\ 0 & .7 & .3 \\ 0 & .2 & .8 \end{bmatrix}$

5. $\begin{bmatrix} \frac{1}{8} & \frac{1}{4} & \frac{5}{8} \\ 0 & 1 & 0 \\ 0 & 0 & 1 \end{bmatrix}$

6. $\begin{bmatrix} 1 & 0 & 0 & 0 \\ 0 & \frac{5}{8} & \frac{1}{8} & \frac{1}{4} \\ 0 & 0 & 1 & 0 \\ 0 & \frac{1}{6} & 0 & \frac{5}{6} \end{bmatrix}$

7. $\begin{bmatrix} 1 & 0 & 0 & 0 \\ 0 & 1 & 0 & 0 \\ .3 & .2 & .1 & .4 \\ 0 & 0 & .5 & .5 \end{bmatrix}$

8. $\begin{bmatrix} 1 & 0 & 0 & 0 \\ 0 & 1 & 0 & 0 \\ 0 & 0 & .2 & .8 \\ 0 & 0 & .6 & .4 \end{bmatrix}$

In Exercises 9–14, rewrite each of the given absorbing stochastic matrices so that the absorbing states appear first, partition the resulting matrix, and identify the submatrices R and S.

9. $\begin{bmatrix} .6 & .4 \\ 0 & 1 \end{bmatrix}$

10. $\begin{bmatrix} \frac{1}{4} & \frac{1}{4} & \frac{1}{2} \\ 0 & 1 & 0 \\ 0 & 0 & 1 \end{bmatrix}$

11. $\begin{bmatrix} 0 & .5 & .5 \\ .2 & .4 & .4 \\ 0 & 0 & 1 \end{bmatrix}$

12. $\begin{bmatrix} .5 & 0 & .5 \\ 0 & 1 & 0 \\ .3 & .1 & .6 \end{bmatrix}$

13. $\begin{bmatrix} .4 & .2 & 0 & .4 \\ .2 & .3 & .3 & .2 \\ 0 & 0 & 1 & 0 \\ 0 & 0 & 0 & 1 \end{bmatrix}$

14. $\begin{bmatrix} .1 & .2 & .3 & .4 \\ 0 & 1 & 0 & 0 \\ 0 & 0 & 1 & 0 \\ 0 & .2 & 0 & .8 \end{bmatrix}$

In Exercises 15–24, compute the steady-state matrix of each of the given stochastic matrices.

15. $\begin{bmatrix} .55 & .45 \\ 0 & 1 \end{bmatrix}$

16. $\begin{bmatrix} 1 & 0 & 0 \\ 0 & 1 & 0 \\ .2 & .3 & .5 \end{bmatrix}$

17. $\begin{bmatrix} 1 & 0 & 0 \\ .2 & .4 & .4 \\ .3 & .2 & .5 \end{bmatrix}$

18. $\begin{bmatrix} \frac{1}{5} & 0 & \frac{4}{5} \\ 0 & 1 & 0 \\ 0 & \frac{3}{8} & \frac{5}{8} \end{bmatrix}$

19. $\begin{bmatrix} \frac{1}{2} & \frac{1}{2} & 0 & 0 \\ 0 & 1 & 0 & 0 \\ \frac{1}{3} & 0 & \frac{2}{3} & 0 \\ 0 & 0 & 0 & 1 \end{bmatrix}$

20. $\begin{bmatrix} 1 & 0 & 0 & 0 \\ \frac{1}{8} & \frac{1}{8} & \frac{1}{4} & \frac{1}{2} \\ \frac{1}{3} & 0 & \frac{2}{3} & 0 \\ 0 & 0 & 0 & 1 \end{bmatrix}$

21. $\begin{bmatrix} 1 & 0 & 0 & 0 \\ 0 & 1 & 0 & 0 \\ \frac{1}{4} & \frac{1}{4} & \frac{1}{2} & 0 \\ \frac{1}{3} & \frac{1}{3} & 0 & \frac{1}{3} \end{bmatrix}$

22. $\begin{bmatrix} 1 & 0 & 0 & 0 \\ 0 & 1 & 0 & 0 \\ .2 & .4 & 0 & .4 \\ .1 & .2 & .4 & .3 \end{bmatrix}$

23. $\begin{bmatrix} 1 & 0 & 0 & 0 & 0 \\ 0 & 1 & 0 & 0 & 0 \\ 0 & 0 & 1 & 0 & 0 \\ .2 & .1 & .3 & .2 & .2 \\ .1 & .2 & .1 & .2 & .4 \end{bmatrix}$

24. $\begin{bmatrix} 1 & 0 & 0 & 0 & 0 \\ 0 & 1 & 0 & 0 & 0 \\ \frac{1}{4} & 0 & \frac{1}{4} & \frac{1}{2} & 0 \\ \frac{1}{3} & \frac{1}{3} & \frac{1}{3} & 0 & 0 \\ 0 & \frac{1}{2} & 0 & \frac{1}{2} & 0 \end{bmatrix}$

25. As more and more old cars are taken off the road and replaced by late models that use unleaded fuel, the consumption of leaded gasoline will continue to drop. Suppose that the following transition matrix describes this Markov process,

$$A = \begin{array}{c} \\ L \\ UL \end{array} \begin{array}{c} L \quad UL \\ \begin{bmatrix} .80 & .20 \\ 0 & 1 \end{bmatrix} \end{array}$$

where L denotes leaded gasoline and UL denotes unleaded gasoline.

a. Show that A is an absorbing stochastic matrix and rewrite it so that the absorbing state appears first. Partition the resulting matrix and identify the submatrices R and S.

b. Compute the steady-state matrix of A and interpret your results.

26. Diane has decided to play the following game of chance. She places a $1 bet on each repeated play of the game in which the probability of her winning $1 is 0.5. She has further decided to continue playing the game until she has either accumulated a total of $3 or has lost all her money. What is the probability that Diane will eventually leave the game a winner if she started with a capital of $1? of $2?

27. Refer to Exercise 26. Suppose that Diane has decided to stop playing only after she has accumulated a sum of $4 or has lost all her money. All other conditions being the same, what is the probability that Diane will leave the game a winner if she started with a capital of $1? of $2? of $3?

28. Because of a proliferation of more affordable automated office equipment, more and more companies are turning to them as replacements for obsolete equipment. The following transition matrix describes the Markov process. Here M stands for manual typewriter, E stands for electric typewriter, and W stands for electric typewriters with some form of word-processing capability.

$$A = \begin{array}{c} \\ M \\ E \\ W \end{array} \begin{array}{c} M \quad E \quad W \\ \begin{bmatrix} .10 & .70 & .20 \\ 0 & .60 & .40 \\ 0 & 0 & 1 \end{bmatrix} \end{array}$$

a. Show that A is an absorbing stochastic matrix and rewrite it so that the absorbing state appears first. Partition the resulting matrix and identify the submatrices R and S.

b. Compute the steady-state matrix A and interpret your results.

29. The registrar of the Computronics Institute has compiled the following statistics on the progress of the school's students in their two-year computer programming course leading to an associate degree: 75 percent of beginning students in a particular year successfully complete their first year of studies and move on to the second year, whereas 25 percent drop out of the program; 90 percent of second-year students in a particular year go on to graduate at the end of the year, whereas 10 percent drop out of the program.

a. Construct the transition matrix associated with this Markov process.

b. Compute the steady-state matrix.

c. Determine the probability that a beginning student enrolled in the program will complete the course successfully.

30. Refer to Example 14. If the offspring are crossed successively with plants of genotype aa only, show that in the long run all the flowers produced by the plants will be white.

Solutions to Self-Check Exercises

9.3

1. a. State 2 is an absorbing state. States 1 and 3 are not absorbing, but each has a possibility (with probability 0.3 and 0.6) for an object to go from these states to state 2. Therefore, the matrix T is an absorbing stochastic matrix.

b. Reordering the states as indicated, we rewrite

$$
\begin{array}{c c c c}
 & 1 & 2 & 3 \\
\begin{array}{c} 1 \\ 2 \\ 3 \end{array} &
\left[\begin{array}{ccc}
.2 & .3 & .5 \\
0 & 1 & 0 \\
0 & .6 & .4
\end{array}\right]
\end{array}
$$

in the form

$$
\begin{array}{c c c c}
 & 2 & 3 & 1 \\
\begin{array}{c} 2 \\ 3 \\ 1 \end{array} &
\left[\begin{array}{c c c}
1 & 0 & 0 \\
\hline
.6 & .4 & 0 \\
.3 & .5 & .2
\end{array}\right]
\end{array}.
$$

We see that

$$
R = \begin{bmatrix} .6 \\ .3 \end{bmatrix} \quad \text{and} \quad S = \begin{bmatrix} .4 & 0 \\ .5 & .2 \end{bmatrix}
$$

c. We compute

$$
I - S = \begin{bmatrix} 1 & 0 \\ 0 & 1 \end{bmatrix} - \begin{bmatrix} .4 & 0 \\ .5 & .2 \end{bmatrix} = \begin{bmatrix} .6 & 0 \\ -.5 & .8 \end{bmatrix}
$$

and, using the techniques of Section 2.5,

$$
(I - S)^{-1} = \begin{bmatrix} 1.67 & 0 \\ 1.04 & 1.25 \end{bmatrix}
$$

and so

$$
(I - S)^{-1} R = \begin{bmatrix} 1.67 & 0 \\ 1.04 & 1.25 \end{bmatrix} \begin{bmatrix} .6 \\ .3 \end{bmatrix} = \begin{bmatrix} 1 \\ 1 \end{bmatrix}.
$$

Therefore, the steady-state matrix of T is

$$\begin{bmatrix} 1 & 0 & 0 \\ 1 & 0 & 0 \\ 1 & 0 & 0 \end{bmatrix}.$$

2. We want to compute the steady-state matrix of T. Note that T is in the form

$$\begin{bmatrix} I & O \\ R & S \end{bmatrix}$$

where

$$R = \begin{bmatrix} .3 \\ .2 \end{bmatrix} \quad \text{and} \quad S = \begin{bmatrix} .6 & .1 \\ .3 & .5 \end{bmatrix}.$$

We compute

$$I - S = \begin{bmatrix} 1 & 0 \\ 0 & 1 \end{bmatrix} - \begin{bmatrix} .6 & .1 \\ .3 & .5 \end{bmatrix} = \begin{bmatrix} .4 & -.1 \\ -.3 & .5 \end{bmatrix}$$

and using the technique of Section 2.5,

$$(I - S)^{-1} = \begin{bmatrix} 2.94 & 0.59 \\ 1.77 & 2.36 \end{bmatrix}$$

so

$$(I - S)^{-1} R = \begin{bmatrix} 2.94 & 0.59 \\ 1.77 & 2.36 \end{bmatrix} \begin{bmatrix} .3 \\ .2 \end{bmatrix} = \begin{bmatrix} 1 \\ 1 \end{bmatrix}.$$

Therefore, the steady-state matrix of T is

$$\begin{array}{c} \\ \text{CAT} \\ \text{ER} \\ \text{MT} \end{array} \begin{array}{ccc} \text{CAT} & \text{ER} & \text{MT} \\ \begin{bmatrix} 1 & 0 & 0 \\ 1 & 0 & 0 \\ 1 & 0 & 0 \end{bmatrix}. \end{array}$$

Interpreting the steady-state matrix of T, we see that in the long run all courts in this state will use computer-aided transcription.

9.4

Game Theory and Strictly Determined Games

▶ Two-Person Games

▶ Optimal Strategies

The theory of games is a relatively new branch of mathematics and owes much of its development to John von Neuman (1903–1957), one of the mathematical giants of this century. Basically, the theory of games combines matrix methods with the theory of probability to determine the optimal strategies to be employed by two or more opponents involved in a competitive situation, each opponent seeking to maximize his or her "gains," or equivalently to minimize his or her "losses." As such, the players may be poker players, managers of rival corporations seeking to extend their share of the market, campaign managers, or generals of opposing armies, to name a few.

For simplicity, we will limit our discussion to games with two players. Such games are, naturally enough, called two-person games.

▶ Two-Person Games

EXAMPLE

15 Richie and Chuck play a coin-matching game in which each player selects a side of a penny without prior knowledge of the other's choice. Then upon a predetermined signal both players disclose their choices simultaneously. Chuck agrees to pay Richie a sum of $3 if both choose heads; if Richie chooses heads and Chuck chooses tails, then Richie pays Chuck $6; if Richie chooses tails and Chuck chooses heads, then Chuck pays Richie $2; finally, if both Richie and Chuck choose tails, then Chuck pays Richie $1. In this game, the objective of each player is to discover a strategy that will ensure that his winnings are maximized (equivalently, his losses are minimized). ◀

The coin-matching game is an example of a **zero-sum game**; that is, a game in which the payoff to one party results in an equal loss to the other. For such games, the sum of the payments made by both players at the end of each play add up to zero.

In order to facilitate the analysis of the problem, we represent the given data in the form of a matrix.

$$
\begin{array}{c}
 & & C\text{'s moves} \\
 & & \text{heads} \quad \text{tails} \\
\text{R's moves} \begin{array}{c} \text{heads} \\ \text{tails} \end{array} & \begin{bmatrix} 3 & -6 \\ 2 & 1 \end{bmatrix}
\end{array}
$$

Each row of the matrix corresponds to one of the two possible moves by Richie (referred to as the row player, R), whereas each column corresponds to one of the two possible moves by Chuck (referred to as the column player, C). Each entry in the matrix represents the payoff from C to R. For example, the entry $a_{11} = 3$ represents a $3 payoff from Chuck to Richie (C to R) when Richie chooses to play row 1 (heads) and Chuck chooses to play column 1 (heads). On the other hand, the entry $a_{12} = -6$ represents (because it's negative) a $6 payoff to C (from R) when R chooses to play row 1 (heads) and C chooses to play column 2 (tails). (Interpret the meaning of $a_{21} = 2$ and $a_{22} = 1$ for yourself.)

More generally, suppose we are given a two-person game with two players R and C. Furthermore, suppose that R has m possible moves $R_1, R_2, \ldots, R_m$ and that C has n possible moves $C_1, C_2, \ldots, C_n$. Then the game has a representation in terms of an $m \times n$ matrix in which each row of the matrix represents one of the m possible moves of R and each column of the matrix represents one of the n possible moves of C:

$$
\begin{array}{c}
\text{C's moves} \\
\begin{array}{c}
 \\
\text{R's moves}
\end{array}
\begin{array}{c}
\begin{array}{cccccc}
C_1 & C_2 & \ldots & C_j & \ldots & C_n
\end{array} \\
\begin{array}{c}
R_1 \\
R_2 \\
\vdots \\
R_i \\
\vdots \\
R_m
\end{array}
\left[
\begin{array}{cccccc}
a_{11} & a_{12} & \ldots & a_{1j} & \ldots & a_{1n} \\
a_{21} & a_{22} & \ldots & a_{2j} & \ldots & a_{2n} \\
\vdots & \vdots & & \vdots & & \vdots \\
a_{i1} & a_{i2} & \ldots & a_{ij} & \ldots & a_{in} \\
\vdots & \vdots & & \vdots & & \vdots \\
a_{m1} & a_{m2} & \ldots & a_{mj} & \ldots & a_{mn}
\end{array}
\right]
\end{array}
\end{array}
$$

The entry a_{ij} in the ith row and jth column of the (payoff) matrix represents the payoff to R when R chooses move R_i and C chooses move C_j. In this context, note that a payoff to R means, in actuality, a payoff to C in the event that the value of a_{ij} is negative.

EXAMPLE

16   The payoff matrix associated with a game is given by

$$
\begin{array}{c}
\text{C's moves} \\
\begin{array}{cc}
 & \begin{array}{ccc} C_1 & C_2 & C_3 \end{array} \\
\text{R's moves} \quad \begin{array}{c} R_1 \\ R_2 \end{array} & \left[\begin{array}{ccc} 1 & -2 & 3 \\ 4 & -5 & -1 \end{array} \right]
\end{array}
\end{array}
$$

Give an interpretation of this payoff matrix.

SOLUTION   In this two-person game, player R has two possible moves, whereas player C has three possible moves. The payoffs are determined as follows:
If R chooses R_1, then

a. R wins 1 unit if C chooses C_1

b. R loses 2 units if C chooses C_2, and

c. R wins 3 units if C chooses C_3

If R chooses R_2, then

a. R wins 4 units if C chooses C_1

b. R loses 5 units if C chooses C_2, and

c. R loses 1 unit if C chooses C_3

◀

▶ Optimal Strategies

Let us return to the payoff matrix of Example 15 and see how it may be used to help us determine the "best" strategy for each of the two players R and C. For convenience, this matrix is reproduced here:

C's moves

$$
\begin{array}{cc}
 & C_1 \quad C_2 \\
\text{R's moves} \quad \begin{array}{c} R_1 \\ R_2 \end{array} & \begin{bmatrix} 3 & -6 \\ 2 & 1 \end{bmatrix}
\end{array}
$$

Let us first consider the game from R's point of view. Since the entries in the payoff matrix represent payoffs to him, his initial reaction might be to seek out the largest entry in the matrix and consider the row containing such an entry as a possible move. Thus he is led to the consideration of R_1 as a possible move. Let us examine this choice a little more closely. To be sure, R would realize the largest possible payoff to himself ($3) if C chose C_1; but if C chose C_2, then R would lose $6! Since R does not know beforehand what C's move will be, a more prudent approach on his part would be to assume that no matter what row he chooses, C will counter with a move (column) that will result in the smallest payoff to him. To maximize the payoff to himself under these circumstances, R would then select from among the moves (rows) the one in which the smallest payoff is as large as possible. This strategy for R called, for obvious reasons, the **maximin strategy** may be summarized as follows:

Maximin Strategy

1. For each row of the payoff matrix, find the smallest entry in that row.

2. Choose the row for which the entry found in Step 1 is as large as possible. This row constitutes R's "best" move.

For the problem under consideration, the work involved may be organized as follows:

Row
minima

$$
\begin{bmatrix} 3 & -6 \\ 2 & 1 \end{bmatrix} \quad \begin{array}{l} -6 \\ ① \end{array} \leftarrow \text{Larger of the row minima}
$$

From these results, it is seen that R's "best" move is row 2. By choosing this move R stands to win at least $1.

Next, let us consider the game from C's point of view. His objective is to minimize the payoff to R. This is accomplished by choosing the column whose largest payoff is as small as possible. This strategy for C, called the **minimax strategy**, may be summarized as follows:

Minimax Strategy

1. For each column of the payoff matrix, find the largest entry in that column.

2. Choose the column for which the entry found in Step 1 is as small as possible. This column constitutes C's "best" move.

The work involved in the determination of C's "best" move may be organized as follows.

$$\begin{bmatrix} 3 & -6 \\ 2 & 1 \end{bmatrix}$$

Column maxima 3 ①
                   ↑
Smaller of the column maxima

From these results, it is seen that C's "best" move is column 2. By choosing this move C stands to lose at most $1.

EXAMPLE

17 For the game with the following payoff matrix, determine the maximin and minimax strategies for each player.

$$\begin{bmatrix} -3 & -2 & 4 \\ -2 & 0 & 3 \\ 6 & -1 & 1 \end{bmatrix}$$

SOLUTION We determine the minimum of each row and the maximum of each column of the payoff matrix and then display these numbers by circling the largest of the row minima and the smallest of the column maxima:

Row
minima

$$\begin{bmatrix} -3 & -2 & 4 \\ -2 & 0 & 3 \\ 6 & -1 & 1 \end{bmatrix} \quad \begin{matrix} -3 \\ -2 \\ Ⓐ \end{matrix}$$ ← Largest of the row minima

Column maxima 6 ⓪ 4
                   ↑
Smallest of the column maxima

From these results we conclude that the maximin strategy (for the row player) is to play row 3, whereas the minimax strategy (for the column player) is to play column 2. ◄

EXAMPLE

18   Determine the maximin and minimax strategies for each player in a game whose payoff matrix is given by

$$\begin{bmatrix} 3 & 4 & -4 \\ 2 & -1 & -3 \end{bmatrix}$$

SOLUTION   Proceeding as in the last example, we obtain the following:

$$\begin{bmatrix} 3 & 4 & -4 \\ 2 & -1 & -3 \end{bmatrix} \quad \begin{array}{c} \text{Row} \\ \text{minima} \\ -4 \\ \boxed{-3} \end{array} \leftarrow \text{Larger of the row minima}$$

Column maxima   3   4 $\boxed{-3}$

↑
Smallest of the column maxima

from which we conclude that the maximin strategy for the row player is to play row 2, whereas the minimax strategy for the column player is to play column 3.   ◀

In arriving at the maximin and minimax strategies for the respective players, we have made the assumption that each player has no prior knowledge of his opponent's choice of move. Therefore one is forced to adopt a conservative approach, in essence, making the most of the worst.

This raises the following question: Suppose a game is played repeatedly and one of the players realizes that the opponent is employing his maximin (or minimax) strategy. Can this knowledge be used to the player's advantage? To obtain a partial answer to this question, let us consider the game posed in Example 17. There, the minimax strategy for the column player is to play column 2. Suppose, in repeated plays of the game, a player consistently plays that column and this strategy becomes known to the row player. The row player may then change the strategy from playing row 3 (the maximin strategy) to playing row 2, thereby reducing losses from one unit to zero. Thus, at least for this game, the knowledge that a player is employing the maximin or minimax strategy can be used to the opponent's advantage.

There is, however, a class of games in which the knowledge that a player is using the maximin (or minimax) strategy proves of no help to the opponent. Consider, for example, the game of Example 18. There, the row player's maximin strategy is to play row 2, and the column player's minimax strategy is to play column 3. Suppose, in repeated plays of the game, R (the row player) has discovered that C (the column player) consistently chooses to play column 3 (the minimax strategy). Can this knowledge be used to R's advantage? Now, other than playing row 2 (the maximin strategy), R may choose to play row 1. But if R makes this choice, then he would lose 4 units instead of 3 units! Clearly, in this case, the knowledge that C is using the minimax strategy cannot be used to advantage. R's *optimal* (best) *strategy* is the maximin strategy.

Optimal Strategy

The **optimal strategy** in a game is the strategy that is most profitable to a particular player.

Next, suppose that in repeated plays of the game, C has discovered that R consistently plays row 2 (his optimal strategy). Can this knowledge be used to his advantage? Another glance at the payoff matrix reveals that by playing column 1, C stands to lose 2 units and by playing column 2 he stands to win 1 unit as compared to winning 3 units by playing column 3 as called for by the minimax strategy. Thus, as in the case of R, C does not benefit from the knowledge of his opponent's move. Furthermore, his optimal strategy coincides with the minimax strategy.

This game, in which the row player cannot benefit from the knowledge that his opponent is using the minimax strategy and the column player cannot benefit from the knowledge that his opponent is using the maximin strategy, is said to be *strictly determined*.

A Strictly Determined Game

A **strictly determined game** is characterized by the following properties:

1. There is an entry in the payoff matrix that is *simultaneously* the smallest entry in its row and the largest entry in its column. Such an entry is called a **saddle point** for the game.

2. The optimal strategy for the row player is precisely the maximin strategy and is the row containing the saddle point. The optimal strategy for the column player is the minimax strategy and is the column containing the saddle point.

The saddle point of a strictly determined game is also referred to as the **value of the game.** If the value of a strictly determined game is positive, then the game favors the row player. If the value is negative, it favors the column player. If the value of the game is zero, the game is called a **fair game.**

Returning to the coin-matching game discussed earlier, we conclude that Richie's optimal strategy consists of playing row 2 repeatedly, whereas Chuck's optimal strategy consists of playing column 2 repeatedly. Furthermore, the value of the game is 1, implying that the game favors the row player Richie.

EXAMPLE

19   A two-person, zero-sum game is defined by the payoff matrix

$$A = \begin{bmatrix} 1 & 2 & -3 \\ -1 & 2 & -2 \\ 2 & 3 & -4 \end{bmatrix}$$

a. Show that the game is strictly determined and find the saddle point(s) for the game.

b. What is the optimal strategy for each player?

c. What is the value of the game? Does the game favor one player over the other?

SOLUTION

a. First of all, we determine the minimum of each row and the maximum of each column of the payoff matrix A and display these minima and maxima as follows:

Row
minima

$$\begin{bmatrix} 1 & 2 & -3 \\ -1 & 2 & \boxed{-2} \\ 2 & 3 & -4 \end{bmatrix} \quad \begin{array}{l} -3 \\ -2 \quad \leftarrow \text{Largest of the row minima} \\ -4 \end{array}$$

Column maxima   2   3   −2
                        ↑
            Smallest of the column maxima

From these results, we see that the circled entry, -2, is simultaneously the smallest entry in its row and the largest entry in its column. Therefore, the game is strictly determined, with the entry $a_{23} = -2$ as its saddle point.

b. From these results, we see that the optimal strategy for the row player is to make the move represented by the second row of the matrix, and the optimal strategy for the column player is to make the move represented by the third column.

c. The value of the game is -2, which implies that if both players adopt their best strategy the column player will win 2 units in a play. Consequently, the game favors the column player.   ◄

A game may have more than one saddle point, as the next example shows.

EXAMPLE

20 A two-person, zero-sum game is defined by the payoff matrix

$$A = \begin{bmatrix} 4 & 5 & 4 \\ -2 & -5 & -3 \\ 4 & 6 & 8 \end{bmatrix}$$

a. Show that the game is strictly determined and find the saddle points for the game.

b. Discuss the optimal strategies for the players.

c. Does the game favor one player over the other?

SOLUTION

a. Proceeding as in the previous example, we obtain the following information:

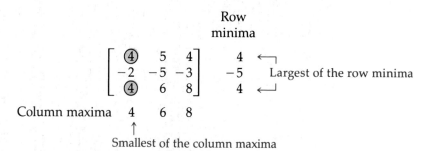

We see that each of the circled entries, 4, is simultaneously the smallest entry in the row and the largest entry in the column containing it. Therefore, the game is strictly determined, and in this case it has two saddle points: the entry $a_{11} = 4$ and the entry $a_{31} = 4$. In general it can be shown that every saddle point of a payoff matrix must have the same value.

b. Since the game has two saddle points, both lying in the first column and in the first and third rows of the payoff matrix, we see that the row player's optimal strategy consists of playing either row 1 or row 3 consistently, whereas the column player's optimal strategy consists of playing column 1 repeatedly.

c. The value of the game is 4, which implies that it favors the row player. ◀

EXAMPLE

21 Two television subscription companies, UBS and Telerama, are planning to extend their operations to a certain city. Each has the option of making its services available to prospective subscribers with a special introductory subscription rate. It is estimated that if both UBS and Telerama offer the special subscription rate, each will get 50 percent of the market, whereas if UBS offers the special subscription rate and Telerama does not, UBS will get 70 percent of the market. If Telerama offers the special subscription rate and UBS does not, it is estimated that UBS will get 40 percent of the market. If both companies elect not to offer the special subscription rate, it is estimated that UBS will get 60 percent of the market.

a. Construct the payoff matrix for the game.

b. Show that the game is strictly determined.

c. Determine the optimal strategy for each company and find the value of the game.

SOLUTION

a. The required payoff matrix is given by

Telerama

		Intro. rate	Usual rate
UBS	Intro. rate	$\begin{bmatrix} .50 \end{bmatrix}$	$.70$
	Usual rate	$.40$	$.60 \end{bmatrix}$

b. The entry $a_{11} = 0.50$ is the smaller entry in its row and the larger entry in its column. Therefore, the entry $a_{11} = 0.50$ is a saddle point of the game and the game is strictly determined.

c. The entry $a_{11} = 0.50$ is the only saddle point of the game, so UBS's optimal strategy is to choose row 1, and Telerama's optimal strategy is to choose column 1. In other words, both companies should offer their potential customers their respective introductory subscription rates. ◄

Self-Check Exercises

9.4

1. A two-person, zero-sum game is defined by the payoff matrix

$$A = \begin{bmatrix} -2 & 1 & 3 \\ 3 & 2 & 2 \\ 2 & -1 & 4 \end{bmatrix}$$

a. Show that the game is strictly determined and find the saddle point(s) for the game.

b. What is the optimal strategy for each player?

c. What is the value of the game? Does the game favor one player over the other?

2. The management of Delta Corporation, a construction and development company, is deciding whether to go ahead with the construction of a large condominium complex. A financial analysis of the project indicates that if the company goes ahead with the development and the home mortgage rate drops a point or more by next year, when the complex is expected to be completed, it will stand to make a profit of $750,000. If the company goes ahead with the development and the mortgage rate stays within one point of the current rate by next year, it will stand to make a profit of $600,000. If the company goes ahead with the development and the mortgage rate increases a point or more by next year, it will stand to make a profit of $350,000. If the company does not go ahead with the development and the mortgage rate drops a point or more by next year, it will stand to make a profit of $400,000. If the company does not go ahead with the development and the mortgage rate stays within one point of the current rate by next year, it will stand to make a profit of $350,000. Finally, if the company does not go ahead with the development

and the mortgage rate increases a point or more by next year, it stands to make $250,000.

    a. Represent this information in the form of a payoff matrix.

    b. Assuming that the home mortgage rate trend is volatile over the next year, determine whether or not the company should go ahead with the project.

Solutions to Self-Check Exercises 9.4 can be found on page 534.

▶

Exercises 9.4

In Exercises 1–8, determine the maximin and minimax strategies for each of the given two-person, zero-sum matrix games.

1. $\begin{bmatrix} 2 & 3 \\ 4 & 1 \end{bmatrix}$ **2.** $\begin{bmatrix} -1 & 3 \\ 2 & 5 \end{bmatrix}$ **3.** $\begin{bmatrix} 1 & 3 & 2 \\ 0 & -1 & 4 \end{bmatrix}$

4. $\begin{bmatrix} 1 & 4 & -2 \\ 4 & 6 & -3 \end{bmatrix}$ **5.** $\begin{bmatrix} 3 & 2 & 1 \\ 1 & -2 & 3 \\ 6 & 4 & 1 \end{bmatrix}$ **6.** $\begin{bmatrix} 1 & 4 \\ 2 & -2 \\ 3 & 0 \end{bmatrix}$

7. $\begin{bmatrix} 4 & 2 & 1 \\ 1 & 0 & -1 \\ 2 & 1 & 3 \end{bmatrix}$ **8.** $\begin{bmatrix} -1 & 1 & 2 \\ 3 & 1 & 1 \\ -1 & 1 & 2 \\ 3 & 2 & -1 \end{bmatrix}$

In Exercises 9–18, determine whether the given two-person, zero-sum matrix game is strictly determined. If a game is strictly determined,

    a. Find the saddle point(s) of the game.

    b. Find the optimal strategy for each player.

    c. Find the value of the game.

    d. Determine whether the game favors one player over the other.

9. $\begin{bmatrix} 2 & 3 \\ 1 & -4 \end{bmatrix}$ **10.** $\begin{bmatrix} 1 & 0 \\ 0 & -1 \end{bmatrix}$ **11.** $\begin{bmatrix} 1 & 3 & 2 \\ -1 & 4 & -6 \end{bmatrix}$

12. $\begin{bmatrix} 3 & 2 \\ -1 & -2 \\ 4 & 1 \end{bmatrix}$ **13.** $\begin{bmatrix} 1 & 3 & 4 & 2 \\ 0 & 2 & 6 & -4 \\ -1 & -3 & -2 & 1 \end{bmatrix}$ **14.** $\begin{bmatrix} 2 & 4 & 2 \\ 0 & 3 & 0 \\ -1 & -2 & 1 \end{bmatrix}$

15. $\begin{bmatrix} 1 & 2 \\ 0 & 3 \\ -1 & 2 \\ 2 & -2 \end{bmatrix}$ **16.** $\begin{bmatrix} -1 & 2 & 4 \\ 2 & 3 & 5 \\ 0 & 1 & -3 \\ -2 & 4 & -2 \end{bmatrix}$

17. $\begin{bmatrix} 1 & -1 & 3 & 2 \\ 1 & 0 & 2 & 2 \\ -2 & 2 & 3 & -1 \end{bmatrix}$ **18.** $\begin{bmatrix} 3 & -1 & 0 & -4 \\ 2 & 1 & 0 & 2 \\ -3 & 1 & -2 & 1 \\ -1 & -1 & -2 & 1 \end{bmatrix}$

19. Robin and Cathy play a game of matching fingers. On a predetermined signal, both players simultaneously extend 1, 2, or 3 fingers from a closed fist. If the sum of the number of fingers extended is even, then Robin receives an amount in dollars equal to that sum from Cathy. If the sum of the number of fingers extended is odd, then Cathy receives an amount in dollars equal to that sum from Robin.

a. Construct the payoff matrix for the game.

b. Find the maximin and the minimax strategies for Robin and Cathy, respectively.

c. Is the game strictly determined?

d. If the answer to (c) is yes, what is the value of the game?

20. Brady's, a conventional department store, and Value-Mart, a discount department store, are each considering opening new stores at one of two possible sites: the Civic Center and North Shore Plaza. The strategies available to the management of each store are given in the following payoff matrix, where each entry represents the amounts (in hundreds of thousands of dollars) either gained or lost by one business from or to the other as a result of the sites selected.

$$\begin{array}{c} \text{Value-Mart} \\ \begin{array}{cc} \text{Center} & \text{Plaza} \end{array} \\ \text{Brady's} \begin{array}{c} \text{Civic Center} \\ \text{North Shore Plaza} \end{array} \begin{bmatrix} 2 & -2 \\ 3 & -4 \end{bmatrix} \end{array}$$

a. Show that the game is strictly determined.

b. What is the value of the game?

c. Determine the best strategy for the management of each store (that is, determine the ideal locations for each store).

21. The management of the Acrosonic Company is faced with the problem of deciding whether to expand the production of its line of electrostatic loudspeaker systems. It has been estimated that an expansion will result in an annual profit of $200,000 for Acrosonic if the general economic climate is good. On the other hand, an expansion during a period of economic recession will cut its annual profit to $120,000. As an alternative, Acrosonic may hold the production of its electrostatic loudspeaker systems at the current level and expand its line of conventional loudspeaker systems. In this event, the company is expected to make a profit of $50,000 in an expanding economy (because many potential customers will be expected to buy electrostatic loudspeaker systems from other competitors) and a profit of $150,000 in a recessionary economy.

a. Construct the payoff matrix for this game. [Hint: The row player is the management of the company and the column player is the economy.]

b. Should management recommend expanding the company's line of electrostatic loudspeaker systems?

22. The proprietor of the Belvedere Restaurant is faced with the problem of deciding whether to expand his restaurant facilities now or to wait until some future date to do so. If he expands the facilities now and the economy experiences a period of growth during the coming year, he will make a net profit of $442,000; if he expands now and a period of zero growth follows, then he will make a net profit of $40,000; and if he expands now and an economic recession follows, he will suffer a net loss of $108,000. If he does not expand the restaurant now and the economy experiences a period of growth during the coming year, he will make a net profit of $280,000; if he does not expand now and a period of zero growth follows, he will make a net profit of $190,000.

Finally, if he does not expand now and an economic recession follows, he will make a net profit of $100,000.

a. Represent this information in the form of a payoff matrix.

b. Assuming that the state of the economy during the coming year is uncertain, determine whether or not the owner of the restaurant should expand his facilities at this time.

23. Two barber shops, Roland's Barber Shop and Charley's Barber Shop, are both located in the business district of a certain town. Roland estimates that if he raises the price of a haircut by $1, he will increase his market share by 3 percent if Charley raises his price by the same amount; he will decrease his market share by 1 percent if Charley holds his price at the same level; and he will decrease his market share by 3 percent if Charley lowers his price by $1. If Roland keeps his price the same, he will increase his market share by 2 percent if Charley raises his price by $1; he will keep the same market share if Charley holds the price at the same level; and he will decrease his market share by 2 percent if Charley lowers his price by $1. Finally, if Roland lowers the price he charges by $1, his market share will increase by 5 percent if Charley raises his price by the same amount; he will increase his market share by 2 percent if Charley holds his price at the same level; and he will increase his market share by 1 percent if Charley lowers his price by $1.

a. Construct the payoff matrix for this game.

b. Show that the game is strictly determined.

c. If neither party is willing to lower the price he charges for a haircut, show that both should keep their present price structures.

Solutions to Self-Check Exercises

9.4

1. *a.* Displaying the minimum of each row and the maximum of each column of the payoff matrix A, we obtain

$$
\begin{array}{ccc}
& & \text{Row minima} \\
\begin{bmatrix} -2 & 1 & 3 \\ 3 & ② & 2 \\ 2 & -1 & 4 \end{bmatrix} & \begin{array}{c} -2 \\ 2 \\ -1 \end{array} & \leftarrow \text{Largest of the row minima}
\end{array}
$$

Column maxima   3   2   4

↑
Smallest of the column maxima

From these results, we see that the circled entry, 2, is simultaneously the smallest entry in its row and the largest entry in its column. Therefore, the game is strictly determined with the entry $a_{22} = 2$ as its saddle point.

b. From these results, we see that the optimal strategy for the row player is to make the move represented by the second row of the matrix, and the optimal strategy for the column player is to make the move represented by the second column.

c. The value of the game is 2, which implies that if both players adopt their best strategy, the row player will win 2 units in a play. Consequently, the game favors the row player.

2. *a.* We may view this situation as a game in which the row player is Delta Corporation and the column player is the home mortgage rate. The required payoff matrix is

Mortgage Rate

		Decrease	Steady	Increase
Delta Corp.	Project	750	600	350
	No project	400	350	250

(All figures are in thousands of dollars.)

b. From (a), the payoff matrix under consideration is

$$\begin{bmatrix} 750 & 600 & 350 \\ 400 & 350 & 250 \end{bmatrix}$$

Proceeding in the usual manner, we find

Row minima

$$\begin{bmatrix} 750 & 600 & (350) \\ 450 & 300 & 250 \end{bmatrix} \qquad \begin{matrix} 350 \\ 250 \end{matrix} \qquad \leftarrow \text{Larger of the row minima}$$

Column maxima   750   600   350

↑

Smallest of the column maxima

From these results, we see that the entry $a_{13} = 350$ is a saddle point and the game is strictly determined. We can also conclude that the company should go ahead with the project.

▶
9.5

Games with Mixed Strategies

▶ Mixed Strategies

▶ Expected Value of a Game

In Section 9.4 we discussed strictly determined games and found that the optimal strategy for the row player is to select the row containing a saddle point for the game, and the optimal strategy for the column player is to select the column containing a saddle point. Furthermore, in repeated plays of the game, each player's optimal strategy consists of making the same move over and over again, since the discovery of the opponent's optimal strategy cannot be used to advantage. Such strategies are called **pure strategies.** In this section, we will look at games that are not strictly determined and the strategies associated with such games.

▶ Mixed Strategies

As a simple example of a game that is not strictly determined, let us consider the following slightly modified version of the coin-matching game played by Richie and Chuck (see Example 15). Suppose Richie wins $3 if both parties choose heads, $1 if both choose tails, and loses $2 if one chooses heads and the other tails. Then the payoff matrix for this game is given by

$$
\begin{array}{c}
& & \text{C's moves} \\
& & \begin{array}{cc} C_1 & C_2 \\ \text{(heads)} & \text{(tails)} \end{array} \\
\text{R's moves} & \begin{array}{c} R_1 \text{ (heads)} \\ R_2 \text{ (tails)} \end{array} & \begin{bmatrix} 3 & -2 \\ -2 & 1 \end{bmatrix}
\end{array}
$$

A quick examination of this matrix reveals that it contains no entry that is simultaneously the smallest entry in its row and the largest entry in its column; that is, the game has no saddle point and is, therefore, not strictly determined. What strategy might Richie adopt for the game? Offhand, it would seem that he should consistently select row 1, since he stands to win $3 by playing this row and only $1 by playing row 2 at a risk, in either case, of losing $2. However, if Chuck discovers that Richie is playing row 1 consistently, he would counter this strategy by playing column 2, causing Richie to lose $2 on each play! In view of this, Richie is led to consider a strategy whereby he chooses row 1 some of the time and row 2 at other times. A similar analysis of the game from Chuck's point of view suggests, too, that he might consider choosing column 1 some of the time and column 2 at other times. Such strategies are called **mixed strategies.**

From a practical point of view there are many ways in which a player may choose moves in a game with mixed strategies. For example, in the game just mentioned, if Richie decides to play heads half the time and tails the other half of the time, he could toss an unbiased coin before each move and let the outcome of the toss determine which move he would make. Another more general but less practical way of deciding on the choice of a move is: Having determined beforehand the proportion of the time row 1 is to be chosen (and therefore the proportion of the time row 2 is to be chosen), Richie might construct a spinner (see Figure 9.3) in which the areas of the two sectors reflect these proportions

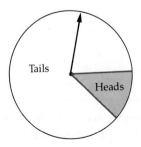

Tails

Heads

Figure 9.3

and let the move be decided by the outcome of a spin. These two methods for determining a player's move in a game with mixed strategies will guarantee that the strategy will not fall into a pattern that will be discovered by the opponent.

From the mathematical point of view, we may describe the mixed strategy of a row player in terms of a row vector whose dimension coincides with the number of possible moves the player has. For example, if Richie had decided on a strategy in which he chose to play row 1 half the time and row 2 the other half of the time, then this strategy is represented by the row vector

$$[0.5 \quad 0.5].$$

Similarly, the mixed strategy for a column vector may be represented by a column vector of appropriate dimension. For example, returning to our illustration, suppose Chuck has decided that he would choose column one 20 percent of the time and column two 80 percent of the time. This strategy is represented by the column vector

$$\begin{bmatrix} 0.2 \\ 0.8 \end{bmatrix}.$$

▶ Expected Value of a Game

For the purpose of comparing the merits of a player's different mixed strategies in a game, it is convenient to introduce a number called the *expected value* of the game. The expected value measures the average payoff to the row player when both players adopt a particular set of mixed strategies. We now explain this notion using a 2×2 matrix game whose payoff matrix has the general form

$$A = \begin{bmatrix} a_{11} & a_{12} \\ a_{21} & a_{22} \end{bmatrix}.$$

Suppose that in repeated plays of the game the row player R adopts the mixed strategy

$$P = [p_1 \quad p_2].$$

That is, the player selects row 1 with probability p_1 and row 2 with probability p_2 and the column player C adopts the mixed strategy

$$Q = \begin{bmatrix} q_1 \\ q_2 \end{bmatrix}.$$

That is, the column player selects column 1 with probability q_1 and column 2 with probability q_2. Now, in each play of the game, there are four possible outcomes that may be represented by the ordered pairs

(row 1, column 1), (row 1, column 2), (row 2, column 1), and (row 2, column 2)

where the first number of each ordered pair represents R's selection and the second member of each ordered pair represents C's selection. Since the choice of moves is made by one player without knowing the other's choice, each pair of events (for example, the events "row 1" and "column 1") constitutes a pair of independent events. Therefore, the probability of R choosing row 1 and C choosing column 1, P(row 1, column 1) is given by

$$P(\text{row 1, column 1}) = P(\text{row 1}) \cdot P(\text{column 1})$$
$$= p_1 q_1.$$

In a similar manner, we compute the probability of each of the other three outcomes. These calculations together with the payoffs associated with each of the four possible outcomes may be summarized as follows:

Outcome	Probability	Payoff
(row 1, column 1)	$p_1 q_1$	a_{11}
(row 1, column 2)	$p_1 q_2$	a_{12}
(row 2, column 1)	$p_2 q_1$	a_{21}
(row 2, column 2)	$p_2 q_2$	a_{22}

Then the *expected payoff* E of the game is the sum of the products of the payoffs and the corresponding probabilities (see Section 8.2). Thus,

$$E = p_1 q_1 a_{11} + p_1 q_2 a_{12} + p_2 q_1 a_{21} + p_2 q_2 a_{22}.$$

In terms of the matrices P, A, and Q, we have the following relatively simple expression for E, namely,

$$E = PAQ,$$

which you may verify (see Problem 22, Exercises 9.5). This result may be generalized as follows:

Expected Value of a Game

Let

$$P = [p_1\ p_2 \cdots p_m] \quad \text{and} \quad Q = \begin{bmatrix} q_1 \\ q_2 \\ \vdots \\ q_n \end{bmatrix}$$

be the vectors representing the mixed strategies for the row player R and the column player C, respectively, in a game with an $m \times n$ payoff matrix

$$A = \begin{bmatrix} a_{11} & a_{12} & \cdots & a_{1n} \\ a_{21} & a_{22} & \cdots & a_{2n} \\ \vdots & \vdots & & \vdots \\ a_{m1} & a_{m2} & \cdots & a_{mn} \end{bmatrix}$$

Then the expected value of the game is given by

$$E = PAQ = [p_1\, p_2 \cdots p_m] \begin{bmatrix} a_{11} & a_{12} & \cdots & a_{1n} \\ a_{21} & a_{22} & \cdots & a_{2n} \\ \vdots & \vdots & & \vdots \\ a_{m1} & a_{m2} & \cdots & a_{mn} \end{bmatrix} \begin{bmatrix} q_1 \\ q_2 \\ \vdots \\ q_n \end{bmatrix}$$

We now look at several examples involving the computation of the expected value of a game.

EXAMPLE

22 Consider a coin-matching game played by Richie and Chuck with a payoff matrix given by

$$A = \begin{bmatrix} 3 & -2 \\ -2 & 1 \end{bmatrix}$$

Compute the expected payoff of the game if Richie adopts the mixed strategy P and Chuck adopts the mixed strategy Q when

a. $$P = [0.5 \quad 0.5] \quad \text{and} \quad Q = \begin{bmatrix} 0.5 \\ 0.5 \end{bmatrix}$$

b. $$P = [0.8 \quad 0.2] \quad \text{and} \quad Q = \begin{bmatrix} 0.1 \\ 0.9 \end{bmatrix}$$

SOLUTION *a.* We compute

$$E = PAQ = [0.5 \quad 0.5] \begin{bmatrix} 3 & -2 \\ -2 & 1 \end{bmatrix} \begin{bmatrix} 0.5 \\ 0.5 \end{bmatrix}$$

$$= [0.5 \quad -0.5] \begin{bmatrix} 0.5 \\ 0.5 \end{bmatrix}$$

$$= 0.$$

Thus, in repeated plays of the game, it may be expected in the long run that the game will end in a draw.

b. We compute

$$E = PAQ = [0.8 \quad 0.2] \begin{bmatrix} 3 & -2 \\ -2 & 1 \end{bmatrix} \begin{bmatrix} 0.1 \\ 0.9 \end{bmatrix}$$

$$= [2 \quad -1.4] \begin{bmatrix} 0.1 \\ 0.9 \end{bmatrix}$$

$$= -1.06.$$

That is, in the long run Richie may be expected to lose $1.06 on the average in each play. ◀

EXAMPLE

23 The payoff matrix for a certain game is given by

$$A = \begin{bmatrix} 1 & -2 \\ -1 & 2 \\ 3 & -3 \end{bmatrix}$$

a. Find the expected payoff to the row player if the row player R uses her maximin pure strategy and the column player C uses her minimax pure strategy.

b. Find the expected payoff to the row player if R uses her maximin strategy 50 percent of the time and chooses each of the other two rows 25 percent of the time, while C chooses each column 50 percent of the time.

SOLUTION
a. The maximin and minimax strategies for the row and column player, respectively, may be found using the method of the last section. Thus,

Row
minima

$$\begin{bmatrix} 1 & -2 \\ -1 & 2 \\ 3 & -3 \end{bmatrix} \quad \begin{matrix} -2 \\ \fbox{-1} \\ -3 \end{matrix} \quad \leftarrow \text{Largest of the row minima}$$

Column maxima 3 ②

↑
Smaller of the column maxima

From these results, we see that R's optimal pure strategy is to choose row 2, whereas C's optimal pure strategy is to choose column 2. Furthermore, if both players use their respective optimal strategy, then the expected payoff to R is 2 units.

b. In this case, R's mixed strategy may be represented by the row vector

$$P = [0.25 \quad 0.50 \quad 0.25],$$

and C's mixed strategy may be represented by the column vector

$$Q = \begin{bmatrix} 0.5 \\ 0.5 \end{bmatrix}.$$

The expected payoff to the row player will then be given by

$$E = PAQ = [0.25 \quad 0.50 \quad 0.25] \begin{bmatrix} 1 & -2 \\ -1 & 2 \\ 3 & -3 \end{bmatrix} \begin{bmatrix} 0.5 \\ 0.5 \end{bmatrix}$$

$$= [0.25 \quad 0.50 \quad 0.25] \begin{bmatrix} -0.5 \\ 0.5 \\ 0 \end{bmatrix}$$

$$= 0.125. \quad \blacktriangleleft$$

In the last section we studied optimal strategies associated with strictly determined games and found them to be precisely the maximin and minimax pure strategies adopted by the row and column players, respectively. We will now look at optimal mixed strategies associated with matrix games that are not strictly determined. In particular, we will state, without proof, the optimal mixed strategies to be adopted by the players in a 2×2 matrix game.

As we saw earlier, a player in a nonstrictly determined game should adopt a mixed strategy, since a pure strategy will soon be detected by the opponent, who may then use this knowledge to his advantage in devising a counter-strategy. Since there are infinitely many mixed strategies for each player in such a game, the question arises as to how an optimal mixed strategy may be discovered for each player. Let us recall that an optimal mixed strategy for a player is one arrived at subject to the condition that the row player seeks to maximize his expected payoff, whereas the column player simultaneously seeks to minimize it.

Without going into unnecessary details, let us note that the problem of finding the optimal mixed strategies for the players in a nonstrictly determined game is equivalent to one of solving the related linear programming problem. However, for a 2×2 nonstrictly determined game, the optimal mixed strategies for the players may be found by employing the formulas contained in the following result, which we state without proof.

Optimal Strategies for Nonstrictly Determined Games

Let

$$\begin{bmatrix} a & b \\ c & d \end{bmatrix}$$

be the payoff matrix for a nonstrictly determined game. Then the optimal mixed strategy for the row player is given by

$$P = [p_1 \quad p_2]$$

where

$$p_1 = \frac{d - c}{a + d - b - c} \quad \text{and} \quad p_2 = 1 - p_1,$$

and the optimal mixed strategy for the column player is given by

$$Q = \begin{bmatrix} q_1 \\ q_2 \end{bmatrix}$$

where

$$q_1 = \frac{d - b}{a + d - b - c} \quad \text{and} \quad q_2 = 1 - q_1.$$

Furthermore, the value of the game is given by the expected value of the game $E = PAQ$ where P and Q are the optimal mixed strategies for the row and column players, respectively. Thus

$$E = PAQ$$
$$= \frac{ad - bc}{a + d - b - c}.$$

The next example illustrates the use of these formulas for finding the optimal mixed strategies and for finding the value of a 2 × 2 (nonstrictly determined) game.

EXAMPLE

24 Consider the coin-matching game played by Richie and Chuck with the payoff matrix

$$A = \begin{bmatrix} 3 & -2 \\ -2 & 1 \end{bmatrix} \quad \text{(See Example 22.)}$$

a. Find the optimal mixed strategies for both Richie and Chuck.

b. Find the value of the game. Does it favor one player over the other?

SOLUTION

a. The game under consideration has no saddle point and is accordingly nonstrictly determined. Using the formulas for determining the optimal mixed strategies for a 2 × 2 game with $a = 3$, $b = -2$, $c = -2$, and $d = 1$, we find that

$$p_1 = \frac{d - c}{a + d - b - c} = \frac{1 - (-2)}{3 + 1 - (-2) - (-2)} = \frac{3}{8}$$

and

$$p_2 = 1 - p_1$$

$$= 1 - \frac{3}{8}$$

$$= \frac{5}{8},$$

so Richie's optimal mixed strategy is given by

$$P = [p_1 \quad p_2]$$

$$= [\tfrac{3}{8} \quad \tfrac{5}{8}].$$

Next, we compute

$$q_1 = \frac{d - b}{a + d - b - c} = \frac{1 - (-2)}{3 + 1 - (-2) - (-2)} = \frac{3}{8}$$

and

$$q_2 = 1 - q_1$$

$$= 1 - \frac{3}{8}$$

$$= \frac{5}{8},$$

giving Chuck's optimal mixed strategy as

$$Q = \begin{bmatrix} \tfrac{3}{8} \\ \tfrac{5}{8} \end{bmatrix}.$$

b. The value of the game may be found by computing the matrix product PAQ where P and Q are the vectors found in (a). Equivalently, we may compute it using the formula

$$\begin{aligned} E &= \frac{ad - bc}{a + d - b - c} \\ &= \frac{(3)(1) - (-2)(-2)}{3 + 1 - (-2) - (-2)} \\ &= -\frac{1}{8}. \end{aligned}$$

Since the value of the game is negative, we conclude that the coin-matching game with the particular given payoff matrix favors Chuck (the column player) over Richie. Over the long run, in repeated plays of the game where each player uses his optimal strategy, Chuck is expected to win 1/8, or 12.5 cents, on the average per play. ◀

EXAMPLE

25 As part of their investment strategy, the Carringtons have earmarked $40,000 for short-term investments in the stock market and the money market. The performance of the investments depends on the prime rate (that is, the interest rate that banks charge their best customers); an increase in the prime rate generally favors their investment in the money market, whereas a decrease in the prime rate generally favors their investment in the stock market. Suppose the following payoff matrix gives the expected percentage increase or decrease in the value of each investment for each state of the prime rate:

	Prime rate Up	Prime rate Down
Money market investment	15	10
Stock market investment	−5	25

a. Determine the optimal investment strategy for the Carrington's short-term investment of $40,000.

b. What short-term profit can the Carringtons expect to make on their investments?

SOLUTION
a. We treat the problem as a matrix game in which the Carringtons are the row player. Letting $p = [p_1 \quad p_2]$ denote their optimal strategy, we find

$$p_1 = \frac{d - c}{a + d - b - c} = \frac{25 - (-5)}{15 + 25 - 10 - (-5)} = \frac{30}{35} \approx 0.86$$

and $\quad p_2 = 1 - p_1 = 1 - 0.86 = 0.14.$

Thus, the Carringtons should put $(0.86)(40,000)$, or \$34,400, into the money market and \$5,600 into the stock market.

b. The expected value of the game is given by

$$E = \frac{ad - bc}{a + d - b - c}$$

$$= \frac{(15)(25) - (10)(-5)}{15 + 25 - 10 - (-5)} = \frac{425}{35}$$

$$\approx 12.14.$$

Thus, the Carringtons can expect to make a short-term profit of 12.14 percent on their total investment of \$40,000, that is, a profit of $(0.1214)(40,000)$, or \$4,856. ◀

Self-Check Exercises

9.5

1. The payoff matrix for a game is given by

$$A = \begin{bmatrix} 2 & 3 & -1 \\ -3 & 2 & -2 \\ 3 & -2 & 2 \end{bmatrix}.$$

a. Find the expected payoff to the row player if the row player R uses the maximin pure strategy and the column player C uses the minimax pure strategy.

b. Find the expected payoff to the row player if R uses the maximin strategy 40 percent of the time and chooses each of the other two rows 30 percent of the time, while C uses the minimax strategy 50 percent of the time and chooses each of the other two columns 25 percent of the time.

c. Which pair of strategies favors the row player?

2. A farmer has allocated 2000 acres of his farm for planting two crops. Crop A is more susceptible to frost than crop B. If there is no frost in the growing season, then he can expect to make \$40 per acre from crop A and \$25 per acre from crop B. If there is mild frost, the expected profits are \$20 per acre from crop A and \$30 per acre from crop B. How many acres of each crop should the farmer cultivate in order to maximize his profits? What profit could he expect to make using this optimal strategy?

Solutions to Self-Check Exercises 9.5 can be found on page 548.

▶

Exercises 9.5

In Exercises 1–6, find the expected payoff E of each of the following games whose payoff matrix and strategies P and Q (for the row and column player, respectively) are given.

1. $\begin{bmatrix} 3 & 1 \\ -4 & 2 \end{bmatrix}$, $P = [\frac{1}{2} \ \frac{1}{2}]$, $Q = \begin{bmatrix} \frac{3}{5} \\ \frac{2}{5} \end{bmatrix}$

2. $\begin{bmatrix} -1 & 4 \\ 3 & -2 \end{bmatrix}$, $P = [0.8 \ 0.2]$, $Q = \begin{bmatrix} 0.6 \\ 0.4 \end{bmatrix}$

3. $\begin{bmatrix} -4 & 3 \\ 2 & 1 \end{bmatrix}$, $P = [\frac{1}{3} \ \frac{2}{3}]$ $Q = \begin{bmatrix} \frac{3}{4} \\ \frac{1}{4} \end{bmatrix}$

4. $\begin{bmatrix} 1 & 0 & -1 \\ -1 & 2 & 3 \\ -3 & 1 & 2 \end{bmatrix}$, $P = [\frac{1}{2} \ \frac{1}{2} \ 0]$, $Q = \begin{bmatrix} \frac{1}{3} \\ \frac{1}{3} \\ \frac{1}{3} \end{bmatrix}$

5. $\begin{bmatrix} 2 & 0 & -2 \\ 1 & -1 & 3 \\ 2 & 1 & -4 \end{bmatrix}$, $P = [0.2 \ 0.6 \ 0.2]$, $Q = \begin{bmatrix} 0.2 \\ 0.6 \\ 0.2 \end{bmatrix}$

6. $\begin{bmatrix} 1 & -4 & 2 \\ 2 & 1 & -1 \\ 2 & -2 & 0 \end{bmatrix}$, $P = [.2 \ .3 \ .5]$ $Q = \begin{bmatrix} .6 \\ .2 \\ .2 \end{bmatrix}$

7. The payoff matrix for a game is given by

$$\begin{bmatrix} 1 & -2 \\ -2 & 3 \end{bmatrix}.$$

Compute the expected payoffs of the game for the following pairs of strategies and deduce from the results of the computation which pair of strategies is most advantageous to R.

a. $P = [1 \ 0]$, $Q = \begin{bmatrix} 1 \\ 0 \end{bmatrix}$ b. $P = [0 \ 1]$, $Q = \begin{bmatrix} 1 \\ 0 \end{bmatrix}$

c. $P = [\frac{1}{2} \ \frac{1}{2}]$, $Q = \begin{bmatrix} \frac{1}{2} \\ \frac{1}{2} \end{bmatrix}$ d. $P = [0.5 \ 0.5]$, $Q = \begin{bmatrix} 0.8 \\ 0.2 \end{bmatrix}$

8. The payoff matrix for a game is given by

$$\begin{bmatrix} 3 & 1 & 1 \\ 0 & 2 & 0 \\ -1 & 0 & 2 \end{bmatrix}.$$

Compute the expected payoffs of the game for the following pairs of strategies and deduce from the results of the computation which pair of strategies is most advantageous to R.

a. $P = [\frac{1}{3} \ \frac{1}{3} \ \frac{1}{3}]$ $Q = \begin{bmatrix} \frac{1}{3} \\ \frac{1}{3} \\ \frac{1}{3} \end{bmatrix}$

b. $P = [\frac{1}{4} \ \frac{1}{2} \ \frac{1}{4}]$ $Q = \begin{bmatrix} \frac{1}{8} \\ \frac{3}{8} \\ \frac{1}{2} \end{bmatrix}$

c. $P = [.4 \quad .3 \quad .3]$ $\qquad Q = \begin{bmatrix} .6 \\ .2 \\ .2 \end{bmatrix}$

d. $P = [.1 \quad .5 \quad .4]$ $\qquad Q = \begin{bmatrix} .3 \\ .3 \\ .4 \end{bmatrix}$

9. The payoff matrix for a game is given by

$$\begin{bmatrix} -3 & 3 & 2 \\ -3 & 1 & 1 \\ 1 & -2 & 1 \end{bmatrix}.$$

a. Find the expected payoff to the row player if the row player R uses the maximin pure strategy and the column player C uses the minimax pure strategy.

b. Find the expected payoff to the row player if R uses the maximin strategy 50 percent of the time and chooses each of the other two rows 25 percent of the time, while C uses the minimax strategy 60 percent of the time and chooses each of the other columns 20 percent of the time.

c. Which pair of strategies favors the row player?

In Exercises 10–15, find the optimal strategies, P and Q, for the row and column player, respectively. Also compute the expected payoff E of each matrix game and determine which player it favors, if any.

10. $\begin{bmatrix} 4 & 1 \\ 2 & 3 \end{bmatrix}$ $\qquad$ 11. $\begin{bmatrix} 2 & 5 \\ 3 & -6 \end{bmatrix}$ $\qquad$ 12. $\begin{bmatrix} -1 & 2 \\ 1 & -3 \end{bmatrix}$

13. $\begin{bmatrix} -1 & 3 \\ 2 & 0 \end{bmatrix}$ $\qquad$ 14. $\begin{bmatrix} -2 & -6 \\ -8 & -4 \end{bmatrix}$ $\qquad$ *15. $\begin{bmatrix} 2 & 5 \\ -2 & 4 \end{bmatrix}$

16. Consider the coin-matching game played by Richie and Chuck (see Examples 22 and 24) with the payoff matrix

$$A = \begin{bmatrix} 4 & -2 \\ -2 & 1 \end{bmatrix}.$$

a. Find the optimal strategies for Richie and Chuck.

b. Find the value of the game. Does it favor one player over the other?

17. As part of their investment strategy, the Carringtons have decided to put $100,000 into stock market investments and also into purchasing precious metals. The performance of the investments depends on the state of the economy in the next year. In an expanding economy it is expected that their stock market investment will outperform their investment in precious metals, whereas an economic recession will have precisely the opposite effect. Suppose the following payoff matrix gives the expected percentage increase or decrease in the value of each investment for each state of the economy:

	Expanding economy	Economic recession
Stock market investment	$\begin{bmatrix} 20 \end{bmatrix}$	-5
Commodity investment	10	15

a. Determine the optimal investment strategy for the Carringtons' investment of $100,000.

b. What profit can the Carringtons expect to make on their investments over the year?

18. The Maxwells have decided to invest $40,000 in the common stocks of two companies listed on the New York Stock Exchange. One of the companies derives its revenue mainly from its worldwide operation of a chain of hotels, whereas the other company is a major brewery in the country. It is expected that if the economy is in a state of growth, then the hotel stock should outperform the brewery stock; however, the brewery stock is expected to hold its own better than the hotel stock in a recessionary period. Suppose that the following payoff matrix gives the expected percentage increase or decrease in the value of each investment for each state of the economy:

$$
\begin{array}{c}
\text{Expanding} \quad \text{Economic} \\
\text{economy} \quad\quad \text{recession}
\end{array}
$$

$$
\begin{array}{l}
\text{Investment in hotel stock} \\
\text{Investment in brewery stock}
\end{array}
\begin{bmatrix}
25 & -5 \\
10 & 15
\end{bmatrix}
$$

a. Determine the optimal investment strategy for the Maxwell's investment of $40,000.

b. What profit can the Maxwells expect to make on their investments?

19. Mr. Robinson and Mr. Carson are running for a seat in the U.S. Senate. If both candidates campaign only in the major cities of the state, then Mr. Robinson is expected to get 60 percent of the votes; if both candidates campaign only in the rural areas, then Mr. Robinson is expected to get 55 percent of the votes; if Mr. Robinson campaigns exclusively in the city and Mr. Carson campaigns exclusively in the rural areas, then Mr. Robinson is expected to get 40 percent of the votes; finally, if Mr. Robinson campaigns exclusively in the rural areas and Mr. Carson campaigns exclusively in the city, then Mr. Robinson is expected to get 45 percent of the votes.

a. Construct the payoff matrix for the game and show that it is not strictly determined.

b. Find the optimal strategy for both Mr. Robinson and Mr. Carson.

20. Two dentists, Dr. Russell and Dr. Carlton, are planning to establish practices in a newly developed community. Both have allocated approximately the same total budget for advertising in the local newspaper and for the distribution of fliers announcing their practices. Because of the location of their offices, Dr. Russell is expected to get 48 percent of the business if both dentists advertise only in the local newspaper; if both dentists advertise through fliers, then Dr. Russell is expected to get 45 percent of the business; if Dr. Russell advertises exclusively in the local newspaper and Dr. Carlton advertises exclusively through fliers, then Dr. Russell is expected to get 65 percent of the business. Finally, if Dr. Russell advertises through fliers exclusively and Dr. Carlton advertises exclusively in the local newspaper, then Dr. Russell is expected to get 50 percent of the business.

a. Construct the payoff matrix for the game and show that it is not strictly determined.

b. Find the optimal strategy for both Dr. Russell and Dr. Carlton.

21. Let

$$
\begin{bmatrix}
a_{11} & a_{12} \\
a_{21} & a_{22}
\end{bmatrix}
$$

be the payoff matrix associated with a 2 × 2 matrix game and let

$$P = [p_1 \quad p_2] \quad \text{and} \quad Q = \begin{bmatrix} q_1 \\ q_2 \end{bmatrix}$$

be the mixed strategies for the row and column players, respectively. Show by direct computation that the expected value of the game is given by $E = PAQ$.

*22. Let

$$\begin{bmatrix} a & b \\ c & d \end{bmatrix}$$

be the payoff matrix associated with a nonstrictly determined 2 × 2 matrix game. Prove that the expected payoff of the game is given by

$$E = \frac{ad - bc}{a + d - b - c}.$$

[*Hint:* Compute $E = PAQ$ where P and Q are the optimal strategies for the row and column players, respectively.]

Solutions to Self-Check Exercises

9.5

1. a. From the following calculations

Row minima

$$\begin{bmatrix} 2 & 3 & -1 \\ -3 & 2 & -2 \\ 3 & -2 & 2 \end{bmatrix} \quad \begin{matrix} \text{(-1)} \\ -3 \\ -2 \end{matrix} \quad \leftarrow \text{Largest of the row minima}$$

Column maxima   3   3   ②
↑
Smallest of the column maxima

we see that R's optimal pure strategy is to choose row 1, whereas C's optimal pure strategy is to choose column 3. Furthermore, if both players use their respective optimal strategies, then the expected payoff to R is -1 unit.

b. R's mixed strategy may be represented by the row vector

$$P = [0.4 \quad 0.3 \quad 0.3],$$

and C's mixed strategy may be represented by the column vector

$$Q = \begin{bmatrix} 0.25 \\ 0.25 \\ 0.50 \end{bmatrix}.$$

The expected payoff to the row player will then be given by

$$E = PAQ = [0.4 \quad 0.3 \quad 0.3] \begin{bmatrix} 2 & 3 & -1 \\ -3 & 2 & -2 \\ 3 & -2 & 2 \end{bmatrix} \begin{bmatrix} 0.25 \\ 0.25 \\ 0.50 \end{bmatrix}$$

$$= [0.4 \quad 0.3 \quad 0.3] \begin{bmatrix} 0.75 \\ -1.25 \\ 1.25 \end{bmatrix}$$

$$= 0.3.$$

c. From the results of (a) and (b) we see that the mixed strategies of (b) will be better for R.

2. We may view this problem as a matrix game with the farmer as the row player and the weather as the column player. The payoff matrix for the game is

$$\begin{array}{cc} & \begin{array}{cc} \text{No} & \text{Mild} \\ \text{frost} & \text{frost} \end{array} \\ \begin{array}{c} \text{Crop A} \\ \text{Crop B} \end{array} & \begin{bmatrix} 40 & 20 \\ 25 & 30 \end{bmatrix} \end{array}$$

The game under consideration has no saddle point and is accordingly nonstrictly determined. Letting $p = [p_1 \quad p_2]$ denote the farmer's optimal strategy and using the formula for determining the optimal mixed strategies for a 2×2 game with $a = 40$, $b = 20$, $c = 25$, and $d = 30$, we find

$$p_1 = \frac{d - c}{a + d - b - c} = \frac{30 - 25}{40 + 30 - 20 - 25} = \frac{5}{25} = \frac{1}{5}$$

and

$$p_2 = 1 - p_1 = 1 - \frac{1}{5} = \frac{4}{5}.$$

Therefore, the farmer should cultivate (1/5)(2000), or 400, acres of crop A and 1600 acres of crop B. By using his optimal strategy, the farmer could expect to realize a profit of

$$E = \frac{ad - bc}{a + d - b - c}$$

$$= \frac{(40)(30) - (20)(25)}{40 + 30 - 20 - 25}$$

$$= 28,$$

or \$28 per acre; that is, a total profit of (28)(2,000), or \$56,000.

▶

Chapter 9 Review Exercises

In Exercises 1–4, determine which of the following matrices are regular.

1. $\begin{bmatrix} 1 & 0 \\ -2 & -8 \end{bmatrix}$
2. $\begin{bmatrix} .3 & .7 \\ 1 & 0 \end{bmatrix}$

3. $\begin{bmatrix} \frac{1}{2} & 0 & \frac{1}{2} \\ 0 & 0 & 1 \\ \frac{1}{3} & \frac{1}{3} & \frac{1}{3} \end{bmatrix}$ 4. $\begin{bmatrix} .3 & .2 & .1 \\ 0 & 1 & 0 \\ .5 & 0 & .5 \end{bmatrix}$

In Exercises 5–8, find the steady-state matrix for the given transition matrix.

5. $\begin{bmatrix} .6 & .4 \\ .3 & .7 \end{bmatrix}$ 6. $\begin{bmatrix} .5 & .5 \\ .4 & .6 \end{bmatrix}$

7. $\begin{bmatrix} .6 & .2 & .2 \\ .4 & .2 & .4 \\ .3 & .2 & .5 \end{bmatrix}$ 8. $\begin{bmatrix} .1 & .3 & .6 \\ .2 & .4 & .4 \\ .6 & .2 & .2 \end{bmatrix}$

9. A study conducted by the State Department of Agriculture in a sun-belt state reveals an increasing trend toward urbanization of the farmland within the state. Ten years ago, 50 percent of the land within the state was used for agricultural purposes (A), 15 percent had been urbanized (U), and the remaining 35 percent was neither agricultural nor urban (N). Since that time, 10 percent of the agricultural land has been converted to urban land, 5 percent has been used for other purposes, and the remaining 85 percent is still agricultural. Ninety-five percent of the urban land has remained urban, whereas 5 percent of it has been used for nonagricultural purposes. Of the land that was neither agricultural nor urban, 10 percent has been converted to agricultural land, 5 percent has been urbanized, and the remaining 85 percent remains unchanged.

   a. Construct the transition matrix for the Markov chain that describes the shift in land use within the state.

   b. Find the probability vector describing the distribution of land within the state ten years ago.

   c. Assuming that this trend continues, find the probability vector describing the distribution of land within the state ten years from now.

10. *Auto Trend* magazine conducted a survey among automobile owners in a certain area of the country to determine what type of car they now own and what type of car they expect to own four years from now. For purposes of classification, automobiles mentioned in the survey were placed into three categories: large, intermediate, and small. Results of the survey follow:

Future car

		Large	Inter.	Small
Present car	Large	.3	.3	.4
	Intermediate	.1	.5	.4
	Small	.1	.2	.7

Assuming that these results indicate the long-term buying trend of car owners within the area, what will the distribution of cars (relative to size) be in this area over the long run?

In Exercises 11–14, determine whether each of the games within the given payoff matrix is strictly determined. If so, give the optimal pure strategies for the row player and the column player and also give the value of the game.

11. $\begin{bmatrix} 1 & 2 \\ 3 & 5 \\ 4 & 6 \end{bmatrix}$ 12. $\begin{bmatrix} 1 & 0 & 3 \\ 2 & -1 & -2 \end{bmatrix}$

13. $\begin{bmatrix} 1 & 3 & 6 \\ -2 & 4 & 3 \\ -5 & -4 & -2 \end{bmatrix}$

14. $\begin{bmatrix} 4 & 3 & 2 \\ -6 & 3 & -1 \\ 2 & 3 & 4 \end{bmatrix}$

In Exercises 15–17, find the expected payoff E of each of the following games whose payoff matrix and strategies P and Q (for the row and column player, respectively) are given.

15. $\begin{bmatrix} 4 & 8 \\ 6 & -12 \end{bmatrix}$, $\quad P = [\frac{1}{2} \ \frac{1}{2}]$, $\quad Q = \begin{bmatrix} \frac{1}{4} \\ \frac{3}{4} \end{bmatrix}$

16. $\begin{bmatrix} 3 & 0 & -3 \\ 2 & 1 & 2 \end{bmatrix}$, $\quad P = [\frac{1}{3} \ \frac{2}{3}]$, $\quad Q = \begin{bmatrix} \frac{1}{3} \\ \frac{1}{3} \\ \frac{1}{3} \end{bmatrix}$

17. $\begin{bmatrix} 3 & -1 & 2 \\ 1 & 2 & 4 \\ -2 & 3 & 6 \end{bmatrix}$, $\quad P = [0.2 \ \ 0.4 \ \ 0.4]$, $\quad Q = \begin{bmatrix} 0.2 \\ 0.6 \\ 0.2 \end{bmatrix}$

In Exercises 18–21, find the optimal strategies, P and Q, for the row player and the column player, respectively. Also compute the expected payoff E of each matrix game and determine which player it favors, if any.

18. $\begin{bmatrix} 1 & -2 \\ 0 & 3 \end{bmatrix}$

19. $\begin{bmatrix} 4 & -7 \\ -5 & 6 \end{bmatrix}$

20. $\begin{bmatrix} 3 & -6 \\ 1 & 2 \end{bmatrix}$

21. $\begin{bmatrix} 12 & 10 \\ 6 & 14 \end{bmatrix}$

22. Two competing record stores, Disco-Mart and Record World, each have the option of selling a certain popular record label at a price of either $7 or $8 per record. If both sell the label at the same price, they are each expected to get 50 percent of the business. If Disco-Mart sells the label at $7 per record and Record World sells the label at $8 per record, Disco-Mart is expected to get 70 percent of the business; whereas if Disco-Mart sells the label at $8 per record and Record World sells the label at $7 per record, Disco-Mart is expected to get 40 percent of the business.

    a. Represent this information in the form of a payoff matrix.

    b. Determine the "optimal" price that each company should sell the record label for to ensure that it captures the largest share of the market.

23. The management of a division of the National Motor Corporation that produces compact and subcompact cars has estimated that the quantity demanded of their compact models is 1500 units a week if the price of oil increases at a higher than normal rate, whereas the quantity demanded of their subcompact models is 2500 units a week under similar conditions. On the other hand, the quantity demanded of their compact models and subcompact models is 3000 units and 2000 units a week, respectively, if the price of oil increases at a normal rate. Determine the percentages of compact and subcompact cars the division should plan to manufacture.

Introduction
to Logic

One of the earliest records of man's effort to understand the reasoning process can be found in Aristotle's work ORGANON (third century B.C.) in which he presented his ideas on logical arguments. Up until the present day, Aristotelian logic has been the basis for traditional logic, the systematic study of valid inferences.

The field of mathematics known as **symbolic logic**, in which symbols are used to replace ordinary language, had its beginnings in the eighteenth and nineteenth centuries with the works of the German mathematician Gottfried Wilhelm Leibniz (1646–1716) and the English mathematician George Boole (1815–1864). Boole introduced algebraic-type operations to the field of logic, thereby providing us with a systematic method of combining statements. Boolean Algebra is the basis of modern-day computer technology, as well as being central to the study of pure mathematics.

In this appendix, we will introduce the fundamental concepts of symbolic logic. Beginning with the definitions of statements and their truth values, we will proceed to a discussion of the combination of statements and valid arguments. In Section A.6, we present an application of symbolic logic that is widely used in computer technology—switching networks.

▶APPENDIX

A

A.1

Propositions and Connectives

We use deductive reasoning in many of the things we do. Whether in a formal debate or in an expository article, we use language to express our thoughts. In English, sentences are used to express assertions, questions, commands, and wishes. In our study of logic, we will be concerned with only one type of sentence, the declarative sentence.

Definition
A **proposition,** or *statement,* is a declarative sentence that can be classified as either true or false, but not both.

Commands, requests, questions, or exclamations are examples of sentences that are not propositions.

EXAMPLE

1   Which of the following are propositions?

a. Toronto is the capital of Ontario.

b. Close the door!

c. There are a hundred trillion connections among the neurons in the human brain.

d. Who is the chief justice of the Supreme Court?

e. The new television sit-com is a successful show.

f. The 1984 Summer Olympic Games were held in Montreal.

g. $x + 3 = 8$

h. How wonderful!

i. Either seven is an odd number or it is even.

SOLUTION   Statements (a), (c), (e), (f), and (i) are propositions. Statement (c) would be difficult to verify, but the validity of the statement can, at least theoretically, be determined. Statement (e) is a proposition if we assume that the meaning of the term *successful* has been defined. For example, one might use the Nielsen ratings to determine the success of a new show. Statement (f) is a proposition that is false.

Statements (b), (d), (g), and (h) are *not* propositions. Statement (b) is a command, (d) is a question, and (h) is an exclamation. Statement (g) is an open sentence that cannot be classified as true or false. For example, if $x = 5$, then $5 + 3 = 8$ and the sentence is true. On the other hand, if $x = 4$, then $4 + 3 \neq 8$ and the sentence is false. ◄

Having considered what is meant by a proposition, we now discuss the ways in which propositions may be combined. For example, the two propositions

a. Toronto is the capital of Ontario

and

b. Toronto is the largest city in Canada

may be joined to form the proposition

c. Toronto is the largest city in Canada and is the capital of Ontario.

Propositions (a) and (b) are called **prime,** or *simple,* propositions, because they are simple statements expressing a single complete thought. We will use the lowercase letters p, q, r, . . . to denote prime propositions in the ensuing discussion.

Propositions that are combinations of two or more propositions, such as proposition (c), are called compound propositions. The words used to combine propositions are called *logical connectives.* The connectives that we will consider are given in Table A.1 along with their symbols. We will discuss the first three in this section and the remaining two in Section A.3.

Name	Logical Connective	Symbol
Conjunction	and	$\wedge$
Negation	not	$\sim$
Disjunction	or	$\vee$
Conditional	if . . . then	$\rightarrow$
Biconditional	if and only if	$\leftrightarrow$

Table A.1

Definition

A *conjunction* is a statement of the form "*p* and *q*" and is represented symbolically by

$$p \wedge q.$$

The conjunction $p \wedge q$ is true if *both* p and q are true; it is false otherwise.

We have already encountered a conjunction in an earlier example. The two propositions

p: Toronto is the capital of Ontario.

and q: Toronto is the largest city in Canada.

were combined to form the conjunction

$p \wedge q$: Toronto is the capital of Ontario and is the largest city in Canada.

Definition

A **disjunction** is a proposition of the form "*p* or *q*" and is represented symbolically by

$$p \vee q.$$

The disjunction $p \vee q$ is false if *both* statements p and q are false; it is true in all other cases.

The use of the word *or* in this definition is meant to convey the meaning "one or the other, or both." Since this disjunction is true when *both* p and q are true, as well as when either p or q is true, it is sometimes referred to as an **inclusive disjunction.**

EXAMPLE

2 Consider the propositions

p: Dorm residents can purchase meal plans.

q: Dorm residents can purchase à la carte cards.

The disjunction is

$p \vee q$: Dorm residents can purchase meal plans or à la carte cards.   ◀

The disjunction in Example 2 is true when dorm residents can purchase either meal plans or à la carte cards or both meal plans and à la carte cards.

Definition

An **exclusive disjunction** is a proposition of the form "*p* or *q*" and is denoted by

$$p \veebar q.$$

The disjunction $p \veebar q$ is true if *either p or q* is true.

In contrast to an inclusive disjunction, an exclusive disjunction is *false* when both *p* and *q* are true. In Example 2, the exclusive disjunction would not be true if dorm residents could buy both meal plans and à la carte cards. The difference between exclusive and inclusive disjunction is further illustrated in the next example.

EXAMPLE

3   Consider the propositions

p: The base price of each condominium unit includes a private deck.

q: The base price of each condominium unit includes a private patio.

Find the exclusive disjunction $p \veebar q$.

SOLUTION   The exclusive disjunction is

$p \veebar q$: The base price of each condominium unit includes a private deck or a patio.

Observe that this statement is not true when both *p* and *q* are true. In other words, the base price of each condominium unit does not include both a private deck and a patio. ◀

In our everyday use of language, the meaning of the word *or* is not always clear. In legal documents the two cases are distinguished by the words *and/or* (inclusive) and *either/or* (exclusive). In mathematics, we use the word *or* in the inclusive sense, unless otherwise specified.

Definition

A **negation** is a proposition of the form "not p" and is represented symbolically by

$$\sim p$$

The proposition $\sim p$ is true if p is false and vice versa.

EXAMPLE

4 Form the negation of the proposition

p: Prices rose on the New York Stock Exchange today.

SOLUTION

The negation is

$\sim p$: Prices did not rise on the New York Stock Exchange today. ◀

EXAMPLE

5 Consider the two propositions

p: The birth rate declined in the United States last year.

q: The population of the United States increased last year.

Write the following statements in symbolic form:

a. Last year the birth rate declined and the population increased in the United States.

b. Either the birth rate declined or the population increased in the United States last year.

c. It is not true that the birth rate declined and the population increased in the United States last year.

SOLUTION

The symbolic form of each statement is given by

a. $p \wedge q$

b. $p \veebar q$

c. $\sim (p \wedge q)$ ◀

►

Exercises A.1

1. Determine which of the following are propositions.

    a. The defendant was convicted of grand larceny.

    b. Who won the 1984 presidential election?

    c. The first McDonald's fast food restaurant was opened in California.

    d. $x - 1 \geq 0$

    e. Keep off the grass!

    f. Coughing may be caused by a lack of water in the air.

2. Determine which of the following are propositions.

    a. Exercise protects men and women from sudden heart attacks.

    b. If voters do not pass the proposition, then taxes will be increased.

    c. Don't swim immediately after eating!

    d. What ever happened to Baby Jane?

    e. $\dfrac{x^2 + 1}{x^2 + 1} = 1.$

    f. Several major corporations will enter or expand operations into the home security products market this year.

3. Identify the logical connective that is used in the following statements.

    a. Housing starts in the United States did not increase last month.

    b. Mr. Bieber's will is valid if and only if he was of sound mind and memory when he made the will.

    c. Americans are saving less and spending more this year.

    d. If you traveled on your job and weren't reimbursed, then you can deduct these expenses from your taxable income.

    e. Both loss of appetite and irritability are symptoms of mental stress.

    f. Prices for many imported goods have either stayed flat or dropped slightly this year.

4. Let p and q denote the propositions

    p: Domestic car sales increased over the past year.

    q: Foreign car sales decreased over the past year.

    Express the following compound propositions in words:

    a. $p \vee q$ *b.* $p \wedge q$ *c.* $p \veebar q$ *d.* $\sim p$

    e. $\sim p \vee q$ *f.* $\sim p \vee \sim q$

5. Let p and q denote the propositions

    p: Every employee is required to be fingerprinted.

    q: Every employee is required to take an oath of allegiance.

Express the following compound propositions in words.

a. $p \vee q$ b. $p \wedge q$ c. $\sim p \vee \sim q$ d. $\sim p \wedge \sim q$

e. $p \vee \sim q$

6. State the negation of the following propositions.

    a. New orders for manufactured goods fell last month.

    b. For many men, housecleaning is not a familiar task.

    c. Drinking during pregnancy affects the size and weight of babies.

    d. Not all patients suffering from influenza lose weight.

    e. The commuter airline industry is now undergoing a shakeup.

    f. The Dow Jones industrial average registered its fourth consecutive decline today.

7. Let p and q denote the propositions

    p: The doctor recommended surgery to treat his hyperthyroidism.

    q: The doctor recommended radioactive iodine to treat his hyperthyroidism.

    a. State the exclusive disjunction for these propositions in words.

    b. State the inclusive disjunction for these propositions in words.

8. Let p and q denote the propositions

    p: The SAT verbal scores improved in this school district last year.

    q: The SAT math scores improved in this school district last year.

Express each of the following statements symbolically.

    a. The SAT verbal scores and the SAT math scores improved in this school district last year.

    b. Either the SAT verbal scores or the SAT math scores improved in this school district last year.

    c. Neither the SAT verbal scores nor the SAT math scores improved in this school district last year.

    d. It is not true that the SAT math scores did not improve in this school district last year.

9. Let p and q denote the propositions

    p: Laura purchased a VHS videocassette recorder.

    q: Laura did not purchase a Beta videocassette recorder.

Express each of the following statements symbolically.

    a. Laura purchased either a VHS or a Beta videocassette recorder.

    b. Laura purchased a VHS and a Beta videocassette recorder.

    c. It is not true that Laura purchased a Beta videocassette recorder.

    d. Laura neither purchased a VHS nor a Beta videocassette recorder.

10. Let p, q, and r denote the propositions

p: The popularity of prime-time soaps increased this year.

q: The popularity of prime-time situation comedies increased this year.

r: The popularity of prime-time detective shows decreased this year.

Express each of the following propositions in words.

a. $\sim p \wedge \sim q$

b. $\sim p \vee r$

c. $\sim (\sim r) \vee \sim q$

d. $\sim p \veebar \sim q$

A.2

Truth Tables

A primary objective of studying logic is to determine whether a given proposition is true or false. In other words, we wish to determine the **truth value** of a given statement.

Consider, for example, the conjunction $p \wedge q$ formed from the prime propositions p and q. There are four possible truth values for the given conjunction, namely,

1. p is true and q is true.
2. p is true and q is false.
3. p is false and q is true.
4. p is false and q is false.

By definition, the conjunction is true when both p and q are true and it is false otherwise. We can summarize this information in the form of a truth table as shown in Table A.2.

p	q	$p \wedge q$
T	T	T
T	F	F
F	T	F
F	F	F

Table A.2

In a similar manner we can construct the truth tables for inclusive disjunction, exclusive disjunction, and negation (see Tables A.3(a), (b), and (c), respectively).

p	q	$p \vee q$
T	T	T
T	F	T
F	T	T
F	F	F

p	q	$p \veebar q$
T	T	F
T	F	T
F	T	T
F	F	F

p	$\sim p$
T	F
F	T

Table A.3   (a)   (b)   (c)

In general, when we are given a compound proposition, we are concerned with the problem of finding every possible combination of truth values associated with the compound proposition. To systematize this procedure, we construct a truth table exhibiting all possible truth values for the given proposition. The next several examples illustrate this method.

EXAMPLE

6   Construct the truth table for the proposition $\sim p \vee q$.

SOLUTION

The truth table is constructed in the following manner.

1. The two prime propositions p and q are placed at the head of the first two columns (see Table A.4).

2. The two propositions containing the connectives, $\sim p$ and $\sim p \vee q$, are placed at the head of the next two columns.

3. The possible truth values for p are entered in the column headed by p and then the possible truth values for q are entered in the column headed by q. Notice that *the possible T values are always exhausted first.*

4. The possible truth values for the negation $\sim p$ are entered in the column headed by $\sim p$ and then the possible truth values for the disjunction $\sim p \vee q$ are entered in the column headed by $\sim p \vee q$.

p	q	$\sim p$	$\sim p \vee q$
T	T	F	T
T	F	F	F
F	T	T	T
F	F	T	T

Table A.4   ◀

EXAMPLE

7   Determine the truth table for the proposition $\sim p \vee (p \wedge q)$.

SOLUTION

Proceeding as in the previous example, the truth table is constructed as shown in Table A.5. Since the given proposition is the disjunction of the propositions $\sim p$ and $(p \wedge q)$, columns are introduced for $\sim p$, $p \wedge q$, and $\sim p \vee (p \wedge q)$.

p	q	$\sim p$	$p \wedge q$	$\sim p \vee (p \wedge q)$
T	T	F	T	T
T	F	F	F	F
F	T	T	F	T
F	F	T	F	T

Table A.5 ◀

The following is an example involving three prime propositions.

EXAMPLE

8 Determine the truth table for the proposition $(p \vee q) \vee (r \wedge \sim p)$.

SOLUTION

Following the method of the previous examples, the truth table shown in Table A.6 is constructed.

p	q	r	$p \vee q$	$\sim p$	$r \wedge \sim p$	$(p \vee q) \vee (r \wedge \sim p)$
T	T	T	T	F	F	T
T	T	F	T	F	F	T
T	F	T	T	F	F	T
T	F	F	T	F	F	T
F	T	T	T	T	T	T
F	T	F	T	T	F	T
F	F	T	F	T	T	T
F	F	F	F	T	F	F

Table A.6 ◀

Observe that the truth table in Example 8 contains eight rows, whereas the truth tables in Examples 6 and 7 contain four rows. In general, *if a compound proposition $P(p, q, \ldots)$ contains the n prime propositions p, q, $\ldots$, then the corresponding truth table contains 2^n rows.* For example, since the compound proposition in Example 8 contains three prime propositions, its corresponding truth table contains 2^3, or 8, rows.

▶

Exercises A.2

In Exercises 1–14, construct a truth table for the given compound proposition.

1. $p \vee \sim q$
2. $\sim p \wedge \sim q$
3. $\sim (\sim p)$
4. $\sim (p \wedge q)$
5. $p \vee \sim p$
6. $\sim (\sim p \vee \sim q)$

7. $\sim p \wedge (p \vee q)$ 8. $(p \vee \sim q) \wedge q$

9. $(p \vee q) \wedge (p \wedge \sim q)$ 10. $(p \vee q) \wedge \sim p$

11. $(p \vee q) \wedge (p \vee r)$ 12. $p \wedge (q \vee r)$

13. $(p \wedge \sim q) \vee (p \wedge r)$ 14. $\sim (p \wedge q) \vee (q \wedge r)$

15. If a compound proposition consists of the prime propositions p, q, r, and s, how many rows does its corresponding truth table contain?

▶ A.3

The Conditional and the Biconditional Connectives

In this section we will introduce two other connectives: the conditional and the biconditional. We will also discuss three variations of conditional statements: the inverse, the contrapositive, and the converse.

We often use expressions of the form

If it rains, *then* the baseball game will be postponed.

to specify the conditions under which a statement will be true. The "if . . . then" statement is the building block upon which deductive reasoning is based, and an understanding of its use in forming logical proofs is of fundamental importance.

Definition

A **conditional statement** is a proposition of the form "if p, then q" and is represented symbolically by

$$p \to q$$

The connective "if . . . then" is called the **conditional connective,** the proposition p the **hypothesis,** and the proposition q the **conclusion.** A conditional statement is false if the hypothesis is true and the conclusion is false; it is true in all other cases.

The truth table determined by the conditional $p \to q$ is shown in Table A.7.

p	q	$p \rightarrow q$
T	T	T
T	F	F
F	T	T
F	F	T

Table A.7

One question that is often asked is: Why is a conditional statement true when its hypothesis is false? We can answer this question by considering the following conditional statement made by a mother to her son.

If you do your homework, then you may watch T.V.

Think of the statement consisting of the two prime propositions

p: You do your homework
q: You may watch T.V.

as a *promise* made by the mother to her son. Four cases arise:

1. The son does his homework and his mother lets him watch T.V.

2. The son does his homework and his mother does not let him watch T.V.

3. The son does not do his homework and his mother lets him watch T.V.

4. The son does not do his homework and his mother does not let him watch T.V.

In (1), p and q are both true, the promise has been kept and, consequently, $p \rightarrow q$ is true. In (2), p is true and q is false and the mother has broken her promise. Therefore $p \rightarrow q$ is false. In (3) and (4), p is not true and the promise is *not* broken. Thus, $p \rightarrow q$ is regarded as a true statement. In other words, the conditional statement is only regarded as false if the "promise" is broken.

There are several equivalent expressions for the conditional connective "if . . . then." Among them are

1. p implies q

2. p only if q

3. q if p

4. q whenever p

5. Suppose p, then q

Care should be taken not to confuse the conditional "q if p" with the conditional $q \rightarrow p$, because the two statements have quite different meanings. For example, the conditional $p \rightarrow q$ formed from the two propositions

p: There is a fire.

q: You call the fire department.

is

> If there is a fire, then you call the fire department.

whereas the conditional $q \rightarrow p$ is

> If you call the fire department, then there is a fire.

Obviously, the two statements have quite different meanings.

We refer to statements that are variations of the conditional $p \rightarrow q$ as **logical variants.** We define the three logical variants of the conditional $p \rightarrow q$ as follows:

1. The **converse** is a compound statement of the form "if q, then p" and is represented symbolically by

$$q \rightarrow p$$

2. The **contrapositive** is a compound statement of the form "if not q, then not p" and is represented symbolically by

$$\sim q \rightarrow \sim p$$

3. The **inverse** is a compound statement of the form "if not p, then not q" and is represented symbolically by

$$\sim p \rightarrow \sim q$$

EXAMPLE

9 Given the two propositions

p: You vote in the presidential election.

q: You are a registered voter.

a. State the conditional $p \rightarrow q$.

b. State the converse, the contrapositive, and the inverse of $p \rightarrow q$.

SOLUTION

a. The conditional is "If you vote in the presidential election, then you are a registered voter."

b. The converse is "If you are a registered voter, then you vote in the presidential election." The contrapositive is "If you are not a registered voter, then you do not vote in the presidential election." The inverse is "If you do not vote in the presidential election, then you are not a registered voter." ◀

The truth table for the conditional $p \rightarrow q$ and its three logical variants is shown in Table A.8.

p	q	Conditional $p \rightarrow q$	Converse $q \rightarrow p$	$\sim p$	$\sim q$	Contrapositive $\sim q \rightarrow \sim p$	Inverse $\sim p \rightarrow \sim q$
T	T	T	T	F	F	T	T
T	F	F	T	F	T	F	T
F	T	T	F	T	F	T	F
F	F	T	T	T	T	T	T

Table A.8

Notice that the conditional $p \rightarrow q$ and its contrapositive $\sim q \rightarrow \sim p$ have identical truth tables. In other words, the conditional statement and its contrapositive have the same meaning.

Definition

Two propositions p and q are logically equivalent, denoted by

$$p \Leftrightarrow q,$$

if they have identical truth tables.

Referring once again to the truth table in Table A.8, we see that $p \rightarrow q$ is logically equivalent to its contrapositive. Similarly, the converse of a conditional statement is logically equivalent to the inverse.

It is sometimes easier to prove the contrapositive of a conditional statement than it is to prove the conditional itself. We may use this to our advantage in establishing a proof, as shown in the next example.

EXAMPLE

10   Prove that if n^2 is an odd number, then n is an odd number.

SOLUTION

Let

p: n^2 is odd

q: n is an odd number

Since $p \rightarrow q$ is logically equivalent to $\sim q \rightarrow \sim p$, it suffices to prove $\sim q \rightarrow \sim p$. Thus, we wish to prove that if n is not an odd number, then n^2 is an even number. If n is even, then $n = 2k$ where k is an integer. Therefore,

$$n^2 = (2k)(2k) = 2(2k^2).$$

Since $2(2k^2)$ is a multiple of 2, it is an even number and consequently not odd. Thus, we have shown that the contrapositive $\sim q \rightarrow \sim p$ is true and it follows that the conditional $p \rightarrow q$ is also true.   ◀

We now turn our attention to the last of the five basic connectives.

Definition

Statements of the form "*p* if and only if *q*" are called **biconditional proposi-tions** and are represented symbolically by

$$p \leftrightarrow q.$$

The connective "if and only if" is called the biconditional connective. The biconditional $p \leftrightarrow q$ is true whenever *p* and *q* are *both true* or *both false*.

The truth table for $p \leftrightarrow q$ is shown in Table A.9.

p	q	$p \leftrightarrow q$
T	T	T
T	F	F
F	T	F
F	F	T

Table A.9

EXAMPLE

11 Let

p: Mark is going to the senior prom.

q: Linda is going to the senior prom.

State the biconditional $p \leftrightarrow q$.

SOLUTION The required statement is

Mark is going to the senior prom,
if and only if,
Linda is going to the senior prom.   ◄

As suggested by its name, the biconditional statement actually comprises two conditional statements. For instance, an *equivalent expression* for the bicon-ditional statement in Example 11 is given by the two conditional statements:

Mark is going to the senior prom
if Linda is going to the senior prom.

and   Linda is going to the senior prom
if Mark is going to the senior prom.

Thus, "p if q and q if p" is equivalent to "p if and only if q."

The equivalent expressions that are most commonly used in forming conditional and biconditional statements are summarized in Table A.10.

		Equivalent Form
Conditional Statement	if p, then q	p is sufficient for q p only if q q is necessary for p q, if p
Biconditional Statement	p if and only q	if p then q, if q then p p is necessary and sufficient for q

Table A.10

The words *necessary* and *sufficient* occur frequently in mathematical proofs. When we say that p is *sufficient* for q we mean that *when p is true, q is also true;* that is, $p \rightarrow q$. In like manner, when we say that p is *necessary* for q, we mean that *if p is not true, then q is not true;* that is, $\sim p \rightarrow \sim q$. But this last expression is logically equivalent to $q \rightarrow p$. Thus, when we say that p is necessary and sufficient for q, it follows that $p \rightarrow q$ and $q \rightarrow p$. Example 12 illustrates the use of these expressions.

EXAMPLE

12 Let

> p: The stock market will go up.
>
> q: Interest rates decrease.

Represent the following statements symbolically:

a. If interest rates decrease, then the stock market will go up.

b. If interest rates do not decrease, then the stock market will not go up.

c. The stock market will go up if and only if interest rates decrease.

d. Decreasing interest rates is a sufficient condition for the stock market to go up.

e. Decreasing interest rates is a necessary and sufficient condition for a rising stock market.

SOLUTION

a. $q \rightarrow p$

b. $\sim q \rightarrow \sim p$ or $p \rightarrow q$

c. $q \leftrightarrow p$

d. $q \rightarrow p$

e. $q \leftrightarrow p$

◀

▶

Exercises A.3

In Exercises 1–4, write the converse, the contrapositive, and the inverse of the given conditional statement.

1. $p \to \sim q$
2. $\sim p \to \sim q$
3. $q \to p$
4. $\sim p \to q$

Exercises 5 and 6 refer to the following propositions p and q.

> p: It is snowing.
>
> q: The temperature is below freezing.

5. Express the conditional and the biconditional of p and q in words.

6. Express the converse, contrapositive, and inverse of the conditional $p \to q$ in words.

Exercises 7 and 8 refer to the following propositions p and q.

> p: The company's union and management reach a settlement.
>
> q: The workers will not strike.

7. Express the conditional and the biconditional of p and q in words.

8. Express the converse, contrapositive, and inverse of the conditional $p \to q$ in words.

In Exercises 9–11, determine whether the given statement is true or false.

9. A conditional proposition and its converse are logically equivalent.

10. The converse and the inverse of a conditional proposition are logically equivalent.

11. A conditional proposition and its inverse are logically equivalent.

12. Consider the conditional statement "If the owner lowers the selling price of the house, then I will buy it." Under what conditions is the conditional statement false?

13. Consider the biconditional statement "I will buy the house if and only if the owner lowers the selling price." Under what conditions is the biconditional statement false?

In Exercises 14–23, construct a truth table for the given compound proposition.

14. $\sim (p \to q)$
15. $\sim (q \to \sim p)$
16. $\sim (p \to q) \wedge p$
17. $(p \to q) \vee (q \to p)$
18. $(p \to \sim q) \veebar \sim p$
19. $(p \to q) \wedge (\sim p \vee q)$
20. $(p \to q) \leftrightarrow (\sim q \to \sim p)$
21. $\sim q \to (\sim p \wedge \sim q)$
22. $(p \vee q) \to \sim r$
23. $[(p \to q) \vee (q \to r)] \to (p \to r)$

In Exercises 24–29, determine whether the given compound propositions are logically equivalent.

24. $\sim p \vee q; \; p \to q$
25. $\sim (p \vee q); \; \sim p \wedge \sim q$
26. $q \to p; \; \sim p \to \sim q$
27. $\sim p \to q; \; \sim p \vee q$

28. $(p \to q) \to r; (p \vee q) \vee r$ 29. $p \vee (q \wedge r); (p \vee q) \wedge (p \vee r)$

30. Let p and q denote the following propositions

p: Taxes are increased.

q: The federal deficit increases.

Represent the following statements symbolically.

a. If taxes are increased, then the federal deficit will not increase.

b. If taxes are not increased, then the federal deficit will increase.

c. The federal deficit will not increase if and only if taxes are increased.

d. Increased taxation is a sufficient condition for halting the growth of the federal deficit.

e. Increased taxation is a necessary and sufficient condition for halting the growth of the federal deficit.

A.4

Laws of Logic

Just as the laws of algebra enable us to perform operations with real numbers, the **laws of logic** provide us with a systematic method of connecting statements. These laws are:

Let p, q, and r be any three propositions, then

1. $p \wedge p \Leftrightarrow p$	*Idempotent law for conjunction*
2. $p \vee p \Leftrightarrow p$	*Idempotent law for disjunction*
3. $(p \wedge q) \wedge r \Leftrightarrow p \wedge (q \wedge r)$	*Associative law for conjunction*
4. $(p \vee q) \vee r \Leftrightarrow p \vee (q \vee r)$	*Associative law for disjunction*
5. $p \wedge q \Leftrightarrow q \wedge p$	*Commutative law for conjunction*
6. $p \vee q \Leftrightarrow q \vee p$	*Commutative law for disjunction*
7. $p \wedge (q \vee r) \Leftrightarrow (p \wedge q) \vee (p \wedge r)$	*Distributive law for conjunction*
8. $p \vee (q \wedge r) \Leftrightarrow (p \vee q) \wedge (p \vee r)$	*Distributive law for disjunction*
9. $\sim (p \vee q) \Leftrightarrow \sim p \wedge \sim q$	*DeMorgan's law*
10. $\sim (p \wedge q) \Leftrightarrow \sim p \vee \sim q$	*DeMorgan's law*

In order to verify any of these laws, we need only construct a truth table to show that the given statements are logically equivalent. We illustrate this procedure in the next example.

EXAMPLE

13   Prove the distributive law for conjunction.

SOLUTION

We wish to prove that $p \wedge (q \vee r)$ is logically equivalent to $(p \wedge q) \vee (p \wedge r)$. This is easily done by constructing the associated truth table as shown in Table A.11.

p	q	r	$q \vee r$	$p \wedge q$	$p \wedge r$	$p \wedge (q \vee r)$	$(p \wedge q) \vee (p \wedge r)$
T	T	T	T	T	T	T	T
T	T	F	T	T	F	T	T
T	F	T	T	F	T	T	T
T	F	F	F	F	F	F	F
F	T	T	T	F	F	F	F
F	T	F	T	F	F	F	F
F	F	T	T	F	F	F	F
F	F	F	F	F	F	F	F

Table A.11

Since the entries in the last two columns are the same, we conclude that $p \wedge (q \vee r) \Leftrightarrow (p \wedge q) \vee (p \wedge r)$. ◄

The proof of the other laws is left to you as an exercise.

DeMorgan's laws, (9) and (10), are useful in forming the negation of a statement. For example, if we wish to state the negation of the proposition

Steve plans to major in business administration or economics.

we can represent the statement symbolically by $p \vee q$, where the prime propositions are

p: Steve plans to major in business administration.

q: Steve plans to major in economics.

Then the negation of $p \vee q$ is $\sim (p \vee q)$. Using DeMorgan's laws, we have

$$\sim (p \vee q) \Leftrightarrow \sim p \wedge \sim q.$$

Thus, the required statement is

Steve does not plan to major in business administration *and* he does not plan to major in economics.

Up to this point we have considered propositions that have both true and false entries in their truth tables. Some statements have the property that they are always true; other propositions have the property that they are always false.

Definition

A **tautology** is a statement that is always true.
A **contradiction** is a statement that is always false.

EXAMPLE

14 Show that

a. $p \lor \sim p$ is a tautology.

b. $p \land \sim p$ is a contradiction.

SOLUTION

a. The truth table associated with $p \lor \sim p$ is shown in Table A.12(a). Since all the entries in the last column are T's, the proposition is a tautology.

b. The truth table for $p \land \sim p$ is shown in Table A.12(b). Since all the entries in the last column are F's, the proposition is a contradiction.

p	$\sim p$	$p \lor \sim p$
T	F	T
F	T	T

p	$\sim p$	$p \land \sim p$
T	F	F
F	T	F

Table A.12   (a)   (b)

In Section A.3 we defined logically equivalent statements as statements that had identical truth tables. The definition of equivalence may also be restated in terms of a tautology.

Definition

Two propositions p and q are logically equivalent if the biconditional $p \leftrightarrow q$ is a tautology.

We can see why these two definitions are really the same by looking more closely at the truth table for $p \leftrightarrow q$ (Table A.13).

	p	q	$p \leftrightarrow q$
Case 1	T	T	T
Case 2	T	F	F
Case 3	F	T	F
Case 4	F	F	T

p	q	$p \leftrightarrow q$
T	T	T
F	F	T

Table A.13   (a)   (b)

If the biconditional is always true, then the second and third cases shown in the truth table are excluded and we are left with the truth table in Table A.13(b). Notice that the entries in each row of the p and q columns are identical. In other words, p and q have identical truth values, and hence are logically equivalent.

In addition to the ten laws stated earlier, we have the following four laws involving tautologies and contradictions.

Let t be a tautology and c a contradiction. Then

11. $p \lor \sim p \Leftrightarrow t$
12. $p \land \sim p \Leftrightarrow c$
13. $p \lor t \Leftrightarrow p$
14. $p \land t \Leftrightarrow p$

It now remains only to show how these laws are used to simplify proofs.

EXAMPLE

15   Using the preceding laws, show that $p \lor (\sim p \land q) \Leftrightarrow (p \lor q)$.

SOLUTION

$$\begin{aligned}
p \lor (\sim p \land q) &\Leftrightarrow (p \lor \sim p) \land (p \lor q) &&\text{by Law 8}\\
&\Leftrightarrow \quad\quad t \ \land (p \lor q) &&\text{by Law 11}\\
&\Leftrightarrow p \lor q &&\text{by Law 14}
\end{aligned}$$
◀

EXAMPLE

16   Using the laws of logic, show that $\sim(p \lor q) \lor (\sim p \land q) \Leftrightarrow \sim p$.

SOLUTION

$$\begin{aligned}
\sim(p \lor q) \lor (\sim p \land q) &\Leftrightarrow (\sim p \land \sim q) \lor (\sim p \land q) &&\text{by Law 9}\\
&\Leftrightarrow \sim p \land (\sim q \lor q) &&\text{by Law 7}\\
&\Leftrightarrow \sim p \land t &&\text{by Law 11}\\
&\Leftrightarrow \sim p &&\text{by Law 14}
\end{aligned}$$
◀

▶

Exercises A.4

1. Prove the idempotent law for conjunction, $p \land p \Leftrightarrow p$.
2. Prove the idempotent law for disjunction, $p \lor p \Leftrightarrow p$.
3. Prove the associative law for conjunction, $(p \land q) \land r \Leftrightarrow p \land (q \land r)$.
4. Prove the associative law for disjunction, $(p \lor q) \lor r \Leftrightarrow p \lor (q \lor r)$.
5. Prove the commutative law for conjunction, $p \land q \Leftrightarrow q \land p$.
6. Prove the commutative law for disjunction, $p \lor q \Leftrightarrow q \lor p$.
7. Prove the distributive law for disjunction, $p \lor (q \land r) \Leftrightarrow (p \lor q) \land (p \lor r)$.
8. Prove DeMorgan's laws
   a. $\sim(p \lor q) \Leftrightarrow \sim p \land \sim q$
   b. $\sim(p \land q) \Leftrightarrow \sim p \lor \sim q$

In Exercises 9–16, determine whether the given statement is a tautology, a contradiction, or neither.

9. $(p \rightarrow q) \leftrightarrow (\sim p \vee q)$

10. $(p \veebar q) \wedge (p \leftrightarrow q)$

11. $p \rightarrow (p \vee q)$

12. $(p \rightarrow q) \vee (q \rightarrow p)$

13. $(p \rightarrow q) \leftrightarrow (\sim q \rightarrow \sim p)$

14. $[(p \rightarrow q) \wedge (q \rightarrow r)] \rightarrow (p \rightarrow r)$

15. $[(p \rightarrow q) \vee (q \rightarrow r)] \rightarrow (p \rightarrow r)$

16. $[p \wedge (q \vee r)] \leftrightarrow [(p \wedge q) \vee (p \wedge r)]$

17. Let p and q denote the statements

p: The candidate opposes changes in the social security system.

q: The candidate supports the ERA.

Use DeMorgan's laws to state the negation of $p \wedge q$ and the negation of $p \vee q$.

In Exercises 18–23, use the laws of logic to prove the given proposition.

18. $[p \wedge (q \vee \sim q) \vee (p \wedge q)] \Leftrightarrow p \vee (p \wedge q)$

19. $p \vee (\sim p \wedge \sim q) \Leftrightarrow p \vee \sim q$

20. $(p \wedge \sim q) \vee (p \wedge \sim r) \Leftrightarrow p \wedge (\sim q \vee \sim r)$

21. $(p \vee q) \vee \sim q \Leftrightarrow p$

22. $p \wedge [\sim(q \wedge r)] \Leftrightarrow (p \wedge \sim q) \vee (p \wedge \sim r)$

23. $p \vee (q \vee r) \Leftrightarrow r \vee (q \vee p)$

▶ A.5

Arguments

In this section we will discuss arguments and the methods used to determine the validity of arguments.

Definition
An **argument** or **proof** consists of a set of propositions $p_1, p_2, \ldots, p_n$, called the *premises*, and a proposition q, called the *conclusion*. An argument is *valid* if and only if the conclusion is true whenever the premises are all true. An argument that is *not* valid is called a *fallacy*, or an *invalid* argument.

The next example illustrates the form in which an argument is presented. Notice that the premises are written separately above the horizontal line and the conclusion is written below the line.

EXAMPLE

17   An argument is presented in the following way:

Premises: If Pam studies diligently, she passes her exams.
             Pam studies diligently.

Conclusion: Pam passes her exams.   ◀

To determine the validity of an argument, we first write the argument in symbolic form and then construct the associated truth table containing the prime propositions, the premises, and the conclusion. We then *check the rows in which the premises are all true. If the conclusion in each of these rows is also true, then the argument is valid. Otherwise, it is a fallacy.*

EXAMPLE

18   Determine the validity of the argument in Example 17.

SOLUTION   The symbolic form of the argument is

$$p \to q$$
$$\underline{p}$$
$$\therefore q$$

Observe that the conclusion is preceded by the symbol $\therefore$, which is used to represent the word *therefore*. The truth table associated with this argument is shown in Table A.14. Observe that only row 1 contains true values for both premises. Since the conclusion is also true in this row, we conclude that the argument is valid.

Propositions		Premises		Conclusion
p	q	$p \to q$	p	q
T	T	T	T	T
T	F	F	T	F
F	T	T	F	T
F	F	T	F	F

Table A.14   ◀

EXAMPLE

19   Determine the validity of the argument

If Michael is overtired, then he is grumpy.
Michael is grumpy.

Therefore, Michael is overtired.

SOLUTION

The argument is written symbolically as follows:

$$p \rightarrow q$$
$$q$$
$$\overline{\therefore p}$$

The associated truth table is shown in Table A.15.

p	q	$p \rightarrow q$	q	p
T	T	T	T	T
T	F	F	F	T
F	T	T	T	F
F	F	T	F	F

Table A.15

Observe that the entries for the premises in the third row are both true, but the corresponding entry for the conclusion is false. We conclude that the argument is a fallacy. ◄

When we consider the validity of an argument, we are concerned only with the *form* of the argument and not the truth or falsity of the premises. In other words, the conclusion of an argument may follow validly from the premises, but the premises themselves may be false. The next example demonstrates this point.

EXAMPLE

20 Determine the validity of the argument

Reggie is a wealthy man.
Wealthy men are happy.

Therefore, Reggie is happy.

SOLUTION

The symbolic form of the argument is

$$p \rightarrow q$$
$$q \rightarrow r$$
$$\overline{\therefore p \rightarrow r}$$

The associated truth table is shown in Table A.16.

p	q	r	$p \rightarrow q$	$q \rightarrow r$	$p \rightarrow r$	
T	T	T	T	T	T	(Check)
T	T	F	T	F	F	
T	F	T	F	T	T	
T	F	F	F	T	F	
F	T	T	T	T	T	(Check)
F	T	F	T	F	T	
F	F	T	T	T	T	(Check)
F	F	F	T	T	T	(Check)

Table A.16

Since the conclusion is true in each of the rows that contain true values for both premises, we conclude that the argument is valid. Note that the validity of the argument is not affected by the truth or falsity of the premise "Wealthy men are happy." ◀

A question that may already have arisen in your mind is: How does one determine the validity of an argument if the associated truth table does not contain true values for all the premises? The next example provides the answer to this question.

EXAMPLE

21 Determine the validity of the argument

> The door is locked.
> The door is unlocked.
> _____
> Therefore, the door is locked.

SOLUTION The symbolic form of the argument is

$$
\begin{array}{c}
p \\
\sim p \\
\hline
\therefore\ p
\end{array}
$$

The associated truth table is shown in Table A.17.

p	$\sim p$	p
T	F	T
F	T	F

Table A.17

Observe that no rows contain true values for both premises. Nevertheless, the argument is considered to be valid since the condition that the conclusion is true whenever the premises are all true is not violated. Again, we remind you that the validity of an argument is not determined by the truth or falsity of its premises or conclusion, but rather, it is only determined by the form of the argument.   ◀

The next proposition provides us with an alternative method for determining the validity of an argument.

Proposition

Suppose an argument consists of the premises $p_1, p_2, \ldots, p_n$ and conclusion q. Then the argument is valid if and only if the proposition $[p_1 \wedge p_2 \wedge \ldots \wedge p_n] \to q$ is a tautology.

To prove this proposition we must show that (a) if an argument is valid, then the given proposition is a tautology and (b) if the given proposition is a tautology, then the argument is valid.

Proof of (a): Since the argument is valid, it follows that q is true whenever all of the premises $p_1, p_2, \ldots, p_n$ are true. But the conjunction $[p_1 \wedge p_2 \wedge \ldots \wedge p_n]$ is true only when all the premises are true (by the definition of conjunction). Hence q is true whenever the conjunction is true and we conclude that the conditional $[p_1 \wedge p_2 \wedge \ldots \wedge p_n] \to q$ is true. (Recall that a conditional statement is always true when its hypotheses are false. Hence, to show that a conditional is a tautology, we need only prove that its conclusion is always true when its hypotheses are true.)

Proof of (b): Since the given proposition is a tautology, q is always true. Therefore, q is true when the conjunction $(p_1 \wedge p_2) \wedge \ldots \wedge p_n)$ is true. But the conjunction is true only when all of the premises $p_1, p_2, \ldots p_n$ are true. Hence, q is true whenever the premises are all true and the argument is valid.

Having proved this proposition, we can now use it to determine the validity of an argument. If we are given an argument, we construct a truth table for $[p_1 \wedge p_2 \wedge \ldots \wedge p_n] \to q$. If the truth table contains all the true values in its last column, then the argument is valid; otherwise it is invalid. This method of proof is illustrated in the next example.

EXAMPLE

22   Determine the validity of the argument

Either you return the book on time, or you have to pay a fine.
You do not return the book on time.
_____

Therefore, you have to pay a fine.

SOLUTION   The symbolic form of the argument is

$$p \lor q$$
$$\underline{\sim p}$$
$$\therefore q$$

Following the method described, we construct the truth table for $[(p \lor q) \land \sim p] \to q$ as shown in Table A.18.

p	q	$p \lor q$	$\sim p$	$(p \lor q) \land \sim p$	$[(p \lor q) \land \sim p] \to q$
T	T	T	F	F	T
T	F	T	F	F	T
F	T	T	T	T	T
F	F	F	T	F	T

Table A.18

Since the entries in the last column are all T's, the proposition is a tautology and we conclude that the argument is valid.   ◀

By familiarizing ourselves with a few of the most commonly used argument forms, we can simplify the problem of determining the validity of an argument. Some of the argument forms most commonly used are:

1. *Modus Ponens* (a manner of affirming), or Rule of Detachment

$$p \to q$$
$$\underline{p}$$
$$\therefore q$$

2. *Modus Tollens* (a manner of denying)

$$p \to q$$
$$\underline{\sim q}$$
$$\therefore \sim p$$

3. *Law of Syllogisms*

$$p \to q$$
$$\underline{q \to r}$$
$$\therefore p \to r$$

The truth tables verifying modus ponens and the law of syllogisms have already been constructed (see Tables A.14 and A.16, respectively). The verification of modus tollens is left to you as an exercise. The next two examples illustrate the use of these argument forms.

EXAMPLE

23 Determine the validity of the argument

If the battery is dead, the car will not start.
The car starts.

Therefore, the battery is not dead.

SOLUTION As before, we express the argument in symbolic form. Thus,

$$
\begin{array}{c}
p \to q \\
\sim q \\
\hline
\therefore \sim p
\end{array}
$$

Identifying the form of this argument as modus tollens, we conclude that the given argument is valid. ◀

EXAMPLE

24 Determine the validity of the argument

If mortgage rates are lowered, housing sales increase.
If housing sales increase, then the price of houses increases.

Therefore, if mortgage rates are lowered, the price of houses increases.

SOLUTION The symbolic form of the argument is

$$
\begin{array}{c}
p \to q \\
q \to r \\
\hline
\therefore p \to r
\end{array}
$$

Using the law of syllogisms, we conclude that the argument is valid. ◀

▶

Exercises A.5

In Exercises 1–16, determine whether the given argument is valid.

1. $p \to q$
   $q \to r$
   $\therefore p \to r$

2. $p \lor q$
   $\sim p$
   $\therefore q$

3. $p \land q$
   $\sim p$
   $\therefore q$

4. $p \to q$
   $\sim q$
   $\therefore \sim p$

5. $p \rightarrow q$
   $\dfrac{\sim p}{\therefore \sim q}$

6. $p \rightarrow q$
   $\dfrac{q \wedge r}{\therefore p \vee r}$

7. $p \leftrightarrow q$
   $\dfrac{q}{\therefore p}$

8. $p \wedge q$
   $\dfrac{\sim p \rightarrow \sim q}{\therefore p \wedge \sim q}$

9. $p \rightarrow q$
   $q \rightarrow p$
   $\overline{\therefore p \leftrightarrow q}$

10. $p \rightarrow \sim q$
    $p \vee q$
    $\overline{\therefore \sim q}$

11. $p \leftrightarrow q$
    $q \leftrightarrow r$
    $\overline{\therefore p \leftrightarrow r}$

12. $p \rightarrow q$
    $q \leftrightarrow r$
    $\overline{\therefore p \wedge r}$

13. $p \veebar r$
    $q \wedge r$
    $\overline{\therefore p \rightarrow r}$

14. $p \leftrightarrow q$
    $q \vee r$
    $\sim r$
    $\overline{\therefore p \rightarrow \sim r}$

15. $p \leftrightarrow q$
    $q \vee r$
    $\sim p$
    $\overline{\therefore \sim p \rightarrow \sim r}$

16. $p \rightarrow q$
    $r \rightarrow q$
    $p \wedge q$
    $\overline{\therefore p \vee r}$

In Exercises 17–22, represent the given argument symbolically and determine whether it is a valid argument.

17. If Carla studies, then she passes her exams.
    Carla did not study.

    Therefore, Carla did not pass her exams.

18. If Tony is wealthy, he is either intelligent or a good businessman.
    Tony is intelligent and he is not a good businessman.

    Therefore, Tony is not wealthy.

19. Steve will attend the matinee and/or the evening show.
    If Steve doesn't go to the matinee show, then he will not go to the evening show.

    Therefore, Steve will attend the matinee show.

20. If Mary wins the race, then Stacy loses the race.
    Neither Mary nor Linda won the race.

    Therefore, Stacy won the race.

21. If mortgage rates go up, then housing prices will go up.
    If housing prices go up, more people will rent houses.
    More people are renting houses.

    Therefore, mortgage rates went up.

22. If taxes are cut, then retail sales increase.
    If retail sales increase, then the unemployment rate will decrease.
    If the unemployment rate decreases, then the incumbent will win the election.

    Therefore, if taxes are not cut, the incumbent will not win the election.

23. Given the statement "All good cooks prepare gourmet food," which of the following conclusions follow logically?

    a. If George prepares gourmet food, then he is a good cook.

    b. If Brenda does not prepare gourmet food, then she is not a good cook.

    c. Everyone who prepares gourmet food is a good cook.

    d. Some people who prepare gourmet food are not good cooks.

24. What conclusion can be drawn from the following statements?

    His date is pretty or she is tall and skinny.

    If his date is tall, then she is a brunette.

    His date is not a brunette.

25. Show that modus tollens is a valid form of argument.

▶ A.6

Applications of Logic to Switching Networks

In this section we will see how the principles of logic can be used in the design and analysis of switching networks. A **switching network** is an arrangement of wires and switches connecting two terminals. Such networks are used extensively in digital computers.

A switch may be open or closed. If a switch is closed, current will flow through the wire. If it is open, no current will flow through the wire. Because a switch has exactly two states, it can be represented by a proposition p that is true if the switch is closed and false if the switch is open.

Now let us consider a circuit with two switches p and q. If the circuit is connected as shown in Figure A.1, the switches p and q are said to be **in series**.

Figure A.1

For such a network, current will flow from A to B if and only if both p and q are closed, but no current will flow if one or more of the switches is open. Thinking of p and q as propositions, we have the following truth table (Table A.19).

p	q	
T	T	T
T	F	F
F	T	F
F	F	F

Table A.19

(Recall that T corresponds to the situation in which the switch is closed.) From the truth table we see that two switches p and q connected in series are analogous to the conjunction $p \wedge q$ of the two propositions p and q.

If a circuit is connected as shown in Figure A.2, the switches p and q are said to be connected **in parallel.**

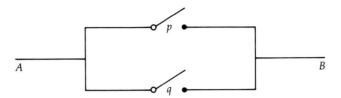

Figure A.2

For such a network, current will flow from A to B if and only if one or more of the switches p or q is closed. Once again, thinking of p and q as propositions, we have the following truth table (Table A.20).

p	q	
T	T	T
T	F	T
F	T	T
F	F	F

Table A.20

From the truth table, we conclude that two switches p and q connected in parallel are analogous to the inclusive disjunction $p \vee q$ of the two propositions p and q.

EXAMPLE

25 Find a logic statement that represents the network shown in Figure A.3. By constructing the truth table for this logic statement, determine the conditions under which current will flow from A to B in the network.

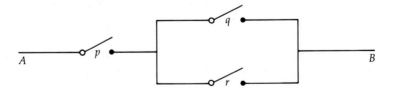

Figure A.3

SOLUTION The required logic statement is $p \wedge (q \vee r)$. Next, we construct the truth table for the logic statement $p \wedge (q \vee r)$ [Table A.21].

p	q	r	$q \vee r$	$p \wedge (q \vee r)$
T	T	T	T	T
T	T	F	T	T
T	F	T	T	T
T	F	F	F	F
F	T	T	T	F
F	T	F	T	F
F	F	T	T	F
F	F	F	F	F

Table A.21

From the truth table, we conclude that current will flow from A to B if and only if one of the following conditions is satisfied:

1. p, q, and r are all closed.

2. p and q are closed but r is open.

3. p and r are closed but q is open.

In other words, current will flow from A to B if and only if p is closed and either q or r or both q and r are closed . ◀

Before looking at some additional examples, let us remark that $\sim p$ represents a switch that is open when p is closed and vice versa. Furthermore, a circuit that is always closed is represented by a tautology, $p \vee \sim p$, whereas a circuit that is always open is represented by a contradiction, $p \wedge \sim p$. (Why?)

EXAMPLE

26 Given the logic statement $(p \vee q) \wedge (r \vee s \vee \sim p)$, draw the corresponding network.

SOLUTION Recalling that the disjunction $p \vee q$ of the propositions p and q represents two switches p and q connected in parallel and the conjunction $p \wedge q$ represents the two switches connected in series, we obtain the following network (Figure A.4).

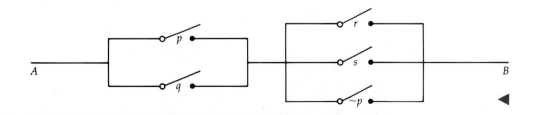

Figure A.4

The theory of networks developed so far, when used in conjunction with the laws of logic, is a useful tool in network analysis. In particular, network analysis enables us to find equivalent, and often simpler, networks, as the following example shows.

EXAMPLE

27 Find a logic statement representing the network shown in Figure A.5. Also, find a simpler but equivalent network.

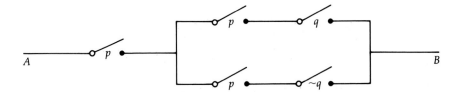

Figure A.5

SOLUTION

The logic statement corresponding to the given network is $p \wedge [(p \wedge q) \vee (p \wedge \sim q)]$.

Next, using the rules of logic to simplify this statement, we obtain

$$p \wedge [(p \wedge q) \vee (p \wedge \sim q)] \Leftrightarrow p \wedge [p \wedge (q \vee \sim q)] \quad \text{(Distributive law)}$$
$$\Leftrightarrow p \wedge p \quad\quad\quad\quad\quad \text{(Tautology)}$$
$$\Leftrightarrow p$$

Thus, the given network is equivalent to the one shown in Figure A.6.

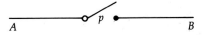

Figure A.6

Exercises A.6

In Exercises 1–5, find a logic statement corresponding to the given network. Determine the conditions under which current will flow from A to B.

1.

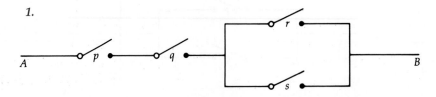

2.

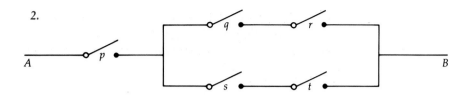

3.

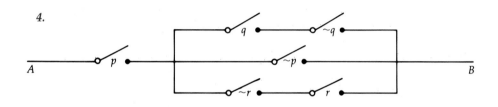

4.

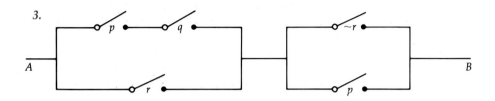

5.

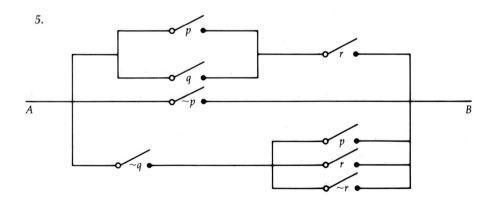

In Exercises 6–11, draw the network corresponding to the given logic statement.

6. $p \vee (q \wedge r)$

7. $(p \wedge q) \wedge r$

8. $[p \vee (q \wedge r)] \wedge \sim q$

9. $(p \vee q) \vee [r \wedge (\sim r \vee \sim p)]$

10. $(\sim p \vee q) \wedge (p \vee \sim q)$

11. $(p \wedge q) \vee [(r \vee \sim q) \wedge (s \vee \sim p)]$

In Exercises 12–15, find a logic statement corresponding to the given network. Then find a simpler but equivalent network.

12.

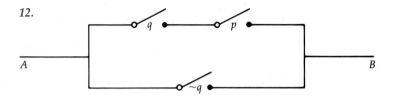

13.

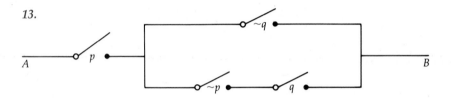

14.

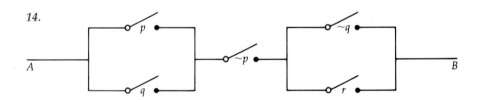

15.

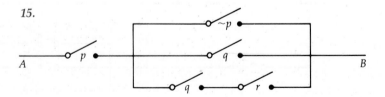

16. A hallway light is to be operated by two switches, one located at the bottom of the staircase and the other located at the top of the staircase. Design a suitable network. [*Hint:* Let *p* and *q* be the switches. Construct a truth table for the associated propositions.]

Appendix B

The System

of Real

Numbers

In this appendix we will briefly review the *system of real numbers*. This system comprises a set of objects called real numbers together with two operations, addition and multiplication, that enable us to combine two or more real numbers to obtain other real numbers. These operations are subjected to certain rules that we will state after first recalling the set of real numbers.

The set of real numbers may be constructed from the set of *natural* (also called *counting*) *numbers*

$$N = \{1, 2, 3, \ldots\}$$

by adjoining other objects (numbers) to it. Thus, the set

$$W = \{0, 1, 2, 3, \ldots\}$$

obtained by adjoining the single number 0 to N is called the set of *whole numbers*. By adjoining the *negatives* of the numbers 1, 2, 3, . . . to the set W of whole numbers, we obtain the set of *integers*

$$I = \{\ldots, -3, -2, -1, 0, 1, 2, 3, \ldots\}$$

Next, consider the set

$$Ra = \left\{ \frac{a}{b} \,\middle|\, a \text{ and } b \text{ are integers with } b \neq 0 \right\}$$

Now, the set I of integers is contained in the set Ra of *rational numbers*. To see this, observe that each integer may be written in the form a/b with $b = 1$, thus qualifying as a member of the set Ra. The converse, however, is false, for the rational numbers (fractions) such as

$$\frac{1}{2}, \frac{23}{25}, \text{ and so on}$$

are clearly not integers.

The sets N, W, I, and Ra constructed thus far have

$$N \subset W \subset I \subset Ra$$

that is, N is a proper subset of W; W is a proper subset of I, and so on.

Finally, consider the set Ir of all numbers that cannot be expressed in the form a/b where a, b are integers ($b \neq 0$). The members of this set, called the set of *irrational numbers*, include $\sqrt{2}$, $\sqrt{3}$, π, and so on. The set

$$R = Ra \cup Ir$$

that is, the set comprising all rational numbers as well as irrational numbers, is called the set of *real numbers* (See Figure B.1).

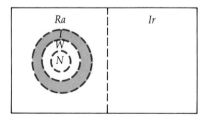

Figure B.1

Note the following important representation of real numbers: Every real number has a decimal representation; a rational number has a representation in terms of a repeated decimal. For example,

$$\frac{1}{7} = 0.142857142857142857\ldots$$ (Note that the block of integers 142857 repeats.)

On the other hand, the irrational number $\sqrt{2}$ has a representation in terms of a nonrepeating decimal. Thus

$$\sqrt{2} = 1.41428\ldots$$

As mentioned earlier, any two real numbers may be combined to obtain another real number. The operation of *addition*, written $+$, enables us to combine any two numbers a and b to obtain their sum, denoted by $a + b$. Another operation, called *multiplication*, and written $\cdot$, enables us to combine any two real numbers a and b to form their product, the number $a \cdot b$, or, written more simply, ab. These two operations are subjected to the following rules of operation:

Given any three real numbers a, b, and c, we have

I Under addition

1. $a + b = b + a$ (Commutative law of addition)

2. $a + (b + c) = (a + b) + c$ (Associative law of addition)

3. $a + 0 = a$ (Identity law of addition)

4. $a + (-a) = 0$ (Inverse law of addition)

II Under multiplication

1. $ab = ba$ (Commutative law of multiplication)

2. $a(bc) = (ab)c$ (Associative law of multiplication)

3. $a \cdot 1 = a$ (Identity law of multiplication)

4. $a(1/a) = 1$ $(a \neq 0)$ (Inverse law of multiplication)

III Under addition and multiplication

1. $a(b + c) = ab + ac$ (Distributive law for addition and multiplication)

Appendix C
Tables

Table 1   **593**

Compound Amount, Present Value, Annuity

Table 1   Compound Amount, Present Value, Annuity

$$i = \tfrac{1}{4}\%$$

n	$(1+i)^n$	$(1+i)^{-n}$	$s_{\overline{n}\rvert i}$	$a_{\overline{n}\rvert i}$	$s_{\overline{n}\rvert i}^{-1}$
1	1.0025 0000	0.9975 0623	1.0000 0000	0.9975 0623	1.0000 0000
2	1.0050 0625	0.9950 1869	2.0025 0000	1.9925 2492	0.4993 7578
3	1.0075 1877	0.9925 3734	3.0075 0625	2.9850 6227	0.3325 0139
4	1.0100 3756	0.9900 6219	4.0150 2502	3.9751 2446	0.2490 6445
5	1.0125 6266	0.9875 9321	5.0250 6258	4.9627 1766	0.1990 0250
6	1.0150 9406	0.9851 3038	6.0376 2523	5.9478 4804	0.1656 2803
7	1.0176 3180	0.9826 7370	7.0527 1930	6.9305 2174	0.1417 8928
8	1.0201 7588	0.9802 2314	8.0703 5110	7.9107 4487	0.1239 1035
9	1.0227 2632	0.9777 7869	9.0905 2697	8.8885 2357	0.1100 0462
10	1.0252 8313	0.9753 4034	10.1132 5329	9.8638 6391	0.0988 8015
11	1.0278 4634	0.9729 0807	11.1385 3642	10.8367 7198	0.0897 7840
12	1.0304 1596	0.9704 8187	12.1663 8277	11.8072 5384	0.0821 9370
13	1.0329 9200	0.9680 6171	13.1967 9872	12.7753 1555	0.0757 7595
14	1.0355 7448	0.9656 4759	14.2297 9072	13.7409 6314	0.0702 7510
15	1.0381 6341	0.9632 3949	15.2653 6520	14.7042 0264	0.0655 0777
16	1.0407 5882	0.9608 3740	16.3035 2861	15.6650 4004	0.0613 3642
17	1.0433 6072	0.9584 4130	17.3442 8743	16.6234 8133	0.0576 5587
18	1.0459 6912	0.9560 5117	18.3876 4815	17.5795 3250	0.0543 8433
19	1.0485 8404	0.9536 6700	19.4336 1727	18.5331 9950	0.0514 5722
20	1.0512 0550	0.9512 8878	20.4822 0131	19.4844 8828	0.0488 2288
21	1.0538 3352	0.9489 1649	21.5334 0682	20.4334 0477	0.0464 3947
22	1.0564 6810	0.9465 5011	22.5872 4033	21.3799 5488	0.0442 7278
23	1.0591 0927	0.9441 8964	23.6437 0843	22.3241 4452	0.0422 9455
24	1.0617 5704	0.9418 3505	24.7028 1770	23.2659 7957	0.0404 8121
25	1.0644 1144	0.9394 8634	25.7645 7475	24.2054 6591	0.0388 1298
26	1.0670 7247	0.9371 4348	26.8289 8619	25.1426 0939	0.0372 7312
27	1.0697 4015	0.9348 0646	27.8960 5865	26.0774 1585	0.0358 4736
28	1.0724 1450	0.9324 7527	28.9657 9880	27.0098 9112	0.0345 2347
29	1.0750 9553	0.9301 4990	30.0382 1330	27.9400 4102	0.0332 9093
30	1.0777 8327	0.9278 3032	31.1133 0883	28.8678 7134	0.0321 4059
31	1.0804 7773	0.9255 1653	32.1910 9210	29.7933 8787	0.0310 6449
32	1.0831 7892	0.9232 0851	33.2715 6983	30.7165 9638	0.0300 5569
33	1.0858 8687	0.9209 0624	34.3547 4876	31.6375 0262	0.0291 0806
34	1.0886 0159	0.9186 0972	35.4406 3563	32.5561 1234	0.0282 1620
35	1.0913 2309	0.9163 1892	36.5292 3722	33.4724 3126	0.0273 7533
36	1.0940 5140	0.9140 3384	37.6205 6031	34.3864 6510	0.0265 8121
37	1.0967 8653	0.9117 5445	38.7146 1171	35.2982 1955	0.0258 3004
38	1.0995 2850	0.9094 8075	39.8113 9824	36.2077 0030	0.0251 1843
39	1.1022 7732	0.9072 1272	40.9109 2673	37.1149 1302	0.0244 4335
40	1.1050 3301	0.9049 5034	42.0132 0405	38.0198 6336	0.0238 0204
41	1.1077 9559	0.9026 9361	43.1182 3706	38.9225 5697	0.0231 9204
42	1.1105 6508	0.9004 4250	44.2260 3265	39.8229 9947	0.0226 1112
43	1.1133 4149	0.8981 9701	45.3365 9774	40.7211 9648	0.0220 5724
44	1.1161 2485	0.8959 5712	46.4499 3923	41.6171 5359	0.0215 2855
45	1.1189 1516	0.8937 2281	47.5660 6408	42.5108 7640	0.0210 2339
46	1.1217 1245	0.8914 9407	48.6849 7924	43.4023 7047	0.0205 4022
47	1.1245 1673	0.8892 7090	49.8066 9169	44.2916 4137	0.0200 7762
48	1.1273 2802	0.8870 5326	50.9312 0842	45.1786 9463	0.0196 3433
49	1.1301 4634	0.8848 4116	52.0585 3644	46.0635 3580	0.0192 0915
50	1.1329 7171	0.8826 3457	53.1886 8278	46.9461 7037	0.0188 0099

Table 1 (continued)

$i = \frac{1}{4}\%$

n	$(1+i)^n$	$(1+i)^{-n}$	$s_{\overline{n}|i}$	$a_{\overline{n}|i}$	$s_{\overline{n}|i}^{-1}$
51	1.1358 0414	0.8804 3349	54.3216 5449	47.8266 0386	0.0184 0886
52	1.1386 4365	0.8782 3790	55.4574 5862	48.7048 4176	0.0180 3184
53	1.1414 9026	0.8760 4778	56.5961 0227	49.5808 8953	0.0176 6906
54	1.1443 4398	0.8738 6312	57.7375 9252	50.4547 5265	0.0173 1974
55	1.1472 0484	0.8716 8391	58.8819 3650	51.3264 3656	0.0169 8314
56	1.1500 7285	0.8695 1013	60.0291 4135	52.1959 4669	0.0166 5858
57	1.1529 4804	0.8673 4178	61.1792 1420	53.0632 8847	0.0163 4542
58	1.1558 3041	0.8651 7883	62.3321 6223	53.9284 6730	0.0160 4308
59	1.1587 1998	0.8630 2128	63.4879 9264	54.7914 8858	0.0157 5101
60	1.1616 1678	0.8608 6911	64.6467 1262	55.6523 5769	0.0154 6869
61	1.1645 2082	0.8587 2230	65.8083 2940	56.5110 7999	0.0151 9564
62	1.1674 3213	0.8565 8085	66.9728 5023	57.3676 6083	0.0149 3142
63	1.1703 5071	0.8544 4474	68.1402 8235	58.2221 0557	0.0146 7561
64	1.1732 7658	0.8523 1395	69.3106 3306	59.0744 1952	0.0144 2780
65	1.1762 0977	0.8501 8848	70.4839 0964	59.9246 0800	0.0141 8764
66	1.1791 5030	0.8480 6831	71.6601 1942	60.7726 7631	0.0139 5476
67	1.1820 9817	0.8459 5343	72.8392 6971	61.6186 2974	0.0137 2886
68	1.1850 5342	0.8438 4382	74.0213 6789	62.4624 7355	0.0135 0961
69	1.1880 1605	0.8417 3947	75.2064 2131	63.3042 1302	0.0132 9674
70	1.1909 8609	0.8396 4037	76.3944 3736	64.1438 5339	0.0130 8996
71	1.1939 6356	0.8375 4650	77.5854 2345	64.9813 9989	0.0128 8902
72	1.1969 4847	0.8354 5786	78.7793 8701	65.8168 5774	0.0126 9368
73	1.1999 4084	0.8333 7442	79.9763 3548	66.6502 3216	0.0125 0370
74	1.2029 4069	0.8312 9618	81.1762 7632	67.4815 2834	0.0123 1887
75	1.2059 4804	0.8292 2312	82.3792 1701	68.3107 5146	0.0121 3898
76	1.2089 6291	0.8271 5523	83.5851 6505	69.1379 0670	0.0119 6385
77	1.2119 8532	0.8250 9250	84.7941 2797	69.9629 9920	0.0117 9327
78	1.2150 1528	0.8230 3491	86.0061 1329	70.7860 3411	0.0116 2708
79	1.2180 5282	0.8209 8246	87.2211 2857	71.6070 1657	0.0114 6511
80	1.2210 9795	0.8189 3512	88.4391 8139	72.4259 5169	0.0113 0721
81	1.2241 5070	0.8168 9289	89.6602 7934	73.2428 4458	0.0111 5321
82	1.2272 1108	0.8148 5575	90.8844 3004	74.0577 0033	0.0110 0298
83	1.2302 7910	0.8128 2369	92.1116 4112	74.8705 2402	0.0108 5639
84	1.2333 5480	0.8107 9670	93.3419 2022	75.6813 2072	0.0107 1330
85	1.2364 3819	0.8087 7476	94.5752 7502	76.4900 9548	0.0105 7359
86	1.2395 2928	0.8067 5787	95.8117 1321	77.2968 5335	0.0104 3714
87	1.2426 2811	0.8047 4600	97.0512 4249	78.1015 9935	0.0103 0384
88	1.2457 3468	0.8027 3915	98.2938 7060	78.9043 3850	0.0101 7357
89	1.2488 490i	0.8007 3731	99.5396 0527	79.7050 7581	0.0100 4625
90	1.2519 7114	0.7987 4046	100.7884 5429	80.5038 1627	0.0099 2177
91	1.2551 0106	0.7967 4859	102.0404 2542	81.3005 6486	0.0098 0004
92	1.2582 3882	0.7947 6168	103.2955 2649	82.0953 2654	0.0096 8096
93	1.2613 8441	0.7927 7973	104.5537 6530	82.8881 0628	0.0095 6446
94	1.2645 3787	0.7908 0273	105.8151 4972	83.6789 0900	0.0094 5044
95	1.2676 9922	0.7888 3065	107.0796 8759	84.4677 3966	0.0093 3884
96	1.2708 6847	0.7868 6349	108.3473 8681	85.2546 0315	0.0092 2957
97	1.2740 4564	0.7849 0124	109.6182 5528	86.0395 0439	0.0091 2257
98	1.2772 3075	0.7829 4388	110.8923 0091	86.8224 4827	0.0090 1776
99	1.2804 2383	0.7809 9140	112.1695 3167	87.6034 3967	0.0089 1508
100	1.2836 2489	0.7790 4379	113.4499 5550	88.3824 8346	0.0088 1446

Table 1

595

Compound Amount, Present Value, Annuity

Table 1 (continued)

$$i = \tfrac{1}{2}\%$$

n	$(1+i)^n$	$(1+i)^{-n}$	$s_{\overline{n}|i}$	$a_{\overline{n}|i}$	$s_{\overline{n}|i}^{-1}$
1	1.0050 0000	0.9950 2488	1.0000 0000	0.9950 2488	1.0000 0000
2	1.0100 2500	0.9900 7450	2.0050 0000	1.9850 9938	0.4987 5312
3	1.0150 7513	0.9851 4876	3.0150 2500	2.9702 4814	0.3316 7221
4	1.0201 5050	0.9802 4752	4.0301 0013	3.9504 9566	0.2481 3279
5	1.0252 5125	0.9753 7067	5.0502 5063	4.9258 6633	0.1980 0997
6	1.0303 7751	0.9705 1808	6.0755 0188	5.8963 8441	0.1645 9546
7	1.0355 2940	0.9656 8963	7.1058 7939	6.8620 7404	0.1407 2854
8	1.0407 0704	0.9608 8520	8.1414 0879	7.8229 5924	0.1228 2886
9	1.0459 1058	0.9561 0468	9.1821 1583	8.7790 6392	0.1089 0736
10	1.0511 4013	0.9513 4794	10.2280 2641	9.7304 1186	0.0977 7057
11	1.0563 9583	0.9466 1489	11.2791 6654	10.6770 2673	0.0886 5903
12	1.0616 7781	0.9419 0534	12.3355 6237	11.6189 3207	0.0810 6643
13	1.0669 8620	0.9372 1924	13.3972 4018	12.5561 5131	0.0746 4224
14	1.0723 2113	0.9325 5646	14.4642 2639	13.4887 0777	0.0691 3609
15	1.0776 8274	0.9279 1688	15.5365 4752	14.4166 2465	0.0643 6436
16	1.0830 7115	0.9233 0037	16.6142 3026	15.3399 2502	0.0601 8937
17	1.0884 8651	0.9187 0684	17.6973 0141	16.2586 3186	0.0565 0579
18	1.0939 2894	0.9141 3616	18.7857 8791	17.1727 6802	0.0532 3173
19	1.0993 9858	0.9095 8822	19.8797 1685	18.0823 5624	0.0503 0253
20	1.1048 9558	0.9050 6290	20.9791 1544	18.9874 1915	0.0476 6645
21	1.1104 2006	0.9005 6010	22.0840 1101	19.8879 7925	0.0452 8163
22	1.1159 7216	0.8960 7971	23.1944 3107	20.7840 5896	0.0431 1380
23	1.1215 5202	0.8916 2160	24.3104 0322	21.6756 8055	0.0411 3465
24	1.1271 5978	0.8871 8567	25.4319 5524	22.5628 6622	0.0393 2061
25	1.1327 9558	0.8827 7181	26.5591 1502	23.4456 3803	0.0376 5186
26	1.1384 5955	0.8783 7991	27.6919 1059	24.3240 1794	0.0361 1163
27	1.1441 5185	0.8740 0986	28.8303 7015	25.1980 2780	0.0346 8565
28	1.1498 7261	0.8696 6155	29.9745 2200	26.0676 8936	0.0333 6167
29	1.1556 2197	0.8653 3488	31.1243 9461	26.9330 2423	0.0321 2914
30	1.1614 0008	0.8610 2973	32.2800 1658	27.7940 5397	0.0309 7892
31	1.1672 0708	0.8567 4600	33.4414 1666	28.6507 9997	0.0299 0304
32	1.1730 4312	0.8524 8358	34.6086 2375	29.5032 8355	0.0288 9453
33	1.1789 0833	0.8482 4237	35.7816 6686	30.3515 2592	0.0279 4727
34	1.1848 0288	0.8440 2226	36.9605 7520	31.1955 4818	0.0270 5586
35	1.1907 2689	0.8398 2314	38.1453 7807	32.0353 7132	0.0262 1550
36	1.1966 8052	0.8356 4492	39.3361 0496	32.8710 1624	0.0254 2194
37	1.2026 6393	0.8314 8748	40.5327 8549	33.7025 0372	0.0246 7139
38	1.2086 7725	0.8273 5073	41.7354 4942	34.5298 5445	0.0239 6045
39	1.2147 2063	0.8232 3455	42.9441 2666	35.3530 8900	0.0232 8607
40	1.2207 9424	0.8191 3886	44.1588 4730	36.1722 2786	0.0226 4552
41	1.2268 9821	0.8150 6354	45.3796 4153	36.9872 9141	0.0220 3631
42	1.2330 3270	0.8110 0850	46.6065 3974	37.7982 9991	0.0214 5622
43	1.2391 9786	0.8069 7363	47.8395 7244	38.6052 7354	0.0209 0320
44	1.2453 9385	0.8029 5884	49.0787 7030	39.4082 3238	0.0203 7541
45	1.2516 2082	0.7989 6402	50.3241 6415	40.2071 9640	0.0198 7117
46	1.2578 7892	0.7949 8907	51.5757 8497	41.0021 8547	0.0193 8894
47	1.2641 6832	0.7910 3390	52.8336 6390	41.7932 1937	0.0189 2733
48	1.2704 8916	0.7870 9841	54.0978 3222	42.5803 1778	0.0184 8503
49	1.2768 4161	0.7831 8250	55.3683 2138	43.3635 0028	0.0180 6087
50	1.2832 2581	0.7792 8607	56.6451 6299	44.1427 8635	0.0176 5376

Table 1 (*continued*)

$$i = \tfrac{1}{2}\%$$

n	$(1+i)^n$	$(1+i)^{-n}$	$s_{\overline{n}|i}$	$a_{\overline{n}|i}$	$s_{\overline{n}|i}^{-1}$
51	1.2896 4194	0.7754 0902	57.9283 8880	44.9181 9537	0.0172 6269
52	1.2960 9015	0.7715 5127	59.2180 3075	45.6897 4664	0.0168 8675
53	1.3025 7060	0.7677 1270	60.5141 2090	46.4574 5934	0.0165 2507
54	1.3090 8346	0.7638 9324	61.8166 9150	47.2213 5258	0.0161 7686
55	1.3156 2887	0.7600 9277	63.1257 7496	47.9814 4535	0.0158 4139
56	1.3222 0702	0.7563 1122	64.4414 0384	48.7377 5657	0.0155 1797
57	1.3288 1805	0.7525 4847	65.7636 1086	49.4903 0505	0.0152 0598
58	1.3354 6214	0.7488 0445	67.0924 2891	50.2391 0950	0.0149 0481
59	1.3421 3946	0.7450 7906	68.4278 9105	50.9841 8855	0.0146 1392
60	1.3488 5015	0.7413 7220	69.7700 3051	51.7255 6075	0.0143 3280
61	1.3555 9440	0.7376 8378	71.1188 8066	52.4632 4453	0.0140 6096
62	1.3623 7238	0.7340 1371	72.4744 7507	53.1972 5824	0.0137 9796
63	1.3691 8424	0.7303 6190	73.8368 4744	53.9276 2014	0.0135 4337
64	1.3760 3016	0.7267 2826	75.2060 3168	54.6543 4839	0.0132 9681
65	1.3829 1031	0.7231 1269	76.5820 6184	55.3774 6109	0.0130 5789
66	1.3898 2486	0.7195 1512	77.9649 7215	56.0969 7621	0.0128 2627
67	1.3967 7399	0.7159 3544	79.3547 9701	56.8129 1165	0.0126 0163
68	1.4037 5785	0.7123 7357	80.7515 7099	57.5252 8522	0.0123 8366
69	1.4107 7664	0.7088 2943	82.1553 2885	58.2341 1465	0.0121 7206
70	1.4178 3053	0.7053 0291	83.5661 0549	58.9394 1756	0.0119 6657
71	1.4249 1968	0.7017 9394	84.9839 3602	59.6412 1151	0.0117 6693
72	1.4320 4428	0.6983 0243	86.4088 5570	60.3395 1394	0.0115 7289
73	1.4392 0450	0.6948 2829	87.8408 9998	61.0343 4222	0.0113 8422
74	1.4464 0052	0.6913 7143	89.2801 0448	61.7257 1366	0.0112 0070
75	1.4536 3252	0.6879 3177	90.7265 0500	62.4136 4543	0.0110 2214
76	1.4609 0069	0.6845 0923	92.1801 3752	63.0981 5466	0.0108 4832
77	1.4682 0519	0.6811 0371	93.6410 3821	63.7792 5836	0.0106 7908
78	1.4755 4622	0.6777 1513	95.1092 4340	64.4569 7350	0.0105 1423
79	1.4829 2395	0.6743 4342	96.5847 8962	65.1313 1691	0.0103 5360
80	1.4903 3857	0.6709 8847	98.0677 1357	65.8023 0538	0.0101 9704
81	1.4977 9026	0.6676 5022	99.5580 5214	66.4699 5561	0.0100 4439
82	1.5052 7921	0.6643 2858	101.0558 4240	67.1342 8419	0.0098 9552
83	1.5128 0561	0.6610 2346	102.5611 2161	67.7953 0765	0.0097 5028
84	1.5203 6964	0.6577 3479	104.0739 2722	68.4530 4244	0.0096 0855
85	1.5279 7148	0.6544 6248	105.5942 9685	69.1075 0491	0.0094 7021
86	1.5356 1134	0.6512 0644	107.1222 6834	69.7587 1135	0.0093 3513
87	1.5432 8940	0.6479 6661	108.6578 7968	70.4066 7796	0.0092 0320
88	1.5510 0585	0.6447 4290	110.2011 6908	71.0514 2086	0.0090 7431
89	1.5587 6087	0.6415 3522	111.7521 7492	71.6929 5608	0.0089 4837
90	1.5665 5468	0.6383 4350	113.3109 3580	72.3312 9958	0.0088 2527
91	1.5743 8745	0.6351 6766	114.8774 9048	72.9664 6725	0.0087 0493
92	1.5822 5939	0.6320 0763	116.4518 7793	73.5984 7487	0.0085 8724
93	1.5901 7069	0.6288 6331	118.0341 3732	74.2273 3818	0.0084 7213
94	1.5981 2154	0.6257 3464	119.6243 0800	74.8530 7282	0.0083 5950
95	1.6061 1215	0.6226 2153	121.2224 2954	75.4756 9434	0.0082 4930
96	1.6141 4271	0.6195 2391	122.8285 4169	76.0952 1825	0.0081 4143
97	1.6222 1342	0.6164 4170	124.4426 8440	76.7116 5995	0.0080 3583
98	1.6303 2449	0.6133 7483	126.0648 9782	77.3250 3478	0.0079 3242
99	1.6384 7611	0.6103 2321	127.6952 2231	77.9353 5799	0.0078 3115
100	1.6466 6849	0.6072 8678	129.3336 9842	78.5426 4477	0.0077 3194

Table 1   **597**

Compound Amount, Present Value, Annuity

Table 1 *(continued)*

$$i = \tfrac{3}{4}\%$$

n	$(1+i)^n$	$(1+i)^{-n}$	$s_{\overline{n}|i}$	$a_{\overline{n}|i}$	$s_{\overline{n}|i}^{-1}$
1	1.0075 0000	0.9925 5583	1.0000 0000	0.9925 5583	1.0000 0000
2	1.0150 5625	0.9851 6708	2.0075 0000	1.9777 2291	0.4981 3200
3	1.0226 6917	0.9778 3333	3.0225 5625	2.9555 5624	0.3308 4579
4	1.0303 3919	0.9705 5417	4.0452 2542	3.9261 1041	0.2472 0501
5	1.0380 6673	0.9633 2920	5.0755 6461	4.8894 3961	0.1970 2242
6	1.0458 5224	0.9561 5802	6.1136 3135	5.8455 9763	0.1635 6891
7	1.0536 9613	0.9490 4022	7.1594 8358	6.7946 3785	0.1396 7488
8	1.0615 9885	0.9419 7540	8.2131 7971	7.7366 1325	0.1217 5552
9	1.0695 6084	0.9349 6318	9.2747 7856	8.6715 7642	0.1078 1929
10	1.0775 8255	0.9280 0315	10.3443 3940	9.5995 7958	0.0966 7123
11	1.0856 6441	0.9210 9494	11.4219 2194	10.5206 7452	0.0875 5094
12	1.0938 0690	0.9142 3815	12.5075 8636	11.4349 1267	0.0799 5148
13	1.1020 1045	0.9074 3241	13.6013 9325	12.3423 4508	0.0735 2188
14	1.1102 7553	0.9006 7733	14.7034 0370	13.2430 2242	0.0680 1146
15	1.1186 0259	0.8939 7254	15.8136 7923	14.1369 9395	0.0632 3639
16	1.1269 9211	0.8873 1766	16.9322 8183	15.0243 1261	0.0590 5879
17	1.1354 4455	0.8807 1231	18.0592 7394	15.9050 2492	0.0553 7321
18	1.1439 6039	0.8741 5614	19.1947 1849	16.7791 8107	0.0520 9766
19	1.1525 4009	0.8676 4878	20.3386 7888	17.6468 2984	0.0491 6740
20	1.1611 8414	0.8611 8985	21.4912 1897	18.5080 1969	0.0465 3063
21	1.1698 9302	0.8547 7901	22.6524 0312	19.3627 9870	0.0441 4543
22	1.1786 6722	0.8484 1589	23.8222 9614	20.2112 1459	0.0419 7748
23	1.1875 0723	0.8421 0014	25.0009 6336	21.0533 1473	0.0399 9846
24	1.1964 1353	0.8358 3140	26.1884 7059	21.8891 4614	0.0381 8474
25	1.2053 8663	0.8296 0933	27.3848 8412	22.7187 5547	0.0365 1650
26	1.2144 2703	0.8234 3358	28.5902 7075	23.5421 8905	0.0349 7693
27	1.2235 3523	0.8173 0380	29.8046 9778	24.3594 9286	0.0335 5176
28	1.2327 1175	0.8112 1966	31.0282 3301	25.1707 1251	0.0322 2871
29	1.2419 5709	0.8051 8080	32.2609 4476	25.9758 9331	0.0309 9723
30	1.2512 7176	0.7991 8690	33.5029 0184	26.7750 8021	0.0298 4816
31	1.2606 5630	0.7932 3762	34.7541 7361	27.5683 1783	0.0287 7452
32	1.2701 1122	0.7873 3262	36.0148 2991	28.3556 5045	0.0277 6634
33	1.2796 3706	0.7814 7158	37.2849 4113	29.1371 2203	0.0268 2048
34	1.2892 3434	0.7756 5418	38.5645 7819	29.9127 7621	0.0259 3053
35	1.2989 0359	0.7698 8008	39.8538 1253	30.6826 5629	0.0250 9170
36	1.3086 4537	0.7641 4896	41.1527 1612	31.4468 0525	0.0242 9973
37	1.3184 6021	0.7584 6051	42.4613 6149	32.2052 6576	0.0235 5082
38	1.3283 4866	0.7528 1440	43.7798 2170	32.9580 8016	0.0228 4157
39	1.3383 1128	0.7472 1032	45.1081 7037	33.7052 9048	0.0221 6893
40	1.3483 4861	0.7416 4796	46.4464 8164	34.4469 3844	0.0215 3016
41	1.3584 6123	0.7361 2701	47.7948 3026	35.1830 6545	0.0209 2276
42	1.3686 4969	0.7306 4716	49.1532 9148	35.9137 1260	0.0203 4452
43	1.3789 1456	0.7252 0809	50.5219 4117	36.6389 2070	0.0197 9338
44	1.3892 5642	0.7198 0952	51.9008 5573	37.3587 3022	0.0192 6751
45	1.3996 7584	0.7144 5114	53.2901 1215	38.0731 8136	0.0187 6521
46	1.4101 7341	0.7091 3264	54.6897 8799	38.7823 1401	0.0182 8495
47	1.4207 4971	0.7038 5374	56.0999 6140	39.4861 6774	0.0178 2532
48	1.4314 0533	0.6986 1414	57.5207 1111	40.1847 8189	0.0173 8504
49	1.4421 4087	0.6934 1353	58.9521 1644	40.8781 9542	0.0169 6292
50	1.4529 5693	0.6882 5165	60.3942 5732	41.5664 4707	0.0165 5787

Table 1 (*continued*)

$$i = \tfrac{3}{4}\%$$

n	$(1+i)^n$	$(1+i)^{-n}$	$s_{\overline{n}|i}$	$a_{\overline{n}|i}$	$s_{\overline{n}|i}^{-1}$
51	1.4638 5411	0.6831 2819	61.8472 1424	42.2495 7525	0.0161 6888
52	1.4748 3301	0.6780 4286	63.3110 6835	42.9276 1812	0.0157 9503
53	1.4858 9426	0.6729 9540	64.7859 0136	43.6006 1351	0.0154 3546
54	1.4970 3847	0.6679 8551	66.2717 9562	44.2685 9902	0.0150 8938
55	1.5082 6626	0.6630 1291	67.7688 3409	44.9316 1193	0.0147 5605
56	1.5195 7825	0.6580 7733	69.2771 0035	45.5896 8926	0.0144 3478
57	1.5309 7509	0.6531 7849	70.7966 7860	46.2428 6776	0.0141 2496
58	1.5424 5740	0.6483 1612	72.3276 5369	46.8911 8388	0.0138 2597
59	1.5540 2583	0.6434 8995	73.8701 1109	47.5346 7382	0.0135 3727
60	1.5656 8103	0.6386 9970	75.4241 3693	48.1733 7352	0.0132 5836
61	1.5774 2363	0.6339 4511	76.9898 1795	48.8073 1863	0.0129 8873
62	1.5892 5431	0.6292 2592	78.5672 4159	49.4365 4455	0.0127 2795
63	1.6011 7372	0.6245 4185	80.1564 9590	50.0610 8640	0.0124 7560
64	1.6131 8252	0.6198 9266	81.7576 6962	50.6809 7906	0.0122 3127
65	1.6252 8139	0.6152 7807	83.3708 5214	51.2962 5713	0.0119 9460
66	1.6374 7100	0.6106 9784	84.9961 3353	51.9069 5497	0.0117 6524
67	1.6497 5203	0.6061 5170	86.6336 0453	52.5131 0667	0.0115 4286
68	1.6621 2517	0.6016 3940	88.2833 5657	53.1147 4607	0.0113 2716
69	1.6745 9111	0.5971 6070	89.9454 8174	53.7119 0677	0.0111 1785
70	1.6871 5055	0.5927 1533	91.6200 7285	54.3046 2210	0.0109 1464
71	1.6998 0418	0.5883 0306	93.3072 2340	54.8929 2516	0.0107 1728
72	1.7125 5271	0.5839 2363	95.0070 2758	55.4768 4880	0.0105 2554
73	1.7253 9685	0.5795 7681	96.7195 8028	56.0564 2561	0.0103 3917
74	1.7383 3733	0.5752 6234	98.4449 7714	56.6316 8795	0.0101 5796
75	1.7513 7486	0.5709 7999	100.1833 1446	57.2026 6794	0.0099 8170
76	1.7645 1017	0.5667 2952	101.9346 8932	57.7693 9746	0.0098 1020
77	1.7777 4400	0.5625 1069	103.6991 9949	58.3319 0815	0.0096 4328
78	1.7910 7708	0.5583 2326	105.4769 4349	58.8902 3141	0.0094 8074
79	1.8045 1015	0.5541 6701	107.2680 2056	59.4443 9842	0.0093 2244
80	1.8180 4398	0.5500 4170	109.0725 3072	59.9944 4012	0.0091 6821
81	1.8316 7931	0.5459 4710	110.8905 7470	60.5403 8722	0.0090 1790
82	1.8454 1691	0.5418 8297	112.7222 5401	61.0822 7019	0.0088 7136
83	1.8592 5753	0.5378 4911	114.5676 7091	61.6201 1930	0.0087 2847
84	1.8732 0196	0.5338 4527	116.4269 2845	62.1539 6456	0.0085 8908
85	1.8872 5098	0.5298 7123	118.3001 3041	62.6838 3579	0.0084 5308
86	1.9014 0536	0.5259 2678	120.1873 8139	63.2097 6257	0.0083 2034
87	1.9156 6590	0.5220 1169	122.0887 8675	63.7317 7427	0.0081 9076
88	1.9300 3339	0.5181 2575	124.0044 5265	64.2499 0002	0.0080 6423
89	1.9445 0865	0.5142 6873	125.9344 8604	64.7641 6875	0.0079 4064
90	1.9590 9246	0.5104 4043	127.8789 9469	65.2746 0918	0.0078 1989
91	1.9737 8565	0.5066 4063	129.8380 8715	65.7812 4981	0.0077 0190
92	1.9885 8905	0.5028 6911	131.8118 7280	66.2841 1892	0.0075 8657
93	2.0035 0346	0.4991 2567	133.8004 6185	66.7832 4458	0.0074 7382
94	2.0185 2974	0.4954 1009	135.8039 6531	67.2786 5467	0.0073 6356
95	2.0336 6871	0.4917 2217	137.8224 9505	67.7703 7685	0.0072 5571
96	2.0489 2123	0.4880 6171	139.8561 6377	68.2584 3856	0.0071 5020
97	2.0642 8814	0.4844 2850	141.9050 8499	68.7428 6705	0.0070 4696
98	2.0797 7030	0.4808 2233	143.9693 7313	69.2236 8938	0.0069 4592
99	2.0953 6858	0.4772 4301	146.0491 4343	69.7009 3239	0.0068 4701
100	2.1110 8384	0.4736 9033	148.1445 1201	70.1746 2272	0.0067 5017

Table 1

599

Compound Amount, Present Value, Annuity

Table 1 (*continued*)

$i = 1\%$

n	$(1+i)^n$	$(1+i)^{-n}$	$s_{\overline{n}|i}$	$a_{\overline{n}|i}$	$s_{\overline{n}|i}^{-1}$
1	1.0100 0000	0.9900 9901	1.0000 0000	0.9900 9901	1.0000 0000
2	1.0201 0000	0.9802 9605	2.0100 0000	1.9703 9506	0.4975 1244
3	1.0303 0100	0.9705 9015	3.0301 0000	2.9409 8521	0.3300 2211
4	1.0406 0401	0.9609 8034	4.0604 0100	3.9019 6555	0.2462 8109
5	1.0510 1005	0.9514 6569	5.1010 0501	4.8534 3124	0.1960 3980
6	1.0615 2015	0.9420 4524	6.1520 1506	5.7954 7647	0.1625 4837
7	1.0721 3535	0.9327 1805	7.2135 3521	6.7281 9453	0.1386 2828
8	1.0828 5671	0.9234 8322	8.2856 7056	7.6516 7775	0.1206 9029
9	1.0936 8527	0.9143 3982	9.3685 2727	8.5660 1758	0.1067 4036
10	1.1046 2213	0.9052 8695	10.4622 1254	9.4713 0453	0.0955 8208
11	1.1156 6835	0.8963 2372	11.5668 3467	10.3676 2825	0.0864 5408
12	1.1268 2503	0.8874 4923	12.6825 0301	11.2550 7747	0.0788 4879
13	1.1380 9328	0.8786 6260	13.8093 2804	12.1337 4007	0.0724 1482
14	1.1494 7421	0.8699 6297	14.9474 2132	13.0037 0304	0.0669 0117
15	1.1609 6896	0.8613 4947	16.0968 9554	13.8650 5252	0.0621 2378
16	1.1725 7864	0.8528 2126	17.2578 6449	14.7178 7378	0.0579 4460
17	1.1843 0443	0.8443 7749	18.4304 4314	15.5622 5127	0.0542 5806
18	1.1961 4748	0.8360 1731	19.6147 4757	16.3982 6858	0.0509 8205
19	1.2081 0895	0.8277 3992	20.8108 9504	17.2260 0850	0.0480 5175
20	1.2201 9004	0.8195 4447	22.0190 0399	18.0455 5297	0.0454 1531
21	1.2323 9194	0.8114 3017	23.2391 9403	18.8569 8313	0.0430 3075
22	1.2447 1586	0.8033 9621	24.4715 8598	19.6603 7934	0.0408 6372
23	1.2571 6302	0.7954 4179	25.7163 0183	20.4558 2113	0.0388 8584
24	1.2697 3465	0.7875 6613	26.9734 6485	21.2433 8726	0.0370 7347
25	1.2824 3200	0.7797 6844	28.2431 9950	22.0231 5570	0.0354 0675
26	1.2952 5631	0.7720 4796	29.5256 3150	22.7952 0366	0.0338 6888
27	1.3082 0888	0.7644 0392	30.8208 8781	23.5596 0759	0.0324 4553
28	1.3212 9097	0.7568 3557	32.1290 9669	24.3164 4316	0.0311 2444
29	1.3345 0388	0.7493 4215	33.4503 8766	25.0657 8530	0.0298 9502
30	1.3478 4892	0.7419 2292	34.7848 9153	25.8077 0822	0.0287 4811
31	1.3613 2740	0.7345 7715	36.1327 4045	26.5422 8537	0.0276 7573
32	1.3749 4068	0.7273 0411	37.4940 6785	27.2695 8947	0.0266 7089
33	1.3886 9009	0.7201 0307	38.8690 0853	27.9896 9255	0.0257 2744
34	1.4025 7699	0.7129 7334	40.2576 9862	28.7026 6589	0.0248 3997
35	1.4166 0276	0.7059 1420	41.6602 7560	29.4085 8009	0.0240 0368
36	1.4307 6878	0.6989 2495	43.0768 7836	30.1075 0504	0.0232 1431
37	1.4450 7647	0.6920 0490	44.5076 4714	30.7995 0994	0.0224 6805
38	1.4595 2724	0.6851 5337	45.9527 2361	31.4846 6330	0.0217 6150
39	1.4741 2251	0.6783 6967	47.4122 5085	32.1630 3298	0.0210 9160
40	1.4888 6373	0.6716 5314	48.8863 7336	32.8346 8611	0.0204 5560
41	1.5037 5237	0.6650 0311	50.3752 3709	33.4996 8922	0.0198 5102
42	1.5187 8989	0.6584 1892	51.8789 8946	34.1581 0814	0.0192 7563
43	1.5339 7779	0.6518 9992	53.3977 7936	34.8100 0806	0.0187 2737
44	1.5493 1757	0.6454 4546	54.9317 5715	35.4554 5352	0.0182 0441
45	1.5648 1075	0.6390 5492	56.4810 7472	36.0945 0844	0.0177 0505
46	1.5804 5885	0.6327 2764	58.0458 8547	36.7272 3608	0.0172 2775
47	1.5962 6344	0.6264 6301	59.6263 4432	37.3536 9909	0.0167 7111
48	1.6122 2608	0.6202 6041	61.2226 0777	37.9739 5949	0.0163 3384
49	1.6283 4834	0.6141 1921	62.8348 3385	38.5880 7871	0.0159 1474
50	1.6446 3182	0.6080 3882	64.4631 8218	39.1961 1753	0.0155 1273

Table 1 (*continued*)

$i = 1\%$

n	$(1+i)^n$	$(1+i)^{-n}$	$s_{\overline{n}|i}$	$a_{\overline{n}|i}$	$s_{\overline{n}|i}^{-1}$
51	1.6610 7814	0.6020 1864	66.1078 1401	39.7981 3617	0.0151 2680
52	1.6776 8892	0.5960 5806	67.7688 9215	40.3941 9423	0.0147 5603
53	1.6944 6581	0.5901 5649	69.4465 8107	40.9843 5072	0.0143 9956
54	1.7114 1047	0.5843 1336	71.1410 4688	41.5686 6408	0.0140 5658
55	1.7285 2457	0.5785 2808	72.8524 5735	42.1471 9216	0.0137 2637
56	1.7458 0982	0.5728 0008	74.5809 8192	42.7199 9224	0.0134 0824
57	1.7632 6792	0.5671 2879	76.3267 9174	43.2871 2102	0.0131 0156
58	1.7809 0060	0.5615 1365	78.0900 5966	43.8486 3468	0.0128 0573
59	1.7987 0960	0.5559 5411	79.8709 6025	44.4045 8879	0.0125 2020
60	1.8166 9670	0.5504 4962	81.6696 6986	44.9550 3841	0.0122 4445
61	1.8348 6367	0.5449 9962	83.4863 6655	45.5000 3803	0.0119 7800
62	1.8532 1230	0.5396 0358	85.3212 3022	46.0396 4161	0.0117 2041
63	1.8717 4443	0.5342 6097	87.1744 4252	46.5739 0258	0.0114 7125
64	1.8904 6187	0.5289 7126	89.0461 8695	47.1028 7385	0.0112 3013
65	1.9093 6649	0.5237 3392	90.9366 4882	47.6266 0777	0.0109 9667
66	1.9284 6015	0.5185 4844	92.8460 1531	48.1451 5621	0.0107 7052
67	1.9477 4475	0.5134 1429	94.7744 7546	48.6585 7050	0.0105 5136
68	1.9672 2220	0.5083 3099	96.7222 2021	49.1669 0149	0.0103 3889
69	1.9868 9442	0.5032 9801	98.6894 4242	49.6701 9949	0.0101 3280
70	2.0067 6337	0.4983 1486	100.6763 3684	50.1685 1435	0.0099 3282
71	2.0268 3100	0.4933 8105	102.6831 0021	50.6618 9539	0.0097 3870
72	2.0470 9931	0.4884 9609	104.7099 3121	51.1503 9148	0.0095 5019
73	2.0675 7031	0.4836 5949	106.7570 3052	51.6340 5097	0.0093 6706
74	2.0882 4601	0.4788 7078	108.8246 0083	52.1129 2175	0.0091 8910
75	2.1091 2847	0.4741 2949	110.9128 4684	52.5870 5124	0.0090 1609
76	2.1302 1975	0.4694 3514	113.0219 7530	53.0564 8637	0.0088 4784
77	2.1515 2195	0.4647 8726	115.1521 9506	53.5212 7364	0.0086 8416
78	2.1730 3717	0.4601 8541	117.3037 1701	53.9814 5905	0.0085 2488
79	2.1947 6754	0.4556 2912	119.4767 5418	54.4370 8817	0.0083 6983
80	2.2167 1522	0.4511 1794	121.6715 2172	54.8882 0611	0.0082 1885
81	2.2388 8237	0.4466 5142	123.8882 3694	55.3348 5753	0.0080 7179
82	2.2612 7119	0.4422 2913	126.1271 1931	55.7770 8666	0.0079 2851
83	2.2838 8390	0.4378 5063	128.3883 9050	56.2149 3729	0.0077 8887
84	2.3067 2274	0.4335 1547	130.6722 7440	56.6484 5276	0.0076 5273
85	2.3297 8997	0.4292 2324	132.9789 9715	57.0776 7600	0.0075 1998
86	2.3530 8787	0.4249 7350	135.3087 8712	57.5026 4951	0.0073 9050
87	2.3766 1875	0.4207 6585	137.6618 7499	57.9234 1535	0.0072 6417
88	2.4003 8494	0.4165 9985	140.0384 9374	58.3400 1520	0.0071 4089
89	2.4243 8879	0.4124 7510	142.4388 7868	58.7524 9030	0.0070 2056
90	2.4486 3267	0.4083 9119	144.8632 6746	59.1608 8148	0.0069 0306
91	2.4731 1900	0.4043 4771	147.3119 0014	59.5652 2919	0.0067 8832
92	2.4978 5019	0.4003 4427	149.7850 1914	59.9655 7346	0.0066 7624
93	2.5228 2869	0.3963 8046	152.2828 6933	60.3619 5392	0.0065 6673
94	2.5480 5698	0.3924 5590	154.8056 9803	60.7544 0982	0.0064 5971
95	2.5735 3755	0.3885 7020	157.3537 5501	61.1429 8002	0.0063 5511
96	2.5992 7293	0.3847 2297	159.9272 9256	61.5277 0299	0.0062 5284
97	2.6252 6565	0.3809 1383	162.5265 6548	61.9086 1682	0.0061 5284
98	2.6515 1831	0.3771 4241	165.1518 3114	62.2857 5923	0.0060 5503
99	2.6780 3349	0.3734 0832	167.8033 4945	62.6591 6755	0.0059 5936
100	2.7048 1383	0.3697 1121	170.4813 8294	63.0288 7877	0.0058 6574

Table 1

601

Compound Amount, Present Value, Annuity

Table 1 (*continued*)

$$i = 1\tfrac{1}{4}\%$$

n	$(1+i)^n$	$(1+i)^{-n}$	$s_{\overline{n}|i}$	$a_{\overline{n}|i}$	$s_{\overline{n}|i}^{-1}$
1	1.0125 0000	0.9876 5432	1.0000 0000	0.9876 5432	1.0000 0000
2	1.0251 5625	0.9754 6106	2.0125 0000	1.9631 1538	0.4968 9441
3	1.0379 7070	0.9634 1833	3.0376 5625	2.9265 3371	0.3292 0117
4	1.0509 4534	0.9515 2428	4.0756 2695	3.8780 5798	0.2453 6102
5	1.0640 8215	0.9397 7706	5.1265 7229	4.8178 3504	0.1950 6211
6	1.0773 8318	0.9281 7488	6.1906 5444	5.7460 0992	0.1615 3381
7	1.0908 5047	0.9167 1593	7.2680 3762	6.6627 2585	0.1375 8872
8	1.1044 8610	0.9053 9845	8.3588 8809	7.5681 2429	0.1196 3314
9	1.1182 9218	0.8942 2069	9.4633 7420	8.4623 4498	0.1056 7055
10	1.1322 7083	0.8831 8093	10.5816 6637	9.3455 2591	0.0945 0307
11	1.1464 2422	0.8722 7746	11.7139 3720	10.2178 0337	0.0853 6839
12	1.1607 5452	0.8615 0860	12.8603 6142	11.0793 1197	0.0777 5831
13	1.1752 6395	0.8508 7269	14.0211 1594	11.9301 8466	0.0713 2100
14	1.1899 5475	0.8403 6809	15.1963 7988	12.7705 5275	0.0658 0515
15	1.2048 2918	0.8299 9318	16.3863 3463	13.6005 4592	0.0610 2646
16	1.2198 8955	0.8197 4635	17.5911 6382	14.4202 9227	0.0568 4672
17	1.2351 3817	0.8096 2602	18.8110 5336	15.2299 1829	0.0531 6023
18	1.2505 7739	0.7996 3064	20.0461 9153	16.0295 4893	0.0498 8479
19	1.2662 0961	0.7897 5866	21.2967 6893	16.8193 0759	0.0469 5548
20	1.2820 3723	0.7800 0855	22.5629 7854	17.5993 1613	0.0443 2039
21	1.2980 6270	0.7703 7881	23.8450 1577	18.3696 9495	0.0419 3748
22	1.3142 8848	0.7608 6796	25.1430 7847	19.1305 6291	0.0397 7238
23	1.3307 1709	0.7514 7453	26.4573 6695	19.8820 3744	0.0377 9666
24	1.3473 5105	0.7421 9707	27.7880 8403	20.6242 3451	0.0359 8665
25	1.3641 9294	0.7330 3414	29.1354 3508	21.3572 6865	0.0343 2247
26	1.3812 4535	0.7239 8434	30.4996 2802	22.0812 5299	0.0327 8729
27	1.3985 1092	0.7150 4626	31.8808 7337	22.7962 9925	0.0313 6677
28	1.4159 9230	0.7062 1853	33.2793 8429	23.5025 1778	0.0300 4863
29	1.4336 9221	0.6974 9978	34.6953 7659	24.2000 1756	0.0288 2228
30	1.4516 1336	0.6888 8867	36.1290 6880	24.8889 0623	0.0276 7854
31	1.4697 5853	0.6803 8387	37.5806 8216	25.5692 9010	0.0266 0942
32	1.4881 3051	0.6719 8407	39.0504 4069	26.2412 7418	0.0256 0791
33	1.5067 3214	0.6636 8797	40.5385 7120	26.9049 6215	0.0246 6786
34	1.5255 6629	0.6554 9429	42.0453 0334	27.5604 5644	0.0237 8387
35	1.5446 3587	0.6474 0177	43.5708 6983	28.2078 5822	0.0229 5111
36	1.5639 4382	0.6394 0916	45.1155 0550	28.8472 6737	0.0221 6533
37	1.5834 9312	0.6315 1522	46.6794 4932	29.4787 8259	0.0214 2270
38	1.6032 8678	0.6237 1873	48.2926 4243	30.1025 0133	0.0207 1983
39	1.6233 2787	0.6160 1850	49.8862 2921	30.7185 1983	0.0200 5365
40	1.6436 1946	0.6084 1334	51.4895 5708	31.3269 3316	0.0194 2141
41	1.6641 6471	0.6009 0206	53.1331 7654	31.9278 3522	0.0188 2063
42	1.6849 6677	0.5934 8352	54.7973 4125	32.5213 1874	0.0182 4906
43	1.7060 2885	0.5861 5656	56.4823 0801	33.1074 7530	0.0177 0466
44	1.7273 5421	0.5789 2006	58.1883 3687	33.6863 9536	0.0171 8557
45	1.7489 4614	0.5717 7290	59.9156 9108	34.2581 6825	0.0166 9012
46	1.7708 0797	0.5647 1397	61.6646 3721	34.8228 8222	0.0162 1675
47	1.7929 4306	0.5577 4219	63.4354 4518	35.3806 2442	0.0157 6406
48	1.8153 5485	0.5508 5649	65.2283 8824	35.9314 8091	0.0153 3075
49	1.8380 4679	0.5440 5579	67.0437 4310	36.4755 3670	0.0149 1563
50	1.8610 2237	0.5373 3905	68.8817 8989	37.0128 7574	0.0145 1763

Table 1 (continued)

$$i = 1\tfrac{1}{4}\%$$

n	$(1+i)^n$	$(1+i)^{-n}$	$s_{\overline{n}|i}$	$a_{\overline{n}|i}$	$s_{\overline{n}|i}^{-1}$
51	1.8842 8515	0.5307 0524	70.7428 1226	37.5435 8099	0.0141 3571
52	1.9078 3872	0.5241 5332	72.6270 9741	38.0677 3431	0.0137 6897
53	1.9316 8670	0.5176 8229	74.5349 3613	38.5854 1660	0.0134 1653
54	1.9558 3279	0.5112 9115	76.4666 2283	39.0967 0776	0.0130 7760
55	1.9802 8070	0.5049 7892	78.4224 5562	39.6016 8667	0.0127 5145
56	2.0050 3420	0.4987 4461	80.4027 3631	40.1004 3128	0.0124 3739
57	2.0300 9713	0.4925 8727	82.4077 7052	40.5930 1855	0.0121 3478
58	2.0554 7335	0.4865 0594	84.4378 6765	41.0795 2449	0.0118 4303
59	2.0811 6676	0.4804 9970	86.4933 4099	41.5600 2419	0.0115 6158
60	2.1071 8135	0.4745 6760	88.5745 0776	42.0345 9179	0.0112 8993
61	2.1335 2111	0.4687 0874	90.6816 8910	42.5033 0054	0.0110 2758
62	2.1601 9013	0.4629 2222	92.8152 1022	42.9662 2275	0.0107 7410
63	2.1871 9250	0.4572 0713	94.9754 0034	43.4234 2988	0.0105 2904
64	2.2145 3241	0.4515 6259	97.1625 9285	43.8749 9247	0.0102 9203
65	2.2422 1407	0.4459 8775	99.3771 2526	44.3209 8022	0.0100 6268
66	2.2702 4174	0.4404 8173	101.6193 3933	44.7614 6195	0.0098 4065
67	2.2986 1976	0.4350 4368	103.8895 8107	45.1965 0563	0.0096 2560
68	2.3273 5251	0.4296 7277	106.1882 0083	45.6261 7840	0.0094 1724
69	2.3564 4442	0.4243 6817	108.5155 5334	46.0505 4656	0.0092 1527
70	2.3858 9997	0.4191 2905	110.8719 9776	46.4696 7562	0.0090 1941
71	2.4157 2372	0.4139 5462	113.2578 9773	46.8836 3024	0.0088 2941
72	2.4459 2027	0.4088 4407	115.6736 2145	47.2924 7431	0.0086 4501
73	2.4764 9427	0.4037 9661	118.1195 4172	47.6962 7093	0.0084 6600
74	2.5074 5045	0.3988 1147	120.5960 3599	48.0950 8240	0.0082 9215
75	2.5387 9358	0.3938 8787	123.1034 8644	48.4889 7027	0.0081 2325
76	2.5705 2850	0.3890 2506	125.6422 8002	48.8779 9533	0.0079 5910
77	2.6026 6011	0.3842 2228	128.2128 0852	49.2622 1761	0.0077 9953
78	2.6351 9336	0.3794 7879	130.8154 6863	49.6416 9640	0.0076 4435
79	2.6681 3327	0.3747 9387	133.4506 6199	50.0164 9027	0.0074 9341
80	2.7014 8494	0.3701 6679	136.1187 9526	50.3866 5706	0.0073 4652
81	2.7352 5350	0.3655 9683	138.8202 8020	50.7522 5389	0.0072 0356
82	2.7694 4417	0.3610 8329	141.5555 3370	51.1133 3717	0.0070 6437
83	2.8040 6222	0.3566 2547	144.3249 7787	51.4699 6264	0.0069 2881
84	2.8391 1300	0.3522 2268	147.1290 4010	51.8221 8532	0.0067 9675
85	2.8746 0191	0.3478 7426	149.9681 5310	52.1700 5958	0.0066 6808
86	2.9105 3444	0.3435 7951	152.8427 5501	52.5136 3909	0.0065 4267
87	2.9469 1612	0.3393 3779	155.7532 8945	52.8529 7688	0.0064 2041
88	2.9837 5257	0.3351 4843	158.7002 0557	53.1881 2531	0.0063 0119
89	3.0210 4948	0.3310 1080	161.6839 5814	53.5191 3611	0.0061 8490
90	3.0588 1260	0.3269 2425	164.7050 0762	53.8460 6035	0.0060 7146
91	3.0970 4775	0.3228 8814	167.7638 2021	54.1689 4850	0.0059 6076
92	3.1357 6085	0.3189 0187	170.8608 6796	54.4878 5037	0.0058 5271
93	3.1749 5786	0.3149 6481	173.9966 2881	54.8028 1518	0.0057 4724
94	3.2146 4483	0.3110 7636	177.1715 8667	55.1138 9154	0.0056 4425
95	3.2548 2789	0.3072 3591	180.3862 3151	55.4211 2744	0.0055 4366
96	3.2955 1324	0.3034 4287	183.6410 5940	55.7245 7031	0.0054 4540
97	3.3367 0716	0.2996 9666	186.9365 7264	56.0242 6698	0.0053 4941
98	3.3784 1600	0.2959 9670	190.2732 7980	56.3202 6368	0.0052 5560
99	3.4206 4620	0.2923 4242	193.6516 9580	56.6126 0610	0.0051 6391
100	3.4634 0427	0.2887 3326	197.0723 4200	56.9013 3936	0.0050 7428

Table 1

603

Compound Amount, Present Value, Annuity

Table 1 (*continued*)

$$i = 1\tfrac{1}{2}\%$$

n	$(1+i)^n$	$(1+i)^{-n}$	$s_{\overline{n}\rvert i}$	$a_{\overline{n}\rvert i}$	$s_{\overline{n}\rvert i}^{-1}$
1	1.0150 0000	0.9852 2167	1.0000 0000	0.9852 2167	1.0000 0000
2	1.0302 2500	0.9706 6175	2.0150 0000	1.9558 8342	0.4962 7792
3	1.0456 7838	0.9563 1699	3.0452 2500	2.9122 0042	0.3283 8296
4	1.0613 6355	0.9421 8423	4.0909 0338	3.8543 8465	0.2444 4478
5	1.0772 8400	0.9282 6033	5.1522 6693	4.7826 4497	0.1940 8932
6	1.0934 4326	0.9145 4219	6.2295 5093	5.6971 8717	0.1605 2521
7	1.1098 4491	0.9010 2679	7.3229 9419	6.5982 1396	0.1365 5616
8	1.1264 9259	0.8877 1112	8.4328 3911	7.4859 2508	0.1185 8402
9	1.1433 8998	0.8745 9224	9.5593 3169	8.3605 1732	0.1046 0982
10	1.1605 4083	0.8616 6723	10.7027 2167	9.2221 8455	0.0934 3418
11	1.1779 4894	0.8489 3323	11.8632 6249	10.0711 1779	0.0842 9384
12	1.1956 1817	0.8363 8742	13.0412 1143	10.9075 0521	0.0766 7999
13	1.2135 5244	0.8240 2702	14.2368 2960	11.7315 3222	0.0702 4036
14	1.2317 5573	0.8118 4928	15.4503 8205	12.5433 8150	0.0647 2332
15	1.2502 3207	0.7998 5150	16.6821 3778	13.3432 3301	0.0599 4436
16	1.2689 8555	0.7880 3104	17.9323 6984	14.1312 6405	0.0557 6508
17	1.2880 2033	0.7763 8526	19.2013 5539	14.9076 4931	0.0520 7966
18	1.3073 4064	0.7649 1159	20.4893 7572	15.6725 6089	0.0488 0578
19	1.3269 5075	0.7536 0747	21.7967 1636	16.4261 6837	0.0458 7847
20	1.3468 5501	0.7424 7042	23.1236 6710	17.1686 3879	0.0432 4574
21	1.3670 5783	0.7314 9795	24.4705 2211	17.9001 3673	0.0408 6550
22	1.3875 6370	0.7206 8763	25.8375 7994	18.6208 2437	0.0387 0331
23	1.4083 7715	0.7100 3708	27.2251 4364	19.3308 6145	0.0367 3075
24	1.4295 0281	0.6995 4392	28.6335 2080	20.0304 0537	0.0349 2410
25	1.4509 4535	0.6892 0583	30.0630 2361	20.7196 1120	0.0332 6345
26	1.4727 0953	0.6790 2052	31.5139 6896	21.3986 3172	0.0317 3196
27	1.4948 0018	0.6689 8574	32.9866 7850	22.0676 1746	0.0303 1527
28	1.5172 2218	0.6590 9925	34.4814 7867	22.7267 1671	0.0290 0108
29	1.5399 8051	0.6493 5887	35.9987 0085	23.3760 7558	0.0277 7878
30	1.5630 8022	0.6397 6243	37.5386 8137	24.0158 3801	0.0266 3919
31	1.5865 2642	0.6303 0781	39.1017 6159	24.6461 4582	0.0255 7430
32	1.6103 2432	0.6209 9292	40.6882 8801	25.2671 3874	0.0245 7710
33	1.6344 7918	0.6118 1568	42.2986 1233	25.8789 5442	0.0236 4144
34	1.6589 9637	0.6027 7407	43.9330 9152	26.4817 2849	0.0227 6189
35	1.6838 8132	0.5938 6608	45.5920 8789	27.0755 9458	0.0219 3363
36	1.7091 3954	0.5850 8974	47.2759 6921	27.6606 8431	0.0211 5240
37	1.7347 7663	0.5764 4309	48.9851 0874	28.2371 2740	0.0204 1437
38	1.7607 9828	0.5679 2423	50.7198 8538	28.8050 5163	0.0197 1613
39	1.7872 1025	0.5595 3126	52.4806 8366	29.3645 8288	0.0190 5463
40	1.8140 1841	0.5512 6232	54.2678 9391	29.9158 4250	0.0184 2710
41	1.8412 2868	0.5431 1559	56.0819 1232	30.4589 6079	0.0178 3106
42	1.8688 4712	0.5350 8925	57.9231 4100	30.9940 5004	0.0172 6426
43	1.8968 7982	0.5271 8153	59.7919 8812	31.5212 3157	0.0167 2465
44	1.9253 3302	0.5193 9067	61.6888 6794	32.0406 2223	0.0162 1038
45	1.9542 1301	0.5117 1494	63.6142 0096	32.5523 3718	0.0157 1976
46	1.9835 2621	0.5041 5265	65.5684 1398	33.0564 8983	0.0152 5125
47	2.0132 7910	0.4967 0212	67.5519 4018	33.5531 9195	0.0148 0342
48	2.0434 7829	0.4893 6170	69.5652 1929	34.0425 5365	0.0143 7500
49	2.0741 3046	0.4821 2975	71.6086 9758	34.5246 8339	0.0139 6478
50	2.1052 4242	0.4750 0468	73.6828 2804	34.9996 8807	0.0135 7168

Table 1 (*continued*)

$$i = 1\tfrac{1}{2}\%$$

n	$(1+i)^n$	$(1+i)^{-n}$	$s_{\overline{n}|i}$	$a_{\overline{n}|i}$	$s_{\overline{n}|i}^{-1}$
51	2.1368 2106	0.4679 8491	75.7880 7046	35.4676 7298	0.0131 9469
52	2.1688 7337	0.4610 6887	77.9248 9152	35.9287 4185	0.0128 3287
53	2.2014 0647	0.4542 5505	80.0937 6489	36.3829 9690	0.0124 8537
54	2.2344 2757	0.4475 4192	82.2951 7136	36.8305 3882	0.0121 5138
55	2.2679 4398	0.4409 2800	84.5295 9893	37.2714 6681	0.0118 3018
56	2.3019 6314	0.4344 1182	86.7975 4292	37.7058 7863	0.0115 2106
57	2.3364 9259	0.4279 9194	89.0995 0606	38.1338 7058	0.0112 2341
58	2.3715 3998	0.4216 6694	91.4359 9865	38.5555 3751	0.0109 3661
59	2.4071 1308	0.4154 3541	93.8075 3863	38.9709 7292	0.0106 6012
60	2.4432 1978	0.4092 9597	96.2146 5171	39.3802 6889	0.0103 9343
61	2.4798 6807	0.4032 4726	98.6578 7149	39.7835 1614	0.0101 3604
62	2.5170 6609	0.3972 8794	101.1377 3956	40.1808 0408	0.0098 8751
63	2.5548 2208	0.3914 1669	103.6548 0565	40.5722 2077	0.0096 4741
64	2.5931 4442	0.3856 3221	106.2096 2774	40.9578 5298	0.0094 1534
65	2.6320 4158	0.3799 3321	108.8027 7215	41.3377 8618	0.0091 9094
66	2.6715 2221	0.3743 1843	111.4348 1374	41.7121 0461	0.0089 7386
67	2.7115 9504	0.3687 8663	114.1063 3594	42.0808 9125	0.0087 6376
68	2.7522 6896	0.3633 3658	116.8179 3098	42.4442 2783	0.0085 6033
69	2.7935 5300	0.3579 6708	119.5701 9995	42.8021 9490	0.0083 6329
70	2.8354 5629	0.3526 7692	122.3637 5295	43.1548 7183	0.0081 7235
71	2.8779 8814	0.3474 6495	125.1992 0924	43.5023 3678	0.0079 8727
72	2.9211 5796	0.3423 3000	128.0771 9738	43.8446 6677	0.0078 0779
73	2.9649 7533	0.3372 7093	130.9983 5534	44.1819 3771	0.0076 3368
74	3.0094 4996	0.3322 8663	133.9633 3067	44.5142 2434	0.0074 6473
75	3.0545 9171	0.3273 7599	136.9727 8063	44.8416 0034	0.0073 0072
76	3.1004 1059	0.3225 3793	140.0273 7234	45.1641 3826	0.0071 4146
77	3.1469 1674	0.3177 7136	143.1277 8292	45.4819 0962	0.0069 8676
78	3.1941 2050	0.3130 7523	146.2746 9967	45.7949 8485	0.0068 3645
79	3.2420 3230	0.3084 4850	149.4688 2016	46.1034 3335	0.0066 9036
80	3.2906 6279	0.3038 9015	152.7108 5247	46.4073 2349	0.0065 4832
81	3.3400 2273	0.2993 9916	156.0015 1525	46.7067 2265	0.0064 1019
82	3.3901 2307	0.2949 7454	159.3415 3798	47.0016 9720	0.0062 7583
83	3.4409 7492	0.2906 1531	162.7316 6105	47.2923 1251	0.0061 4509
84	3.4925 8954	0.2863 2050	166.1726 3597	47.5786 3301	0.0060 1784
85	3.5449 7838	0.2820 8917	169.6652 2551	47.8607 2218	0.0058 9396
86	3.5981 5306	0.2779 2036	173.2102 0389	48.1386 4254	0.0057 7333
87	3.6521 2535	0.2738 1316	176.8083 5695	48.4124 5571	0.0056 5584
88	3.7069 0723	0.2697 6666	180.4604 8230	48.6822 2237	0.0055 4138
89	3.7625 1084	0.2657 7997	184.1673 8954	48.9480 0234	0.0054 2984
90	3.8189 4851	0.2618 5218	187.9299 0038	49.2098 5452	0.0053 2113
91	3.8762 3273	0.2579 8245	191.7488 4889	49.4678 3696	0.0052 1516
92	3.9343 7622	0.2541 6990	195.6250 8162	49.7220 0686	0.0051 1182
93	3.9933 9187	0.2504 1369	199.5594 5784	49.9724 2055	0.0050 1104
94	4.0532 9275	0.2467 1300	203.5528 4971	50.2191 3355	0.0049 1273
95	4.1140 9214	0.2430 6699	207.6061 4246	50.4622 0054	0.0048 1681
96	4.1758 0352	0.2394 7487	211.7202 3459	50.7016 7541	0.0047 2321
97	4.2384 4057	0.2359 3583	215.8960 3811	50.9376 1124	0.0046 3186
98	4.3020 1718	0.2324 4909	220.1344 7868	51.1700 6034	0.0045 4268
99	4.3665 4744	0.2290 1389	224.4364 9586	51.3990 7422	0.0044 5560
100	4.4320 4565	0.2256 2944	228.8030 4330	51.6247 0367	0.0043 7057

Table 1

605

Compound Amount, Present Value, Annuity

Table 1 (*continued*)

$$i = 1\tfrac{3}{4}\%$$

n	$(1+i)^n$	$(1+i)^{-n}$	$s_{\overline{n}\rvert i}$	$a_{\overline{n}\rvert i}$	$s_{\overline{n}\rvert i}^{-1}$
1	1.0175 0000	0.9828 0098	1.0000 0000	0.9828 0098	1.0000 0000
2	1.0353 0625	0.9658 9777	2.0175 0000	1.9486 9875	0.4956 6295
3	1.0534 2411	0.9492 8528	3.0528 0625	2.8979 8403	0.3275 6746
4	1.0718 5903	0.9329 5851	4.1062 3036	3.8309 4254	0.2435 3237
5	1.0906 1656	0.9169 1254	5.1780 8938	4.7478 5508	0.1931 2142
6	1.1097 0235	0.9011 4254	6.2687 0596	5.6489 9762	0.1595 2256
7	1.1291 2215	0.8856 4378	7.3784 0831	6.5346 4139	0.1355 3059
8	1.1488 8178	0.8704 1157	8.5075 3045	7.4050 5297	0.1175 4292
9	1.1689 8721	0.8554 4135	9.6564 1224	8.2604 9432	0.1035 5813
10	1.1894 4449	0.8407 2860	10.8253 9945	9.1012 2291	0.0923 7534
11	1.2102 5977	0.8262 6889	12.0148 4394	9.9274 9181	0.0832 3038
12	1.2314 3931	0.8120 5788	13.2251 0371	10.7395 4969	0.0756 1377
13	1.2529 8950	0.7980 9128	14.4565 4303	11.5376 4097	0.0691 7283
14	1.2749 1682	0.7843 6490	15.7095 3253	12.3220 0587	0.0636 5562
15	1.2972 2786	0.7708 7459	16.9844 4935	13.0928 8046	0.0588 7739
16	1.3199 2935	0.7576 1631	18.2816 7721	13.8504 9677	0.0546 9958
17	1.3430 2811	0.7445 8605	19.6016 0656	14.5950 8282	0.0510 1623
18	1.3665 3111	0.7317 7990	20.9446 3468	15.3268 6272	0.0477 4492
19	1.3904 4540	0.7191 9401	22.3111 6578	16.0460 5673	0.0448 2061
20	1.4147 7820	0.7068 2458	23.7016 1119	16.7528 8130	0.0421 9122
21	1.4395 3681	0.6946 6789	25.1163 8938	17.4475 4919	0.0398 1464
22	1.4647 2871	0.6827 2028	26.5559 2620	18.1302 6948	0.0376 5638
23	1.4903 6146	0.6709 7817	28.0206 5490	18.8012 4764	0.0356 8796
24	1.5164 4279	0.6594 3800	29.5110 1637	19.4606 8565	0.0338 8565
25	1.5429 8054	0.6480 9632	31.0274 5915	20.1087 8196	0.0322 2952
26	1.5699 8269	0.6369 4970	32.5704 3969	20.7457 3166	0.0307 0269
27	1.5974 5739	0.6259 9479	34.1404 2238	21.3717 2644	0.0292 9079
28	1.6254 1290	0.6152 2829	35.7378 7977	21.9869 5474	0.0279 8151
29	1.6538 5762	0.6046 4697	37.3632 9267	22.5916 0171	0.0267 6424
30	1.6828 0013	0.5942 4764	39.0171 5029	23.1858 4934	0.0256 2975
31	1.7122 4913	0.5840 2716	40.6999 5042	23.7698 7650	0.0245 7005
32	1.7422 1349	0.5739 8247	42.4121 9955	24.3438 5897	0.0235 7812
33	1.7727 0223	0.5641 1053	44.1544 1305	24.9079 6951	0.0226 4779
34	1.8037 2452	0.5544 0839	45.9271 1527	25.4623 7789	0.0217 7363
35	1.8352 8970	0.5448 7311	47.7308 3979	26.0072 5100	0.0209 5082
36	1.8674 0727	0.5355 0183	49.5661 2949	26.5427 5283	0.0201 7507
37	1.9000 8689	0.5262 9172	51.4335 3675	27.0690 4455	0.0194 4257
38	1.9333 3841	0.5172 4002	53.3336 2365	27.5862 8457	0.0187 4990
39	1.9671 7184	0.5083 4400	55.2669 6206	28.0946 2857	0.0180 9399
40	2.0015 9734	0.4996 0098	57.2341 3390	28.5942 2955	0.0174 7209
41	2.0366 2530	0.4910 0834	59.2357 3124	29.0852 3789	0.0168 8170
42	2.0722 6624	0.4825 6348	61.2723 5654	29.5678 0135	0.0163 2057
43	2.1085 3090	0.4742 6386	63.3446 2278	30.0420 6522	0.0157 8666
44	2.1454 3019	0.4661 0699	65.4531 5367	30.5081 7221	0.0152 7810
45	2.1829 7522	0.4580 9040	67.5985 8386	30.9662 6261	0.0147 9321
46	2.2211 7728	0.4502 1170	69.7815 5908	31.4164 7431	0.0143 3043
47	2.2600 4789	0.4424 6850	72.0027 3637	31.8589 4281	0.0138 8836
48	2.2995 9872	0.4348 5848	74.2627 8425	32.2938 0129	0.0134 6569
49	2.3398 4170	0.4273 7934	76.5623 8298	32.7211 8063	0.0130 6124
50	2.3807 8893	0.4200 2883	78.9022 2468	33.1412 0946	0.0126 7391

Table 1 *(continued)*

$$i = 1\tfrac{3}{4}\%$$

n	$(1+i)^n$	$(1+i)^{-n}$	$s_{\overline{n}|i}$	$a_{\overline{n}|i}$	$s_{\overline{n}|i}^{-1}$
51	2.4224 5274	0.4128 0475	81.2830 1361	33.5540 1421	0.0123 0269
52	2.4648 4566	0.4057 0492	83.7054 6635	33.9597 1913	0.0119 4665
53	2.5079 8046	0.3987 2719	86.1703 1201	34.3584 4633	0.0116 0492
54	2.5518 7012	0.3918 6947	88.6782 9247	34.7503 1579	0.0112 7672
55	2.5965 2785	0.3851 2970	91.2301 6259	35.1354 4550	0.0109 6129
56	2.6419 6708	0.3785 0585	93.8266 9043	35.5139 5135	0.0106 5795
57	2.6882 0151	0.3719 9592	96.4686 5752	35.8859 4727	0.0103 6606
58	2.7352 4503	0.3655 9796	99.1568 5902	36.2515 4523	0.0100 8503
59	2.7831 1182	0.3593 1003	101.8921 0405	36.6108 5526	0.0098 1430
60	2.8318 1628	0.3531 3025	104.6752 1588	36.9639 8552	0.0095 5336
61	2.8813 7306	0.3470 5676	107.5070 3215	37.3110 4228	0.0093 0172
62	2.9317 9709	0.3410 8772	110.3884 0522	37.6521 3000	0.0090 5892
63	2.9831 0354	0.3352 2135	113.3202 0231	37.9873 5135	0.0088 2455
64	3.0343 0785	0.3294 5587	116.3033 0585	38.3168 0723	0.0085 9821
65	3.0884 2574	0.3237 8956	119.3386 1370	38.6405 9878	0.0083 7952
66	3.1424 7319	0.3182 2069	122.4270 3944	38.9588 1748	0.0081 6813
67	3.1974 6647	0.3127 4761	125.5695 1263	39.2715 6509	0.0079 6372
68	3.2534 2213	0.3073 6866	128.7669 7910	39.5789 3375	0.0077 6596
69	3.3103 5702	0.3020 8222	132.0204 0124	39.8810 1597	0.0075 7459
70	3.3682 8827	0.2968 8670	135.3307 5826	40.1779 0267	0.0073 8930
71	3.4272 3331	0.2917 8054	138.6990 4653	40.4696 8321	0.0072 0985
72	3.4872 0990	0.2867 6221	142.1262 7984	40.7564 4542	0.0070 3600
73	3.5482 3607	0.2818 3018	145.6134 8974	41.0382 7560	0.0068 6750
74	3.6103 3020	0.2769 8298	149.1617 2581	41.3152 5857	0.0067 0413
75	3.6735 1098	0.2722 1914	152.7720 5601	41.5874 7771	0.0065 4570
76	3.7377 9742	0.2675 3724	156.4455 6699	41.8550 1495	0.0063 9200
77	3.8032 0888	0.2629 3586	160.1833 6441	42.1179 5081	0.0062 4284
78	3.8697 6503	0.2584 1362	163.9865 7329	42.3763 6443	0.0060 9806
79	3.9374 8592	0.2539 6916	167.8563 3832	42.6303 3359	0.0059 5748
80	4.0063 9192	0.2496 0114	171.7938 2424	42.8799 3474	0.0058 2093
81	4.0765 0378	0.2453 0825	175.8002 1617	43.1252 4298	0.0056 8828
82	4.1478 4260	0.2410 8919	179.8767 1995	43.3663 3217	0.0055 5936
83	4.2204 2984	0.2369 4269	184.0245 6255	43.6032 7486	0.0054 3406
84	4.2942 8737	0.2328 6751	188.2449 9239	43.8361 4237	0.0053 1223
85	4.3694 3740	0.2288 6242	192.5392 7976	44.0650 0479	0.0051 9375
86	4.4459 0255	0.2249 2621	196.9087 1716	44.2899 3099	0.0050 7850
87	4.5237 0584	0.2210 5770	201.3546 1971	44.5109 8869	0.0049 6636
88	4.6028 7070	0.2172 5572	205.8783 2555	44.7282 4441	0.0048 5724
89	4.6834 2093	0.2135 1914	210.4811 9625	44.9417 6355	0.0047 5102
90	4.7653 8080	0.2098 4682	215.1646 1718	45.1516 1037	0.0046 4760
91	4.8487 7496	0.2062 3766	219.9299 9798	45.3578 4803	0.0045 4690
92	4.9336 2853	0.2026 9057	224.7787 7295	45.5605 3860	0.0044 4882
93	5.0199 6703	0.1992 0450	229.7124 0148	45.7597 4310	0.0043 5327
94	5.1078 1645	0.1957 7837	234.7323 6850	45.9555 2147	0.0042 6017
95	5.1972 0324	0.1924 1118	239.8401 8495	46.1479 3265	0.0041 6944
96	5.2881 5429	0.1891 0190	245.0373 8819	46.3370 3455	0.0040 8101
97	5.3806 9699	0.1858 4953	250.3255 4248	46.5228 8408	0.0039 9480
98	5.4748 5919	0.1826 5310	255.7062 3947	46.7055 3718	0.0039 1074
99	5.5706 6923	0.1795 1165	261.1810 9866	46.8850 4882	0.0038 2876
100	5.6681 5594	0.1764 2422	266.7517 6789	47.0614 7304	0.0037 4880

Table 1

607

Compound Amount, Present Value, Annuity

Table 1 (*continued*)

$$i = 2\%$$

n	$(1+i)^n$	$(1+i)^{-n}$	$s_{\overline{n}|i}$	$a_{\overline{n}|i}$	$s_{\overline{n}|i}^{-1}$
1	1.0200 0000	0.9803 9216	1.0000 0000	0.9803 9216	1.0000 0000
2	1.0404 0000	0.9611 6878	2.0200 0000	1.9415 6094	0.4950 4950
3	1.0612 0800	0.9423 2233	3.0604 0000	2.8838 8327	0.3267 5467
4	1.0824 3216	0.9238 4543	4.1216 0800	3.8077 2870	0.2426 2375
5	1.1040 8080	0.9057 3081	5.2040 4016	4.7134 5951	0.1921 5839
6	1.1261 6242	0.8879 7138	6.3081 2096	5.6014 3089	0.1585 2581
7	1.1486 8567	0.8705 6018	7.4342 8338	6.4719 9107	0.1345 1196
8	1.1716 5938	0.8534 9037	8.5829 6905	7.3254 8144	0.1165 0980
9	1.1950 9257	0.8367 5527	9.7546 2843	8.1622 3671	0.1025 1544
10	1.2189 9442	0.8203 4830	10.9497 2100	8.9825 8501	0.0913 2653
11	1.2433 7431	0.8042 6304	12.1687 1542	9.7868 4805	0.0821 7794
12	1.2682 4179	0.7884 9318	13.4120 8973	10.5733 4122	0.0745 5960
13	1.2936 0663	0.7730 3253	14.6803 3152	11.3483 7375	0.0681 1835
14	1.3194 7876	0.7578 7502	15.9739 3815	12.1062 4877	0.0626 0197
15	1.3458 6834	0.7430 1473	17.2934 1692	12.8492 6350	0.0578 2547
16	1.3727 8571	0.7284 4581	18.6392 8525	13.5777 0931	0.0536 5013
17	1.4002 4142	0.7141 6256	20.0120 7096	14.2918 7188	0.0499 6984
18	1.4282 4625	0.7001 5937	21.4123 1238	14.9920 3125	0.0467 0210
19	1.4568 1117	0.6864 3076	22.8405 5863	15.6784 6201	0.0437 8177
20	1.4859 4740	0.6729 7133	24.2973 6980	16.3514 3334	0.0411 5672
21	1.5156 6634	0.6597 7582	25.7833 1719	17.0112 0916	0.0387 8477
22	1.5459 7967	0.6468 3904	27.2989 8354	17.6580 4820	0.0366 3140
23	1.5768 9926	0.6341 5592	28.8449 6321	18.2922 0412	0.0346 6810
24	1.6084 3725	0.6217 2149	30.4218 6247	18.9139 2560	0.0328 7110
25	1.6406 0599	0.6095 3087	32.0302 9972	19.5234 5647	0.0312 2044
26	1.6734 1811	0.5975 7928	33.6709 0572	20.1210 3576	0.0296 9923
27	1.7068 8648	0.5858 6204	35.3443 2383	20.7068 9780	0.0282 9309
28	1.7410 2421	0.5743 7455	37.0512 1031	21.2812 7236	0.0269 8967
29	1.7758 4469	0.5631 1231	38.7922 3451	21.8443 8466	0.0257 7836
30	1.8113 6158	0.5520 7089	40.5680 7921	22.3964 5555	0.0246 4992
31	1.8475 8882	0.5412 4597	42.3794 4079	22.9377 0152	0 0235 9635
32	1.8845 4059	0.5306 3330	44.2270 2961	23.4683 3482	0.0226 1061
33	1.9222 3140	0.5202 2873	46.1115 7020	23.9885 6355	0.0216 8653
34	1.9606 7603	0.5100 2817	48.0338 0160	24.4985 9172	0.0208 1867
35	1.9998 8955	0.5000 2761	49.9944 7763	24.9986 1933	0.0200 0221
36	2.0398 8734	0.4902 2315	51.9943 6719	25.4888 4248	0.0192 3285
37	2.0806 8509	0.4806 1093	54.0342 5453	25.9694 5341	0.0185 0678
38	2.1222 9879	0.4711 8719	56.1149 3962	26.4406 4060	0.0178 2057
39	2.1647 4477	0.4619 4822	58.2372 3841	26.9025 8883	0.0171 7114
40	2.2080 3966	0.4528 9042	60.4019 8318	27.3554 7924	0.0165 5575
41	2.2522 0046	0.4440 1021	62.6100 2284	27.7994 8945	0.0159 7188
42	2.2972 4447	0.4353 0413	64.8622 2330	28.2347 9358	0.0154 1729
43	2.3431 8936	0.4267 6875	67.1594 6777	28.6615 6233	0.0148 8993
44	2.3900 5314	0.4184 0074	69.5026 5712	29.0799 6307	0.0143 8794
45	2.4378 5421	0.4101 9680	71.8927 1027	29.4901 5987	0.0139 0962
46	2.4866 1129	0.4021 5373	74.3305 6447	29.8923 1360	0.0134 5342
47	2.5363 4351	0.3942 6836	76.8171 7576	30.2865 8196	0.0130 1792
48	2.5870 7039	0.3865 3761	79.3535 1927	30.6731 1957	0.0126 0184
49	2.6388 1179	0.3789 5844	81.9405 8966	31.0520 7801	0.0122 0396
50	2.6915 8803	0.3715 2788	84.5794 0145	31.4236 0589	0.0118 2321

Table 1 (*continued*)

$i = 2\%$

n	$(1+i)^n$	$(1+i)^{-n}$	$s_{\overline{n}|i}$	$a_{\overline{n}|i}$	$s_{\overline{n}|i}^{-1}$
51	2.7454 1979	0.3642 4302	87.2709 8948	31.7878 4892	0.0114 5856
52	2.8003 2819	0.3571 0100	90.0164 0927	32.1449 4992	0.0111 0909
53	2.8563 3475	0.3500 9902	92.8167 3746	32.4950 4894	0.0107 7392
54	2.9134 6144	0.3432 3433	95.6730 7221	32.8382 8327	0.0104 5226
55	2.9717 3067	0.3365 0425	98.5865 3365	33.1747 8752	0.0101 4337
56	3.0311 6529	0.3299 0613	101.5582 6432	33.5046 9365	0.0098 4656
57	3.0917 8859	0.3234 3738	104.5894 2961	33.8281 3103	0.0095 6120
58	3.1536 2436	0.3170 9547	107.6812 1820	34.1452 2650	0.0092 8667
59	3.2166 9685	0.3108 7791	110.8348 4257	34.4561 0441	0.0090 2243
60	3.2810 3079	0.3047 8227	114.0515 3942	34.7608 8668	0.0087 6797
61	3.3466 5140	0.2988 0614	117.3325 7021	35.0596 9282	0.0085 2278
62	3.4135 8443	0.2929 4720	120.6792 2161	35.3526 4002	0.0082 8643
63	3.4818 5612	0.2872 0314	124.0928 0604	35.6398 4316	0.0080 5848
64	3.5514 9324	0.2815 7170	127.5746 6216	35.9214 1486	0.0078 3855
65	3.6225 2311	0.2760 5069	131.1261 5541	36.1974 6555	0.0076 2624
66	3.6949 7357	0.2706 3793	134.7486 7852	36.4681 0348	0.0074 2122
67	3.7688 7304	0.2653 3130	138.4436 5209	36.7334 3478	0.0072 2316
68	3.8442 5050	0.2601 2873	142.2125 2513	36.9935 6351	0.0070 3173
69	3.9211 3551	0.2550 2817	146.0567 7563	37.2485 9168	0.0068 4665
70	3.9995 5822	0.2500 2761	149.9779 1114	37.4986 1929	0.0066 6765
71	4.0795 4939	0.2451 2511	153.9774 6937	37.7437 4441	0.0064 9446
72	4.1611 4038	0.2403 1874	158.0570 1875	37.9840 6314	0.0063 2683
73	4.2443 6318	0.2356 0661	162.2181 5913	38.2196 6975	0.0061 6454
74	4.3292 5045	0.2309 8687	166.4625 2231	38.4506 5662	0.0060 0736
75	4.4158 3546	0.2264 5771	170.7917 7276	38.6771 1433	0.0058 5508
76	4.5041 5216	0.2220 1737	175.2076 0821	38.8991 3170	0.0057 0751
77	4.5942 3521	0.2176 6408	179.7117 6038	39.1167 9578	0.0055 6447
78	4.6861 1991	0.2133 9616	184.3059 9558	39.3301 9194	0.0054 2576
79	4.7798 4231	0.2092 1192	188.9921 1549	39.5394 0386	0.0052 9123
80	4.8754 3916	0.2051 0973	193.7719 5780	39.7445 1359	0.0051 6071
81	4.9729 4794	0.2010 8797	198.6473 9696	39.9456 0156	0.0050 3405
82	5.0724 0690	0.1971 4507	203.6203 4490	40.1427 4663	0.0049 1110
83	5.1738 5504	0.1932 7948	208.6927 5180	40.3360 2611	0.0047 9173
84	5.2773 3214	0.1894 8968	213.8666 0683	40.5255 1579	0.0046 7581
85	5.3828 7878	0.1857 7420	219.1439 3897	40.7112 8999	0.0045 6321
86	5.4905 3636	0.1821 3157	224.5268 1775	40.8934 2156	0.0044 5381
87	5.6003 4708	0.1785 6036	230.0173 5411	41.0719 8192	0.0043 4750
88	5.7123 5402	0.1750 5918	235.6177 0119	41.2470 4110	0.0042 4416
89	5.8266 0110	0.1716 2665	241.3300 5521	41.4186 6774	0.0041 4370
90	5.9431 3313	0.1682 6142	247.1566 5632	41.5869 2916	0.0040 4602
91	6.0619 9579	0.1649 6217	253.0997 8944	41.7518 9133	0.0039 5101
92	6.1832 3570	0.1617 2762	259.1617 8523	41.9136 1895	0.0038 5859
93	6.3069 0042	0.1585 5649	265.3450 2094	42.0721 7545	0.0037 6868
94	6.4330 3843	0.1554 4754	271.6519 2135	42.2276 2299	0.0036 8118
95	6.5616 9920	0.1523 9955	278.0849 5978	42.3800 2254	0.0035 9602
96	6.6929 3318	0.1494 1132	284.6466 5898	42.5294 3386	0.0035 1313
97	6.8267 9184	0.1464 8169	291.3395 9216	42.6759 1555	0.0034 3242
98	6.9633 2768	0.1436 0950	298.1663 8400	42.8195 2505	0.0033 5383
99	7.1025 9423	0.1407 9363	305.1297 1168	42.9603 1867	0.0032 7729
100	7.2446 4612	0.1380 3297	312.2323 0591	43.0983 5164	0.0032 0274

Table 1 **609**

Compound Amount, Present Value, Annuity

Table 1 (continued)

$$i = 2\tfrac{1}{4}\%$$

n	$(1+i)^n$	$(1+i)^{-n}$	$s_{\overline{n}\rvert i}$	$a_{\overline{n}\rvert i}$	$s_{\overline{n}\rvert i}^{-1}$
1	1.0225 0000	0.9779 9511	1.0000 0000	0.9779 9511	1.0000 0000
2	1.0455 0625	0.9564 7444	2.0225 0000	1.9344 6955	0.4944 3758
3	1.0690 3014	0.9354 2732	3.0680 0625	2.8698 9687	0.3259 4458
4	1.0930 8332	0.9148 4335	4.1370 3639	3.7847 4021	0.2417 1893
5	1.1176 7769	0.8947 1232	5.2301 1971	4.6794 5253	0.1912 0021
6	1.1428 2544	0.8750 2427	6.3477 9740	5.5544 7680	0.1575 3496
7	1.1685 3901	0.8557 6946	7.4906 2284	6.4102 4626	0.1335 0025
8	1.1948 3114	0.8369 3835	8.6591 6186	7.2471 8461	0.1154 8462
9	1.2217 1484	0.8185 2161	9.8539 9300	8.0657 0622	0.1014 8170
10	1.2492 0343	0.8005 1013	11.0757 0784	8.8662 1635	0.0902 8768
11	1.2773 1050	0.7828 9499	12.3249 1127	9.6491 1134	0.0811 3649
12	1.3060 4999	0.7656 6748	13.6022 2177	10.4147 7882	0.0735 1740
13	1.3354 3611	0.7488 1905	14.9082 7176	11.1635 9787	0.0670 7686
14	1.3654 8343	0.7323 4137	16.2437 0788	11.8959 3924	0.0615 6230
15	1.3962 0680	0.7162 2628	17.6091 9130	12.6121 6551	0.0567 8852
16	1.4276 2146	0.7004 6580	19.0053 9811	13.3126 3131	0.0526 1663
17	1.4597 4294	0.6850 5212	20.4330 1957	13.9976 8343	0.0489 4039
18	1.4925 8716	0.6699 7763	21.8927 6251	14.6676 6106	0.0456 7720
19	1.5261 7037	0.6552 3484	23.3853 4966	15.3228 9590	0.0427 6182
20	1.5605 0920	0.6408 1647	24.9115 2003	15.9637 1237	0.0401 4207
21	1.5956 2066	0.6267 1538	26.4720 2923	16.5904 2775	0.0377 7572
22	1.6315 2212	0.6129 2457	28.0676 4989	17.2033 5232	0.0356 2821
23	1.6682 3137	0.5994 3724	29.6991 7201	17.8027 8955	0.0336 7097
24	1.7057 6658	0.5862 4668	31.3674 0338	18.3890 3624	0.0318 8023
25	1.7441 4632	0.5733 4639	33.0731 6996	18.9623 8263	0.0302 3599
26	1.7833 8962	0.5607 2997	34.8173 1628	19.5231 1260	0.0287 2134
27	1.8235 1588	0.5483 9117	36.6007 0590	20.0715 0376	0.0273 2188
28	1.8645 4499	0.5363 2388	38.4242 2178	20.6078 2764	0.0260 2525
29	1.9064 9725	0.5245 2213	40.2887 6677	21.1323 4977	0.0248 2081
30	1.9493 9344	0.5129 8008	42.1952 6402	21.6453 2985	0.0236 9934
31	1.9932 5479	0.5016 9201	44.1446 5746	22.1470 2186	0.0226 5280
32	2.0381 0303	0.4906 5233	46.1379 1226	22.6376 7419	0.0216 7415
33	2.0839 6034	0.4798 5558	48.1760 1528	23.1175 2977	0.0207 5722
34	2.1308 4945	0.4692 9641	50.2599 7563	23.5868 2618	0.0198 9655
35	2.1787 9356	0.4589 6960	52.3908 2508	24.0457 9577	0.0190 8731
36	2.2278 1642	0.4488 7002	54.5696 1864	24.4946 6579	0.0183 2522
37	2.2779 4229	0.4389 9268	56.7974 3506	24.9336 5848	0.0176 0643
38	2.3291 9599	0.4293 3270	59.0753 7735	25.3629 9118	0.0169 2753
39	2.3816 0290	0.4198 8528	61.4045 7334	25.7828 7646	0.0162 8543
40	2.4351 8897	0.4106 4575	63.7861 7624	26.1935 2221	0.0156 7738
41	2.4899 8072	0.4016 0954	66.2213 6521	26.5951 3174	0.0151 0087
42	2.5460 0528	0.3927 7216	68.7113 4592	26.9879 0390	0.0145 5364
43	2.6032 9040	0.3841 2925	71.2573 5121	27.3720 3316	0.0140 3364
44	2.6618 6444	0.3756 7653	73.8606 4161	27.7477 0969	0.0135 3901
45	2.7217 5639	0.3674 0981	76.5225 0605	28.1151 1950	0.0130 6805
46	2.7829 9590	0.3593 2500	79.2442 6243	28.4744 4450	0.0126 1921
47	2.8456 1331	0.3514 1809	82.0272 5834	28.8258 6259	0.0121 9107
48	2.9096 3961	0.3436 8518	84.8728 7165	29.1695 4777	0.0117 8233
49	2.9751 0650	0.3361 2242	87.7825 1126	29.5056 7019	0.0113 9179
50	3.0420 4640	0.3287 2608	90.7576 1776	29.8343 9627	0.0110 1836

Table 1 (*continued*)

$$i = 2\tfrac{1}{4}\%$$

n	$(1+i)^n$	$(1+i)^{-n}$	$s_{\overline{n}|i}$	$a_{\overline{n}|i}$	$s_{\overline{n}|i}^{-1}$
51	3.1104 9244	0.3214 9250	93.7996 6416	30.1558 8877	0.0106 6102
52	3.1804 7852	0.3144 1810	96.9101 5661	30.4703 0687	0.0103 1884
53	3.2520 3929	0.3074 9936	100.0906 3513	30.7778 0623	0.0099 9094
54	3.3252 1017	0.3007 3287	103.3426 7442	31.0785 3910	0.0096 7654
55	3.4000 2740	0.2941 1528	106.6678 8460	31.3726 5438	0.0093 7489
56	3.4765 2802	0.2876 4330	110.0679 1200	31.6602 9768	0.0090 8530
57	3.5547 4990	0.2813 1374	113.5444 4002	31.9416 1142	0.0088 0712
58	3.6347 3177	0.2751 2347	117.0991 8992	32.2167 3489	0.0085 3977
59	3.7165 1324	0.2690 6940	120.7339 2169	32.4858 0429	0.0082 8268
60	3.8001 3479	0.2631 4856	124.4504 3493	32.7489 5285	0.0080 3533
61	3.8856 3782	0.2573 5801	128.2505 6972	33.0063 1086	0.0077 9724
62	3.9730 6467	0.2516 9487	132.1362 0754	33.2580 0573	0.0075 6795
63	4.0624 5862	0.2461 5635	136.1092 7221	33.5041 6208	0.0073 4704
64	4.1538 6394	0.2407 3971	140.1717 3083	33.7449 0179	0.0071 3411
65	4.2473 2588	0.2354 4226	144.3255 9477	33.9803 4405	0.0069 2878
66	4.3428 9071	0.2302 6138	148.5729 2066	34.2106 0543	0.0067 3070
67	4.4406 0576	0.2251 9450	152.9158 1137	34.4357 9993	0.0065 3955
68	4.5405 1939	0.2202 3912	157.3564 1713	34.6560 3905	0.0063 5500
69	4.6426 8107	0.2153 9278	161.8969 3651	34.8714 3183	0.0061 7677
70	4.7471 4140	0.2106 5309	166.5396 1758	35.0820 8492	0.0060 0458
71	4.8539 5208	0.2060 1769	171.2867 5898	35.2881 0261	0.0058 3816
72	4.9631 6600	0.2014 8429	176.1407 1106	35.4895 8691	0.0056 7728
73	5.0748 3723	0.1970 5065	181.1038 7705	35.6866 3756	0.0055 2169
74	5.1890 2107	0.1927 1458	186.1787 1429	35.8793 5214	0.0053 7118
75	5.3057 7405	0.1884 7391	191.3677 3536	36.0678 2605	0.0052 2554
76	5.4251 5396	0.1843 2657	196.6735 0941	36.2521 5262	0.0050 8457
77	5.5472 1993	0.1802 7048	202.0986 6337	36.4324 2310	0.0049 4808
78	5.6720 3237	0.1763 0365	207.6458 8329	36.6087 2675	0.0048 1589
79	5.7996 5310	0.1724 2411	213.3179 1567	36.7811 5085	0.0046 8784
80	5.9301 4530	0.1686 2993	219.1175 6877	36.9497 8079	0.0045 6376
81	6.0635 7357	0.1649 1925	225.0477 1407	37.1147 0004	0.0044 4350
82	6.2000 0397	0.1612 9022	231.1112 8763	37.2759 9026	0.0043 2692
83	6.3395 0406	0.1577 4105	237.3112 9160	37.4337 3130	0.0042 1387
84	6.4821 4290	0.1542 6997	243.6507 9567	37.5880 0127	0.0041 0423
85	6.6279 9112	0.1508 7528	250.1329 3857	37.7388 7655	0.0039 9787
86	6.7771 2092	0.1475 5528	256.7609 2969	37.8864 3183	0.0038 9467
87	6.9296 0614	0.1443 0835	263.5380 5060	38.0307 4018	0.0037 9452
88	7.0855 2228	0.1411 3286	270.4676 5674	38.1718 7304	0.0036 9730
89	7.2449 4653	0.1380 2724	277.5531 7902	38.3099 0028	0.0036 0291
90	7.4079 5782	0.1349 8997	284.7981 2555	38.4448 9025	0.0035 1126
91	7.5746 3688	0.1320 1953	292.2060 8337	38.5769 0978	0.0034 2224
92	7.7450 6621	0.1291 1445	299.7807 2025	38.7060 2423	0.0033 3577
93	7.9193 3020	0.1262 7331	307.5257 8645	38.8322 9754	0.0032 5176
94	8.0975 1512	0.1234 9468	315.4451 1665	38.9557 9221	0.0031 7012
95	8.2797 0921	0.1207 7719	323.5426 3177	39.0765 6940	0.0030 9078
96	8.4660 0267	0.1181 1950	331.8223 4099	39.1946 8890	0.0030 1366
97	8.6564 8773	0.1155 2029	340.2883 4366	39.3102 0920	0.0029 3868
98	8.8512 5871	0.1129 7828	348.9448 3139	39.4231 8748	0.0028 6578
99	9.0504 1203	0.1104 9221	357.7960 9010	39.5336 7968	0.0027 9489
100	9.2540 4630	0.1080 6084	366.8465 0213	39.6417 4052	0.0027 2594

Table 1

611

Compound Amount, Present Value, Annuity

Table 1 (*continued*)

$$i = 2\tfrac{1}{2}\%$$

n	$(1+i)^n$	$(1+i)^{-n}$	$s_{\overline{n}|i}$	$a_{\overline{n}|i}$	$s_{\overline{n}|i}^{-1}$
1	1.0250 0000	0.9756 0976	1.0000 0000	0.9756 0976	1.0000 0000
2	1.0506 2500	0.9518 1440	2.0250 0000	1.9274 2415	0.4938 2716
3	1.0768 9063	0.9285 9941	3.0756 2500	2.8560 2356	0.3251 3717
4	1.1038 1289	0.9059 5064	4.1525 1563	3.7619 7421	0.2408 1788
5	1.1314 0821	0.8838 5429	5.2563 2852	4.6458 2850	0.1902 4686
6	1.1596 9342	0.8622 9687	6.3877 3673	5.5081 2536	0.1565 4997
7	1.1886 8575	0.8412 6524	7.5474 3015	6.3493 9060	0.1324 9543
8	1.2184 0290	0.8207 4657	8.7361 1590	7.1701 3717	0.1144 6735
9	1.2488 6297	0.8007 2836	9.9545 1880	7.9708 6553	0.1004 5689
10	1.2800 8454	0.7811 9840	11.2033 8177	8.7520 6393	0.0892 5876
11	1.3120 8666	0.7621 4478	12.4834 6631	9.5142 0871	0.0801 0596
12	1.3448 8882	0.7435 5589	13.7955 5297	10.2577 6460	0.0724 8713
13	1.3785 1104	0.7254 2038	15.1404 4179	10.9831 8497	0.0660 4827
14	1.4129 7382	0.7077 2720	16.5189 5284	11.6909 1217	0.0605 3653
15	1.4482 9817	0.6904 6556	17.9319 2666	12.3813 7773	0.0557 6646
16	1.4845 0562	0.6736 2493	19.3802 2483	13.0550 0266	0.0515 9899
17	1.5216 1826	0.6571 9506	20.8647 3045	13.7121 9772	0.0479 2777
18	1.5596 5872	0.6411 6591	22.3863 4871	14.3533 6363	0.0446 7008
19	1.5986 5019	0.6255 2772	23.9460 0743	14.9788 9134	0.0417 6062
20	1.6386 1644	0.6102 7094	25.5446 5761	15.5891 6229	0.0391 4713
21	1.6795 8185	0.5953 8629	27.1832 7405	16.1845 4857	0.0367 8733
22	1.7215 7140	0.5808 6467	28.8628 5590	16.7654 1324	0.0346 4661
23	1.7646 1068	0.5666 9724	30.5844 2730	17.3321 1048	0.0326 9638
24	1.8087 2595	0.5528 7535	32.3490 3798	17.8849 8583	0.0309 1282
25	1.8539 4410	0.5393 9059	34.1577 6393	18.4243 7642	0.0292 7592
26	1.9002 9270	0.5262 3472	36.0117 0803	18.9506 1114	0.0277 6875
27	1.9478 0002	0.5133 9973	37.9120 0073	19.4640 1087	0.0263 7687
28	1.9964 9502	0.5008 7778	39.8598 0075	19.9648 8866	0.0250 8793
29	2.0464 0739	0.4886 6125	41.8562 9577	20.4535 4991	0.0238 9127
30	2.0975 6758	0.4767 4269	43.9027 0316	20.9302 9259	0.0227 7764
31	2.1500 0677	0.4651 1481	46.0002 7074	21.3954 0741	0.0217 3900
32	2.2037 5694	0.4537 7055	48.1502 7751	21.8491 7796	0.0207 6831
33	2.2588 5086	0.4427 0298	50.3540 3445	22.2918 8094	0.0198 5938
34	2.3153 2213	0.4319 0534	52.6128 8531	22.7237 8628	0.0190 0675
35	2.3732 0519	0.4213 7107	54.9282 0744	23.1451 5734	0.0182 0558
36	2.4325 3532	0.4110 9372	57.3014 1263	23.5562 5107	0.0174 5158
37	2.4933 4870	0.4010 6705	59.7339 4794	23.9573 1812	0.0167 4090
38	2.5556 8242	0.3912 8492	62.2272 9664	24.3486 0304	0.0160 7012
39	2.6195 7448	0.3817 4139	64.7829 7906	24.7303 4443	0.0154 3615
40	2.6850 6384	0.3724 3062	67.4025 5354	25.1027 7505	0.0148 3623
41	2.7521 9043	0.3633 4695	70.0876 1737	25.4661 2200	0.0142 6786
42	2.8209 9520	0.3544 8483	72.8398 0781	25.8206 0683	0.0137 2876
43	2.8915 2008	0.3458 3886	75.6608 0300	26.1664 4569	0.0132 1688
44	2.9638 0808	0.3374 0376	78.5523 2308	26.5038 4945	0.0127 3037
45	3.0379 0328	0.3291 7440	81.5161 3116	26.8330 2386	0.0122 6752
46	3.1138 5086	0.3211 4576	84.5540 3443	27.1541 6962	0.0118 2676
47	3.1916 9713	0.3133 1294	87.6678 8530	27.4674 8255	0.0114 0669
48	3.2714 8956	0.3056 7116	90.8595 8243	27.7731 5371	0.0110 0599
49	3.3532 7680	0.2982 1576	94.1310 7199	28.0713 6947	0.0106 2348
50	3.4371 0872	0.2909 4221	97.4843 4879	28.3623 1168	0.0102 5806

Table 1 (*continued*)

$$i = 2\tfrac{1}{2}\%$$

n	$(1+i)^n$	$(1+i)^{-n}$	$s_{\overline{n}|i}$	$a_{\overline{n}|i}$	$s_{\overline{n}|i}^{-1}$
51	3.5230 3644	0.2838 4606	100.9214 5751	28.6461 5774	0.0099 0870
52	3.6111 1235	0.2769 2298	104.4444 9395	28.9230 8072	0.0095 7446
53	3.7013 9016	0.2701 6876	108.0556 0629	29.1932 4948	0.0092 5449
54	3.7939 2491	0.2635 7928	111.7569 9645	29.4568 2876	0.0089 4799
55	3.8887 7303	0.2571 5052	115.5509 2136	29.7139 7928	0.0086 5419
56	3.9859 9236	0.2508 7855	119.4396 9440	29.9648 5784	0.0083 7243
57	4.0856 4217	0.2447 5956	123.4256 8676	30.2096 1740	0.0081 0204
58	4.1877 8322	0.2387 8982	127.5113 2893	30.4484 0722	0.0078 4244
59	4.2924 7780	0.2329 6568	131.6991 1215	30.6813 7290	0.0075 9307
60	4.3997 8975	0.2272 8359	135.9915 8995	30.9086 5649	0.0073 5340
61	4.5097 8449	0.2217 4009	140.3913 7970	31.1303 9657	0.0071 2294
62	4.6225 2910	0.2163 3179	144.9011 6419	31.3467 2836	0.0069 0126
63	4.7380 9233	0.2110 5541	149.5236 9330	31.5577 8377	0.0066 8790
64	4.8565 4464	0.2059 0771	154.2617 8563	31.7636 9148	0.0064 8249
65	4.9779 5826	0.2008 8557	159.1183 3027	31.9645 7705	0.0062 8463
66	5.1024 0721	0.1959 8593	164.0962 8853	32.1605 6298	0.0060 9398
67	5.2299 6739	0.1912 0578	169.1986 9574	32.3517 6876	0.0059 1021
68	5.3607 1658	0.1865 4223	174.4286 6314	32.5383 1099	0.0057 3300
69	5.4947 3449	0.1819 9241	179.7893 7971	32.7203 0340	0.0055 6206
70	5.6321 0286	0.1775 5358	185.2841 1421	32.8978 5698	0.0053 9712
71	5.7729 0543	0.1732 2300	190.9162 1706	33.0710 7998	0.0052 37C 0
72	5.9172 2806	0.1689 9805	196.6891 2249	33.2400 7803	0.0050 8417
73	6.0651 5876	0.1648 7615	202.6063 5055	33.4049 5417	0.0049 3568
74	6.2167 8773	0.1608 5478	208.6715 0931	33.5658 0895	0.0047 9222
75	6.3722 0743	0.1569 3149	214.8882 9705	33.7227 4044	0.0046 5358
76	6.5315 1261	0.1531 0389	221.2605 0447	33.8758 4433	0.0045 1956
77	6.6948 0043	0.1493 6965	227.7920 1709	34.0252 1398	0.0043 8997
78	6.8621 7044	0.1457 2649	234.4868 1751	34.1709 4047	0.0042 6463
79	7.0337 2470	0.1421 7218	241.3489 8795	34.3131 1265	0.0041 4338
80	7.2095 6782	0.1387 0457	248.3827 1265	34.4518 1722	0.0040 2605
81	7.3898 0701	0.1353 2153	255.5922 8047	34.5871 3875	0.0039 1248
82	7.5745 5219	0.1320 2101	262.9820 8748	34.7191 5976	0.0038 0254
83	7.7639 1599	0.1288 0098	270.5566 3966	34.8479 6074	0.0036 9608
84	7.9580 1389	0.1256 5949	278.3205 5566	34.9736 2023	0.0035 9298
85	8.1569 6424	0.1225 9463	286.2785 6955	35.0962 1486	0.0034 9310
86	8.3608 8834	0.1196 0452	294.4355 3379	35.2158 1938	0.0033 9633
87	8.5699 1055	0.1166 8733	302.7964 2213	35.3325 0671	0.0033 0255
88	8.7841 5832	0.1138 4130	311.3663 3268	35.4463 4801	0.0032 1165
89	9.0037 6228	0.1110 6468	320.1504 9100	35.5574 1269	0.0031 2353
90	9.2288 5633	0.1083 5579	329.1542 5328	35.6657 6848	0.0030 3809
91	9.4595 7774	0.1057 1296	338.3831 0961	35.7714 8144	0.0029 5523
92	9.6960 6718	0.1031 3460	347.8426 8735	35.8746 1604	0.0028 7486
93	9.9384 6886	0.1006 1912	357.5387 5453	35.9752 3516	0.0027 9690
94	10.1869 3058	0.0981 6500	367.4772 2339	36.0734 0016	0.0027 2126
95	10.4416 0385	0.0957 7073	377.6641 5398	36.1691 7089	0.0026 4786
96	10.7026 4395	0.0934 3486	388.1057 5783	36.2626 0574	0.0025 7662
97	10.9702 1004	0.0911 5596	398.8084 0177	36.3537 6170	0.0025 0747
98	11.2444 6530	0.0889 3264	409.7786 1182	36.4426 9434	0.0024 4034
99	11.5255 7693	0.0867 6355	421.0230 7711	36.5294 5790	0.0023 7517
100	11.8137 1635	0.0846 4737	432.5486 5404	36.6141 0526	0.0023 1188

Table 1   **613**

Compound Amount, Present Value, Annuity

Table 1 (*continued*)

$$i = 2\tfrac{3}{4}\%$$

n	$(1+i)^n$	$(1+i)^{-n}$	$s_{\overline{n}|i}$	$a_{\overline{n}|i}$	$s_{\overline{n}|i}^{-1}$
1	1.0275 0000	0.9732 3601	1.0000 0000	0.9732 3601	1.0000 0000
2	1.0557 5625	0.9471 8833	2.0275 0000	1.9204 2434	0.4932 1825
3	1.0847 8955	0.9218 3779	3.0832 5625	2.8422 6213	0.3243 3243
4	1.1146 2126	0.8971 6573	4.1680 4580	3.7394 2787	0.2399 2059
5	1.1452 7334	0.8731 5400	5.2826 6706	4.6125 8186	0.1892 9832
6	1.1767 6836	0.8497 8491	6.4279 4040	5.4623 6678	0.1555 7083
7	1.2091 2949	0.8270 4128	7.6047 0876	6.2894 0806	0.1314 9747
8	1.2423 8055	0.8049 0635	8.8138 3825	7.0943 1441	0.1134 5795
9	1.2765 4602	0.7833 6385	10.0562 1880	7.8776 7826	0.0994 4095
10	1.3116 5103	0.7623 9791	11.3327 6482	8.6400 7616	0.0882 3972
11	1.3477 2144	0.7419 9310	12.6444 1585	9.3820 6926	0.0790 8629
12	1.3847 8378	0.7221 3440	13.9921 3729	10.1042 0366	0.0714 6871
13	1.4228 6533	0.7028 0720	15.3769 2107	10.8070 1086	0.0650 3252
14	1.4619 9413	0.6839 9728	16.7997 8639	11.4910 0814	0.0595 2457
15	1.5021 9896	0.6656 9078	18.2617 8052	12.1566 9892	0.0547 5917
16	1.5435 0944	0.6478 7424	19.7639 7948	12.8045 7315	0.0505 9710
17	1.5859 5595	0.6305 3454	21.3074 8892	13.4351 0769	0.0469 3186
18	1.6295 6973	0.6136 5892	22.8934 4487	14.0487 6661	0.0436 8063
19	1.6743 8290	0.5972 3496	24.5230 1460	14.6460 0157	0.0407 7802
20	1.7204 2843	0.5812 5057	26.1973 9750	15.2272 5213	0.0381 7173
21	1.7677 4021	0.5656 9398	27.9178 2593	15.7929 4612	0.0358 1941
22	1.8163 5307	0.5505 5375	29.6855 6615	16.3434 9987	0.0036 8640
23	1.8663 0278	0.5358 1874	31.5019 1921	16.8793 1861	0.0317 4410
24	1.9176 2610	0.5214 7809	33.3682 2199	17.4007 9670	0.0299 6863
25	1.9703 6082	0.5075 2126	35.2858 4810	17.9083 1795	0.0283 3997
26	2.0245 4575	0.4939 3796	37.2562 0892	18.4022 5592	0.0268 4116
27	2.0802 2075	0.4807 1821	39.2807 5467	18.8829 7413	0.0254 5776
28	2.1374 2682	0.4678 5227	41.3609 7542	19.3508 2640	0.0241 7738
29	2.1962 0606	0.4553 3068	43.4984 0224	19.8061 5708	0.0229 8935
30	2.2566 0173	0.4431 4421	45.6946 0830	20.2493 0130	0.0218 8442
31	2.3186 5828	0.4312 8391	47.9512 1003	20.6805 8520	0.0208 5453
32	2.3824 2138	0.4197 4103	50.2698 6831	21.1003 2623	0.0198 9263
33	2.4479 3797	0.4085 0708	52.6522 8969	21.5088 3332	0.0189 9253
34	2.5152 5626	0.3975 7380	55.1002 2765	21.9064 0712	0.0181 4875
35	2.5844 2581	0.3869 3314	57.6154 8391	22.2933 4026	0.0173 5645
36	2.6554 9752	0.3765 7727	60.1999 0972	22.6699 1753	0.0166 1132
37	2.7285 2370	0.3664 9856	62.8554 0724	23.0364 1609	0.0159 0953
38	2.8035 5810	0.3566 8959	65.5839 3094	23.3931 0568	0.0152 4764
39	2.8806 5595	0.3471 4316	68.3874 8904	23.7402 4884	0.0146 2256
40	2.9598 7399	0.3378 5222	71.2681 4499	24.0781 0106	0.0140 3151
41	3.0412 7052	0.3288 0995	74.2280 1898	24.4069 1101	0.0134 7200
42	3.1249 0546	0.3200 0968	77.2692 8950	24.7269 2069	0.0129 4175
43	3.2108 4036	0.3114 4495	80.3941 9496	25.0383 6563	0.0124 3871
44	3.2991 3847	0.3031 0944	83.6050 3532	25.3414 7507	0.0119 6100
45	3.3898 6478	0.2949 9702	86.9041 7379	25.6364 7209	0.0115 0693
46	3.4830 8606	0.2871 0172	90.2940 3857	25.9235 7381	0.0110 7493
47	3.5788 7093	0.2794 1773	93.7771 2463	26.2029 9154	0.0106 6358
48	3.6772 8988	0.2719 3940	97.3559 9556	26.4749 3094	0.0102 7158
49	3.7784 1535	0.2646 6122	101.0332 8544	26.7395 9215	0.0098 9773
50	3.8823 2177	0.2575 7783	104.8117 0079	26.9971 6998	0.0095 4092

Table 1 (*continued*)

$$i = 2\tfrac{3}{4}\%$$

n	$(1+i)^n$	$(1+i)^{-n}$	$s_{\overline{n}|i}$	$a_{\overline{n}|i}$	$s_{\overline{n}|i}^{-1}$
51	3.9890 8562	0.2506 8402	108.6940 2256	27.2478 5400	0.0092 0014
52	4.0987 8547	0.2439 7471	112.6831 0818	27.4918 2871	0.0088 7444
53	4.2115 0208	0.2374 4497	116.7818 9365	27.7292 7368	0.0085 6297
54	4.3273 1838	0.2310 9000	120.9933 9573	27.9603 6368	0.0082 6491
55	4.4463 1964	0.2249 0511	125.3207 1411	28.1852 6879	0.0079 7953
56	4.5685 9343	0.2188 8575	129.7670 3375	28.4041 5454	0.0077 0612
57	4.6942 2975	0.2130 2749	134.3356 2718	28.6171 8203	0.0074 4404
58	4.8233 2107	0.2073 2603	139.0298 5692	28.8245 0806	0.0071 9270
59	4.9559 6239	0.2017 7716	143.8531 7799	29.0262 8522	0.0069 5153
60	5.0922 5136	0.1963 7679	148.8091 4038	29.2226 6201	0.0067 2002
61	5.2322 8827	0.1911 2097	153.9013 9174	29.4137 8298	0.0064 9767
62	5.3761 7620	0.1860 0581	159.1336 8002	29.5997 8879	0.0062 8402
63	5.5240 2105	0.1810 2755	164.5098 5622	29.7808 1634	0.0060 7866
64	5.6759 3162	0.1761 8253	170.0338 7726	29.9569 9887	0.0058 8118
65	5.8320 1974	0.1714 6718	175.7098 0889	30.1284 6605	0.0056 9120
66	5.9924 0029	0.1668 7804	181.5418 2863	30.2953 4409	0.0055 0837
67	6.1571 9130	0.1624 1172	187.5342 2892	30.4577 5581	0.0053 3236
68	6.3265 1406	0.1580 6493	193.6914 2021	30.6158 2074	0.0051 6285
69	6.5004 9319	0.1538 3448	200.0179 3427	30.7696 5522	0.0049 9955
70	6.6792 5676	0.1497 1726	206.5184 2746	30.9193 7247	0.0048 4218
71	6.8629 3632	0.1457 1023	213.1976 8422	31.0650 8270	0.0046 9048
72	7.0516 6706	0.1418 1044	220.0606 2054	31.2068 9314	0.0045 4420
73	7.2455 8791	0.1380 1503	227.1122 8760	31.3449 0816	0.0044 0311
74	7.4448 4158	0.1343 2119	234.3578 7551	31.4792 2936	0.0042 6698
75	7.6495 7472	0.1307 2622	241.8027 1709	31.6099 5558	0.0041 3560
76	7.8599 3802	0.1272 2747	249.4522 9181	31.7371 8304	0.0040 0878
77	8.0760 8632	0.1238 2235	257.3122 2983	31.8610 0540	0.0038 8633
78	8.2981 7869	0.1205 0837	265.3883 1615	31.9815 1377	0.0037 6806
79	8.5263 7861	0.1172 8309	273.6864 9485	32.0987 9685	0.0036 5382
80	8.7608 5402	0.1141 4412	282.2128 7345	32.2129 4098	0.0035 4342
81	9.0017 7751	0.1110 8917	290.9737 2747	32.3240 3015	0.0034 3674
82	9.2493 2639	0.1081 1598	299.9755 0498	32.4321 4613	0.0033 3361
83	9.5036 8286	0.1052 2237	309.2248 3137	32.5373 6850	0.0032 3389
84	9.7650 3414	0.1024 0620	318.7285 1423	32.6397 7469	0.0031 3747
85	10.0335 7258	0.0996 6540	328.4935 4837	32.7394 4009	0.0030 4420
86	10.3094 9583	0.0969 9795	338.5271 2095	32.8364 3804	0.0029 5397
87	10.5930 0696	0.0944 0190	348.8366 1678	32.9308 3994	0.0028 6667
88	10.8843 1465	0.0918 7533	359.4296 2374	33.0227 1527	0.0027 8219
89	11.1836 3331	0.0894 1638	370.3139 3839	33.1121 3165	0.0027 0041
90	11.4911 8322	0.0870 2324	381.4975 7170	33.1991 5489	0.0026 2125
91	11.8071 9076	0.0846 9415	392.9887 5492	33.2838 4905	0.0025 4460
92	12.1318 8851	0.0824 2740	404.7959 4568	33.3662 7644	0.0024 7038
93	12.4655 1544	0.0802 2131	416.9278 3418	33.4464 9776	0.0023 9850
94	12.8083 1711	0.0780 7427	429.3933 4962	33.5245 7202	0.0023 2887
95	13.1605 4584	0.0759 8469	442.2016 6674	33.6005 5671	0.0022 6141
96	13.5224 6085	0.0739 5104	455.3622 1257	33.6745 0775	0.0021 9605
97	13.8943 2852	0.0719 7181	468.8846 7342	33.7464 7956	0.0021 3272
98	14.2764 2255	0.0700 4556	482.7790 0194	33.8165 2512	0.0020 7134
99	14.6690 2417	0.0681 7086	497.0554 2449	33.8846 9598	0.0020 1185
100	15.0724 2234	0.0663 4634	511.7244 4867	33.9510 4232	0.0019 5418

Table 1

615

Compound Amount, Present Value, Annuity

Table 1 (continued)

$$i = 3\%$$

n	$(1+i)^n$	$(1+i)^{-n}$	$s_{\overline{n}|i}$	$a_{\overline{n}|i}$	$s_{\overline{n}|i}^{-1}$
1	1.0300 0000	0.9708 7379	1.0000 0000	0.9708 7379	1.0000 0000
2	1.0609 0000	0.9425 9591	2.0300 0000	1.9134 6970	0.4926 1084
3	1.0927 2700	0.9151 4166	3.0909 0000	2.8286 1135	0.3235 3036
4	1.1255 0881	0.8884 8705	4.1836 2700	3.7170 9840	0.2390 2705
5	1.1592 7407	0.8626 0878	5.3091 3581	4.5797 0719	0.1883 5457
6	1.1940 5230	0.8374 8426	6.4684 0988	5.4171 9144	0.1545 9750
7	1.2298 7387	0.8130 9151	7.6624 6218	6.2302 8296	0.1305 0635
8	1.2667 7008	0.7894 0923	8.8923 3605	7.0196 9219	0.1124 5639
9	1.3047 7318	0.7664 1673	10.1591 0613	7.7861 0892	0.0984 3386
10	1.3439 1638	0.7440 9391	11.4638 7931	8.5302 0284	0.0872 3051
11	1.3842 3387	0.7224 2128	12.8077 9569	9.2526 2411	0.0780 7745
12	1.4257 6089	0.7013 7988	14.1920 2956	9.9540 0399	0.0704 6209
13	1.4685 3371	0.6809 5134	15.6177 9045	10.6349 5533	0.0640 2954
14	1.5125 8972	0.6611 1781	17.0863 2416	11.2960 7314	0.0585 2634
15	1.5579 6742	0.6418 6195	18.5989 1389	11.9379 3509	0.0537 6658
16	1.6047 0644	0.6231 6694	20.1568 8130	12.5611 0203	0.0496 1085
17	1.6528 4763	0.6050 1645	21.7615 8774	13.1661 1847	0.0459 5253
18	1.7024 3306	0.5873 9461	23.4144 3537	13.7535 1308	0.0427 0870
19	1.7535 0605	0.5702 8603	25.1168 6844	14.3237 9911	0.0398 1388
20	1.8061 1123	0.5536 7575	26.8703 7449	14.8774 7486	0.0372 1571
21	1.8602 9457	0.5375 4928	28.6764 8572	15.4150 2414	0.0348 7178
22	1.9161 0341	0.5218 9250	30.5367 8030	15.9369 1664	0.0327 4739
23	1.9735 8651	0.5066 9175	32.4528 8370	16.4436 0839	0.0308 1390
24	2.0327 9411	0.4919 3374	34.4264 7022	16.9355 4212	0.0290 4742
25	2.0937 7793	0.4776 0557	36.4592 6432	17.4131 4769	0.0274 2787
26	2.1565 9127	0.4636 9473	38.5530 4225	17.8768 4242	0.0259 3829
27	2.2212 8901	0.4501 8906	40.7096 3352	18.3270 3147	0.0245 6421
28	2.2879 2768	0.4370 7675	42.9309 2252	18.7641 0823	0.0232 9323
29	2.3565 6551	0.4243 4636	45.2188 5020	19.1884 5459	0.0221 1467
30	2.4272 6247	0.4119 8676	47.5754 1571	19.6004 4135	0.0210 1926
31	2.5000 8035	0.3999 8715	50.0026 7818	20.0004 2849	0.0199 9893
32	2.5750 8276	0.3883 3703	52.5027 5852	20.3887 6553	0.0190 4662
33	2.6523 3524	0.3770 2625	55.0778 4128	20.7657 9178	0.0181 5612
34	2.7319 0530	0.3660 4490	57.7301 7652	21.1318 3668	0.0173 2196
35	2.8138 6245	0.3553 8340	60.4620 8181	21.4872 2007	0.0165 3929
36	2.8982 7833	0.3450 3243	63.2759 4427	21.8322 5250	0.0158 0379
37	2.9852 2668	0.3349 8294	66.1742 2259	22.1672 3544	0.0151 1162
38	3.0747 8348	0.3252 2615	69.1594 4927	22.4924 6159	0.0144 5934
39	3.1670 2698	0.3157 5355	72.2342 3275	22.8082 1513	0.0138 4385
40	3.2620 3779	0.3065 5684	75.4012 5973	23.1147 7197	0.0132 6238
41	3.3598 9893	0.2976 2800	78.6632 9753	23.4123 9997	0.0127 1241
42	3.4606 9589	0.2889 5922	82.0231 9645	23.7013 5920	0.0121 9167
43	3.5645 1677	0.2805 4294	85.4838 9234	23.9819 0213	0.0116 9811
44	3.6714 5227	0.2723 7178	89.0484 0911	24.2542 7392	0.0112 2985
45	3.7815 9584	0.2644 3862	92.7198 6139	24.5187 1254	0.0107 8518
46	3.8950 4372	0.2567 3653	96.5014 5723	24.7754 4907	0.0103 6254
47	4.0118 9503	0.2492 5876	100.3965 0095	25.0247 0783	0.0099 6051
48	4.1322 5188	0.2419 9880	104.4083 9598	25.2667 0664	0.0095 7777
49	4.2562 1944	0.2349 5029	108.5406 4785	25.5016 5693	0.0092 1314
50	4.3839 0602	0.2281 0708	112.7968 6729	25.7297 6401	0.0088 6550

Table 1 (*continued*)

$i = 3\%$

n	$(1+i)^n$	$(1+i)^{-n}$	$s_{\overline{n}|i}$	$a_{\overline{n}|i}$	$s_{\overline{n}|i}^{-1}$
51	4.5154 2320	0.2214 6318	117.1807 7331	25.9512 2719	0.0085 3382
52	4.6508 8590	0.2150 1280	121.6961 9651	26.1662 3999	0.0082 1718
53	4.7904 1247	0.2087 5029	126.3470 8240	26.3749 9028	0.0079 1471
54	4.9341 2485	0.2026 7019	131.1374 9488	26.5776 6047	0.0076 2558
55	5.0821 4859	0.1967 6717	136.0716 1972	26.7744 2764	0.0073 4907
56	5.2346 1305	0.1910 3609	141.1537 6831	26.9654 6373	0.0070 8447
57	5.3916 5144	0.1854 7193	146.3883 8136	27.1509 3566	0.0068 3114
58	5.5534 0098	0.1800 6984	151.7800 3280	27.3310 0549	0.0065 8848
59	5.7200 0301	0.1748 2508	157.3334 3379	27.5058 3058	0.0063 5593
60	5.8916 0310	0.1697 3309	163.0534 3680	27.6755 6367	0.0061 3296
61	6.0683 5120	0.1647 8941	168.9450 3991	27.8403 5307	0.0059 1908
62	6.2504 0173	0.1599 8972	175.0133 9110	28.0003 4279	0.0057 1385
63	6.4379 1379	0.1553 2982	181.2637 9284	28.1556 7261	0.0055 1682
64	6.6310 5120	0.1508 0565	187.7017 0662	28.3064 7826	0.0053 2760
65	6.8299 8273	0.1464 1325	194.3327 5782	28.4528 9152	0.0051 4581
66	7.0348 8222	0.1421 4879	201.1627 4055	28.5950 4031	0.0049 7110
67	7.2459 2868	0.1380 0853	208.1976 2277	28.7330 4884	0.0048 0313
68	7.4633 0654	0.1339 8887	215.4435 5145	28.8670 3771	0.0046 4159
69	7.6872 0574	0.1300 8628	222.9068 5800	28.9971 2399	0.0044 8618
70	7.9178 2191	0.1262 9736	230.5940 6374	29.1234 2135	0.0043 3663
71	8.1553 5657	0.1226 1880	238.5118 8565	29.2460 4015	0.0041 9266
72	8.4000 1727	0.1190 4737	246.6672 4222	29.3650 8752	0.0040 5405
73	8.6520 1778	0.1155 7998	255.0672 5949	29.4806 6750	0.0039 2053
74	8.9115 7832	0.1122 1357	263.7192 7727	29.5928 8106	0.0037 9191
75	9.1789 2567	0.1089 4521	272.6308 5559	29.7018 2628	0.0036 6796
76	9.4542 9344	0.1057 7205	281.8097 8126	29.8075 9833	0.0035 4849
77	9.7379 2224	0.1026 9131	291.2640 7469	29.9102 8964	0.0034 3331
78	10.0300 5991	0.0997 0030	301.0019 9693	30.0099 8994	0.0033 2224
79	10.3309 6171	0.0967 9641	311.0320 5684	30.1067 8635	0.0032 1510
80	10.6408 9056	0.0939 7710	321.3630 1855	30.2007 6345	0.0031 1175
81	10.9601 1727	0.0912 3990	332.0039 0910	30.2920 0335	0.0030 1201
82	11.2889 2079	0.0885 8243	342.9640 2638	30.3805 8577	0.0029 1576
83	11.6275 8842	0.0860 0236	354.2529 4717	30.4665 8813	0.0028 2284
84	11.9764 1607	0.0834 9743	365.8805 3558	30.5500 8556	0.0027 3313
85	12.3357 0855	0.0810 6547	377.8569 5165	30.6311 5103	0.0026 4650
86	12.7057 7981	0.0787 0434	390.1926 6020	30.7098 5537	0.0025 6284
87	13.0869 5320	0.0764 1198	402.8984 4001	30.7862 6735	0.0024 8202
88	13.4795 6180	0.0741 8639	415.9853 9321	30.8604 5374	0.0024 0393
89	13.8839 4865	0.0720 2562	429.4649 5500	30.9324 7936	0.0023 2848
90	14.3004 6711	0.0699 2779	443.3489 0365	31.0024 0714	0.0022 5556
91	14.7294 8112	0.0678 9105	457.6493 7076	31.0702 9820	0.0021 8508
92	15.1713 6556	0.0659 1364	472.3788 5189	31.1362 1184	0.0021 1694
93	15.6265 0652	0.0639 9383	487.5502 1744	31.2002 0567	0.0020 5107
94	16.0953 0172	0.0621 2993	503.1767 2397	21.2623 3560	0.0019 8737
95	16.5781 6077	0.0603 2032	519.2720 2569	31.3226 5592	0.0019 2577
96	17.0755 0559	0.0585 6342	535.8501 8645	31.3812 1934	0.0018 6619
97	17.5877 7076	0.0568 5769	552.9256 9205	31.4380 7703	0.0018 0856
98	18.1154 0388	0.0552 0164	570.5134 6281	31.4932 7867	0.0017 5281
99	18.6588 6600	0.0535 9383	588.6288 6669	31.5468 7250	0.0016 9886
100	19.2186 3198	0.0520 3284	607.2877 3270	31.5989 0534	0.0016 4667

Table 1

617

Compound Amount, Present Value, Annuity

Table 1 (continued)

$i = 3\frac{1}{2}\%$

n	$(1+i)^n$	$(1+i)^{-n}$	$s_{\overline{n}|i}$	$a_{\overline{n}|i}$	$s_{\overline{n}|i}^{-1}$
1	1.0350 0000	0.9661 8357	1.0000 0000	0.9661 8357	1.0000 0000
2	1.0712 2500	0.9335 1070	2.0350 0000	1.8996 9428	0.4914 0049
3	1.1087 1788	0.9019 4271	3.1062 2500	2.8016 3698	0.3219 3418
4	1.1475 2300	0.8714 4223	4.2149 4288	3.6730 7921	0.2372 5114
5	1.1876 8631	0.8419 7317	5.3624 6588	4.5150 5238	0.1864 8137
6	1.2292 5533	0.8135 0064	6.5501 5218	5.3285 5302	0.1526 6821
7	1.2722 7926	0.7859 9096	7.7794 0751	6.1145 4398	0.1285 4449
8	1.3168 0904	0.7594 1156	9.0516 8677	6.8739 5554	0.1104 7665
9	1.3628 9735	0.7337 3097	10.3684 9581	7.6076 8651	0.0964 4601
10	1.4105 9876	0.7089 1881	11.7313 9316	8.3166 0532	0.0852 4137
11	1.4599 6972	0.6849 4571	13.1419 9192	9.0015 5104	0.0760 9197
12	1.5110 6866	0.6617 8330	14.6019 6164	9.6633 3433	0.0684 8395
13	1.5639 5606	0.6394 0415	16.1130 3030	10.3027 3849	0.0620 6157
14	1.6186 9452	0.6177 8179	17.6769 8636	10.9205 2028	0.0565 7073
15	1.6753 4883	0.5968 9062	19.2956 8088	11.5174 1090	0.0518 2507
16	1.7339 8604	0.5767 0591	20.9710 2971	12.0941 1681	0.0476 8483
17	1.7946 7555	0.5572 0378	22.7050 1575	12.6513 2059	0.0440 4313
18	1.8574 8920	0.5383 6114	24.4996 9130	13.1896 8173	0.0408 1684
19	1.9225 0132	0.5201 5569	26.3571 8050	13.7098 3742	0.0379 4033
20	1.9897 8886	0.5025 6588	28.2796 8181	14.2124 0330	0.0353 6108
21	2.0594 3147	0.4855 7090	30.2694 7068	14.6979 7420	0.0330 3659
22	2.1315 1158	0.4691 5063	32.3289 0215	15.1671 2484	0.0309 3207
23	2.2061 1448	0.4532 8563	34.4604 1373	15.6204 1047	0.0290 1880
24	2.2833 2849	0.4379 5713	36.6665 2821	16.0583 6760	0.0272 7283
25	2.3632 4498	0.4231 4699	38.9498 5669	16.4815 1459	0.0256 7404
26	2.4459 5856	0.4088 3767	41.3131 0168	16.8903 5226	0.0242 0540
27	2.5315 6711	0.3950 1224	43.7590 6024	17.2853 6451	0.0228 5241
28	2.6201 7196	0.3816 5434	46.2906 2734	17.6670 1885	0.0216 0265
29	2.7118 7798	0.3687 4815	48.9107 9930	18.0357 6700	0.0204 4538
30	2.8067 9370	0.3562 7841	51.6226 7728	18.3920 4541	0.0193 7133
31	2.9050 3148	0.3442 3035	54.4294 7098	18.7362 7576	0.0183 7240
32	3.0067 0759	0.3325 8971	57.3345 0247	19.0688 6547	0.0174 4150
33	3.1119 4235	0.3213 4271	60.3412 1005	19.3902 0818	0.0165 7242
34	3.2208 6033	0.3104 7605	63.4531 5240	19.7006 8423	0.0157 5966
35	3.3335 9045	0.2999 7686	66.6740 1274	20.0006 6110	0.0149 9835
36	3.4502 6611	0.2898 3272	70.0076 0318	20.2904 9381	0.0142 8416
37	3.5710 2543	0.2800 3161	73.4578 6930	20.5705 2542	0.0136 1325
38	3.6960 1132	0.2705 6194	77.0288 9472	20.8410 8736	0.0129 8214
39	3.8253 7171	0.2614 1250	80.7249 0604	21.1024 9987	0.0123 8775
40	3.9592 5972	0.2525 7247	84.5502 7775	21.3550 7234	0.0118 2728
41	4.0978 3381	0.2440 3137	88.5095 3747	21.5991 0371	0.0112 9822
42	4.2412 5799	0.2357 7910	92.6073 7128	21.8348 8281	0.0107 9828
43	4.3897 0202	0.2278 0590	96.8486 2928	22.0626 8870	0.0103 2539
44	4.5433 4160	0.2201 0231	101.2383 3130	22.2827 9102	0.0098 7768
45	4.7023 5855	0.2126 5924	105.7816 7290	22.4954 5026	0.0094 5343
46	4.8669 4110	0.2054 6787	110.4840 3145	22.7009 1813	0.0090 5108
47	5.0372 8404	0.1985 1968	115.3509 7255	22.8994 3780	0.0086 6919
48	5.2135 8898	0.1918 0645	120.3882 5659	23.0912 4425	0.0083 0646
49	5.3960 6459	0.1853 2024	125.6018 4557	23.2765 6450	0.0079 6167
50	5.5849 2686	0.1790 5337	130.9979 1016	23.4556 1787	0.0076 3371

Table 1 *(continued)*

$$i = 3\tfrac{1}{2}\%$$

n	$(1+i)^n$	$(1+i)^{-n}$	$s_{\overline{n}|i}$	$a_{\overline{n}|i}$	$s_{\overline{n}|i}^{-1}$
51	5.7803 9930	0.1729 9843	136.5828 3702	23.6286 1630	0.0073 2156
52	5.9827 1327	0.1671 4824	142.3632 3631	23.7957 6454	0.0070 2429
53	6.1921 0824	0.1614 9589	148.3459 4958	23.9572 6043	0.0067 4100
54	6.4088 3202	0.1560 3467	154.5380 5782	24.1132 9510	0.0064 7090
55	6.6331 4114	0.1507 5814	160.9468 8984	24.2640 5323	0.0062 1323
56	6.8653 0108	0.1456 6004	167.5800 3099	24.4097 1327	0.0059 6730
57	7.1055 8662	0.1407 3433	174.4453 3207	24.5504 4760	0.0057 3245
58	7.3542 8215	0.1359 7520	181.5509 1869	24.6864 2281	0.0055 0810
59	7.6116 8203	0.1313 7701	188.9052 0085	24.8177 9981	0.0052 9366
60	7.8780 9090	0.1269 3431	196.5168 8288	24.9447 3412	0.0050 8862
61	8.1538 2408	0.1226 4184	204.3949 7378	25.0673 7596	0.0048 9249
62	8.4392 0793	0.1184 9453	212.5487 9786	25.1858 7049	0.0047 0480
63	8.7345 8020	0.1144 8747	220.9880 0579	25.3003 5796	0.0045 2513
64	9.0402 9051	0.1106 1591	229.7225 8599	25.4109 7388	0.0043 5308
65	9.3567 0068	0.1068 7528	238.7628 7650	25.5178 4916	0.0041 8826
66	9.6841 8520	0.1032 6114	248.1195 7718	25.6211 1030	0.0040 3031
67	10.0231 3168	0.0997 6922	257.8037 6238	25.7208 7951	0.0038 7892
68	10.3739 4129	0.0963 9538	267.8268 9406	25.8172 7489	0.0037 3375
69	10.7370 2924	0.0931 3563	278.2008 3535	25.9104 1052	0.0035 9453
70	11.1128 2526	0.0899 8612	288.9378 6459	26.0003 9664	0.0034 6095
71	11.5017 7414	0.0869 4311	300.0506 8985	26.0873 3975	0.0033 3277
72	11.9043 3624	0.0840 0300	311.5524 6400	26.1713 4275	0.0032 0973
73	12.3209 8801	0.0811 6232	323.4568 0024	26.2525 0508	0.0030 9160
64	12.7522 2259	0.0784 1770	335.7777 8824	26.3309 2278	0.0029 7816
75	13.1985 5038	0.0757 6590	348.5300 1083	26.4066 8868	0.0028 6919
76	13.6604 9964	0.0732 0376	361.7285 6121	26.4798 9244	0.0027 6450
77	14.1386 1713	0.0707 2827	375.3890 6085	26.5506 2072	0.0026 6390
78	14.6334 6873	0.0683 3650	389.5276 7798	26.6189 5721	0.0025 6721
79	15.1456 4013	0.0660 2560	404.1611 4671	26.6849 8281	0.0024 7426
80	15.6757 3754	0.0637 9285	419.3067 8685	26.7487 7567	0.0023 8489
81	16.2243 8835	0.0616 3561	434.9825 2439	26.8104 1127	0.0022 9894
82	16.7922 4195	0.0595 5131	451.2069 1274	26.8699 6258	0.0022 1628
83	17.3799 7041	0.0575 3750	467.9991 5469	26.9275 0008	0.0021 3676
84	17.9882 6938	0.0555 9178	485.3791 2510	26.9830 9186	0.0020 6025
85	18.6178 5881	0.0537 1187	503.3673 9448	27.0368 0373	0.0019 8662
86	19.2694 8387	0.0518 9553	521.9852 5329	27.0886 9926	0.0019 1576
87	19.9439 1580	0.0501 4060	541.2547 3715	27.1388 3986	0.0018 4756
88	20.6419 5285	0.0484 4503	561.1986 5295	27.1872 8489	0.0017 8190
89	21.3644 2120	0.0468 0679	581.8406 0581	27.2340 9168	0.0017 1868
90	22.1121 7595	0.0452 2395	603.2050 2701	27.2793 1564	0.0016 5781
91	22.8861 0210	0.0436 9464	625.3172 0295	27.3230 1028	0.0015 9919
92	23.6871 1568	0.0422 1704	648.2033 0506	27.3652 2732	0.0015 4273
93	24.5161 6473	0.0407 8941	671.8904 2073	27.4060 1673	0.0014 8834
94	25.3742 3049	0.0394 1006	696.4065 8546	27.4454 2680	0.0014 3594
95	26.2623 2856	0.0380 7735	721.7808 1595	27.4835 0415	0.0013 8546
96	27.1815 1006	0.0367 8971	748.0431 4451	27.5202 9387	0.0013 3682
97	28.1328 6291	0.0355 4562	775.2246 5457	27.5558 3948	0.0012 8995
98	29.1175 1311	0.0343 4359	803.3575 1748	27.5901 8308	0.0012 4478
99	30.1366 2607	0.0331 8221	832.4750 3059	27.6233 6529	0.0012 0124
100	31.1914 0798	0.0320 6011	862.6116 5666	27.6554 2540	0.0011 5927

Table 1

Compound Amount, Present Value, Annuity

619

Table 1 (*continued*)

$i = 4\%$

n	$(1+i)^n$	$(1+i)^{-n}$	$s_{\overline{n}|i}$	$a_{\overline{n}|i}$	$s_{\overline{n}|i}^{-1}$
1	1.0400 0000	0.9615 3846	1.0000 0000	0.9615 3846	1.0000 0000
2	1.0816 0000	0.9245 5621	2.0400 0000	1.8860 9467	0.4901 9608
3	1.1248 6400	0.8889 9636	3.1216 0000	2.7750 9103	0.3203 4854
4	1.1698 5856	0.8548 0419	4.2464 6400	3.6298 9522	0.2354 9005
5	1.2166 5290	0.8219 2711	5.4163 2256	4.4518 2233	0.1846 2711
6	1.2653 1902	0.7903 1453	6.6329 7546	5.2421 3686	0.1507 6190
7	1.3159 3178	0.7599 1781	7.8982 9448	6.0020 5467	0.1266 0961
8	1.3685 6905	0.7306 9021	9.2142 2626	6.7327 4487	0.1085 2783
9	1.4233 1181	0.7025 8674	10.5827 9531	7.4353 3161	0.0944 9299
10	1.4802 4428	0.6755 6417	12.0061 0712	8.1108 9578	0.0832 9094
11	1.5394 5406	0.6495 8093	13.4863 5141	8.7604 7671	0.0741 4904
12	1.6010 3222	0.6245 9705	15.0258 0546	9.3850 7376	0.0665 5217
13	1.6650 7351	0.6005 7409	16.6268 3768	9.9856 4785	0.0601 4373
14	1.7316 7645	0.5774 7508	18.2919 1119	10.5631 2293	0.0546 6897
15	1.8009 4351	0.5552 6450	20.0235 8764	11.1183 8743	0.0499 4110
16	1.8729 8125	0.5339 0818	21.8245 3114	11.6522 9561	0.0458 2000
17	1.9479 0050	0.5133 7325	23.6975 1239	12.1656 6885	0.0421 9852
18	2.0258 1652	0.4936 2812	25.6454 1288	12.6592 9697	0.0389 9333
19	2.1068 4918	0.4746 4242	27.6712 2940	13.1339 3940	0.0361 3862
20	2.1911 2314	0.4563 8695	29.7780 7858	13.5903 2634	0.0335 8175
21	2.2787 6807	0.4388 3360	31.9692 0172	14.0291 5995	0.0312 8011
22	2.3699 1879	0.4219 5539	34.2479 6979	14.4511 1533	0.0291 9881
23	2.4647 1554	0.4057 2633	36.6178 8858	14.8568 4167	0.0273 0906
24	2.5633 0416	0.3901 2147	39.0826 0412	15.2469 6314	0.0255 8683
25	2.6658 3633	0.3751 1680	41.6459 0829	15.6220 7994	0.0240 1196
26	2.7724 6978	0.3606 8923	44.3117 4462	15.9827 6918	0.0225 6738
27	2.8833 6858	0.3468 1657	47.0842 1440	16.3295 8575	0.0212 3854
28	2.9987 0332	0.3334 7747	49.9675 8298	16.6630 6322	0.0200 1298
29	3.1186 5145	0.3206 5141	52.9662 8630	16.9837 1463	0.0188 7993
30	3.2433 9751	0.3083 1867	56.0849 3775	17.2920 3330	0.0178 3010
31	3.3731 3341	0.2964 6026	59.3283 3526	17.5884 9356	0.0168 5535
32	3.5080 5875	0.2850 5794	62.7014 6867	17.8735 5150	0.0159 4859
33	3.6483 8110	0.2740 9417	66.2095 2742	18.1476 4567	0.0151 0357
34	3.7943 1634	0.2635 5209	69.8579 0851	18.4111 9776	0.0143 1477
35	3.9460 8899	0.2534 1547	73.6522 2486	18.6646 1323	0.0135 7732
36	4.1039 3255	0.2436 6872	77.5983 1385	18.9082 8195	0.0128 8688
37	4.2680 8986	0.2342 9685	81.7022 4640	19.1425 7880	0.0122 3957
38	4.4388 1345	0.2252 8543	85.9703 3626	19.3678 6423	0.0116 3192
39	4.6163 6599	0.2166 2061	90.4091 4971	19.5844 8484	0.0110 6083
40	4.8010 2063	0.2082 8904	95.0255 1570	19.7927 7388	0.0105 2349
41	4.9930 6145	0.2002 7793	99.8265 3633	19.9930 5181	0.0100 1738
42	5.1927 8391	0.1925 7493	104.8195 9778	20.1856 2674	0.0095 4020
43	5.4004 9527	0.1851 6820	110.0123 8169	20.3707 9494	0.0090 8989
44	5.6165 1508	0.1780 4635	115.4128 7696	20.5488 4129	0.0086 6454
45	5.8411 7568	0.1711 9841	121.0293 9204	20.7200 3970	0.0082 6246
46	6.0748 2271	0.1646 1386	126.8705 6772	20.8846 5356	0.0078 8205
47	6.3178 1562	0.1582 8256	132.9453 9043	21.0429 3612	0.0075 2189
48	6.5705 2824	0.1521 9476	139.2632 0604	21.1951 3088	0.0071 8065
49	6.8333 4937	0.1463 4112	145.8337 3429	21.3414 7200	0.0068 5712
50	7.1066 8335	0.1407 1262	152.6670 8366	21.4821 8462	0.0065 5020

Table 1 (*continued*)

$i = 4\%$

n	$(1+i)^n$	$(1+i)^{-n}$	$s_{\overline{n}|i}$	$a_{\overline{n}|i}$	$s_{\overline{n}|i}^{-1}$
51	7.3909 5068	0.1353 0059	159.7737 6700	21.6174 8521	0.0062 5885
52	7.6865 8871	0.1300 9672	167.1647 1768	21.7475 8193	0.0059 8212
53	7.9940 5226	0.1250 9300	174.8513 0639	21.8726 7493	0.0057 1915
54	8.3138 1435	0.1202 8173	182.8453 5865	21.9929 5667	0.0054 6910
55	8.6463 6692	0.1156 5551	191.1591 7299	22.1086 1218	0.0052 3124
56	8.9922 2160	0.1112 0722	199.8055 3991	22.2198 1940	0.0050 0487
57	9.3519 1046	0.1069 3002	208.7977 6151	22.3267 4943	0.0047 8932
58	9.7259 8688	0.1028 1733	218.1496 7197	22.4295 6676	0.0045 8401
59	10.1150 2635	0.0988 6282	227.8756 5885	22.5284 2957	0.0043 8836
60	10.5196 2741	0.0950 6040	237.9906 8520	22.6234 8997	0.0042 0185
61	10.9404 1250	0.0914 0423	248.5103 1261	22.7148 9421	0.0040 2398
62	11.3780 2900	0.0878 8868	259.4507 2511	22.8027 8289	0.0038 5430
63	11.8331 5016	0.0845 0835	270.8287 5412	22.8872 9124	0.0036 9237
64	12.3064 7617	0.0812 5803	282.6619 0428	22.9685 4927	0.0035 3780
65	12.7987 3522	0.0781 3272	294.9683 8045	23.0466 8199	0.0033 9019
66	13.3106 8463	0.0751 2762	307.7671 1567	23.1218 0961	0.0032 4921
67	13.8431 1201	0.0722 3809	321.0778 0030	23.1940 4770	0.0031 1451
68	14.3968 3649	0.0694 5970	334.9209 1231	23.2635 0740	0.0029 8578
69	14.9727 0995	0.0667 8818	349.3177 4880	23.3302 9558	0.0028 6272
70	15.5716 1835	0.0642 1940	364.2904 5876	23.3945 1498	0.0027 4506
71	16.1944 8308	0.0617 4942	379.8620 7711	23.4562 6440	0.0026 3253
72	16.8422 6241	0.0593 7445	396.0565 6019	23.5156 3885	0.0025 2489
73	17.5159 5290	0.0570 9081	412.8988 2260	23.5727 2966	0.0024 2190
74	18.2165 9102	0.0548 9501	430.4147 7550	23.6276 2468	0.0023 2334
75	18.9452 5466	0.0527 8367	448.6313 6652	23.6804 0834	0.0022 2900
76	19.7030 6485	0.0507 5353	467.5766 2118	23.7311 6187	0.0021 2869
77	20.4911 8744	0.0488 0147	487.2796 8603	23.7799 6333	0.0020 5221
78	21.3108 3494	0.0469 2449	507.7708 7347	23.8268 8782	0.0019 6939
79	22.1632 6834	0.0451 1970	529.0817 0841	23.8720 0752	0.0018 9007
80	23.0497 9907	0.0433 8433	551.2449 7675	23.9153 9185	0.0018 1408
81	23.9717 9103	0.0417 1570	574.2947 7582	23.9571 0754	0.0017 4127
82	24.9306 6267	0.0401 1125	598.2665 6685	23.9972 1879	0.0016 7150
83	25.9278 8918	0.0385 6851	623.1972 2952	24.0357 8730	0.0016 0463
84	26.9650 0475	0.0370 8510	649.1251 1870	24.0728 7240	0.0015 4054
85	28.0436 0494	0.0356 5875	676.0901 2345	24.1085 3116	0.0014 7909
86	29.1653 4914	0.0342 8726	704.1337 2839	24.1428 1842	0.0014 2018
87	30.3319 6310	0.0329 6852	733.2990 7753	24.1757 8694	0.0013 6370
88	31.5452 4163	0.0317 0050	763.6310 4063	24.2074 8745	0.0013 0953
89	32.8070 5129	0.0304 8125	795.1762 8225	24.2379 6870	0.0012 5758
90	34.1193 3334	0.0293 0890	827.9833 3354	24.2672 7759	0.0012 0775
91	35.4841 0668	0.0281 8163	862.1026 6688	24.2954 5923	0.0011 5995
92	36.9034 7094	0.0270 9772	897.5867 7356	24.3225 5695	0.0011 1410
93	38.3796 0978	0.0260 5550	934.4902 4450	24.3486 1245	0.0010 7010
94	39.9147 9417	0.0250 5337	972.8698 5428	24.3736 6582	0.0010 2789
95	41.5113 8594	0.0240 8978	1012.7846 4845	24.3977 5559	0.0009 8738
96	43.1718 4138	0.0231 6325	1054.2960 3439	24.4209 1884	0.0009 4850
97	44.8987 1503	0.0222 7235	1097.4678 7577	24.4431 9119	0.0009 1119
98	46.6946 6363	0.0214 1572	1142.3665 9080	24.4646 0692	0.0008 7538
99	48.5624 5018	0.0205 9204	1189.0612 5443	24.4851 9896	0.0008 4100
100	50.5049 4818	0.0198 0004	1237.6237 0461	24.5049 9900	0.0008 0800

Table 1

621

Compound Amount, Present Value, Annuity

Table 1 (continued)

$$i = 4\tfrac{1}{2}\%$$

n	$(1+i)^n$	$(1+i)^{-n}$	$s_{\overline{n}|i}$	$a_{\overline{n}|i}$	$s_{\overline{n}|i}^{-1}$
1	1.0450 0000	0.9569 3780	1.0000 0000	0.9569 3780	1.0000 0000
2	1.0920 2500	0.9157 2995	2.0450 0000	1.8726 6775	0.4889 9756
3	1.1411 6613	0.8762 9660	3.1370 2500	2.7489 6435	0.3187 7336
4	1.1925 1860	0.8385 6134	4.2781 9113	3.5875 2570	0.2337 4365
5	1.2461 8194	0.8024 5105	5.4707 0973	4.3899 7674	0.1827 9164
6	1.3022 6012	0.7678 9574	6.7168 9166	5.1578 7248	0.1488 7839
7	1.3608 6183	0.7348 2846	8.0191 5179	5.8927 0094	0.1247 0147
8	1.4221 0061	0.7031 8513	9.3800 1362	6.5958 8607	0.1066 0965
9	1.4860 9514	0.6729 0443	10.8021 1423	7.2687 9050	0.0925 7447
10	1.5529 6942	0.6439 2768	12.2882 0937	7.9127 1818	0.0813 7882
11	1.6228 5305	0.6161 9874	13.8411 7879	8.5289 1692	0.0722 4818
12	1.6958 8143	0.5896 6386	15.4640 3184	9.1185 8078	0.0646 6619
13	1.7721 9610	0.5642 7164	17.1599 1327	9.6828 5242	0.0582 7535
14	1.8519 4492	0.5399 7286	18.9321 0937	10.2228 2528	0.0528 2032
15	1.9352 8244	0.5167 2044	20.7840 5429	10.7395 4573	0.0481 1381
16	2.0223 7015	0.4944 6932	22.7193 3673	11.2340 1505	0.0440 1537
17	2.1133 7681	0.4731 7639	24.7417 0689	11.7071 9143	0.0404 1758
18	2.2084 7877	0.4528 0037	26.8550 8370	12.1599 9180	0.0372 3690
19	2.3078 6031	0.4333 0179	29.0635 6246	12.5932 9359	0.0344 0734
20	2.4117 1402	0.4146 4286	31.3714 2277	13.0079 3645	0.0318 7614
21	2.5202 4116	0.3967 8743	33.7831 3680	13.4047 2388	0.0296 0057
22	2.6336 5201	0.3797 0089	36.3033 7795	13.7844 2476	0.0275 4565
23	2.7521 6635	0.3633 5013	38.9370 2996	14.1477 7489	0.0256 8249
24	2.8760 1383	0.3477 0347	41.6891 9631	14.4954 7837	0.0239 8703
25	3.0054 3446	0.3327 3060	44.5652 1015	14.8282 0896	0.0224 3903
26	3.1406 7901	0.3184 0248	47.5706 4460	15.1466 1145	0.0210 2137
27	3.2820 0956	0.3046 9137	50.7113 2361	15.4513 0282	0.0197 1946
28	3.4296 9999	0.2915 7069	53.9933 3317	15.7428 7351	0.0185 2081
29	3.5840 3649	0.2790 1502	57.4230 3316	16.0218 8853	0.0174 1461
30	3.7453 1813	0.2670 0002	61.0070 6966	16.2888 8854	0.0163 9154
31	3.9138 5745	0.2555 0241	64.7523 8779	16.5443 9095	0.0154 4345
32	4.0899 8104	0.2444 9991	68.6662 4524	16.7888 9086	0.0145 6320
33	4.2740 3018	0.2339 7121	72.7562 2628	17.0228 6207	0.0137 4453
34	4.4663 6154	0.2238 9589	77.0302 5646	17.2467 5796	0.0129 8191
35	4.6673 4781	0.2142 5444	81.4966 1800	17.4610 1240	0.0122 7045
36	4.8773 7846	0.2050 2817	86.1639 6581	17.6660 4058	0.0116 0578
37	5.0968 6049	0.1961 9921	91.0413 4427	17.8622 3979	0.0109 8402
38	5.3262 1921	0.1877 5044	96.1382 0476	18.0499 9023	0.0104 0169
39	5.5658 9908	0.1796 6549	101.4644 2398	18.2296 5572	0.0098 5567
40	5.8163 6454	0.1719 2870	107.0303 2306	18.4015 8442	0.0093 4315
41	6.0781 0094	0.1645 2507	112.8466 8760	18.5661 0949	0.0088 6158
42	6.3516 1548	0.1574 4026	118.9247 8854	18.7235 4975	0.0084 0868
43	6.6374 3818	0.1506 6054	125.2764 0402	18.8742 1029	0.0079 8235
44	6.9361 2290	0.1441 7276	131.9138 4220	19.0183 8305	0.0075 8071
45	7.2482 4843	0.1379 6437	138.8499 6510	19.1563 4742	0.0072 0202
46	7.5744 1961	0.1320 2332	146.0982 1353	19.2883 7074	0.0068 4471
47	7.9152 6849	0.1263 3810	153.6726 3314	19.4147 0884	0.0065 0734
48	8.2714 5557	0.1208 9771	161.5879 0163	19.5356 0654	0.0061 8858
49	8.6436 7107	0.1156 9158	169.8593 5720	19.6512 9813	0.0058 8722
50	9.0326 3627	0.1107 0965	178.5030 2828	19.7620 0778	0.0056 0215

Table 1 (continued)

$$i = 4\tfrac{1}{2}\%$$

n	$(1+i)^n$	$(1+i)^{-n}$	$s_{\overline{n}|i}$	$a_{\overline{n}|i}$	$s_{\overline{n}|i}^{-1}$
51	9.4391 0490	0.1059 4225	187.5356 6455	19.8679 5003	0.0053 3232
52	9.8638 6463	0.1013 8014	196.9747 6946	19.9693 3017	0.0050 7679
53	10.3077 3853	0.0970 1449	206.8386 3408	20.0663 4466	0.0048 3469
54	10.7715 8677	0.0928 3683	217.1463 7262	20.1591 8149	0.0046 0519
55	11.2563 0817	0.0888 3907	227.9179 5938	20.2480 2057	0.0043 8754
56	11.7628 4204	0.0850 1347	239.1742 6756	20.3330 3404	0.0041 8105
57	12.2921 6993	0.0813 5260	250.9371 0960	20.4143 8664	0.0039 8506
58	12.8453 1758	0.0778 4938	263.2292 7953	20.4922 3602	0.0037 9897
59	13.4233 5687	0.0744 9701	276.0745 9711	20.5667 3303	0.0036 2221
60	14.0274 0793	0.0712 8901	289.4979 5398	20.6380 2204	0.0034 5426
61	14.6586 4129	0.0682 1915	303.5253 6190	20.7062 4118	0.0032 9462
62	15.3182 8014	0.0652 8148	318.1840 0319	20.7715 2266	0.0031 4284
63	16.0076 0275	0.0624 7032	333.5022 8333	20.8339 9298	0.0029 9848
64	16.7279 4487	0.0597 8021	349.5098 8608	20.8937 7319	0.0028 6115
65	17.4807 0239	0.0572 0594	366.2378 3096	20.9509 7913	0.0027 3047
66	18.2673 3400	0.0547 4253	383.7185 3335	21.0057 2165	0.0026 0608
67	19.0893 6403	0.0523 8519	401.9858 6735	21.0581 0684	0.0024 8765
68	19.9483 8541	0.0501 2937	421.0752 3138	21.1082 3621	0.0023 7487
69	20.8460 6276	0.0479 7069	441.0236 1679	21.1562 0690	0.0022 6745
70	21.7841 3558	0.0459 0497	461.8696 7955	21.2021 1187	0.0021 6511
71	22.7644 2168	0.0439 2820	483.6538 1513	21.2460 4007	0.0020 6759
72	23.7888 2066	0.0420 3655	506.4182 3681	21.2880 7662	0.0019 7465
73	24.8593 1759	0.0402 2637	530.2070 5747	21.3283 0298	0.0018 8606
74	25.9779 8688	0.0384 9413	555.0663 7505	21.3667 9711	0.0018 0159
75	27.1469 9629	0.0368 3649	581.0443 6193	21.4036 3360	0.0017 2104
76	28.3686 1112	0.0352 5023	608.1913 5822	21.4388 8383	0.0016 4422
77	29.6451 9862	0.0337 3228	636.5599 6934	21.4726 1611	0.0015 7094
78	30.9792 3256	0.0322 7969	666.2051 6796	21.5048 9579	0.0015 0104
79	32.3732 9802	0.0308 8965	697.1844 0052	21.5357 8545	0.0014 3434
80	33.8300 9643	0.0295 5948	729.5576 9854	21.5653 4493	0.0013 7069
81	35.3524 5077	0.0282 8658	763.3877 9497	21.5936 3151	0.0013 0995
82	36.9433 1106	0.0270 6850	798.7402 4575	21.6207 0001	0.0012 5197
83	38.6057 6006	0.0259 0287	835.6835 5680	21.6466 0288	0.0011 9663
84	40.3430 1926	0.0247 8744	874.2893 1686	21.6713 9032	0.0011 4379
85	42.1584 5513	0.0237 2003	914.6323 3612	21.6951 1035	0.0010 9334
86	44.0555 8561	0.0226 9860	956.7907 9125	21.7178 0895	0.0010 4516
87	46.0380 8696	0.0217 2115	1000.8463 7685	21.7395 3009	0.0009 9915
88	48.1098 0087	0.0207 8579	1046.8844 6381	21.7603 1588	0.0009 5522
89	50.2747 4191	0.0198 9070	1094.9942 6468	21.7802 0658	0.0009 1325
90	52.5371 0530	0.0190 3417	1145.2690 0659	21.7992 4075	0.0008 7316
91	54.9012 7503	0.0182 1451	1197.8061 1189	21.8174 5526	0.0008 3486
92	57.3718 3241	0.0174 3016	1252.7073 8692	21.8348 8542	0.0007 9827
93	59.9535 6487	0.0166 7958	1310.0792 1933	21.8515 6499	0.0007 6331
94	62.6514 7529	0.0159 6132	1370.0327 8420	21.8675 2631	0.0007 2991
95	65.4707 9168	0.0152 7399	1432.6842 5949	21.8828 0030	0.0006 9799
96	68.4169 7730	0.0146 1626	1498.1550 5117	21.8974 1655	0.0006 6749
97	71.4957 4128	0.0139 8685	1566.5720 2847	21.9114 0340	0.0006 3834
98	74.7130 4964	0.0133 8454	1638.0677 6976	21.9247 8794	0.0006 1048
99	78.0751 3687	0.0128 0817	1712.7808 1939	21.9375 9612	0.0005 8385
100	81.5885 1803	0.0122 5663	1790.8559 5627	21.9498 5274	0.0005 5839

Table 1

623

Compound Amount, Present Value, Annuity

Table 1 (*continued*)

$$i = 5\%$$

n	$(1+i)^n$	$(1+i)^{-n}$	$s_{\overline{n}\rceil i}$	$a_{\overline{n}\rceil i}$	$s_{\overline{n}\rceil i}^{-1}$
1	1.0500 0000	0.9523 8095	1.0000 0000	0.9523 8095	1.0000 0000
2	1.1025 0000	0.9070 2948	2.0500 0000	1.8594 1043	0.4878 0488
3	1.1576 2500	0.8638 3760	2.1525 0000	2.7232 4803	0.3172 0856
4	1.2155 0625	0.8227 0247	4.3101 2500	3.5459 5050	0.2320 1183
5	1.2762 8156	0.7835 2617	5.5256 3125	4.3294 7667	0.1809 7480
6	1.3400 9564	0.7462 1540	6.8019 1281	5.0756 9206	0.1470 1747
7	1.4071 0042	0.7106 8133	8.1420 0845	5.7863 7340	0.1228 1982
8	1.4774 5544	0.6768 3936	9.5491 0888	6.4632 1276	0.1047 2181
9	1.5513 2822	0.6446 0892	11.0265 6432	7.1078 2168	0.0906 9008
10	1.6288 9463	0.6139 1325	12.5778 9254	7.7217 3493	0.0795 0458
11	1.7103 3936	0.5846 7929	14.2067 8716	8.3064 1422	0.0703 8889
12	1.7958 5633	0.5568 3742	15.9171 2652	8.8632 5164	0.0628 2541
13	1.8856 4914	0.5303 2135	17.7129 8285	9.3935 7299	0.0564 5577
14	1.9799 3160	0.5050 6795	19.5986 3199	9.8986 4094	0.0510 2397
15	2.0789 2818	0.4810 1710	21.5785 6359	10.3796 5804	0.0463 4229
16	2.1828 7459	0.4581 1152	23.6574 9177	10.8377 6956	0.0422 6991
17	2.2920 1832	0.4362 9669	25.8403 6636	11.2740 6625	0.0386 9914
18	2.4066 1923	0.4155 2065	28.1323 8467	11.6895 8690	0.0355 4622
19	2.5269 5020	0.3957 3396	30.5390 0391	12.0853 2086	0.0327 4501
20	2.6532 9771	0.3768 8948	33.0659 5410	12.4622 1034	0.0302 4259
21	2.7859 6259	0.3589 4236	35.7192 5181	12.8211 5271	0.0279 9611
22	2.9252 6072	0.3418 4987	38.5052 1440	13.1630 0258	0.0259 7051
23	3.0715 2376	0.3255 7131	41.4304 7512	13.4885 7388	0.0241 3682
24	3.2250 9994	0.3100 6791	44.5019 9887	13.7986 4179	0.0224 7090
25	3.3863˙5494	0.2953 0277	47.7270 9882	14.0939 4457	0.0209 5246
26	3.5556 7269	0.2812 4073	51.1134 5376	14.3751 8530	0.0195 6432
27	3.7334 5632	0.2678 4832	54.6691 2645	14.6430 3362	0.0182 9186
28	3.9201 2914	0.2550 9364	58.4025 8277	14.8981 2726	0.0171 2253
29	4.1161 3560	0.2429 4632	62.3227 1191	15.1410 7358	0.0160 4551
30	4.3219 4238	0.2313 7745	66.4388 4750	15.3724 5103	0.0150 5144
31	4.5380 3949	0.2203 5947	70.7607 8988	15.5928 1050	0.0141 3212
32	4.7649 4147	0.2098 6617	75.2988 2937	15.8026 7667	0.0132 8042
33	5.0031 8854	0.1998 7254	80.0637 7084	16.0025 4921	0.0124 9004
34	5.2533 4797	0.1903 5480	85.0669 5938	16.1929 0401	0.0117 5545
35	5.5160 1537	0.1812 9029	90.3203 0735	16.3741 9429	0.0110 7171
36	5.7918 1614	0.1726 5741	95.8363 2272	16.5468 5171	0.0104 3446
37	6.0814 0694	0.1644 3563	101.6281 3886	16.7112 8734	0.0098 3979
38	6.3854 7729	0.1566 0536	107.7095 4580	16.8678 9271	0.0092 8423
39	6.7047 5115	0.1491 4797	114.0950 2309	17.0170 4067	0.0087 6462
40	7.0399 8871	0.1420 4568	120.7997 7424	17.1590 8635	0.0082 7816
41	7.3919 8815	0.1352 8160	127.8397 6295	17.2943 6796	0.0078 2229
42	7.7615 8756	0.1288 3962	135.2317 5110	17.4232 0758	0.0073 9471
43	8.1496 6693	0.1227 0440	142.9933 3866	17.5459 1198	0.0069 9333
44	8.5571 5028	0.1168 6133	151.1430 0559	17.6627 7331	0.0066 1625
45	8.9850 0779	0.1112 9651	159.7001 5587	17.7740 6982	0.0062 6173
46	9.4342 5818	0.1059 9668	168.6851 6366	17.8800 6650	0.0059 2820
47	9.9059 7109	0.1009 4921	178.1194 2185	17.9810 1571	0.0056 1421
48	10.4012 6965	0.0961 4211	188.0253 9294	18.0771 5782	0.0053 1843
49	10.9213 3313	0.0915 6391	198.4266 6259	18.1697 2173	0.0050 3965
50	11.4673 9979	0.0872 0373	209.3479 9572	18.2559 2546	0.0047 7674

Table 1 (*continued*)

$$i = 6\%$$

n	$(1+i)^n$	$(1+i)^{-n}$	$s_{\overline{n}|i}$	$a_{\overline{n}|i}$	$s_{\overline{n}|i}^{-1}$
1	1.0600 0000	0.9433 9623	1.0000 0000	0.9433 9623	1.0000 0000
2	1.1236 0000	0.8899 9644	2.0600 0000	1.8333 9267	0.4854 3689
3	1.1910 1600	0.8396 1928	3.1836 0000	2.6730 1195	0.3141 0981
4	1.2624 7696	0.7920 9366	4.3746 1600	3.4651 0561	0.2285 9149
5	1.3382 2558	0.7472 5817	5.6370 9296	4.2123 6379	0.1773 9640
6	1.4185 1911	0.7049 6054	6.9753 1854	4.9173 2433	0.1433 6263
7	1.5036 3026	0.6650 5711	8.3938 3765	5.5823 8144	0.1191 3502
8	1.5938 4807	0.6274 1237	9.8974 6791	6.2097 9381	0.1010 3594
9	1.6894 7896	0.5918 9846	11.4913 1598	6.8016 9227	0.0870 2224
10	1 7908 4770	0.5583 9478	13.1807 9494	7.3600 8705	0.0758 6796
11	1.8982 9856	0.5267 8753	14.9716 4264	7.8868 7458	0.0667 9294
12	2.0121 9647	0.4969 6936	16.8699 4120	8.3838 4394	0.0592 7703
13	2.1329 2826	0.4688 3902	18.8821 3767	8.8526 8296	0.0529 6011
14	2.2609 0396	0.4423 0096	21.0150 6593	9.2949 8393	0.0475 8491
15	2.3965 5819	0.4172 6506	23.2759 6988	9.7122 4899	0.0429 6276
16	2.5403 5168	0.3936 4628	25.6725 2808	10.1058 9527	0.0389 5214
17	2.6927 7279	0.3713 6442	28.2128 7976	10.4772 5969	0.0354 4480
18	2.8543 3915	0.3503 4379	30.9056 5255	10.8276 0348	0.0323 5654
19	3.0255 9950	0.3305 1301	33.7599 9170	11.1581 1649	0.0296 2086
20	3.2071 3547	0.3118 0473	36.7855 9120	11.4699 2122	0.0271 8456
21	3.3995 6360	0.2941 5540	39.9927 2668	11.7640 7662	0.0250 0455
22	3.6035 3742	0.2775 0510	43.3922 9028	12.0415 8172	0.0230 4557
23	3.8197 4966	0.2617 9726	46.9958 2769	12.3033 7898	0.0212 7848
24	4.0489 3464	0.2469 7855	50.8155 7735	12.5503 5753	0.0196 7900
25	4.2918 7072	0.2329 9863	54.8645 1200	12.7833 5616	0.0182 2672
26	4.5493 8296	0.2198 1003	59.1563 8272	13.0031 6619	0.0169 0435
27	4.8223 4594	0.2073 6795	63.7057 6568	13.2105 3414	0.0156 9717
28	5.1116 8670	0.1956 3014	68.5281 1162	13.4061 6428	0.0145 9255
29	5.4183 8790	0.1845 5674	73.6397 9832	13.5907 2102	0.0135 7961
30	5.7434 9117	0.1741 1013	79.0581 8622	13.7648 3115	0.0126 4891
31	6.0881 0064	0.1642 5484	84.8016 7739	13.9290 8599	0.0117 9222
32	6.4533 8668	0.1549 5740	90.8897 7803	14.0840 4339	0.0110 0234
33	6.8405 8988	0.1461 8622	97.3431 6471	14.2302 2961	0.0102 7293
34	7.2510 2528	0.1379 1153	104.1837 5460	14.3681 4114	0.0095 9843
35	7.6860 8679	0.1301 0522	111.4347 7987	14.4982 4636	0.0089 7386
36	8.1472 5200	0.1227 4077	119.1208 6666	14.6209 8713	0.0083 9483
37	8.6360 8712	0.1157 9318	127.2681 1866	14.7367 8031	0.0078 5743
38	9.1542 5235	0.1092 3885	135.9042 0578	14.8460 1916	0.0073 5812
39	9.7035 0749	0.1030 5552	145.0584 5813	14.9490 7468	0.0068 9377
40	10.2857 1794	0.0972 2219	154.7619 6562	15.0462 9687	0.0064 6154
41	10.9028 6101	0.0917 1905	165.0476 8356	15.1380 1592	0.0060 5886
42	11.5570 3267	0.0865 2740	175.9505 4457	15.2245 4332	0.0056 8342
43	12.2504 5463	0.0816 2962	187.5075 7724	15.3061 7294	0.0053 3312
44	12.9854 8191	0.0770 0908	199.7580 3188	15.3831 8202	0.0050 0606
45	13.7546 1083	0.0726 5007	212.7435 1379	15.4558 3209	0.0047 0050
46	14.5904 8748	0.0685 3781	226.5081 2462	15.5243 6990	0.0044 1485
47	15.4659 1673	0.0646 5831	241.0986 1210	15.5890 2821	0.0041 4768
48	16.3938 7173	0.0609 9840	256.5645 2882	15.6500 2661	0.0038 9766
49	17.3775 0403	0.0575 4566	272.9584 0055	15.7075 7227	0.0036 6356
50	18.4201 5427	0.0542 8836	290.3359 0458	15.7618 6064	0.0034 4429

Table 1

625

Compound Amount, Present Value, Annuity

Table 1 (continued)

$i = 7\%$

n	$(1+i)^n$	$(1+i)^{-n}$	$s_{\overline{n}|i}$	$a_{\overline{n}|i}$	$s_{\overline{n}|i}^{-1}$
1	1.0700 0000	0.9345 7944	1.0000 0000	0.9345 7944	
2	1.1449 0000	0.8734 3873	2.0700 0000	1.8080 1817	1.0000 0000
3	1.2250 4300	0.8162 9788	2.2149 0000	2.6243 1604	0.4830 9179
4	1.3107 9601	0.7628 9521	4.4399 4300	3.3872 1126	0.3110 5166
5	1.4025 5173	0.7129 8618	5.7507 3901	4.1001 9744	0.2252 2812
					0.1738 9069
6	1.5007 3035	0.6663 4222	7.1532 9074	4.7665 3966	
7	1.6057 8148	0.6227 4974	8.6540 2109	5.3892 8940	0.1397 9580
8	1.7181 8618	0.5820 0910	10.2598 0257	5.9712 9851	0.1155 5322
9	1.8384 5921	0.5439 3374	11.9779 8875	6.5152 3225	0.0974 6776
10	1.9671 5136	0.5083 4929	13.8164 4796	7.0235 8154	0.0834 8647
					0.0723 7750
11	2.1048 5195	0.4750 9280	15.7835 9932	7.4986 7434	
12	2.2521 9159	0.4440 1196	17.8884 5127	7.9426 8630	0.0633 5690
13	2.4098 4500	0.4149 6445	20.1406 4286	8.3576 5074	0.0559 0199
14	2.5785 3415	0.3878 1724	22.5504 8786	8.7454 6799	0.0496 5085
15	2.7590 3154	0.3624 4602	25.1290 2201	9.1079 1401	0.0443 4494
					0.0397 9462
16	2.9521 6375	0.3387 3460	27.8880 5355	9.4466 4860	
17	3.1588 1521	0.3165 7439	20.8402 1730	9.7632 2299	0.0358 5765
18	3.3799 3228	0.2958 6392	33.9990 3251	10.0590 8691	0.0324 2519
19	3.6165 2754	0.2765 0832	37.3789 6479	10.3355 9524	0.0294 1260
20	3.8696 8446	0.2584 1900	40.9954 9232	10.5940 1425	0.0267 5301
					0.0243 9293
21	4.1405 6237	0.2415 1309	44.8651 7678	10.8355 2733	
22	4.4304 0174	0.2257 1317	49.0057 3916	11.0612 4050	0.0222 8900
23	4.7405 2986	0.2109 4688	53.4361 4090	11.2721 8738	0.0204 0577
24	5.0723 6695	0.1971 4662	58.1766 7076	11.4693 3400	0.0187 1393
25	5.4274 3264	0.1842 4918	63.2490 3772	11.6535 8318	0.0171 8902
					0.0158 1052
26	5.8073 5292	0.1721 9549	68.6764 7036	11.8257 7867	
27	6.2138 6763	0.1609 3037	74.4838 2328	11.9867 0904	0.0145 6103
28	6.6488 3836	0.1504 0221	80.6976 9091	12.1371 1125	0.0134 2573
29	7.1142 5705	0.1405 6282	87.3465 2927	12.2776 7407	0.0123 9193
30	7.6122 5504	0.1313 6712	94.4607 8632	12.4090 4118	0.0114 4865
					0.0105 8640
31	8.1451 1290	0.1227 7301	102.0730 4137	12.5318 1419	
32	8.7152 7080	0.1147 4113	110.2181 5426	12.6465 5532	0.0097 9691
33	9.3253 3975	0.1072 3470	118.9334 2506	12.7537 9002	0.0090 7292
34	9.9781 1354	0.1002 1934	128.2587 6481	12.8540 0936	0.0084 0807
35	10.6765 8148	0.0936 6294	138.2368 7835	12.9476 7230	0.0077 9674
					0.0072 3396
36	11.4239 4219	0.0875 3546	148.9134 5984	13.0352 0776	
37	12.2236 1814	0.0818 0884	160.3374 0202	13.1170 1660	0.0067 1531
38	13.0792 7141	0.0764 5686	172.5610 2017	13.1934 7345	0.0062 3685
39	13.9948 2041	0.0714 5501	185.6402 9158	13.2649 2846	0.0057 9505
40	14.9744 5784	0.0667 8038	199.6351 1199	13.3317 0884	0.0053 8676
					0.0050 0914
41	16.0226 6989	0.0624 1157	214.6095 6983	13.3941 2041	
42	17.1442 5678	0.0583 2857	230.6322 3972	13.4524 4898	0.0046 5962
43	18.3443 5475	0.0545 1268	247.7764 9650	13.5069 6167	0.0043 3591
44	19.6284 5959	0.0509 4643	266.1208 5125	13.5579 0810	0.0040 3590
45	21.0024 5176	0.0476 1349	285.7493 1084	13.6055 2159	0.0037 5769
					0.0034 9957
46	22.4726 2338	0.0444 9859	306.7517 6260	13.6500 2018	
47	24.0457 0702	0.0415 8747	329.2243 8598	13.6916 0764	0.0032 5996
48	25.7289 0651	0.0388 6679	353.2700 9300	13.7304 7443	0.0030 3744
49	27.5299 2997	0.0363 2410	378.9989 9951	13.7667 9853	0.0028 3070
50	29.4570 2506	0.0339 4776	406.5289 2947	13.8007 4629	0.0026 3853
					0.0024 5985

Table 1 (*continued*)

$i = 8\%$

n	$(1+i)^n$	$(1+i)^{-n}$	$s_{\overline{n}|i}$	$a_{\overline{n}|i}$	$s_{\overline{n}|i}^{-1}$
1	1.0800 0000	0.9259 2593	1.0000 0000	0.9259 2593	1.0000 0000
2	1.1664 0000	0.8573 3882	2.0800 0000	1.7832 6475	0.4807 6923
3	1.2597 1200	0.7938 3224	3.2464 0000	2.5770 9699	0.3080 3351
4	1.3604 8896	0.7350 2985	4.5061 1200	3.3121 2684	0.2219 2080
5	1.4693 2808	0.6805 8320	5.8666 0096	3.9927 1004	0.1704 5645
6	1.5868 7432	0.6301 6963	7.3359 2904	4.6228 7966	0.1363 1539
7	1.7138 2427	0.5834 9040	8.9228 0336	5.2063 7006	0.1120 7240
8	1.8509 3021	0.5402 6888	10.6366 2763	5.7466 3894	0.0940 1476
9	1.9990 0463	0.5002 4897	12.4875 5784	6.2468 8791	0.0800 7971
10	2.1589 2500	0.4631 9349	14.4865 6247	6.7100 8140	0.0690 2949
11	2.3316 3900	0.4288 8286	16.6454 8746	7.1389 6426	0.0600 7634
12	2.5181 7012	0.3971 1376	18.9771 2646	7.5360 7802	0.0526 9502
13	2.7196 2373	0.3676 9792	21.4952 9658	7.9037 7594	0.0465 2181
14	2.9371 9362	0.3404 6104	24.2149 2030	8.2442 3698	0.0412 9685
15	3.1721 6911	0.3152 4170	27.1521 1393	8.5594 7869	0.0368 2954
16	3.4259 4264	0.2918 9047	30.3242 8304	8.8513 6916	0.0329 7687
17	3.7000 1805	0.2702 6895	33.7502 2569	9.1216 3811	0.0296 2943
18	3.9960 1950	0.2502 4903	37.4502 4374	9.3718 8714	0.0267 0210
19	4.3157 0106	0.2317 1206	41.4462 6324	9.6035 9920	0.0241 2763
20	4.6609 5714	0.2145 4821	45.7619 6430	9.8181 4741	0.0218 5221
21	5.0338 3372	0.1986 5575	50.4229 2144	10.0168 0316	0.0198 3225
22	5.4365 4041	0.1839 4051	55.4567 5516	10.2007 4366	0.0180 3207
23	5.8714 6365	0.1703 1528	60.8932 9557	10.3710 5895	0.0164 2217
24	6.3411 8074	0.1576 9934	66.7647 5922	10.5287 5828	0.0149 7796
25	6.8484 7520	0.1460 1790	73.1059 3995	10.6747 7619	0.0136 7878
26	7.3963 5321	0.1352 0176	79.9544 1515	10.8099 7795	0.0125 0713
27	7.9880 6147	0.1251 8682	87.3507 6836	10.9351 6477	0.0114 4809
28	8.6271 0639	0.1159 1372	95.3388 2983	11.0510 7849	0.0104 8891
29	9.3172 7490	0.1073 2752	103.9659 3622	11.1584 0601	0.0096 1854
30	10.0626 5689	0.0993 7733	113.2832 1111	11.2577 8334	0.0088 2743
31	10.8676 6944	0.0920 1605	123.3458 6800	11.3497 9939	0.0081 0728
32	11.7370 8300	0.0852 0005	134.2135 3744	11.4349 9944	0.0074 5081
33	12.6760 4964	0.0788 8893	145.9506 2044	11.5138 8837	0.0068 5163
34	13.6901 3361	0.0730 4531	158.6266 7007	11.5869 3367	0.0063 0411
35	14.7853 4429	0.0676 3454	172.3168 0368	11.6545 6822	0.0058 0326
36	15.9681 7184	0.0626 2458	187.1021 4797	11.7171 9279	0.0053 4467
37	17.2456 2558	0.0579 8572	203.0703 1981	11.7751 7851	0.0049 2440
38	18.6252 7563	0.0536 9048	220.3159 4540	11.8288 6899	0.0045 3894
39	20.1152 9768	0.0497 1341	238.9412 2103	11.8785 8240	0.0041 8513
40	21.7245 2150	0.0460 3093	259.0565 1871	11.9246 1333	0.0038 6016
41	23.4624 8322	0.0426 2123	280.7810 4021	11.9672 3457	0.0035 6149
42	25.3394 8187	0.0394 6411	304.2435 2342	12.0066 9867	0.0032 8684
43	27.3666 4042	0.0365 4084	329.5830 0530	12.0432 3951	0.0030 3414
44	29.5559 7166	0.0338 3411	356.9496 4572	12.0770 7362	0.0028 0152
45	31.9204 4939	0.0313 2788	386.5056 1738	12.1084 0150	0.0025 8728
46	34.4740 8534	0.0290 0730	418.4260 6677	12.1374 0880	0.0023 8991
47	37.2320 1217	0.0268 5861	452.9001 5211	12.1642 6741	0.0022 0799
48	40.2105 7314	0.0248 6908	490.1321 6428	12.1891 3649	0.0020 4027
49	32.4274 1899	0.0230 2693	530.3427 3742	12.2121 6341	0.0018 8557
50	46.9016 1251	0.0213 2123	573.7701 5642	12.2334 8464	0.0017 4286

Table 2

Binomial Probabilities

627

Table 2 Binomial Probabilities

							p					
n	x	0.05	0.1	0.2	0.3	0.4	0.5	0.6	0.7	0.8	0.9	0.95
2	0	0.902	0.810	0.640	0.490	0.360	0.250	0.160	0.090	0.040	0.010	0.002
	1	0.095	0.180	0.320	0.420	0.480	0.500	0.480	0.420	0.320	0.180	0.095
	2	0.002	0.010	0.040	0.090	0.160	0.250	0.360	0.490	0.640	0.810	0.902
3	0	0.857	0.729	0.512	0.343	0.216	0.125	0.064	0.027	0.008	0.001	
	1	0.135	0.243	0.384	0.441	0.432	0.375	0.288	0.189	0.096	0.027	0.007
	2	0.007	0.027	0.096	0.189	0.288	0.375	0.432	0.441	0.384	0.243	0.135
	3		0.001	0.008	0.027	0.064	0.125	0.216	0.343	0.512	0.729	0.857
4	0	0.815	0.656	0.410	0.240	0.130	0.062	0.026	0.008	0.002		
	1	0.171	0.292	0.410	0.412	0.346	0.250	0.154	0.076	0.026	0.004	
	2	0.014	0.049	0.154	0.265	0.346	0.375	0.346	0.265	0.154	0.049	0.014
	3		0.004	0.026	0.076	0.154	0.250	0.346	0.412	0.410	0.292	0.171
	4			0.002	0.008	0.026	0.062	0.130	0.240	0.410	0.656	0.815
5	0	0.774	0.590	0.328	0.168	0.078	0.031	0.010	0.002			
	1	0.204	0.328	0.410	0.360	0.259	0.156	0.077	0.028	0.006		
	2	0.021	0.073	0.205	0.309	0.346	0.312	0.230	0.132	0.051	0.008	0.001
	3	0.001	0.008	0.051	0.132	0.230	0.312	0.346	0.309	0.205	0.073	0.021
	4			0.006	0.028	0.077	0.156	0.259	0.360	0.410	0.328	0.204
	5				0.002	0.010	0.031	0.078	0.168	0.328	0.590	0.774
6	0	0.735	0.531	0.262	0.118	0.047	0.016	0.004	0.001			
	1	0.232	0.354	0.393	0.303	0.187	0.094	0.037	0.010	0.002		
	2	0.031	0.098	0.246	0.324	0.311	0.234	0.138	0.060	0.015	0.001	
	3	0.002	0.015	0.082	0.185	0.276	0.312	0.276	0.185	0.082	0.015	0.002
	4		0.001	0.015	0.060	0.138	0.234	0.311	0.324	0.246	0.098	0.031
	5			0.002	0.010	0.037	0.094	0.187	0.303	0.393	0.354	0.232
	6				0.001	0.004	0.016	0.047	0.118	0.262	0.531	0.735
7	0	0.698	0.478	0.210	0.082	0.028	0.008	0.002				
	1	0.257	0.372	0.367	0.247	0.131	0.055	0.017	0.004			
	2	0.041	0.124	0.275	0.318	0.261	0.164	0.077	0.025	0.004		
	3	0.004	0.023	0.115	0.227	0.290	0.273	0.194	0.097	0.029	0.003	
	4		0.003	0.029	0.097	0.194	0.273	0.290	0.227	0.115	0.023	0.004
	5			0.004	0.025	0.077	0.164	0.261	0.318	0.275	0.124	0.041
	6				0.004	0.017	0.055	0.131	0.247	0.367	0.372	0.257
	7					0.002	0.008	0.028	0.082	0.210	0.478	0.698
8	0	0.663	0.430	0.168	0.058	0.017	0.004	0.001				
	1	0.279	0.383	0.336	0.198	0.090	0.031	0.008	0.001			
	2	0.051	0.149	0.294	0.296	0.209	0.109	0.041	0.010	0.001		
	3	0.005	0.033	0.147	0.254	0.279	0.219	0.124	0.047	0.009		
	4		0.005	0.046	0.136	0.232	0.273	0.232	0.136	0.046	0.005	
	5			0.009	0.047	0.124	0.219	0.279	0.254	0.147	0.033	0.005
	6			0.001	0.010	0.041	0.109	0.209	0.296	0.294	0.149	0.051
	7				0.001	0.008	0.031	0.090	0.198	0.336	0.383	0.279
	8					0.001	0.004	0.017	0.058	0.168	0.430	0.663

*Table A-2 from ELEMENTARY STATISTICS IN A WORLD OF APPLICATIONS, 1st Ed. by Ramakant Khazanie. Copyright 1979. Reprinted by permission of Scott, Foresman and Company.

Table 2 (*continued*)

								p					
n	x	0.05	0.1	0.2	0.3	0.4	0.5	0.6	0.7	0.8	0.9	0.95
9	0	0.630	0.387	0.134	0.040	0.010	0.002					
	1	0.299	0.387	0.302	0.156	0.060	0.018	0.004				
	2	0.063	0.172	0.302	0.267	0.161	0.070	0.021	0.004			
	3	0.008	0.045	0.176	0.267	0.251	0.164	0.074	0.021	0.003		
	4	0.001	0.007	0.066	0.172	0.251	0.246	0.167	0.074	0.017	0.001	
	5		0.001	0.017	0.074	0.167	0.246	0.251	0.172	0.066	0.007	0.001
	6			0.003	0.021	0.074	0.164	0.251	0.267	0.176	0.045	0.008
	7				0.004	0.021	0.070	0.161	0.267	0.302	0.172	0.063
	8					0.004	0.018	0.060	0.156	0.302	0.387	0.299
	9						0.002	0.010	0.040	0.134	0.387	0.630
10	0	0.599	0.349	0.107	0.028	0.006	0.001					
	1	0.315	0.387	0.268	0.121	0.040	0.010	0.002				
	2	0.075	0.194	0.302	0.233	0.121	0.044	0.011	0.001			
	3	0.010	0.057	0.201	0.267	0.215	0.117	0.042	0.009	0.001		
	4	0.001	0.011	0.088	0.200	0.251	0.205	0.111	0.037	0.006		
	5		0.001	0.026	0.103	0.201	0.246	0.201	0.103	0.026	0.001	
	6			0.006	0.037	0.111	0.205	0.251	0.200	0.088	0.011	0.001
	7			0.001	0.009	0.042	0.117	0.215	0.267	0.201	0.057	0.010
	8				0.001	0.011	0.044	0.121	0.233	0.302	0.194	0.075
	9					0.002	0.010	0.040	0.121	0.268	0.387	0.315
	10						0.001	0.006	0.028	0.107	0.349	0.599
11	0	0.569	0.314	0.086	0.020	0.004						
	1	0.329	0.384	0.236	0.093	0.027	0.005	0.001				
	2	0.087	0.213	0.295	0.200	0.089	0.027	0.005	0.001			
	3	0.014	0.071	0.221	0.257	0.177	0.081	0.023	0.004			
	4	0.001	0.016	0.111	0.220	0.236	0.161	0.070	0.017	0.002		
	5		0.002	0.039	0.132	0.221	0.226	0.147	0.057	0.010		
	6			0.010	0.057	0.147	0.226	0.221	0.132	0.039	0.002	
	7			0.002	0.017	0.070	0.161	0.236	0.220	0.111	0.016	0.001
	8				0.004	0.023	0.081	0.177	0.257	0.221	0.071	0.014
	9				0.001	0.005	0.027	0.089	0.200	0.295	0.213	0.087
	10					0.001	0.005	0.027	0.093	0.236	0.384	0.329
	11							0.004	0.020	0.086	0.314	0.569
12	0	0.540	0.282	0.069	0.014	0.002						
	1	0.341	0.377	0.206	0.071	0.017	0.003					
	2	0.099	0.230	0.283	0.168	0.064	0.016	0.002				
	3	0.017	0.085	0.236	0.240	0.142	0.054	0.012	0.001			
	4	0.002	0.021	0.133	0.231	0.213	0.121	0.042	0.008	0.001		
	5		0.004	0.053	0.158	0.227	0.193	0.101	0.029	0.003		
	6			0.016	0.079	0.177	0.226	0.177	0.079	0.016		
	7				0.003	0.029	0.101	0.193	0.227	0.158	0.053	0.004
	8			0.001	0.008	0.042	0.121	0.213	0.231	0.133	0.021	0.002
	9				0.001	0.012	0.054	0.142	0.240	0.236	0.085	0.017
	10					0.002	0.016	0.064	0.168	0.283	0.230	0.099
	11						0.003	0.017	0.071	0.206	0.377	0.341
	12							0.002	0.014	0.069	0.282	0.540

Table 2

629

Binomial Probabilities

Table 2 (*continued*)

n	x	0.05	0.1	0.2	0.3	0.4	0.5	0.6	0.7	0.8	0.9	0.95
13	0	0.513	0.254	0.055	0.010	0.001						
	1	0.351	0.367	0.179	0.054	0.011	0.002					
	2	0.111	0.245	0.268	0.139	0.045	0.010	0.001				
	3	0.021	0.100	0.246	0.218	0.111	0.035	0.006	0.001			
	4	0.003	0.028	0.154	0.234	0.184	0.087	0.024	0.003			
	5		0.006	0.069	0.180	0.221	0.157	0.066	0.014	0.001		
	6		0.001	0.023	0.103	0.197	0.209	0.131	0.044	0.006		
	7			0.006	0.044	0.131	0.209	0.197	0.103	0.023	0.001	
	8			0.001	0.014	0.066	0.157	0.221	0.180	0.069	0.006	
	9				0.003	0.024	0.087	0.184	0.234	0.154	0.028	0.003
	10				0.001	0.006	0.035	0.111	0.218	0.246	0.100	0.021
	11					0.001	0.010	0.045	0.139	0.268	0.245	0.111
	12						0.002	0.011	0.054	0.179	0.367	0.351
	13							0.001	0.010	0.055	0.254	0.513
14	0	0.488	0.229	0.044	0.007	0.001						
	1	0.359	0.356	0.154	0.041	0.007	0.001					
	2	0.123	0.257	0.250	0.113	0.032	0.006	0.001				
	3	0.026	0.114	0.250	0.194	0.085	0.022	0.003				
	4	0.004	0.035	0.172	0.229	0.155	0.061	0.014	0.001			
	5		0.008	0.086	0.196	0.207	0.122	0.041	0.007			
	6		0.001	0.032	0.126	0.207	0.183	0.092	0.023	0.002		
	7			0.009	0.062	0.157	0.209	0.157	0.062	0.009		
	8			0.002	0.023	0.092	0.183	0.207	0.126	0.032	0.001	
	9				0.007	0.041	0.122	0.207	0.196	0.086	0.008	
	10				0.001	0.014	0.061	0.155	0.229	0.172	0.035	0.004
	11					0.003	0.022	0.085	0.194	0.250	0.114	0.026
	12					0.001	0.006	0.032	0.113	0.250	0.257	0.123
	13						0.001	0.007	0.041	0.154	0.356	0.359
	14							0.001	0.007	0.044	0.229	0.488
15	0	0.463	0.206	0.035	0.005							
	1	0.366	0.343	0.132	0.031	0.005						
	2	0.135	0.267	0.231	0.092	0.022	0.003					
	3	0.031	0.129	0.250	0.170	0.063	0.014	0.002				
	4	0.005	0.043	0.188	0219	0.127	0.042	0.007	0.001			
	5	0.001	0.010	0.103	0.206	0.186	0.092	0.024	0.003			
	6		0.002	0.043	0.147	0.207	0.153	0.061	0.012	0.001		
	7			0.014	0.081	0.177	0.196	0.118	0.035	0.003		
	8			0.003	0.035	0.118	0.196	0.177	0.081	0.014		
	9			0.001	0.012	0.061	0.153	0.207	0.147	0.043	0.002	
	10				0.003	0.024	0.092	0.186	0.206	0.103	0.010	0.001
	11				0.001	0.007	0.042	0.127	0.219	0.188	0.043	0.005
	12					0.002	0.014	0.063	0.170	0.250	0.129	0.031
	13						0.003	0.022	0.092	0.231	0.267	0.135
	14							0.005	0.031	0.132	0.343	0.366
	15								0.005	0.035	0.206	0.463

Table 3   The Standard Normal Distribution

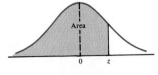

$$F_z(z) = P[Z \le z]$$

z	0.00	0.01	0.02	0.03	0.04	0.05	0.06	0.07	0.08	0.09
−3.4	0.0003	0.0003	0.0003	0.0003	0.0003	0.0003	0.0003	0.0003	0.0003	0.0002
−3.3	0.0005	0.0005	0.0005	0.0004	0.0004	0.0004	0.0004	0.0004	0.0004	0.0003
−3.2	0.0007	0.0007	0.0006	0.0006	0.0006	0.0006	0.0006	0.0005	0.0005	0.0005
−3.1	0.0010	0.0009	0.0009	0.0009	0.0008	0.0008	0.0008	0.0008	0.0007	0.0007
−3.0	0.0013	0.0013	0.0013	0.0012	0.0012	0.0011	0.0011	0.0011	0.0010	0.0010
−2.9	0.0019	0.0018	0.0017	0.0017	0.0016	0.0016	0.0015	0.0015	0.0014	0.0014
−2.8	0.0026	0.0025	0.0024	0.0023	0.0023	0.0022	0.0021	0.0021	0.0020	0.0019
−2.7	0.0035	0.0034	0.0033	0.0032	0.0031	0.0030	0.0029	0.0028	0.0027	0.0026
−2.6	0.0047	0.0045	0.0044	0.0043	0.0041	0.0040	0.0039	0.0038	0.0037	0.0036
−2.5	0.0062	0.0060	0.0059	0.0057	0.0055	0.0054	0.0052	0.0051	0.0049	0.0048
−2.4	0.0082	0.0080	0.0078	0.0075	0.0073	0.0071	0.0069	0.0068	0.0066	0.0064
−2.3	0.0107	0.0104	0.0102	0.0099	0.0096	0.0094	0.0091	0.0089	0.0087	0.0084
−2.2	0.0139	0.0136	0.0132	0.0129	0.0125	0.0122	0.0119	0.0116	0.0113	0.0110
−2.1	0.0179	0.0174	0.0170	0.0166	0.0162	0.0158	0.0154	0.0150	0.0146	0.0143
−2.0	0.0228	0.0222	0.0217	0.0212	0.0207	0.0202	0.0197	0.0192	0.0188	0.0183
−1.9	0.0287	0.0281	0.0274	0.0268	0.0262	0.0256	0.0250	0.0244	0.0239	0.0233
−1.8	0.0359	0.0352	0.0344	0.0336	0.0329	0.0322	0.0314	0.0307	0.0301	0.0294
−1.7	0.0446	0.0436	0.0427	0.0418	0.0409	0.0401	0.0392	0.0384	0.0375	0.0367
−1.6	0.0548	0.0537	0.0526	0.0516	0.0505	0.0495	0.0485	0.0475	0.0465	0.0455
−1.5	0.0668	0.0655	0.0643	0.0630	0.0618	0.0606	0.0594	0.0582	0.0571	0.0559
−1.4	0.0808	0.0793	0.0778	0.0764	0.0749	0.0735	0.0722	0.0708	0.0694	0.0681
−1.3	0.0968	0.0951	0.0934	0.0918	0.0901	0.0885	0.0869	0.0853	0.0838	0.0823
−1.2	0.1151	0.1131	0.1112	0.1093	0.1075	0.1056	0.1038	0.1020	0.1003	0.0985
−1.1	0.1357	0.1335	0.1314	0.1292	0.1271	0.1251	0.1230	0.1210	0.1190	0.1170
−1.0	0.1587	0.1562	0.1539	0.1515	0.1492	0.1469	0.1446	0.1423	0.1401	0.1379
−0.9	0.1841	0.1814	0.1788	0.1762	0.1736	0.1711	0.1685	0.1660	0.1635	0.1611
−0.8	0.2119	0.2090	0.2061	0.2033	0.2005	0.1977	0.1949	0.1922	0.1894	0.1867
−0.7	0.2420	0.2389	0.2358	0.2327	0.2296	0.2266	0.2236	0.2206	0.2177	0.2148
−0.6	0.2743	0.2709	0.2676	0.2643	0.2611	0.2578	0.2546	0.2514	0.2483	0.2451
−0.5	0.3085	0.3050	0.3015	0.2981	0.2946	0.2912	0.2877	0.2843	0.2810	0.2776

Table 3
The Standard Normal Distribution

631

Table 3 (*continued*)

$$F_z(z) = P[Z \le z]$$

z	0.00	0.01	0.02	0.03	0.04	0.05	0.06	0.07	0.08	0.09
−0.4	0.3446	0.3409	0.3372	0.3336	0.3300	0.3264	0.3228	0.3192	0.3156	0.3121
−0.3	0.3821	0.3783	0.3745	0.3707	0.3669	0.3632	0.3594	0.3557	0.3520	0.3483
−0.2	0.4207	0.4168	0.4129	0.4090	0.4052	0.4013	0.3974	0.3936	0.3897	0.3859
−0.1	0.4602	0.4562	0.4522	0.4483	0.4443	0.4404	0.4364	0.4325	0.4286	0.4247
−0.0	0.5000	0.4960	0.4920	0.4880	0.4840	0.4801	0.4761	0.4721	0.4681	0.4641
0.0	0.5000	0.5040	0.5080	0.5120	0.5160	0.5199	0.5239	0.5279	0.5319	0.5359
0.1	0.5398	0.5438	0.5478	0.5517	0.5557	0.5596	0.5636	0.5675	0.5714	0.5753
0.2	0.5793	0.5832	0.5871	0.5910	0.5948	0.5987	0.6026	0.6064	0.6103	0.6141
0.3	0.6179	0.6217	0.6255	0.6293	0.6331	0.6368	0.6406	0.6443	0.6480	0.6517
0.4	0.6554	0.6591	0.6628	0.6664	0.6700	0.6736	0.6772	0.6808	0.6844	0.6879
0.5	0.6915	0.6950	0.6985	0.7019	0.7054	0.7088	0.7123	0.7157	0.7190	0.7224
0.6	0.7257	0.7291	0.7324	0.7357	0.7389	0.7422	0.7454	0.7486	0.7517	0.7549
0.7	0.7580	0.7611	0.7642	0.7673	0.7704	0.7734	0.7764	0.7794	0.7823	0.7852
0.8	0.7881	0.7910	0.7939	0.7967	0.7995	0.8023	0.8051	0.8078	0.8106	0.8133
0.9	0.8159	0.8186	0.8212	0.8238	0.8264	0.8289	0.8315	0.8340	0.8365	0.8389
1.0	0.8413	0.8438	0.8461	0.8485	0.8508	0.8531	0.8554	0.8577	0.8599	0.8621
1.1	0.8643	0.8665	0.8686	0.8708	0.8729	0.8749	0.8770	0.8790	0.8810	0.8830
1.2	0.8849	0.8869	0.8888	0.8907	0.8925	0.8944	0.8962	0.8980	0.8997	0.9015
1.3	0.9032	0.9049	0.9066	0.9082	0.9099	0.9115	0.9131	0.9147	0.9162	0.9177
1.4	0.9192	0.9207	0.9222	0.9236	0.9251	0.9265	0.9278	0.9292	0.9306	0.9319
1.5	0.9332	0.9345	0.9357	0.9370	0.9382	0.9394	0.9406	0.9418	0.9429	0.9441
1.6	0.9452	0.9463	0.9474	0.9484	0.9495	0.9505	0.9515	0.9525	0.9535	0.9545
1.7	0.9554	0.9564	0.9573	0.9582	0.9591	0.9599	0.9608	0.9616	0.9625	0.9633
1.8	0.9641	0.9649	0.9656	0.9664	0.9671	0.9678	0.9686	0.9693	0.9699	0.9706
1.9	0.9713	0.9719	0.9726	0.9732	0.9738	0.9744	0.9750	0.9756	0.9761	0.9767
2.0	0.9772	0.9778	0.9783	0.9788	0.9793	0.9798	0.9803	0.9808	0.9812	0.9817
2.1	0.9821	0.9826	0.9830	0.9834	0.9838	0.9842	0.9846	0.9850	0.9854	0.9857
2.2	0.9861	0.9864	0.9868	0.9871	0.9875	0.9878	0.9881	0.9884	0.9887	0.9890
2.3	0.9893	0.9896	0.9898	0.9901	0.9904	0.9906	0.9909	0.9911	0.9913	0.9916
2.4	0.9918	0.9920	0.9922	0.9925	0.9927	0.9929	0.9931	0.9932	0.9934	0.9936
2.5	0.9938	0.9940	0.9941	0.9943	0.9945	0.9946	0.9948	0.9949	0.9951	0.9952
2.6	0.9953	0.9955	0.9956	0.9957	0.9959	0.9960	0.9961	0.9962	0.9963	0.9964
2.7	0.9965	0.9966	0.9967	0.9968	0.9969	0.9970	0.9971	0.9972	0.9973	0.9974
2.8	0.9974	0.9975	0.9976	0.9977	0.9977	0.9978	0.9979	0.9979	0.9980	0.9981
2.9	0.9981	0.9982	0.9982	0.9983	0.9984	0.9984	0.9985	0.9985	0.9986	0.9986
3.0	0.9987	0.9987	0.9987	0.9988	0.9988	0.9989	0.9989	0.9989	0.9990	0.9990
3.1	0.9990	0.9991	0.9991	0.9991	0.9992	0.9992	0.9992	0.9992	0.9993	0.9993
3.2	0.9993	0.9993	0.9994	0.9994	0.9994	0.9994	0.9994	0.9995	0.9995	0.9995
3.3	0.9995	0.9995	0.9995	0.9996	0.9996	0.9996	0.9996	0.9996	0.9996	0.9997
3.4	0.9997	0.9997	0.9997	0.9997	0.9997	0.9997	0.9997	0.9997	0.9997	0.9998

Answers to

Odd-Numbered

Exercises

Exercises 1.1
page 9

1. (3,3), I   *3.* (2, −2), IV   *5.* (−4, −6), III   *7.* A   *9.* E, F, and G
11. F
13.

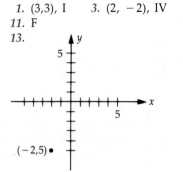

15.

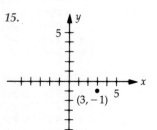

17.

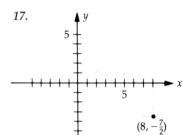

$(8, -\frac{7}{2})$

19.

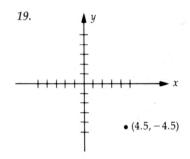

• $(4.5, -4.5)$

21. 5   **23.** $\sqrt{61}$ **25.** $(-8, -6)$ and $(8, -6)$ **29.** 3400 miles   **31.** Route 1

Exercises 1.2
page 25

1. $\frac{1}{2}$ **3.** not defined   **5.** 5   **7.** $\frac{5}{6}$ **9.** $\dfrac{d - b}{c - a}$ $(a \neq c)$ **11.** *a.* 4 *b.* -8

13. parallel   **15.** parallel   **17.** not perpendicular   **19.** yes   **21.** $y = -3$

23. *a.* $2x - y - 10 = 0$ *b.* $x + y - 6 = 0$ *c.* $y - 2 = 0$

25. *a.* $y = 3x + 4$ *b.* $y = -2x - 1$ *c.* $y = 5$

27. $x - 2y + 6 = 0$

29. *a.* $y = -6$ *b.* $3x - 2y = 0$ *c.* $y = b$ **31.** $k = 8$

33. *a.* $4x + 3y - 12 = 0$ *b.* $2x + y + 4 = 0$
    c. $6x - 4y + 3 = 0$.   *d.* $x - 8y - 4 = 0$

35. *a.*

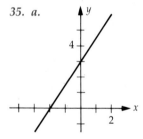

b.

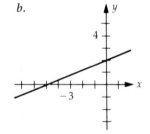

c.

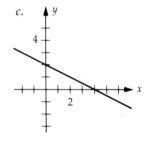

d.

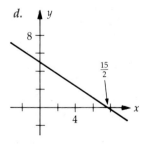

e.

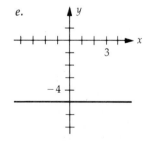

f.

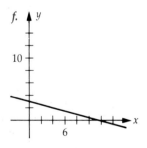

37. *a.* $y = 0.0715x$ *b.* \$0.0715   *c.* \$2502.50

39. a. 88.8 tons   *b.*

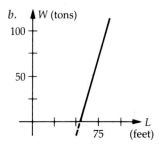

41. a. and *b.*

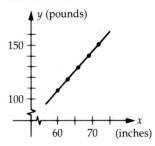

c. $11x - 3y - 336 = 0$
d. 126 lbs

Exercises 1.3
page 42

1. $80; $580   *3.* $836   *5.* 108 days
7. 24% per year   *9.* $172.22 per month   *11.* 10% per year
13. $9850   *15.* $800,000; $700,000
17. a. $6 billion   *b.* $43.5 billion   *c.* $81 billion
19. a. $y = 1.053x$ *b.* $652.86
21. a. $1200/year   *c.*
     b. $V = 60,000 - 12,000n$
     d. $24,000

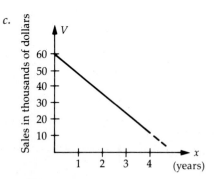

23. $C(x) = 0.6x + 12,100; R(x) = 1.15x; P(x) = 0.55x - 12,100$
25. a.   *b.* $16   *c.* 20,000 units

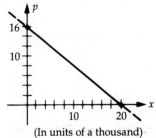

(In units of a thousand)

27. *a.* 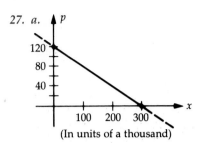   *b.* $120   *c.* 300,000 units

29. $p = -(1/20)x + 100$

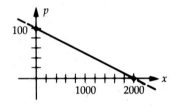

31. $50 per unit

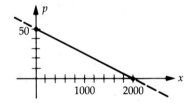

33. *a.* 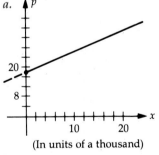   *b.* $18 per unit

35. *a.*   *b.* $20 per unit

37. $p = .015\,x + 300$; 10,000 refrigerators; $300 per unit

Exercises 1.4
page 54

1. (2, 10)

3. (4, 2/3)

5. (−4, −6)

7. 1000 units; $15,000

9. 600 units; $240

11. *a.*

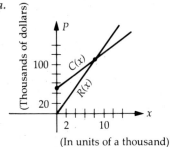

b. 8000 units; $112,000

c.

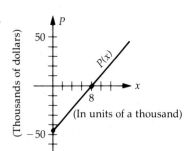

d. (8000, 0)

13. 9259 units; $83,331

15. *a.* $C_1(x) = 18,000 + 15x$
$C_2(x) = 15,000 + 20x$

b.

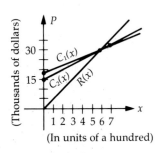

c. machine II; machine II; machine I

d. ($1500); $1500; $4750

17. 7000 units; $6

19. 10,000 units; $3

Exercises 1.5
page 63

1. *a.* $y = 2.3x + 1.5$
b.

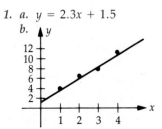

3. *a.* $y = -.77x + 5.73$
b.

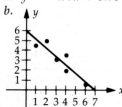

5. *a.* $y = 1.2x + 2$
b.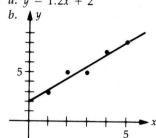

7. *a.* $y = .34x - .9$
b.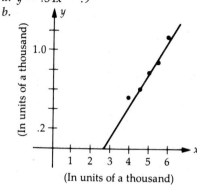
(In units of a thousand)

c. 1276 applications

9. *a.* $y = -2.8x + 440$ *b.*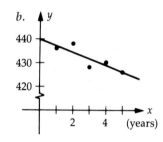

c. 420.4

11. *a.* $y = 49.5x + 158.5$ *b.* \$505,000
13. *a.* $y = 8.83x + 110.93$ *b.* 225.72
15. *a.* $y = 3.05x + 20.06$ *b.* \$56,610

Chapter 1
Review Exercises
page 67

1. $3x + 4y - 18 = 0$
5. 60,000
11. *a.* $C(x) = 6x + 30,000$
    c. $P(x) = 4x - 30,000$
13. $36p - x - 1600 = 0$; \$44.44

3. $3x + 2y + 14 = 0$
9. \$19,550
    b. $R(x) = 10x$
    d. (\$6000); \$2000; \$18,000
15. 2500 units; \$50,000

► CHAPTER

TWO

Exercises 2.1
page 79

1. Unique solution; $(2, 1)$

3. Infinitely many solutions; $\left(t, \frac{2}{5}t - 2 \right)$, t a parameter

5. No solution

7. No solution

9. $\begin{aligned} x + \quad y &= 2000 \\ .06x + .08y &= 144 \end{aligned}$

11. $\begin{aligned} .06x + .08y + .12z &= 21{,}600 \\ z &= 2x \\ .12z &= .08y \end{aligned}$

13. $\begin{aligned} 8000x + 12{,}000y + 16{,}000z &= 1{,}000{,}000 \\ x \qquad\qquad &= 2y \\ x + \quad y + \quad z &= 100 \end{aligned}$

Exercises 2.2
page 94

1. Yes

3. No

5. Yes

7. No

9. Yes

11. $\begin{bmatrix} 1 & 3 & 1 & | & 3 \\ 3 & 8 & 3 & | & 7 \\ 2 & -3 & 1 & | & -10 \end{bmatrix} \xrightarrow[R_3-2R_1]{R_2-3R_1} \begin{bmatrix} 1 & 3 & 1 & | & 3 \\ 0 & -1 & 0 & | & -2 \\ 0 & -9 & -1 & | & -16 \end{bmatrix} \xrightarrow{-R_2} \begin{bmatrix} 1 & 3 & 1 & | & 3 \\ 0 & 1 & 0 & | & 2 \\ 0 & -9 & -1 & | & -16 \end{bmatrix}$

$\xrightarrow[R_3+9R_2]{R_1-3R_2} \begin{bmatrix} 1 & 0 & 1 & | & -3 \\ 0 & 1 & 0 & | & 2 \\ 0 & 0 & -1 & | & 2 \end{bmatrix} \xrightarrow[-R_3]{R_1+R_3} \begin{bmatrix} 1 & 0 & 0 & | & -1 \\ 0 & 1 & 0 & | & 2 \\ 0 & 0 & 1 & | & -2 \end{bmatrix}$

13. $\begin{aligned} x + 3y + \quad z &= \quad 3 \\ 3x + 8y + 3z &= \quad 7; \quad x = -1, \ y = 2, \ z = -2 \\ 2x - 3y + \quad z &= -10 \end{aligned}$

15. a. Yes   b. $(3, -1, 2)$

17. a. Yes   b. $(1, 2, t)$, t, a parameter

19. a. Yes   b. $(2, 4)$ 21. a. Yes   b. $(3, -2, 1, 2)$

23. a. Yes   b. $(3, -1 - t, t, 2)$, t a parameter

25. a. Yes   b. $(2, -1, 2 - t, t)$, t a parameter

27. a. Yes   b. $(2, s, t, u)$, s, t, and u parameters

29. $\left(\dfrac{22}{5}, -\dfrac{9}{5}\right)$ 31. $(2, 1)$ 33. $\left(\dfrac{7}{9}, -\dfrac{1}{9}, -\dfrac{2}{3}\right)$

35. $(19, -7, -15)$ 37. $(2, 0, -3)$

39. No solution   41. $\left(1 - \dfrac{1}{4}s + \dfrac{1}{4}t, s, t\right)$, s, t, parameters

43. $(1, 0, -2, -1)$ 45. $(-256, -33, -12, 167)$

47. $\left(\dfrac{3}{2}t, 1 - \dfrac{1}{2}t, 1 + \dfrac{3}{2}t, t\right)$, t a parameter

49. No solution

51. 96 one-bedroom, 64 two-bedroom, and 32 three-bedroom units

53. $66,667 in high risk, none in medium risk, and $133,333 in low risk

55. 80 doz. sleeveless, 140 doz. short-sleeve, and 160 doz. long-sleeve

57. a. $\begin{aligned} x_1 \qquad\qquad\qquad\qquad + x_6 \qquad &= 1700 \\ x_1 - x_2 \qquad\qquad\qquad + x_7 &= \quad 700 \\ x_2 - x_3 \qquad\qquad\qquad\quad &= \quad 300 \\ - x_3 + x_4 \qquad\qquad\quad &= \quad 400 \\ - x_4 + x_5 \qquad + x_7 &= \quad 700 \\ x_5 + x_6 \qquad\quad &= 1800 \end{aligned}$

b. $\begin{aligned} x_1 &= 1700 - s \\ x_2 &= 1000 - s + t \\ x_3 &= \quad 700 - s + t \\ x_4 &= 1100 - s + t \\ x_5 &= 1800 - s \end{aligned}$

$$x_6 = s$$
$$x_7 = t$$
(900, 1000, 700, 1100, 1000, 800, 800)
(1000, 1100, 800, 1200, 1100, 700, 800)
c. x_6 must be at least 300 cars/hour

Exercises 2.3
page 107

1. $4 \times 4; 4 \times 3; 1 \times 5$

3. 2; 3; 8

5. $\begin{bmatrix} 1 & 6 \\ 6 & -1 \\ 2 & 2 \end{bmatrix}$

7. $\begin{bmatrix} 1 & 1 & -4 \\ -1 & -8 & 1 \\ 6 & 3 & 1 \end{bmatrix}$

9. $\begin{bmatrix} 3 & -4 & -16 \\ 17 & -4 & 16 \end{bmatrix}$

11. $\begin{bmatrix} \frac{7}{2} & 3 & -1 & \frac{10}{3} \\ -\frac{19}{6} & \frac{2}{3} & -\frac{17}{2} & \frac{23}{3} \\ \frac{29}{3} & \frac{17}{6} & -1 & -2 \end{bmatrix}$

13. $x = 6, y = 2, u = 1, z = 0$

15. $x = 20, y = 4, z = -2, u = 2$

19. $\begin{bmatrix} 3 \\ 2 \\ -1 \\ 5 \end{bmatrix}$

21. $\begin{bmatrix} 1 & 3 & 0 \\ -1 & 4 & 1 \\ 2 & 2 & 0 \end{bmatrix}$

23. $\begin{bmatrix} 220 & 215 & 210 & 205 \\ 220 & 210 & 200 & 195 \\ 215 & 205 & 195 & 190 \end{bmatrix}$

25. a. $D = \begin{bmatrix} 2960 & 1510 & 1150 \\ 1100 & 550 & 490 \\ 1230 & 590 & 470 \end{bmatrix}$

b. $E = \begin{bmatrix} 3256 & 1661 & 1265 \\ 1210 & 605 & 539 \\ 1353 & 649 & 517 \end{bmatrix}$

Exercises 2.4
page 118

1. 2×5; not defined

3. $1 \times 1; 7 \times 7$

5. $n = s$ and $m = t$

7. $\begin{bmatrix} -1 \\ 3 \end{bmatrix}$

9. $\begin{bmatrix} 0.57 & 1.93 \\ 0.64 & 1.76 \end{bmatrix}$

11. $\begin{bmatrix} 2 & 9 \\ 5 & 16 \end{bmatrix}$

13. $\begin{bmatrix} 6 & -3 & 0 \\ -2 & 1 & -8 \\ 4 & -4 & 9 \end{bmatrix}$

15. $\begin{bmatrix} -4 & -20 & 4 \\ 4 & 12 & 0 \\ 12 & 32 & 20 \end{bmatrix}$

17. $\begin{bmatrix} 4 & 2 & -10 \\ 6 & 16 & 2 \\ 10 & 26 & 6 \end{bmatrix}$

19. $AB = \begin{bmatrix} 10 & 7 \\ 22 & 15 \end{bmatrix}, \quad BA = \begin{bmatrix} 5 & 8 \\ 13 & 20 \end{bmatrix}$

23. $A = \begin{bmatrix} -2 & -1 \\ 5 & 2 \end{bmatrix}$

25. $A^T = \begin{bmatrix} 2 & 5 \\ 4 & -6 \end{bmatrix}$

27. $AX = B$ where $A = \begin{bmatrix} 2 & 0 \\ 3 & -2 \end{bmatrix}, \quad X = \begin{bmatrix} x \\ y \end{bmatrix}, \quad$ and $\quad B = \begin{bmatrix} 7 \\ 12 \end{bmatrix}$

29. $AX = B$ where $A = \begin{bmatrix} 1 & -2 & 3 \\ 3 & 4 & -2 \\ 2 & -3 & 7 \end{bmatrix}, \quad X = \begin{bmatrix} x \\ y \\ z \end{bmatrix}, \quad$ and $\quad B = \begin{bmatrix} -1 \\ 1 \\ 6 \end{bmatrix}$

31. $\begin{bmatrix} 1575 & 1590 & 1560 & 975 \\ 410 & 405 & 415 & 270 \\ 215 & 205 & 225 & 155 \end{bmatrix}$

33. *a.*
$$A^2 = \begin{bmatrix} 2 & 1 & 1 & 2 & 1 \\ 1 & 3 & 2 & 1 & 2 \\ 1 & 2 & 4 & 2 & 1 \\ 2 & 1 & 2 & 3 & 1 \\ 1 & 2 & 1 & 1 & 2 \end{bmatrix}$$

b. As an example, $a_{11} = 2$, and this gives the 2 one-stop routes from city 1 (Singapore) to city 1; namely, Singapore—HK—Singapore and Singapore—Taipei—Singapore.

Exercises 2.5
page 131

5. $\begin{bmatrix} \frac{1}{7} & -\frac{2}{7} \\ \frac{1}{7} & \frac{3}{14} \end{bmatrix}$

7. $\begin{bmatrix} 2 & -11 & -3 \\ 1 & -6 & -2 \\ 0 & -1 & 0 \end{bmatrix}$

9. does not exist

11. $\begin{bmatrix} -\frac{13}{10} & \frac{7}{5} & \frac{1}{2} \\ \frac{2}{5} & -\frac{1}{5} & 0 \\ -\frac{7}{10} & \frac{3}{5} & \frac{1}{2} \end{bmatrix}$

13. $\begin{bmatrix} 3 & 4 & -6 & 1 \\ -2 & -3 & 5 & -1 \\ -4 & -4 & 7 & -1 \\ -4 & -5 & 8 & -1 \end{bmatrix}$

15. *a.* $x = 24/5$, $y = 23/5$ *b.* $x = 2/5$, $y = 9/5$
17. *a.* $x = -1$, $y = 3$, $z = 2$ *b.* $x = 1$, $y = 8$, $z = -12$
19. *a.* $x = -2/17$, $y = -10/17$, $z = -60/17$ *b.* $x = 1$, $y = 0$, $z = -5$
21. *a.* $x_1 = 1$, $x_2 = -4$, $x_3 = 5$, $x_4 = -1$
    b. $x_1 = 12$, $x_2 = -24$, $x_3 = 21$, $x_4 = -7$
23. *a.*
$$A^{-1} = \begin{bmatrix} -\frac{5}{2} & -\frac{3}{2} \\ 2 & 1 \end{bmatrix}$$

25. *a.*
$$ABC = \begin{bmatrix} 4 & 10 \\ 2 & 3 \end{bmatrix}, \quad A^{-1} = \begin{bmatrix} 3 & -5 \\ 1 & -2 \end{bmatrix}, \quad B^{-1} = \begin{bmatrix} 1 & -3 \\ -1 & 4 \end{bmatrix}, \quad C^{-1} = \begin{bmatrix} \frac{1}{8} & -\frac{3}{8} \\ \frac{1}{4} & \frac{1}{4} \end{bmatrix}$$

27. 600 adults and 400 children on Saturday;
    400 adults and 400 children on Sunday

29.

	Food A	Food B	Food C
Susan	$7\frac{1}{2}$	$3\frac{1}{3}$	$4\frac{7}{12}$
Tom	$7\frac{1}{2}$	$3\frac{1}{3}$	$2\frac{1}{12}$

Exercises 2.6
page 143

1. *a.* $10 million *b.* $160 million
   c. Agricultural; manufacturing and transportation
3. $x = 23.75$ and $y = 21.25$ 5. $x = 42.86$ and $y = 57.14$
9. $45 million and $75 million
11. $34.4 million, $33 million, and $21.6 million

Chapter 2
Review Exercises
page 146

1. $\begin{bmatrix} 2 & 2 \\ -1 & 4 \\ 3 & 3 \end{bmatrix}$

3. $\begin{bmatrix} -6 & -2 \end{bmatrix}$

5. $\begin{bmatrix} 8 & 9 & 11 \\ -10 & -1 & 3 \\ 11 & 12 & 10 \end{bmatrix}$

7. $\begin{bmatrix} 6 & 18 & 6 \\ -12 & 6 & 18 \\ 24 & 0 & 12 \end{bmatrix}$
9. $\begin{bmatrix} -11 & -16 & -15 \\ -4 & -2 & -10 \\ -6 & 14 & 2 \end{bmatrix}$
11. $\begin{bmatrix} -3 & 17 & 8 \\ -2 & 56 & 27 \\ 74 & 78 & 116 \end{bmatrix}$

13. \$2,300,000; \$2,450,000
15. $x = 1, y = -1$

17. $x = 2, y = 2t - 5, z = t; t$ a parameter
19. $x = 1, y = -1, z = 2, w = 2$

21. $\begin{bmatrix} \frac{3}{4} & -\frac{1}{2} \\ -\frac{1}{8} & \frac{1}{4} \end{bmatrix}$
23. $\begin{bmatrix} 0 & -\frac{1}{5} & \frac{2}{5} \\ -2 & 1 & 1 \\ -1 & \frac{1}{5} & \frac{3}{5} \end{bmatrix}$

25. $A^{-1} = \begin{bmatrix} 0 & \frac{1}{7} & \frac{2}{7} \\ -1 & -\frac{4}{7} & \frac{6}{7} \\ -\frac{1}{2} & -\frac{1}{2} & \frac{1}{2} \end{bmatrix}; x = 3, y = -1, z = 2$

27. 30 of each type

► **CHAPTER**

THREE

Exercises 3.1
page 157

1.
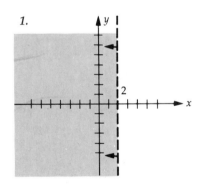

3.
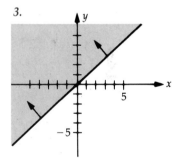

5.
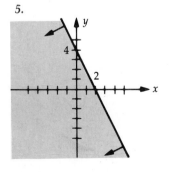

7.
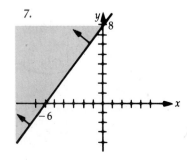

9.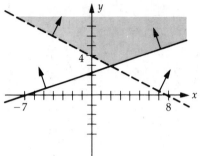

11.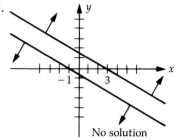

No solution

13.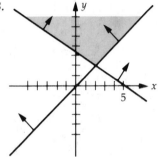

15.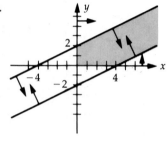

17.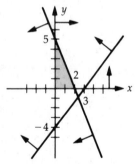

19.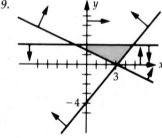

21.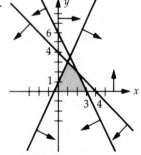

23.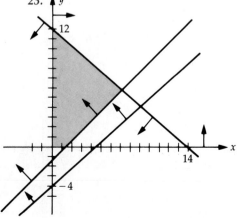

25.

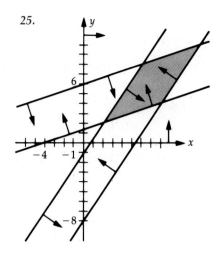

Exercises 3.2
page 166

1. Maximize $P = 3x + 4y$ subject to
$$6x + 9y \leq 300$$
$$5x + 4y \leq 180$$
$$x \geq 0, y \geq 0$$

3. Maximize $P = 2x + (3/2)y$ subject to
$$3x + 4y \leq 1000$$
$$6x + 3y \leq 1200$$
$$x \geq 150, y \geq 0$$

5. Minimize $C = 2x + 5y$ subject to
$$30x + 25y \geq 400$$
$$x + 0.5y \geq 10$$
$$2x + 5y \geq 40$$
$$x \geq 0, y \geq 0$$

7. Minimize $C = 32{,}000 - x - 3y$ subject to
$$x + y \leq 6000$$
$$x + y \geq 2000$$
$$x \leq 3000$$
$$y \leq 4000$$
$$x \geq 0, y \geq 0$$

9. Maximize $P = 200{,}000x + 100{,}000y + 600{,}000z$ subject to
$$3000x + 1000y + 12{,}000z \leq 102{,}000$$
$$x + y + z \leq 25$$
$$z \leq 6$$
$$x \geq 0, y \geq 0, z \geq 0$$

11. Minimize $C = 16x_1 + 20x_2 + 22x_3 + 18x_4 + 16x_5 + 14x_6$ subject to
$$x_1 + x_2 + x_3 \leq 800$$
$$x_4 + x_5 + x_6 \leq 600$$
$$x_1 + x_4 \geq 500$$
$$x_2 + x_5 \geq 400$$
$$x_3 + x_6 \geq 400$$
$$x_1 \geq 0, x_2 \geq 0, \ldots, x_6 \geq 0$$

13. Maximize $P = 3400x + 4000y + 5000z$ subject to
$$6x + 8y + 10z \leq 8{,}200$$
$$24x + 22y + 20z \leq 21{,}800$$
$$18x + 21y + 30z \leq 23{,}700$$
$$x \geq 0, y \geq 0, z \geq 0$$

Exercises 3.3
page 178

1. $x = 0, y = 6, P = 18$ **3.** $x = 14, y = 3, C = 58$
5. Any point (x, y) lying on the line segment joining $(5, 20)$ to $(12, 6)$, $C = 90$
7. $x = 3, y = 3, C = 75$ **9.** $x = 15, y = 17.5, P = 115$
11. $x = 10, y = 38, P = 134$ **13.** max: $x = 6, y = 33/2, P = 258$
       min: $x = 15, y = 3, P = 186$

15. max: $x = 5, y = 15, P = 70$ **17.** 120 model A, 160 model B, $P = \$480$
       min: $x = 0, y = 5, P = 20$
19. $16 million in homeowner loans, $4 million in auto loans; $P = \$2.08$ million
21. 65 acres of crop A, 80 acres of crop B, $P = \$25,750$
23. 2000 at location I, 4000 at location II; $C = \$18,000$
25. 3 oz. of brand A, 4 oz. of brand B; $C = 25¢$

Chapter 3
Review Exercises
page 184

1. $x = 0, y = 4, P = 20$ **3.** $x = 3, y = 10, P = 29$
5. $x = 20, y = 0, C = 40$ **7.** max: $x = 22, y = 0, Q = 22$
       min: $x = 3, y = 5/2, Q = 11/2$

9. $40,000 in each company; $P = \$13,600$

▶ **CHAPTER**

FOUR

Exercises 4.1
page 204

1. In final form; $x = \frac{30}{7}, y = \frac{20}{7}, u = 0, v = 0, P = \frac{220}{7}$

3. Not in final form; pivot element is $\frac{1}{2}$, lying in the first row, second column

5. In final form; $x = \frac{1}{3}, y = 0, z = \frac{13}{3}, u = 0, v = 6, w = 0, P = 17$

7. $x = 30, y = 50, u = 0, v = 0, P = 300$
9. $x = 6, y = 12, u = 0, v = 0, P = 78$
11. $x = 0, y = 10, u = 2, v = 20, P = 120$
13. $x = 15, y = 0, z = 0, u = 0, v = 5, P = 45$
    or $x = 0, y = 15, z = 0, u = 0, v = 5, P = 45$
15. $x = 5, y = 4, z = 0, u = 0, v = 0, w = 0, P = 13$
17. $x = 0, y = 60, z = 20, u = 40, v = 0, w = 0, P = 200$
19. $x = 0, y = 10, z = 0, t = 0, u = 46, v = 86, w = 22, P = 60$
21. 20 units of each product; maximum profit is $140 per shift
23. 65 acres of crop A, 80 acres of crop B; profit is $25,750
25. 22 minutes of morning and 3 minutes of evening advertising time; maximum exposure is 6,200,000 viewers
27. 300 standard units, 300 deluxe units, and 400 luxury units; maximum profit of $4,220,000
29. Does not exist; (Has an unbounded solution.)

Exercises 4.2
page 220

1. $x = 4, y = 0, C = -8$ **3.** $x = 4, y = 3, C = -18$
5. $x = 0, y = 13, z = 18, w = 14, C = -111$

7. Maximize $P = 4u + 6v$ subject to
$u + 3v \leq 2$
$2u + 2v \leq 5;$ $x = 4, y = 0, C = 8$
$u \geq 0, v \geq 0$

9. Maximize $P = 60u + 40v + 30w$ subject to
$6u + 2v + w \leq 6$
$u + v + w \leq 4;$ $x = 10, y = 20, C = 140$
$u \geq 0, v \geq 0, w \geq 0$

11. Maximize $P = 10u + 20v$ subject to
$20u + v \leq 200$
$10u + v \leq 150;$ $x = 0, y = 0, z = 10, C = 1200$
$u + 2v \leq 120$
$u \geq 0, v \geq 0$

13. Maximize $P = 10u + 24v + 16w$ subject to
$u + 2v + w \leq 6$
$2u + v + w \leq 8;$ $x = 8, y = 0, z = 8, C = 80$
$2u + v + w \leq 4$
$u \geq 0, v \geq 0, w \geq 0$

15. Maximize $P = 6u + 2v + 4w$ subject to
$2u + 6v \leq 30$
$4u + 6w \leq 12;$ $x = \dfrac{1}{3}, y = \dfrac{4}{3}, z = 0, C = 26$
$3u + v + 2w \leq 20$
$u \geq 0, v \geq 0, w \geq 0$

17. Loc. I: 500 to Warehouse A, 200 to Warehouse B
Loc. II: 200 to Warehouse B, 400 to Warehouse C
Shipping costs: $20,000

19. 8 oz. orange juice; 6 oz. pink grapefruit juice; 178 calories

Chapter 4 Review Exercises page 225

1. $x = 3, y = 4, u = 0, v = 0, P = 25$
3. $x = 56/5, y = 2/5, z = 0, u = 0, v = 0, P = 23\frac{3}{5}$
5. $x = 3/4, y = 0, z = 7/4, C = 60$
7. 30 units product B, $P = \$180$

▶ **CHAPTER**

FIVE

Exercises 5.1 page 238

1. $1718.19	3. $4974.48	5. $27,566.88	7. $261,751.50
9. $214,983	11. $10\frac{1}{4}\%$	13. 8.3%	15. $29,227.60

17. $30,255.96   19. 13.9 years

21. (a) $37,518.25   (b) $12,518.25

23. $670.84   25. 116%   27. $3.795 million   29. 11.72%

31. $26,267.58   33. $7167.89   35. $15,000

37. *a.* 9%   *b.* 9.202%   *c.* 9.308%   *d.* 9.381%
39. 6.59%

Exercises 5.2
page 249

1. $15,937.40   3. $54,759.60   5. $37,965.60   7. $28,733.19
9. $15,558.62   11. $15,011.28   13. $109,658.81   15. $453.05
17. $44,526.45   19. $9,850.12   21. $608.54
23. between $106,623 and $127,029
25. between $97,372 and $115,465

Exercises 5.3
page 257

1. $14,902.96   3. $444.24   5. $622.13   7. $731.79
9. $1491.19   11. $516.76   13. $172.95   15. $16,274.53
17. *a.* $295 *b.* $1580; $527
19. *a.* $193.60; $152.18 *b.* $969.60; $1304.64
21. $658.31; $17,495.74; $20,212.67; $34,115.46
23. $3135.47   25. $14,497.38
27. $2,090.41; $4,280.21   29. $60,710.01

Exercises 5.4
page 268

1. 30   3. -4.5 5. $-3, 8, 19, 30, 41$
7. $x + 6y$ 9. 795   11. 792   13. 550
15. *a.* 275   *b.* -280 17. 76 days   19. $7.90
21. *b.* $800   23. G.P.; 256; 508   25. Not a G.P.
27. G.P.; 1/3; $364\frac{1}{3}$ 29. 3; 0   31. 293,866   33. $33,667.38
35. annual raise of 8% per year   37. *a.* $20,113.60 *b.* $87,537.33
39. $25,165.80   41. $39,321.60; $110,678.40

Chapter 5
Review Exercises
page 271

1. *a.* $7320.50   *b.* $7387.30 $7422.55   *d.* $7446.75
3. $40,000   5. $5557.65   7. $7,861.70
9. $318.93   11. 7.179%   13. $218.64
15. *a.* $1011.10   *b.* $207,330   *c.* $35,753.43
17. $205.10

▶ **CHAPTER**

SIX

Exercises 6.1
page 287

1. a, b, and d
3. *a.* $\{x \mid x$ is a gold medalist in the 1976 Olympic Games$\}$
   b. $\{x \mid x$ is an integer greater than 2 but less than 8$\}$
   c. $\{x \mid x^2 - 4x + 2 = 0; x$ is a real number$\}$
5. *a.* $\{2, 3, 4, 5, 6\}$ *b.* $\{A, H, I, M, O, P, S, T, U\}$
7. *a.* $\{-2, 3\}$ *b.* $\{-1\}$

9. *a.* T   *b.* F   *c.* F
11. *a.* F   *b.* F
13. *a.* T   *b.* F   *c.* F   *d.* F   *e.* T   *f.* F
15. a, b, c
17. *a.* φ   *b.* φ, {1}   *c.* φ, {1}, {2}, {1, 2}
    d. φ, {1}, {2}, {3}, {1, 2}, {1, 3}, {2, 3}, {1, 2, 3}
    e. φ, {1}, {2}, {3}, {4}, {1, 2}, {1, 3}, {1, 4}, {2, 3}, {2, 4}, {3, 4}, {1, 2, 3}, {1, 2, 4}, {1, 3, 4}, {2, 3, 4}, {1, 2, 3, 4}
19. φ, {IBM}, {U.S. Steel}, {Union Carbide}, {Boeing}, {IBM, U.S. Steel}, {IBM, Union Carbide}, {IBM, Boeing}, {U.S. Steel, Union Carbide}, {U.S. Steel, Boeing}, {Union Carbide, Boeing}, {IBM, U.S. Steel, Union Carbide}, {IBM, U.S. Steel, Boeing}, {IBM, Union Carbide, Boeing}, {U.S. Steel, Union Carbide, Boeing}, {IBM, U.S. Steel, Union Carbide, Boeing} All except the last subset listed above are proper subsets.
21. $A \subset C$
23. *a.*

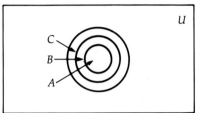

b.

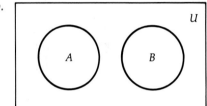

c.

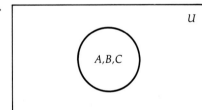

25. *a.*

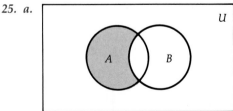

b.

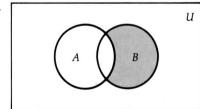

27. *a.*

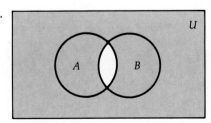

b.

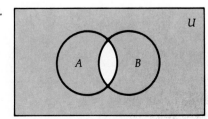

29. a.

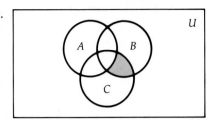

b.

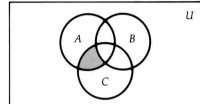

c.

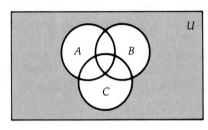

31. a. {2, 4, 6, 8, 10}   *b.* {1, 2, 4, 5, 6, 8, 9, 10}   *c.* U

33. a. C   *b.* ϕ *c.* U

35. a. No   *b.* Yes   *c.* Yes   *d.* No   *e.* No

37. a. The set of all employees at the Universal Life Insurance Company who drink both tea and coffee

  b. The set of all employees at the Universal Life Insurance Company who drink tea but not coffee

  c. The set of all employees at the Universal Life Insurance Company who drink coffee but not tea

39. a. $A \cap B \cap C$ *b.* $(A \cap B \cap C^c) \cup (A \cap B^c \cap C) \cup (A^c \cap B \cap C)$

  c. $(A \cap B) \cup (A \cap C) \cup (B \cap C)$ *d.* $A \cap B^c \cap C$

  e. $(A \cap B \cap C)^c$

41. a. The set of all employees in the hospital who are male nurses

  b. The set of all employees in the hospital who are female doctors

  c. The set of all employees in the hospital who are both doctors and administrators

43. a. $D \cap F$ *b.* $R \cap F^c \cap L^c$

45. a. B^c *b.* $A \cap B$ *c.* $A \cap B \cap C^c$

47. a.

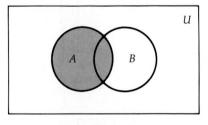

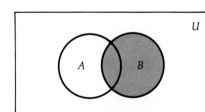

b.

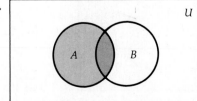

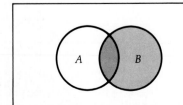

49. a. 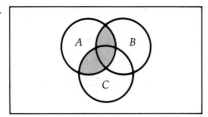   *b.*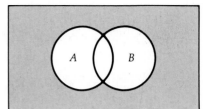

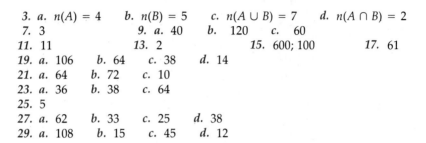

53. a. y, t, s *b.* x, w, z, t, s *c.* x, v, z, u, y, t, s

Exercises 6.2
page 298

3. a. $n(A) = 4$ *b.* $n(B) = 5$ *c.* $n(A \cup B) = 7$ *d.* $n(A \cap B) = 2$
7. 3   *9. a.* 40   *b.* 120   *c.* 60
11. 11   *13.* 2   *15.* 600; 100   *17.* 61
19. a. 106   *b.* 64   *c.* 38   *d.* 14
21. a. 64   *b.* 72   *c.* 10
23. a. 36   *b.* 38   *c.* 64
25. 5
27. a. 62   *b.* 33   *c.* 25   *d.* 38
29. a. 108   *b.* 15   *c.* 45   *d.* 12

Exercises 6.3
page 306

1. 12   *3.* 64   *5.* 24   *7.* 24   *9.* 60
11. 250   *13.* 30   *15. a.* 299   *b.* 208
17. a. 36   *b.*

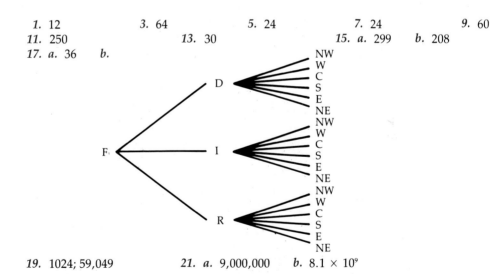

19. 1024; 59,049   *21. a.* 9,000,000   *b.* 8.1×10^9

Exercises 6.4
page 319

1. 20   *3.* 60   *5.* 1   *7.* 35
9. $\dfrac{n!}{2}$ *11.* 84   *13.* $\dfrac{7!}{(7 - r)! \, r!}$ *15.* Combination
17. Permutation   *19.* Combination   *21.* Permutation
23. 720   *25.* 220   *27.* 161,700   *29.* 480
31. a. 40,320   *b.* 384   *c.* 1152
33. a. 20!   *b.* 2880   *35.* 6
37. $C(30, 12) \cdot C(18, 2)$ *39. a.* 3003   *b.* 1485
41. a. 126   *b.* 60   *43. a.* 15,504   *b.* 5400
45. a. 35,960   *b.* 27,405   *47.* 70

Chapter 6
Review Exercises
page 323

1. a. {3}   *b.* {A, E, H, L, S, T}   *c.* {4, 6, 8, 10}   *d.* {−4}
3. ϕ, {e}, {f}, {g}, {e, f}, {e, g}, {f, g}, {e, f, g} all except the last are proper subsets of *A*
7. a. 190   *b.* $\dfrac{n!}{(n-r)!r!}$ *c.* 181,440   *d.* n!   *e.* 120
9. a. 50,400   *b.* 5040
11. a. 446   *b.* 377   *c.* 34   *13. a.* 1287   *b.* 288
15. 1050   *17. a.* 487,635   *b.* 550   *c.* 341,055

▶CHAPTER

SEVEN

Exercises 7.1
page 335

1. ϕ, {a}, {b}, {c}, {a, b}, {a, c}, {b, c}, S
3. {a, b, d, f}; {a}   *5.* {b, c, e}; {a}   *7.* Yes
11. S = {(n, n, n), (n, n, d), (n, d, n), (n, d, d), (d, n, n), (d, n, d), (d, d, n), (d, d, d)}
13. a. {1, 2, 3, 4, 5}   *b.* {2}   *c.* {1, 3, 5}
15. a. {(H, 1), (H, 2), (H, 3), (H, 4), (H, 5), (H, 6), (T, 1), (T, 2), (T, 3), (T, 4),
       (T, 5), (T, 6)}
    b. {(H, 2), (H, 4), (H, 6)}
17. a. ϕ, {1}, {2}, {3}, {1, 2}, {1, 3}, {2, 3}, {1, 2, 3}   *b.* 4   *c.* 4
19. a. $E \cup F$ *b.* $E \cap F$ *c.* $E \cap F^c$
21. a. E^c *b.* $E^c \cap F^c$ *c.* $E \cup F$ *d.* $(E \cap F^c) \cup (E^c \cap F)$
23. a. {x | x > 0}   *b.* {x | 0 < x ≤ 2}   *c.* {x | x > 2}
25. a. S = {0, 1, 2, 3, . . . , 10}   *b.* E = {0, 1, 2, 3}   *c.* F = {5, 6, . . . , 10}
27. a. S = {0, 1, 2, . . . , 20}   *b.* E = {0, 1, 2, . . . , 9}   *c.* F = {20}

Exercises 7.2
page 346

1. {(H, H)}, {(H, T)}, {(T, H)}, {(T, T)}
3. {(D, m)}, {(D, f)}, {(R, m)}, {(R, f)}, {(I, m)}, {(I, f)}
5. {(1, i)}, {(1, d)}, {(1, s)}, {(2, i)}, {(2, d)}, {(2, s)}, . . . , {(5, i)}, {(5, d)}, {(5, s)}
7. 200   *9.* 36

11.

Blood type	A	B	AB	O
Probability	.41	.12	.03	.44

13.

Number of calls received/minute	10	11	12	13	14	15	16	17	18	19
Probability	.05	.125	.1	.025	.1	.3	.2	0	.05	.05

15.

Company	A	B	C	D	E
Probability	.40	.26	.16	.11	.08

17. a. 1/4   *b.* 1/2   *c.* 1/13   *19. a.* 1/6   *b.* 11/36
21. 3/8   *23.* .65   *25. a.* .00001   *b.* .00531
27. No   *29.* Yes
31. a. 3/7   *b.* 3/14   *c.* 1   *33. a.* .35   *b.* .33

Exercises 7.3
page 357

1. 1/2   *3.* 1/36   *5.* 1/9   *7.* 1/52
9. 3/13   *11.* 12/13   *13.* $P(S) \neq 1$
15. Since the five events are not mutually exclusive, property 3 cannot be used—that is, he could win more than one purse.
17. The two events are not mutually exclusive; hence, the probability of the given event is

$$1/6 + 1/6 - 1/36 = 11/36.$$

19. $E^c \cap F^c = \{e\} \neq \phi$ *21.* The space is not necessarily uniform.
23. a. i. $(A \cup B)^c$ *ii.* $A \cap B^c$ *iii.* $A \cap B \cap C^c$
   b. i. .35   *ii.* .25   *iii.* .1
25. a. 0   *b.* .7   *c.* .8   *d.* .3   *e.* 1
27. a. 11/24, 13/24   *b.* 13/24, 11/24   *c.* 1/3
   d. 2/3   *e.* 1/3   *f.* 2/3
29. a. .8   *b.* .4   *c.* .6   *d.* .2   *e.* .4   *f.* .8
31. a. .16   *b.* .38   *c.* .22
33. a. .90   *b.* .40   *c.* .40

Exercises 7.4
page 368

1. a. 1/32   *b.* 5/32   *c.* 31/32   *3. a.* .042   *b.* .420   *c.* .659
5. a. 1/4   *b.* 3/8   *c.* 7/8   *7. a.* .022   *b.* .312
9. a. .005   *b.* .145   *11. a.* .120; .013   *b.* .15; .015
13. .085   *15. a.* 3/5   *b.* 3/10   *c.* 9/10
17. 1/729   *19. a.* .0001   *b.* .0018   *c.* .0504   *d.* .2916
21. .0000154   *23.* .00197   *25.* .0014406
27. a. .618   *b.* .059   *29.* .03

Exercises 7.5
page 385

1. a. .4   *b.* .33   *3. a.* .8   *b.* .75
5. independent   *7. a.* .24   *b.* .76
9. a. 1/4   *b.* 13/51   *c.* 12/51
11. a. 1/2   *b.* 1/2   *c.* 1/2
13. a. .5   *b.* .4   *c.* .2   *d.* .35   *e.* No   *f.* No
15. a. .4   *b.* .3   *c.* .12   *d.* .30   *e.* Yes   *f.* Yes

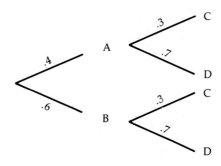

17. $P(A \cap B) \neq P(A) \cdot P(B)$
19. 1/4   *21. a.* 1/21   *b.* 1/3   *23. a.* .28   *b.* .31
25. a. .81   *b.* .162   *27.* not independent
29. a. .092   *b.* .008   *31.* .014

33. a.

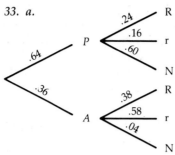

R = recovered within 48 hours
r = recovered after 48 hours
N = never recovered

  b. .24   c. .40
35. a. .16   b. .424   c. .1696

Exercises 7.6
page 397

1.

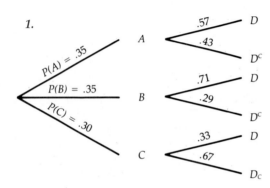

3. a. .45   b. .22
5. a. .48   b. .33
7. a. .08   b. .15   c. .35
9. a. 1/10   b. 2/15   c. 3/10   d. 8/15
11. .0784
13.

15. .53   17. a. 3/4   b. 2/9   19. .856
21. a. .03   b. .29   23. a. .513   b. .390
25. .143   27. a. .4906   b. .62   c. .186
29. a. .543   b. .545   c. .456

Chapter 7   1. a. 0   b. .6   c. .6   d. .4   e. 1
Review Exercises   3. a. .49   b. .39   c. .48   5. a. .019   b. .981
page 404   7. a. .284   b. .984   9. 2/15   11. .457   13. .368

► CHAPTER

EIGHT

Exercises 8.1 page 416

1. any positive integer
3. *a.* See (b).

b.

Outcome	GGG	GGR	GRG	RGG	GRR	RGR	RRG	RRR
Value	3	2	2	2	1	1	1	0

c. {GGG}
5. any positive integer, infinite discrete
7. $0 \le x < \infty$, continuous
9. any positive integer, infinite discrete
11. *a.* .20   *b.* .60   *c.* .30   *d.* 1
13.

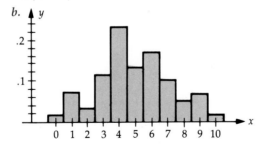

12 13 14 15 16 17 18 19 20

15. *a.*

x	0	1	2	3	4	5	6	7	8	9	10
$P(X = x)$	.017	.067	.033	.117	.233	.133	.167	.1	.05	.067	.017

b.

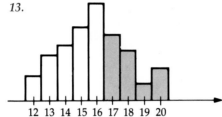

17.

x	1	2	3	4	5	6	7	8	9	10
$P(X = x)$	.007	.029	.021	.079	.164	.15	.20	.207	.114	.029

Exercises 8.2 page 430

1. *a.* 2.6   *b.*

x	0	1	2	3	4
$P(X = x)$	0	.1	.4	.3	.2

; 2.6

3. .86   5. $78.50   7. .91   9. .12
11. 1.73   13. $833.33   15. $163.56
17. *a.* Dahl: 8.52; Farthington: 7.25;   *b.* Farthington Auto Sales
19. $3.50   21. *a.* .357   *b.* .75   *c.* .636
23. −5.3¢   25. −2.7¢
27. *a.* Mean: 74; mode: 85; median: 80   *b.* Mode

Exercises 8.3
page 441

1. $\mu = 2$, $V(X) = 1$, $\sigma = 1$ 3. $\mu = 0$, $V(X) = 1$, $\sigma = 1$
5. $\mu = 518$, $V(X) = 1891$, $\sigma = 43.5$
7. $\mu = 4.5$, $V(X) = 5.25$
11. *a.* Let X = the annual birth rate during the years 1975–1984.

b.

x	14.8	15.3	15.4	15.5	15.7	15.9
$P(X = x)$	.2	.1	.1	.2	.1	.3

c. $\mu = 15.47$, $V(X) = .1541$, $\sigma = .393$
13. $\mu = 5.02$, $V(X) = 3.06$ 15. $V(x) = 3.06$
17. *a.* .64   *b.* .91   19. 7
21. *a.* .96   *b.* 1500   23. 15/16

Exercises 8.4
page 456

1. Yes.   3. No. There are more than 2 outcomes to the experiment.
5. No. The probability of an accident on a clear day is not the same as the probability of an accident on a rainy day.
7. 21/32   9. .0041
11. *a.* $P(X = 0) \approx .08$, $P(X = 1) \approx .26$, $P(X = 2) \approx .35$, $P(X = 3) \approx .23$, $P(X = 4) \approx .08$, $P(X = 5) \approx .01$

b.

x	0	1	2	3	4	5
$P(X = x)$	.08	.26	.35	.23	.08	.01

c. $\mu = 2$, $\sigma = 1.1$

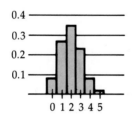

13. No; probability that at most 1 is defective is .74.
15. $\approx .116$ 17. $\approx .0002$ 19. *a.* $\approx .633$ *b.* $\approx .367$
21. *a.* $\approx .273$ *b.* $\approx .650$ 23. .312   25. .913
27. *a.* $\approx .003988$ *b.* .000006   *c.* $\approx 3.997 \times 10^{-9}$
29. $\approx .392$ 31. 7 times   33. *a.* 1200   *b.* ≈ 21.91

Exercises 8.5
page 468

1. .9265   3. .0401   5. .8657
7. *a.*   *b.* .9147

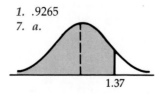

9. *a.*   *b.* .2578

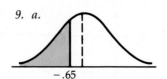

11. a.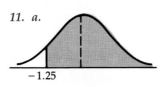

−1.25

b. .8944

13. a.

.68   2.02

b. .2266

15. a. 1.23   b. − .81
17. a. 1.9   b. − 1.9
19. a. .9772   b. .9192   c. .7333

Exercises 8.6
page 479

1. a. .2206   b. .2206   c. .3034
3. a. .0228   b. .0228   c. .4772   d. .7258
5. a. .0038   b. .0918   c. .4082   d. .2514
7. .6247   9. 0.62%   11. A:80; B:73; C:62; D:54
13. a. .5793   b. .2668   c. .0122
15. a. .2877   b. .0008   c. .7287   17. .9265
19. a. .0037   b. drug is very effective   21. 2142

Chapter 8
Review Exercises
page 483

1. $95.88

3. a.

x	0	1	2	3	4
$P(X = x)$	.1296	.3456	.3456	.1536	.0256

b. $\mu = 1.6$, $\sigma \approx .98$
5. a. 2.42   b. − 1.05   c. − 2.03   d. 1.42
7. .2646; .9163   9. At least .75   11. 0.6%   13. .9997

▶ **CHAPTER**
NINE

Exercises 9.1
page 496

1. Yes   3. Yes   5. No   7. Yes   9. No

11. [2/3   1/3]

13. [17/36   23/144   53/144]

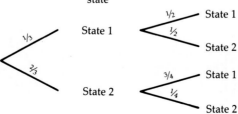

15. [.156 .577 .267]

17. *a.*

$$T = \begin{array}{c} \\ P \\ A \end{array} \begin{array}{cc} P & A \\ \begin{bmatrix} .8 & .2 \\ .3 & .7 \end{bmatrix} \end{array}$$

b.

$$X_0 = \begin{array}{cc} P & A \\ [.2 & .8] \end{array}$$

c. 40%

19. 78.8% city, 21.2% suburb; 77.7% city, 22.3% suburb

21. University: 37%, Campus: 35%, Book Mart: 28%
    University: 34.5%, Campus: 31.35%, Book Mart: 34.15%

23. 36% Business, 23.8% Humanities, 15% Education,
    25.1% Natural Science and others

Exercises 9.2
page 507

1. Regular 3. Not regular 5. Regular 7. Not regular

9. [3/11 8/11] 11. [1/2 1/2] 13. [3/13 8/13 2/13]

15. [3/19 8/19 8/19] 17. 81.8%

19. *a.* 40.8% 1-wage earner; 59.2% 2-wage earners
    b. 30% 1-wage earner; 70% 2-wage earners

21. *a.* 72.5% in single-family homes; 27.5% in condominiums
    b. 70% in single-family homes; 30% in condominiums

23. *a.* 31.7% ABC, 37.35% CBS, 30.95% NBC

    b. $33\frac{1}{3}$% ABC, $33\frac{1}{3}$% CBS, $33\frac{1}{3}$% NBC

25. 25% red, 50% pink, 25% white

Exercises 9.3
page 519

1. Yes 3. No 5. Yes 7. Yes

9. $\begin{bmatrix} 1 & 0 \\ \hline .4 & .6 \end{bmatrix}$, $R = [.4]$, $S = [.6]$

11. $\begin{bmatrix} 1 & 0 & 0 \\ \hline .4 & .4 & .2 \\ .5 & .5 & 0 \end{bmatrix}$, $R = \begin{bmatrix} .4 \\ .5 \end{bmatrix}$, $S = \begin{bmatrix} .4 & .2 \\ .5 & 0 \end{bmatrix}$,

or $\begin{bmatrix} 1 & 0 & 0 \\ \hline .5 & 0 & .5 \\ .4 & .2 & .4 \end{bmatrix}$, $R = \begin{bmatrix} .5 \\ .4 \end{bmatrix}$, $S = \begin{bmatrix} 0 & .5 \\ .2 & .4 \end{bmatrix}$

13. $\begin{bmatrix} 1 & 0 & 0 & 0 \\ 0 & 1 & 0 & 0 \\ \hline .2 & .3 & .3 & .2 \\ .4 & 0 & .2 & .4 \end{bmatrix}$, $R = \begin{bmatrix} .2 & .3 \\ .4 & 0 \end{bmatrix}$, $S = \begin{bmatrix} .3 & .2 \\ .2 & .4 \end{bmatrix}$,

or $\begin{bmatrix} 1 & 0 & 0 & 0 \\ 0 & 1 & 0 & 0 \\ \hline .4 & 0 & .4 & .2 \\ .2 & .3 & .2 & .3 \end{bmatrix}$, $R = \begin{bmatrix} .4 & 0 \\ .2 & 3 \end{bmatrix}$, $S = \begin{bmatrix} .4 & .2 \\ .2 & .3 \end{bmatrix}$, and so forth.

15. $\begin{bmatrix} 1 & 0 \\ \hline 1 & 0 \end{bmatrix}$

17. $\begin{bmatrix} 1 & 0 & 0 \\ \hline 1 & 0 & 0 \\ 1 & 0 & 0 \end{bmatrix}$

19. $\begin{bmatrix} 1 & 0 & 0 & 0 \\ 0 & 1 & 0 & 0 \\ \hline 1 & 0 & 0 & 0 \\ 1 & 0 & 0 & 0 \end{bmatrix}$

21. $\begin{bmatrix} 1 & 0 & 0 & 0 \\ 0 & 1 & 0 & 0 \\ \hline \frac{1}{2} & \frac{1}{2} & 0 & 0 \\ \frac{1}{2} & \frac{1}{2} & 0 & 0 \end{bmatrix}$

23.
$$\begin{bmatrix} 1 & 0 & 0 & | & 0 & 0 \\ 0 & 1 & 0 & | & 0 & 0 \\ 0 & 0 & 1 & | & 0 & 0 \\ \frac{7}{22} & \frac{5}{22} & \frac{10}{22} & | & 0 & 0 \\ \frac{6}{22} & \frac{9}{22} & \frac{7}{22} & | & 0 & 0 \end{bmatrix}$$

25. a.
$$\begin{array}{cc} & \text{UL} \quad \text{L} \\ \begin{array}{c} \text{UL} \\ \text{L} \end{array} & \begin{bmatrix} 1 & | & 0 \\ .2 & | & .8 \end{bmatrix} \end{array}; \qquad R = [.2], \qquad S = [.8]$$

b. $\begin{bmatrix} 1 & | & 0 \\ 1 & | & 0 \end{bmatrix}$; Eventually, only unleaded fuel will be used.

27. .25; .50; .75

29. a.
$$\begin{array}{c} \begin{array}{c} \text{D} \\ \text{G} \\ 1 \\ 2 \end{array} \begin{array}{cccc} \text{D} \quad \text{G} \quad 1 \quad 2 \\ \begin{bmatrix} 1 & 0 & | & 0 & 0 \\ 0 & 1 & | & 0 & 0 \\ .25 & 0 & | & 0 & .75 \\ .1 & .9 & | & 0 & 0 \end{bmatrix} \end{array} \end{array}$$

b. $\begin{bmatrix} 1 & 0 & | & 0 & 0 \\ 0 & 1 & | & 0 & 0 \\ .325 & .675 & | & 0 & 0 \\ .1 & .9 & | & 0 & 0 \end{bmatrix}$ **c.** .675

Exercises 9.4
page 532

1. R: row 1; C: column 2   **3.** R: row 1; C: column 1
5. R: row 1 or row 3; C: column 3   **7.** R: row 1 or row 3; C: column 2
9. strictly determined;   **a.** 2   **b.** R: row 1; C: column 1   **c.** 2
   d. favors row player
11. strictly determined;   **a.** 1   **b.** R: row 1; column 1   **c.** 1
   d. favors row player
13. strictly determined;   **a.** 1   **b.** R: row 1; C: column 1   **c.** 1
   d. favors row player
15. not strictly determined   **17.** not strictly determined
19. a. $\begin{bmatrix} 2 & -3 & 4 \\ -3 & 4 & -5 \\ 4 & -5 & 6 \end{bmatrix}$ **b.** Robin; row 1; Cathy: column 1 or column 2
   c. not strictly determined   **d.** not strictly determined
21. a.   Economy   **b.** Yes
                  good   recess.
   Mgmt. expand $\begin{bmatrix} 200,000 & 120,000 \\ 50,000 & 150,000 \end{bmatrix}$
        not exp.
23. a.   Charley
                    raises holds lowers
        raises $\begin{bmatrix} 3 & -1 & -3 \\ 2 & 0 & -2 \\ 5 & 2 & 1 \end{bmatrix}$
   Roland holds
        lowers

Exercises 9.5
page 545

1. 3/10   **3.** 5/12   **5.** .16
7. a. 1   **b.** −2   **c.** 0   **d.** −3/10; (a) is most advantageous
9. a. 1   **b.** −7/20   **c.** the first strategy

11.
$$\begin{bmatrix} \frac{3}{4} & \frac{1}{4} \end{bmatrix}, \begin{bmatrix} \frac{11}{12} \\ \frac{1}{12} \end{bmatrix};$$ 9/4; favors row player

13.
$$\begin{bmatrix} \frac{1}{3} & \frac{2}{3} \end{bmatrix}, \begin{bmatrix} \frac{1}{2} \\ \frac{1}{2} \end{bmatrix};$$ 1; favors row player

15. R: row 1; C: column 1; 2; favors row player

17. $16,667 in stocks; $83,333 in commodities; $11,667

19. *a.* $\begin{bmatrix} .6 & .4 \\ .45 & .55 \end{bmatrix}$ *b.* $\begin{bmatrix} \frac{1}{3} & \frac{2}{3} \end{bmatrix}$ *c.* $\begin{bmatrix} \frac{1}{2} \\ \frac{1}{2} \end{bmatrix}$

Chapter 9
Review Exercises
page 549

1. not regular   **3.** regular   **5.** [3/7 4/7]   **7.** [.457 .200 .343]

9. *a.*

	A	U	N
A	.85	.10	.05
U	0	.95	.05
N	.10	.05	.85

b.

A	U	N
[.50	.15	.35]

c.

A	U	N
[.424	.262	.314]

11. strictly determined; R: row 3; C: column 1; value is 4

13. strictly determined; R: row 1; C: column 1; value is 1

15. $-1/4$ **17.** 2

19. $P = \begin{bmatrix} \frac{1}{2} & \frac{1}{2} \end{bmatrix}$; $Q = \begin{bmatrix} \frac{13}{22} \\ \frac{9}{22} \end{bmatrix}$; $E = -1/2$; favors column player

21. $P = \begin{bmatrix} \frac{4}{5} & \frac{1}{5} \end{bmatrix}$; $Q = \begin{bmatrix} \frac{2}{5} \\ \frac{3}{5} \end{bmatrix}$; $E = 10.8$; favors row player

23. 25% compact, 75% subcompact

▶ APPENDIX

A

Exercises A.1
page 559

1. (a), (c), (f)

3. *a.* not   *b.* if and only if; and   *c.* and   *d.* and, not; if . . . then
   e. and   *f.* or

5. *a.* Every employee is required to be fingerprinted or to take an oath of allegiance.
   b. Every employee is required to be fingerprinted and to take an oath of allegiance.
   c. Every employee is not required to be fingerprinted or is not required to take an oath of allegiance.
   d. Every employee is neither required to be fingerprinted nor to take an oath of allegiance.
   e. Every employee is required to be fingerprinted or is not required to take an oath of allegiance.

7. *a.* The doctor recommended either surgery or radioactive iodine to treat his hyperthyroidism.

  b. The doctor recommended surgery and/or radioactive iodine to treat his hyperthyroidism.

9. *a.* $p \veebar q$　　　　*b.* $p \wedge q$　　　　*c.* $\sim q$　　　　*d.* $\sim p \wedge \sim q$

Exercises A.2
page 563

1.

p	q	$\sim q$	$p \vee \sim q$
T	T	F	T
T	F	T	T
F	T	F	F
F	F	T	T

3.

p	$\sim p$	$\sim(\sim p)$
T	F	T
F	T	F

5.

p	$\sim p$	$p \vee \sim p$
T	F	T
F	T	T

7.

p	$\sim p$	q	$p \vee q$	$\sim p \wedge (p \vee q)$
T	F	T	T	F
T	F	F	T	F
F	T	T	T	T
F	T	F	F	F

9.

p	q	$\sim q$	$p \vee q$	$p \wedge \sim q$	$(p \vee q) \wedge (p \wedge \sim q)$
T	T	F	T	F	F
T	F	T	T	T	T
F	T	F	T	F	F
F	F	T	F	F	F

11.

p	q	r	$p \vee q$	$p \vee r$	$(p \vee q) \wedge (p \vee r)$
T	T	T	T	T	T
T	T	F	T	T	T
T	F	T	T	T	T
T	F	F	T	T	T
F	T	T	T	T	T
F	T	F	T	F	F
F	F	T	F	T	F
F	F	F	F	F	F

13.

p	q	r	$\sim q$	$p \wedge \sim q$	$p \wedge r$	$(p \wedge \sim q) \vee (p \wedge r)$
T	T	T	F	F	T	T
T	T	F	F	F	F	F
T	F	T	T	T	T	T
T	F	F	T	T	F	T
F	T	T	F	F	F	F
F	T	F	F	F	F	F
F	F	T	T	F	F	F
F	F	F	T	F	F	F

15. 16 rows

Exercises A.3
page 570

1. $\sim q \to p; q \to \sim p; \sim p \to q$ 3. $p \to q; \sim p \to \sim q; \sim q \to \sim p$

5. Conditional: If it is snowing, then the temperature is below freezing.
   Biconditional: It is snowing if and only if the temperature is below freezing.

7. Conditional: If the company's union and management reach a settlement, then the workers will not strike.
   Biconditional: The company's union and management will reach a settlement if and only if the workers do not strike.

9. False   11. False

13. It is false when
    1. I buy the house and the owner does not lower the selling price and
    2. The owner lowers the selling price and I do not buy the house.

15.

p	q	$\sim p$	$q \to \sim p$	$\sim (q \to \sim p)$
T	T	F	F	T
T	F	F	T	F
F	T	T	T	F
F	F	T	T	F

17.

p	q	$p \to q$	$q \to p$	$(p \to q) \lor (q \to p)$
T	T	T	T	T
T	F	F	T	T
F	T	T	F	T
F	F	T	T	T

19.

p	q	$\sim p$	$p \to q$	$\sim p \lor q$	$(p \to q) \land (\sim p \lor q)$
T	T	F	T	T	T
T	F	F	F	F	F
F	T	T	T	T	T
F	F	T	T	T	T

21.

p	q	$\sim p$	$\sim q$	$(\sim p \land \sim q)$	$\sim q \to (\sim p \land \sim q)$
T	T	F	F	F	T
T	F	F	T	F	F
F	T	T	F	F	T
F	F	T	T	T	T

23.

p	q	r	$p \to q$	$q \to r$	$p \to r$	$(p \to q) \lor (q \to r)$	$[(p \to q) \lor (q \to r)] \to (p \to r)$
T	T	T	T	T	T	T	T
T	T	F	T	F	F	T	F
T	F	T	F	T	T	T	T
T	F	F	F	T	F	T	F
F	T	T	T	T	T	T	T
F	T	F	T	F	T	T	T
F	F	T	T	T	T	T	T
F	F	F	T	T	T	T	T

25. logically equivalent   27. not logically equivalent   29. logically equivalent

Exercises A.4
page 574

1.

p	p	$p \land p$
T	T	T
F	F	F

3.

p	q	r	$p \wedge q$	$(p \wedge q) \wedge r$	$q \wedge r$	$p \wedge (q \wedge r)$
T	T	T	T	T	T	T
T	T	F	T	F	F	F
T	F	T	F	F	F	F
T	F	F	F	F	F	F
F	T	T	F	F	T	F
F	T	F	F	F	F	F
F	F	T	F	F	F	F
F	F	F	F	F	F	F

5.

p	q	$p \wedge q$	$q \wedge p$
T	T	T	T
T	F	F	F
F	T	F	F
F	F	F	F

7.

p	q	r	$q \wedge r$	$p \vee (q \wedge r)$	$p \vee q$	$p \vee r$	$(p \vee q) \wedge (p \vee r)$
T	T	T	T	T	T	T	T
T	T	F	F	T	T	T	T
T	F	T	F	T	T	T	T
T	F	F	F	T	T	T	T
F	T	T	T	T	T	T	T
F	T	F	F	F	T	F	F
F	F	T	F	F	F	T	F
F	F	F	F	F	F	F	F

9. Tautology   11. Tautology   13. Tautology   15. Neither

17. $\sim (p \wedge q)$: The candidate does not oppose changes in the social security system or the candidate does not support the ERA.

$\sim (p \vee q)$: The candidate does not oppose changes in the social security system and the candidate does not support the ERA.

19. $p \vee (\sim p \wedge \sim q) \Leftrightarrow (p \vee \sim p) \wedge (p \vee \sim q)$ by Law 8
             $\Leftrightarrow t \wedge (p \vee \sim q)$ by Law 1
             $\Leftrightarrow p \vee \sim q$ by Law 14

21. $(p \vee q) \vee \sim q \Leftrightarrow p \vee (q \vee \sim q)$ by Law 4
             $\Leftrightarrow p \vee t$ by Law 11
             $\Leftrightarrow p$ by Law 13

23. $p \vee (q \vee r) \Leftrightarrow p \vee (r \vee q)$ by Law 6
             $\Leftrightarrow (p \vee r) \vee q$ by Law 4
             $\Leftrightarrow (r \vee p) \vee q$ by Law 6
             $\Leftrightarrow r \vee (p \vee q)$ by Law 4
             $\Leftrightarrow r \vee (q \vee p)$ by Law 6

Exercises A.5
page 581

1. Valid   3. Valid   5. Invalid   7. Valid   9. Valid
11. Valid   13. Valid   15. Invalid

17. $p \rightarrow q$; Invalid
$\dfrac{\sim p}{\therefore \sim q}$

19. $p \vee q$; Valid
$\dfrac{\sim p \rightarrow \sim q}{\therefore p}$

21. $p \rightarrow q$; Invalid
$q \rightarrow r$
$\dfrac{r}{\therefore p}$

23. (b)

25.

p	q	$p \to q$	$\sim q$	$\sim p$
T	T	T	F	F
T	F	F	T	F
F	T	T	F	T
F	F	T	T	T

Exercises A.6
page 586

1. $p \wedge q \wedge (r \vee s)$ 3. $[(p \wedge q) \vee r] \wedge (\sim r \vee p)$

5. $[(p \vee q) \wedge r] \vee (\sim p) \vee [\sim q \wedge (p \vee r \vee \sim r)]$

7.

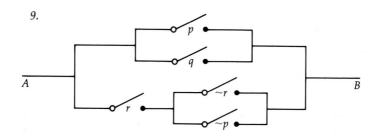

9.

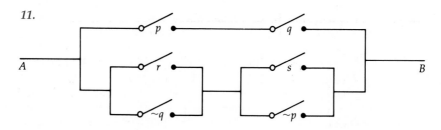

11.

13. $p \wedge [\sim q \vee (\sim p \wedge q)]; p \wedge \sim q$

15. $p \wedge [\sim p \vee q \vee (q \wedge r)]; p \wedge q$

Index

Formulas

Equation of a Straight Line

(a) point-slope form $\quad y - y_1 = m(x - x_1)$

(b) general form $\quad Ax + By + C = 0$

Equation of the Least-Squares Line

$$y = mx + b$$

where m and b satisfy the **normal equations**

$$nb + (x_1 + x_2 + \cdots + x_n)m = y_1 + y_2 + \cdots + y_n$$

$$(x_1 + x_2 + \cdots + x_n)b + (x_1^2 + x_2^2 + \cdots + x_n^2)m = x_1y_1 + x_2y_2 + \cdots + x_ny_n$$

Compound Interest

$$A_n = P(1 + i)^n \qquad (i = r/m)$$

where A_n is the accumulated amount at the end of n conversion periods, P is the principal, r is the interest rate per year, and n is the number of conversion periods.

Effective Rate of Interest

$$R = \left(1 + \frac{r}{m}\right)^m - 1$$

where R is the effective rate of interest, r is the nominal interest rate, and m is the number of conversion periods.

Future Value of an Annuity

$$S = P\left[\frac{(1 + i)^n - 1}{i}\right]$$

Present Value of an Annuity

$$A = P\left[\frac{1 - (1 + i)^{-n}}{i}\right]$$

Amortization Formula

$$P = \frac{Ai}{1 - (1 + i)^{-n}}$$